Mixing in Coagulation and Flocculation

American Water Works Association Research Foundation

Edited by
Appiah Amirtharajah
Mark M. Clark
R. Rhodes Trussell

Disclaimer

This study was funded by the American Water Works Association Research Foundation (AWWARF). AWWARF assumes no responsibility for the opinions or statements of fact expressed in the report. The mention of trade names for commercial products does not represent or imply the approval or endorsement of AWWARF. This report is presented solely for informational purposes.

Copyright © 1991
American Water Works Association Research Foundation
American Water Works Association
6666 West Quincy Ave.
Denver, CO 80235

Printed in USA

ISBN 0-89867-561-8

Contents

Introduction, vi
Acknowledgments, viii

Fundamentals

Chapter 1 Mixing in Water Treatment 3

    Introduction to Rapid Mixing and Slow Mixing, 3
    Mixing for Coagulation, 6
    Mixing for Flocculation, 25
    References, 32

Chapter 2 Mixing in Liquids 35

    Introduction, 35
    Stirred-Tank Reactors, 51
    In-Line Mixers, 68
    References, 77

Chapter 3 Turbulence and Mixing: Modeling Effects on Chemical Reactions 80

    Introduction, 80
    Review of Turbulence and Mixing Theory, 82
    Perspective, 89
    Lagrangian Modeling, 90
    Eulerian Modeling, 95
    Closures for Component and Segregation Balances, 99
    Approximate Models for Local Mixing, 110
    Modeling Complex Chemical Reactions, 116
    Effect of Available Experimental Techniques, 118
    Scale-up of Mixed Reactors, 119
    Relative Costs of Modeling Methods, 120
    Similarity Approximations for Distributed Variables, 121
    Global Variable Correlations From Distributed Modeling, 123
    Conclusion, 123
    References, 123

Chapter 4 Residence Time Distribution 127

    The Simple Closed System, 128
    Mathematical Properties, 128
    Models for Residence Time Distributions, 135
    Measurement Techniques, 146
    Data Analysis, 152
    Chemical Reactions and Micromixing, 158
    Designing for Good Mixing, 164
    Conclusions, 168
    References, 169

**Chapter 5 Micromixing Models and Application to Aluminum
Neutralization Precipitation Reactions** **170**

    What is Micromixing?, 170
    The Mechanism of Mixing, 172
    Micromixing Models, 179
    Precipitation of Sparingly Soluble Salts and Micromixing, 196
    Aluminum Precipitation, 198
    Conclusions, 209
    References, 210

Application

**Chapter 6 Particle Destabilization and Flocculation Reactions
in Turbulent Pipe Flow** **217**

    Introduction, 217
    Structure of Turbulent Pipe Flow, 218
    Types of Pipe Mixers, 220
    Pipe Mixing Performance Criteria, 224
    Design Criteria of Pipe Mixers, 231
    Practical Experience With Pipe Mixers, 244
    Research Needs, 253
    References, 254

Chapter 7 Mixing, Breakup, and Floc Characteristics **256**

    Introduction, 256
    Particle Size and Diameter, 257
    Particle Shape, 258
    Floc Density and Porosity, 258
    Floc Structure, 262
    Floc Strength and Breakup, 266
    Aging of Flocs, 274
    Conclusions and Recommendations, 279
    References, 279

Chapter 8 Mixing and Scale-up **282**

    Introduction, 282
    Scale-Related Effects in Nature and Chemical Processing, 283
    Traditional Scale-up Laws, 284
    Initial Mixing, 287
    Flocculation, 292
    Scale-up Experiments in Flocculation, 297
    Comparison of Batch and Continuous Flocculation, 301
    Summary, Conclusions, and Research Recommendations, 302
    References, 305

Design and Operation

Chapter 9 Design of Impellers for Mixing 309

Impellers, 309
Scale-up Principles, 322
Dynamic Similarity, 325
Other Impeller Types, 327
Flow Patterns in Large Blending Tanks, 327
Flocculation Experiments, 327
Power and Flow Comparisons for Impellers, 337
References, 342

Chapter 10 Pilot-Plant Studies for Design and Operation 343

Introduction, 343
Design of Pilot Studies, 345
Rapid-Mixing Evaluation, 367
Flocculation Evaluation, 368
Full-Scale Testing, 376
Conclusions, 377
References, 379

Chapter 11 Design of Mixers for Water Treatment Plants: Rapid Mixing and Flocculators 380

Introduction, 380
Rapid Mixers, 381
Flocculation, 397
References, 418

Index, 421

Introduction

Appiah Amirtharajah
School of Civil Engineering
Georgia Institute of Technology
Atlanta, GA 30332 USA

Mark M. Clark
Department of Civil Engineering
University of Illinois
Urbana, IL 61801 USA

R. Rhodes Trussell
James M. Montgomery Consulting Engineers, Inc.
Pasadena, CA 91101 USA

This text is a unique attempt at synthesizing the available knowledge on the theory and design of mixing in coagulation and flocculation in a single state-of-the-science document. This book is the culmination of a two-year effort by an international interdisciplinary group of authors who contributed their precious time, willingly and voluntarily. The project was organized by the American Water Works Association Research Foundation.

To produce a state-of-the-science book on mixing for the water treatment industry, the chapters and authors were selectively assembled to provide five specific elements:

    1. An assembly of current knowledge of mixing in coagulation and flocculation in water treatment that would serve as a state-of-the-art reference for academics and practicing engineers.

    2. A review of current ideas on mixing and turbulence in chemical engineering that would indicate new directions for research in environmental engineering by attempting an interdisciplinary transfer of technology and knowledge.

    3. A review of current practice in design and operation of rapid mixing and flocculation units in the water industry.

    4. A mix of international authors to synthesize and critically evaluate ideas and design practices across national boundaries.

    5. A practical guide for the design, operation, modification, and evaluation of mixing processes in water treatment.

The book is divided into three principal sections: fundamentals, application, and design and operation.

Fundamentals: Chapters 1 through 5

Chapter 1 covers mixing in water treatment, with a focus on the chemical processes occurring with coagulants. Mixing in liquids and development of design correlations for stirred tanks, in-line mixers, jet mixers, and static mixers are discussed in chap. 2. Turbulence theory with micromixing and macromixing is analyzed from a chemical engineering perspective in chap. 3. An important aspect of mixing is the residence time distribution generated by dispersion effects in real reactors. The fundamentals of residence time distributions and their measurement and interpretation are covered in chap. 4. Micromixing models and their application to aluminum neutralization and precipitation reactions are discussed in chap. 5.

Chapters 1 through 5 provide the fundamental bases for current understanding of mixing in environmental and chemical engineering. Some of the discussions are quantitative and mathematical. The chapters are written by dual authors from Japan, England, France, and the United States and provide perspectives that are uniquely international and interdisciplinary. For readers chiefly interested in application or design and operation of mixing processes or facilities, a brief overview of the section on fundamentals is all that is necessary before delving in detail into chap. 6 through 11.

Application: Chapters 6 through 8

These three chapters form a bridge between the chapters on fundamentals and the chapters on design and operation. Chapter 6 analyzes the application of turbulent pipe flow for mixing. Pipe mixers are not common in the United States, but some facilities are operating successfully in Germany. Chapter 7 discusses the effect of mixing on the characteristics of flocs that are produced in water treatment. For the design of mixing units, the characteristics of processes generated in bench- and pilot-scale studies need to be scaled-up. Chapter 8 pinpoints some of the difficulties encountered in scale-up of mixing facilities. Some guidance for good practice is suggested. These chapters discuss experiences from Germany, Japan, Belgium, France, and the United States.

Design and Operation: Chapters 9 through 11

These three chapters summarize the principal and best available design and operational practices for mixing. They include a discussion of the impeller design from a major equipment manufacturer; pilot-plant studies for development of design criteria and optimization of the design process; and the engineering of mixing units for rapid mixing and flocculation. The last three chapters are most useful for practical design and operation.

The reader should be cautioned that the nomenclature used throughout the book is not always consistent between chapters. The reader should not compare definitions between chapters.

An important underlying theme throughout the book is an attempt to summarize what is currently known in terms of fundamentals and practice from the chemical engineering field as well as from the environmental process engineering field. Although all the authors met together and shared their mutual understanding of the topics at hand, it will be evident to the careful reader that even the most informed environmental engineers do not fully understand the application of the chemical engineering principles displayed, and chemical engineers could benefit from a better grasp of the specific design issues in environmental processes. Many of the specific design ideas in chemical engineering require adaptation before they can be effectively implemented in environmental engineering. Nevertheless, these chapters represent a unique compilation of the scientific and engineering thought on mixing in water treatment. Careful study will likely lead the reader to even greater insight to design phenomena than many of the contributors now possess—certainly much greater insight than most environmental engineers possess today.

Acknowledgments

This book would not have been possible without the dedication of all the authors. The tremendously long hours they spent in writing, discussing, and editing the contents of each chapter is greatly appreciated.

In addition to the authors, many other individuals were essential to the conception and production of this book. The need for a book on mixing originated from discussions between Vernon Snoeyink and Rhodes Trussell. Vernon Snoeyink, a member of the board of trustees of the American Water Works Association Research Foundation, initiated approval for the project. He sought funding that would enable the authors to meet and discuss the content of their chapters as well as to exchange experiences and expertise.

François Fiessinger provided significant input in selecting authors, structuring the book, and organizing the first authors' meeting.

The project was begun at the AWWA Research Foundation under the leadership of Nancy McTigue; subsequently, Rick Karlin provided excellent overall support and organization, which kept the project on track.

The authors of chap. 8 wish to thank Lyonnaise des Eaux, Le Pecq, France, and the University of Illinois for sponsoring some of the experimental work reported in the chapter. The authors of chap. 11 wish to acknowledge the assistance of Philadelphia Mixer for some of the material included in their chapter.

Finally, all authors and the AWWA Research Foundation wish to thank Mary Kay Kozyra, Candace Bradford, Carrie Dubois, and Sheryl Tongue for their excellent editorial and production skills.

Contributors to this work include:

Appiah Amirtharajah, Ph.D., P.E., Professor, School of Civil Engineering, Georgia Institute of Technology, Atlanta, GA 30332 USA

Mark M. Clark, Ph.D., Assistant Professor, Department of Civil Engineering, University of Illinois at Urbana–Champaign, Urbana, IL 61801-2397 USA

René David, Ph.D., Ingenieur, Directeur de Recherche, Laboratoire des Sciences du Génie Chimique, CNRS-ENSIC— Institut National Polytechnique de Lorraine, 1 rue Grandville, 54001 Nancy Cedex, France

François Fiessinger, Vice President, Zenon Environmental, Inc., 845 Harrington Court, Burlington, Ontario L7N 3P3 Canada

R.J. François, D.Sc., Chemical Engineer, Technical Center Steel Cord—N.V. Bekaert, Kronkelstraat 2, B-8793 Waregem (Sint-Eloois-Vijve), Belgium

J.C. Godfrey, Ph.D., Department of Chemical Engineering, University of Bradford, Bradford, West Yorkshire BD7 1DP, United Kingdom

Rudolf Klute, Dr. Ing., Institut für Siedlungswasserwirtschaft, University of Karlsruhe, Kaiserstrabe 12, D-7500 Karlsruhe 1, Germany

Robert D.G. Monk, P.E., Senior Vice President, Camp Dresser & McKee Inc., 100 Pringle Avenue, Suite 300, Walnut Creek, CA 94596 USA

E. Bruce Nauman, Ph.D., Professor, Department of Chemical Engineering, Rensselaer Polytechnic Institute, Troy, NY 12181-3590 USA

Dale D. Newkirk, P.E., Regional Operations Manager, Metropolitan Water District of Southern California, 1111 Sunset Boulevard, Box 54153, Los Angeles, CA 90054 USA

James Y. Oldshue, Ph.D., P.E., Vice President—Mixing Technology, Lightnin, 135 Mt. Read Boulevard, Rochester, NY 14611 USA

Gary K. Patterson, Ph.D., P.E., Associate Dean of Engineering for Research and Graduate Affairs, Professor of Chemical Engineering, School of Engineering, University of Missouri–Rolla, Rolla, MO 65401 USA

Norihito Tambo, Dr. Eng., Professor, Department of Sanitary and Environmental Engineering, Hokkaido University, Sapporo 060, Japan

R. Rhodes Trussell, Ph.D., P.E., Senior Vice President, James M. Montgomery Consulting Engineers, Inc., 250 North Madison Avenue, Pasadena, CA 91109-7009 USA

Robert P. Zipp, Ph.D., Department of Chemical Engineering, University of Michigan, Ann Arbor, MI 48109 USA

Fundamentals

1

Mixing in Water Treatment

Appiah Amirtharajah, Ph.D., P.E., Professor, School of Civil
  Engineering, Georgia Institute of Technology,
  Atlanta, Ga. USA

Norihito Tambo, Dr. Eng., Professor, Department of Sanitary
  and Environmental Engineering, Hokkaido University,
  Sapporo, Japan

INTRODUCTION TO RAPID MIXING AND SLOW MIXING

In a water treatment plant process train, the mixing unit operation is utilized at several stages during treatment. The three main processes requiring mixing are: (1) initial mixing or rapid mixing, where the chemicals are added for particle destabilization; (2) the slow mixing, or flocculation, stage, where mixing is used to accelerate the formation of aggregates principally by the transport mechanism due to velocity gradients; and (3) the mixing of disinfectants, such as chlorine, for inactivation of microorganisms. This chapter deals principally with the first two mixing processes, which occur during the overall process of coagulation. The coagulation process is defined to include particle destabilization and particle flocculation. The term *flocculation* applies only to the transport aspects of the overall process of coagulation. In the chemical engineering literature, coagulation refers to destabilization by double-layer compression, and flocculation refers to destabilization by the bridging mechanism; both terms exclude particle transport. Figure 1-1 is a schematic of the two major mixing operations in a water treatment plant process

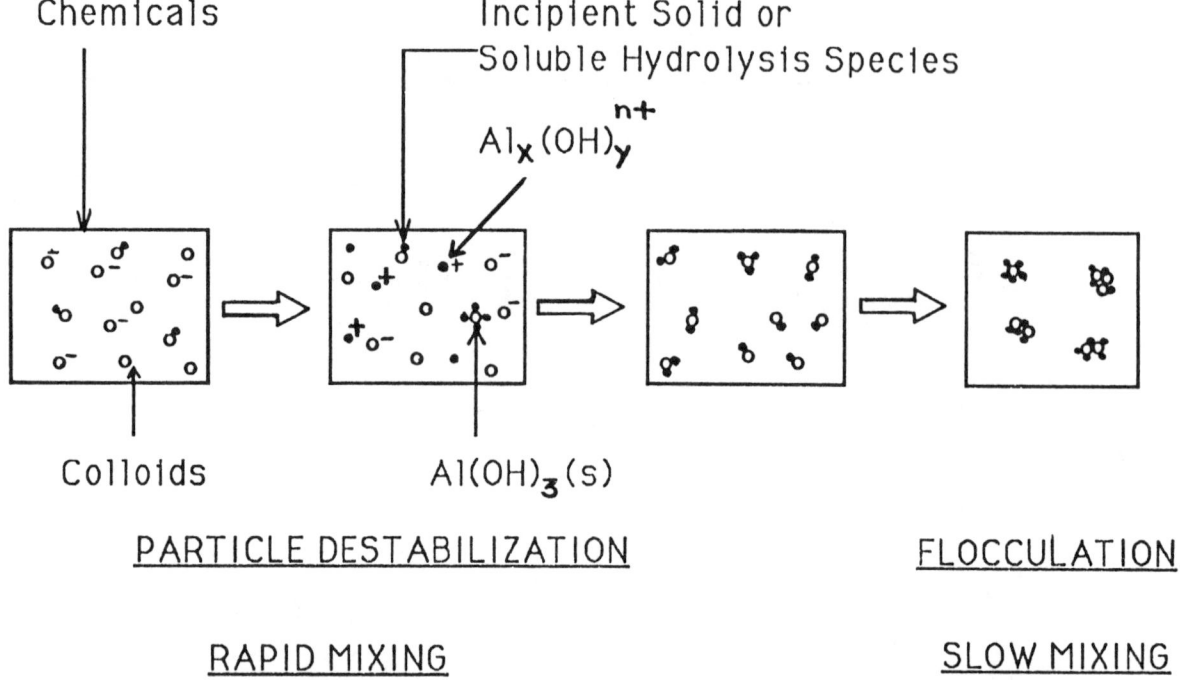

Figure 1-1 The overall process of coagulation with rapid mixing and slow mixing.

train that use alum (aluminum sulfate) as a coagulant. Chemical interactions and particle destabilization occur during the rapid mixing stage, while slow mixing causes flocculation and the growth of aggregates.

In the recent past, significant insights into the process of coagulation have been developed. These developments have resulted from studies that combine the chemical aspects of coagulation with the physical aspects (fluid mechanics) of flocculation. The principal aim of the rapid mixing step in water treatment is to disperse the coagulants uniformly and quickly so as to cause the destabilization of the particles in the raw water. As currently understood, four distinct mechanisms can cause particle destabilization of colloidal suspensions. These mechanisms are: (1) compression of the double layer, described quantitatively by the electrostatic model of Derjaguin, Landau, Verwey, and Overbeek (DLVO theory); (2) adsorption to produce charge neutralization; (3) enmeshment in a precipitate, colorfully described as "sweep coagulation"; and (4) adsorption to permit interparticle bridging. In water treatment, the major mechanisms are charge neutralization with inorganic coagulants, such as aluminum and iron salts or polymeric inorganic coagulants, and sweep coagulation with higher dosages of aluminum and iron salts. When organic polyelectrolytes are used as primary coagulants or coagulant aids, the important mechanisms are charge neutralization and interparticle bridging.

A phenomenological view would indicate that for particles to be destabilized, interactions must occur between the colloids in the raw water and the products of the chemical coagulation reactions. Microscopically, the destabilization step for charge neutralization requires collisions or transport between the colloids and the incipiently forming products of the metal hydrolysis reactions. This initial collisional process is illustrated in the first three steps shown schematically in Figure 1-1. In contrast to charge neutralization, for the sweep coagulation mechanism the chemical conditions for rapid precipitation of the amorphous hydroxides and subsequent flocculation are significantly more important than transport interactions between the colloid and the hydrolysis products during destabilization. The rapid mixing requirements for these two major mechanisms of particle destabilization are analyzed in a subsequent section of this chapter.

Slow Mixing

The principal aim of slow mixing is to transform the destabilized smaller particles into larger aggregates. The transport processes that govern the overall growth of aggregates are called flocculation. Over the last 10 years a reasonable understanding of the flocculation process has been developed at the microscopic level (Tambo et al. 1970; O'Melia 1972; Spielman 1978; Tambo and Watanabe 1979). Flocculation is currently analyzed as being caused by collisions between particles in three mechanisms: (1) Brownian, or perikinetic, flocculation due to the thermal energy of the fluid; (2) velocity gradient, or orthokinetic flocculation, due to bulk fluid motion; and (3) differential settling due to a larger particle overtaking and colliding with a slower settling particle. The velocity gradient mode can be quantitatively described for both laminar-shear and isotropic-turbulence flow fields. Some of the mechanisms have been analyzed (1) incorporating the effects of electrostatic repulsion due to double layers and (2) including the effects of hydrodynamic interactions (Spielman 1978), which tend to retard collisions prior to contact.

The three mechanisms of transport can be used to formulate a general equation for the kinetics of the flocculation process. One of the factors that controls the overall rate of aggregation during orthokinetic flocculation is the energy provided during slow mixing, which is related to the well recognized mean velocity gradient ($\overline{G}$-value) concept of Camp and Stein (1943). The Smoluchowski theory of orthokinetic flocculation is applicable to a laminar flow field that has a well-defined velocity gradient. Camp and Stein (1943) extended the use of this theory for application to turbulent fields by defining the mean velocity gradient $\overline{G}$ in terms of power input per unit volume as

$$\overline{G} = [P/\mu V]^{1/2} \qquad \text{(Eq 1-1)}$$

Where:

P = total power dissipated
μ = viscosity
V = volume

Equation 1-1 may also be expressed as

$$\overline{G} = [\varepsilon/\nu]^{1/2} \qquad \text{(Eq 1-2)}$$

Where:

ε = power input or dissipation per unit mass
ν = kinematic viscosity

It is intuitively obvious that high energy inputs, or $\overline{G}$-values, during the flocculation process may cause breakup or a disaggregation of the flocs formed. Thus, efficient flocculation is bounded by lower and upper limits on the velocity gradients. These topics on the slow mixing or flocculation operation are also discussed in later sections.

MIXING FOR COAGULATION

Rapid Mixing and the Mechanisms of Coagulation

In water treatment, coagulation by inorganic salts occurs predominantly by two mechanisms: (1) adsorption of hydrolysis species on the colloid, causing charge neutralization; and (2) sweep coagulation, where interactions occur between the colloid and the precipitating hydroxide. Figure 1-2 shows the production of the intermediate kinetic hydrolysis species of alum prior to formation of the aluminum hydroxide precipitate and the two mechanisms of coagulation that are dependent on the chemical species formed. Using the mechanisms shown in Figure 1-2 as a basis, Amirtharajah and Mills (1982) developed the alum coagulation diagram shown in Figure 1-3 as a predictive tool to define the coagulant dosage and pH conditions where each of these mechanisms would dominate. The reactions that precede charge

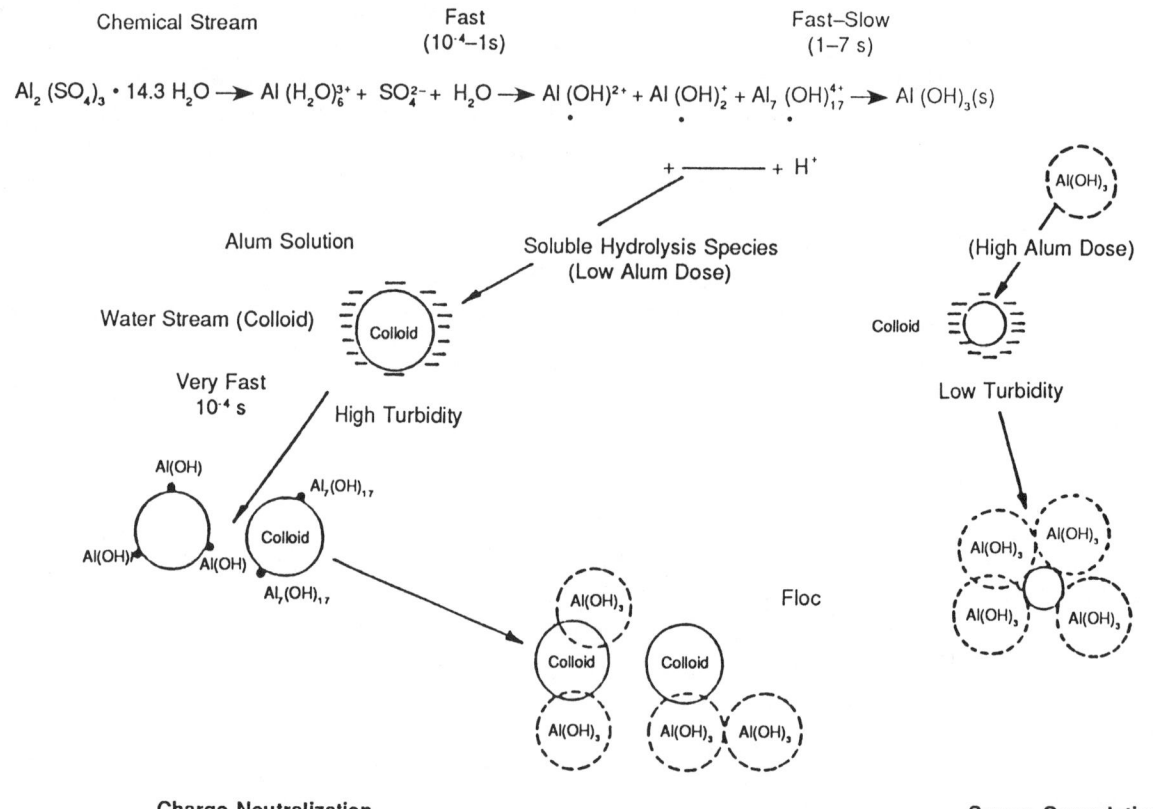

Source: Amirtharajah, A. & Mills, K.M. 1982. *Rapid-Mix Design for Mechanisms of Alum Coagulation. Jour. AWWA, 74:4:210.*

Figure 1-2 Reaction schematics of coagulation.

neutralization with alum are extremely fast; they occur within microseconds without formation of Al(III) hydrolysis polymers and within 1 s if polymers are formed (O'Melia 1972; Hahn and Stumm 1968). The formation of aluminum hydroxide precipitate before sweep coagulation is slower and occurs in the range of 1–7 s (Letterman, Quon, and Gemmell 1973). Because of the competitive nature of the reactions that are involved in the two modes of coagulation, the speedy reaction times imply that for charge neutralization it is imperative that the coagulants be dispersed in the raw water stream as rapidly as possible (less than 0.1 s) so that the hydrolysis products that develop in 0.01–1 s will cause destabilization of the colloid. In contrast, for sweep coagulation, in which the hydroxide formation is in the range of 1–7 s, it is evident that extremely short dispersion times and high intensities of mixing are not as crucial as in charge neutralization. Recently, attempts have been made (Amirtharajah 1981; Amirtharajah and Mills 1982; Amirtharajah and Trusler 1986) to distinguish between the requirements for rapid mixing on the basis of the major mode of coagulation.

Rapid mixing for sweep coagulation. In sweep coagulation, physical interaction occurs between the voluminous precipitates formed (iron or aluminum hydroxide) and the raw water colloids. In typical water treatment practice under sweep coagulation conditions, the water is supersaturated by three to four orders of magnitude, and the hydroxide is precipitated very quickly. Under these circumstances, the chemical conditions for rapid precipitation and subsequent flocculation of particles are significantly more important than transport interactions between the colloid and the hydrolysis products during destabilization. Thus, only the chemical

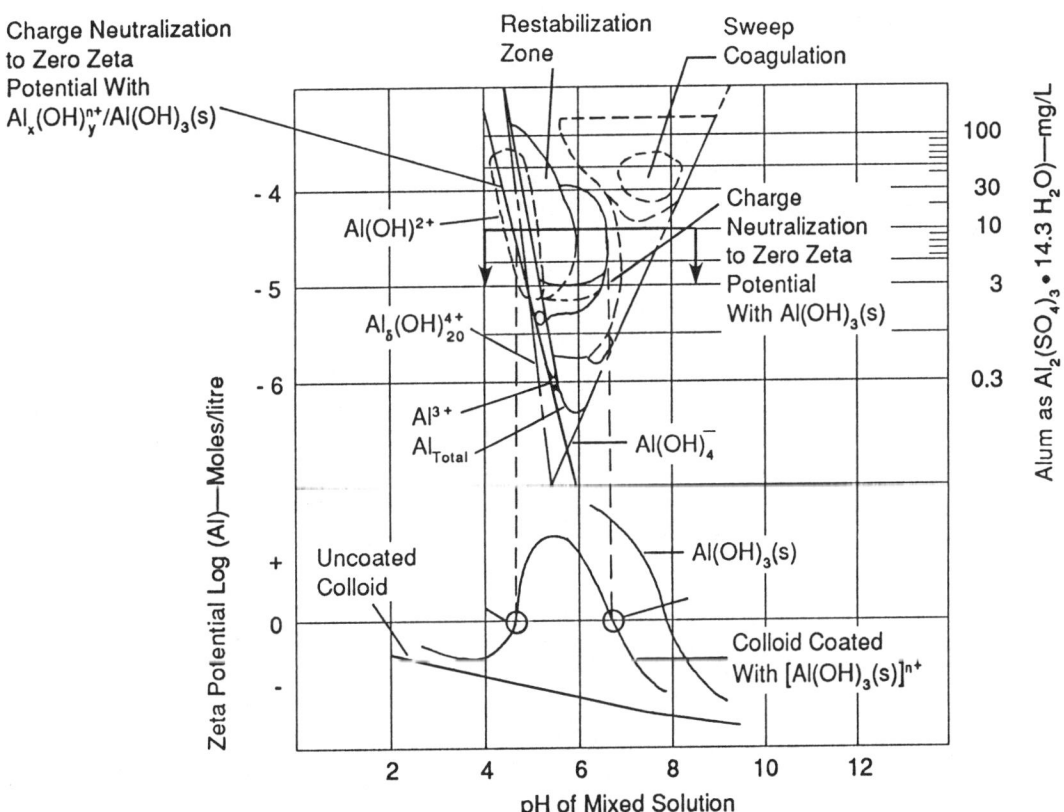

Figure 1-3 The alum coagulation diagram and zeta potential.

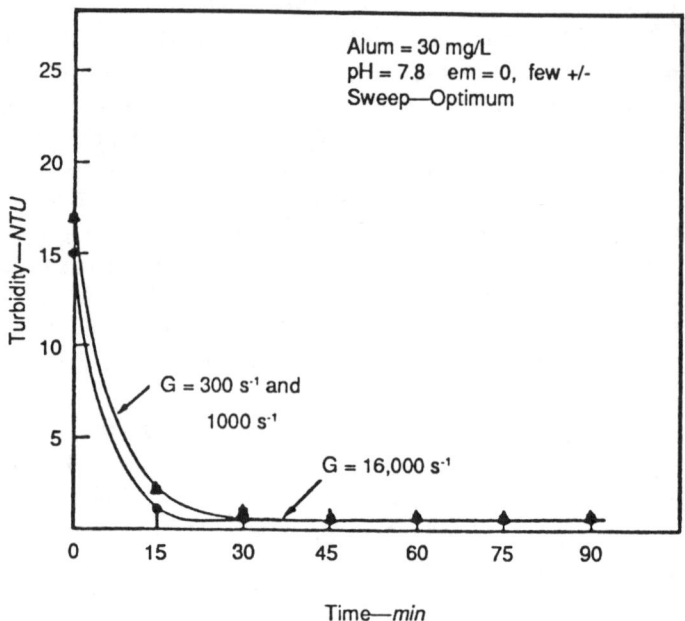

Source: Amirtharajah, A. & Mills, K.M. 1982. Rapid-Mix Design for Mechanisms of Alum Coagulation. Jour. AWWA, 74:4:210.

Figure 1-4 Similar settled-water turbidity curves for optimum sweep coagulation.

aspects of the destabilization step and the transport aspects of flocculation are important. Amirtharajah and Mills (1982) showed that when sweep coagulation is dominant, the results are indifferent to rapid mixing conditions. Figure 1-4 shows a sample of data for different rapid-mix intensities under optimum sweep coagulation conditions defined by the coagulation diagram for alum. It is seen that $\overline{G}$-values from 300 s^{-1} to 16,000 s^{-1} give the same settled-water turbidities.

Since the chemistry of the system and the major mode of coagulation (charge neutralization or sweep coagulation) influences the selection of the rapid-mixing device and its design, it is important to define the chemical conditions for ferric coagulants in addition to those for alum. Figure 1-5 shows the iron coagulation diagram developed by Johnson and Amirtharajah (1983). The diagram may be used to determine the sweep coagulation conditions under which the effectiveness of coagulation will probably be indifferent to rapid mixing conditions. At the lower iron dosages and pH conditions close to the corona region (i.e., dose = 2–15 mg/L and pH = 5.7 to 6.2) charge neutralization will be dominant and high intensities of mixing ($\overline{G}$-values of 3000–4000 s^{-1}) for very short times (< 1 s) would probably be beneficial.

Turbulent rapid mixing for charge neutralization. For colloidal particles to be destabilized in the charge neutralization mechanism of coagulation, there has to be transport or collisions between the colloids and the incipiently forming products of the hydrolysis reactions (see Figure 1-1). This mode of coagulation at low dosages of chemicals often produces small destabilized pinpoint floc that are ideal

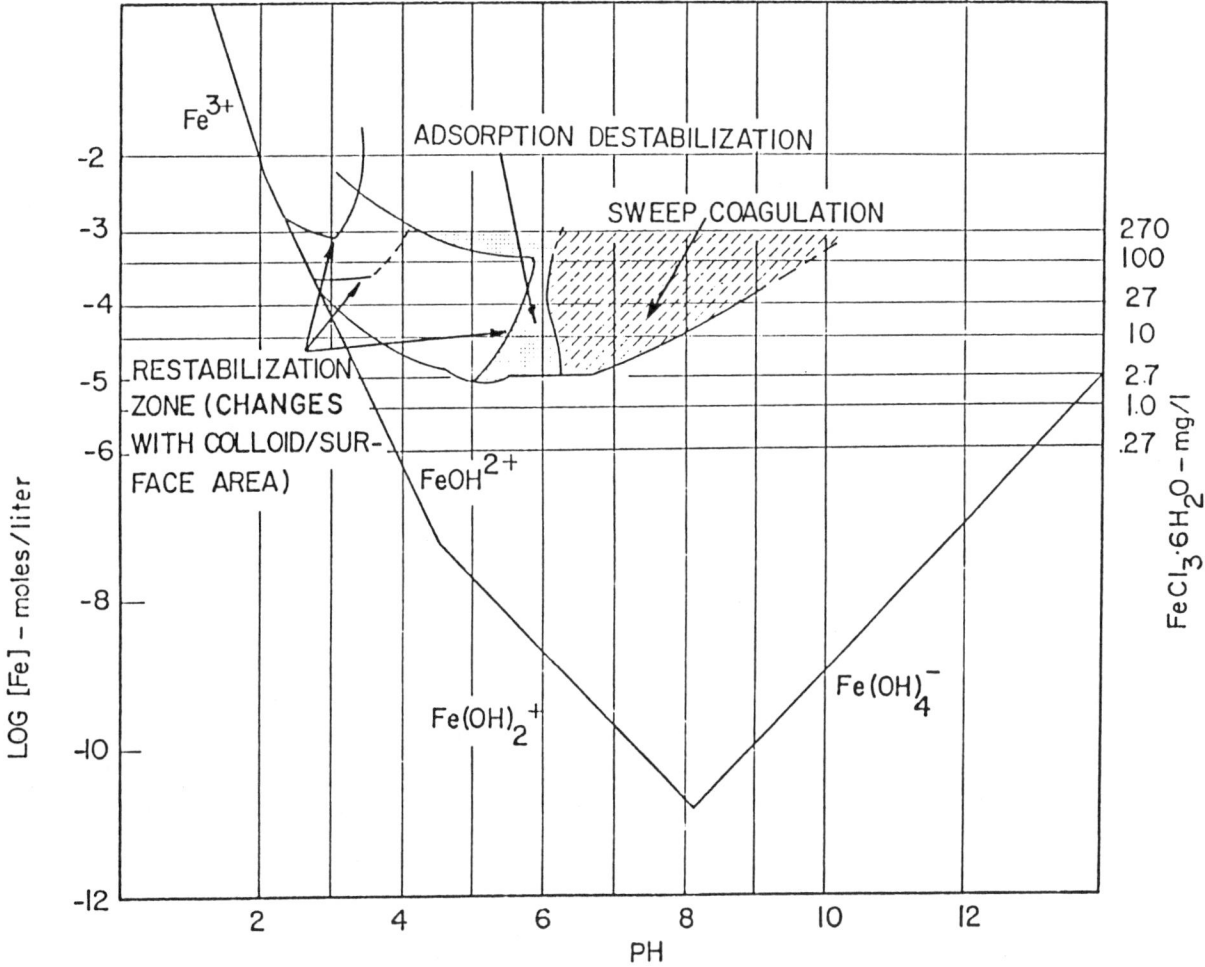

Source: Johnson, P.N. & Amirtharajah, A. 1983. Ferric Chloride and Alum as Single and Dual Coagulants. Jour. AWWA, 75:5:232.

Figure 1-5 Iron (III) coagulation diagram.

for direct filtration. Amirtharajah and Trusler (1986) have applied this concept of transport during the destabilization step to develop a theory for analyzing the required turbulent energy for rapid mixing. The following section presents their theoretical development.

Theory for Particle Destabilization by Inorganic Coagulants in Turbulent Fields

The theory for particle destabilization is developed using accepted models of particle encounters for flocculation in turbulent fields (Saffman and Turner 1956; Delichatsios and Probstein 1975), modified by two novel assumptions.

Flocculation kinetics in turbulent fields. A generalized framework for analysis of flocculation can be developed as follows (presented by O'Melia [1980]). Consider bimolecular collisions between particles of diameters d_1 and d_2 that come into contact by the jth transport mechanism. The rate of contact $N_j(d_1, d_2)$ is expressed as

$$N_j(d_1, d_2) = K_j n(d_1) n(d_2) \qquad \text{(Eq 1-3)}$$

Where:

K_j = rate constant
$n(d_1)$ = number concentration of particles of diameter d_1
$n(d_2)$ = number concentration of particles of diameter d_2

The rate constant for laminar shear K_{LS} (orthokinetic flocculation) is expressed in the very well-known Smoluchowski expression

$$K_{LS} = \frac{(d_1 + d_2)^3}{6} \left(\frac{du}{dz}\right) \qquad \text{(Eq 1-4)}$$

Where:

(du/dz) = laminar velocity gradient

In a turbulent field, collisions due to orthokinetic flocculation tend to predominate for large particles of several microns in diameter. As noted previously, Camp and Stein (1943) obtained a result for orthokinetic flocculation in turbulent systems by a somewhat arbitrary setting of $(du/dz) = \overline{G} = (\varepsilon/\nu)^{1/2}$. A single velocity gradient has no rigorous meaning in a turbulent field composed of systems of eddies. However, with this assumption and combining Eq 1-2, Eq 1-3, and Eq 1-4, the equation for flocculation under conditions of turbulent shear is

$$N_{TS}(d_1, d_2) = \frac{n(d_1) n(d_2)}{6} (d_1 + d_2)^3 \left(\frac{\varepsilon}{\nu}\right)^{1/2}$$

$$= (0.166) n(d_1) n(d_2) (d_1 + d_2)^3 \left(\frac{\varepsilon}{\nu}\right)^{1/2} \qquad \text{(Eq 1-5)}$$

Saffman and Turner (1956) considered the collision frequency to occur by a turbulent flux of fluid into a collisional sphere of diameter $(d_1 + d_2)$. Assuming that at two close points in a turbulent field the relative motion is that of uniform strain, and using Taylor's result for energy dissipation in isotropic turbulence, they derived the following result for the collision rate:

$$N_{TS}(d_1, d_2) = (0.162) n(d_1) n(d_2) (d_1 + d_2)^3 \left(\frac{\varepsilon}{\nu}\right)^{1/2} \qquad \text{(Eq 1-6)}$$

The above result assumed that for collisions to occur in a flow field of uniform strain, the radius of the sphere of influence $(d_1 + d_2)/2$ was small compared to Kolmogoroff's microscale η given by

$$\eta = \left(\frac{\nu^3}{\varepsilon}\right)^{1/4} \qquad \text{(Eq 1-7)}$$

Delichatsios and Probstein (1975) developed a kinetic model for coagulation in isotropic turbulent flow for both conditions where the radius of the collision sphere $(d_1 + d_2)/2$ is smaller than and greater than the microscale η. The following development modified Delichatsios and Probstein's presentation for a monodisperse system to two unequal diameters d_1 and d_2. A simple binary collision mean free path concept was applied to small particles in isotropic turbulence. Their equations for flocculation in the viscous and inertial subranges of turbulence are given by the following expressions.

For the viscous subrange

$$N_{TS1}(d_1, d_2) = (0.051)\, n(d_1)\, n(d_2)\, (d_1 + d_2)^3 \left(\frac{\varepsilon}{\nu}\right)^{1/2}$$

$$\text{for } \frac{(d_1 + d_2)}{2} < \eta \quad\quad\quad \text{(Eq 1-8)}$$

and for the inertial subrange

$$N_{TS2}(d_1, d_2) = (0.427)\, n(d_1)\, n(d_2)\, (d_1 + d_2)^{7/3}\, (\varepsilon)^{1/3}$$

$$\text{for } \frac{(d_1 + d_2)}{2} < \eta \quad\quad\quad \text{(Eq 1-9)}$$

For the case where the particles are smaller than the microscale, it is seen that the three different models, Eq 1-5, Eq 1-6, and Eq 1-8, give the same dependencies on the physical quantities, except for the numerical constants, and can be represented by

$$N_{TS1}(d_1, d_2) = C_3\, (d_1 + d_2)^3 \left(\frac{\varepsilon}{\nu}\right)^{1/2} \quad\quad\quad \text{(Eq 1-10)}$$

Similarly, for particles larger than the microscale

$$N_{TS2}(d_1, d_2) = C_4\, (d_1 + d_2)^{7/3}\, (\varepsilon)^{1/3} \quad\quad\quad \text{(Eq 1-11)}$$

In Eq 1-10 and Eq 1-11, the coefficients C_3 and C_4 include the numerical constants as well as $n(d_1)$ and $n(d_2)$; these coefficients would be constant for a given colloid and a given coagulant dosage.

It needs to be emphasized that these equations are conceptually similar to Smoluchowski's expression for laminar shear (Eq 1-4) and do not incorporate the corrections for electrostatic effects or hydrodynamic retardation. A rationale for the elimination of these two corrections in model formulation is presented elsewhere (Amirtharajah and Trusler 1986). The above expressions are used in the development of the following model.

Particle destabilization in turbulent fields. Consider a colloidal suspension of $n(d_1)$ particles of diameter d_1. Coagulant chemicals at a very low pH (e.g., alum) are added as a pure solution into this suspension. The positively charged hydrolysis species that are formed have to be quickly carried by eddies of turbulence of microscale dimensions so that they interact with the colloidal suspension particles of diameter d_1. These collisions between the negatively charged colloid and the positively charged microspecies cause destabilization of the colloid.

The conceptual framework just presented implies homogeneous nucleation with formation of microsolids followed by interactions with the colloids for particle destabilization. Since heterogeneous nucleation may occur in a system with colloids, the proposed ideas can be easily modified for such systems by assuming that soluble hydroxy species or ions are carried by the turbulent eddies that interact with the colloids to form precipitates on the surfaces of the colloidal particles.

The major assumptions of the theory are shown in a schematic representation of the analytical framework in Figure 1-6. The assumptions are: (1) one of the "particle" dimensions could be replaced by an eddy of the size of the microscale of turbulence; and (2) the power input can also be expressed in terms of the size of the microscale eddies in the collision models.

Viscous subrange. For the viscous subrange, consider Eq 1-10. During rapid mixing and destabilization assume that the collisions occur between the colloid particles (d_1) and eddies of the size of the microscale η, which is the d_2 particle carrying several positively charged species. The rate of destabilization for the viscous subrange $(dD/dt)_v$ will be equal to the rate of collisions.

$$\left(\frac{dD}{dt}\right)_v = N_{TS1}(d_1, d_2) = C_3 (d_1 + \eta)^3 \left(\frac{\varepsilon}{\nu}\right)^{1/2} \quad \text{(Eq 1-12)}$$

In Eq 1-12, substituting for ε from Eq 1-7 gives

$$\left(\frac{dD}{dt}\right)_v = C'_v (d_1 + \eta)^3 \left(\frac{1}{\eta}\right)^2 \quad \text{(Eq 1-13)}$$

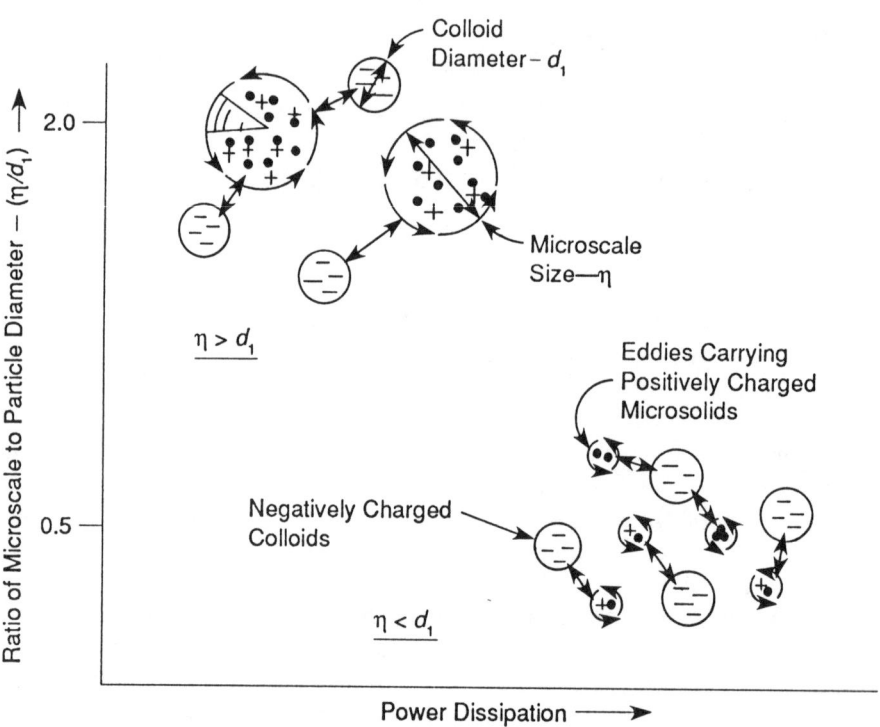

Source: Amirtharajah, A. & Trusler, S.L. 1986. *Destabilization of Particles by Turbulent Rapid Mixing.* Jour. Envir. Engrg., Div. ASCE, 112:6:1085, December 1986. Reprinted with permission of ASCE.

Figure 1-6 Schematic diagram of destabilization due to coagulants carried by microscale eddies.

Where:

C'_v = vC_3, a constant at a given temperature.

Equation 1-13 is one of the final equations of the theory. It gives the particle destabilization as a function of microscale eddy size for the viscous subrange. Expanding the cubic term and differentiation of Eq 1-13 with respect to η will indicate the variations of the rate of destabilization with eddy size.

$$\frac{d}{d\eta}\left[\frac{dD}{dt}\right]_v = C'_v\left[-2\left(\frac{d_1^3}{\eta^3}\right) - 3\left(\frac{d_1^2}{\eta^2}\right) + 1\right] \quad \text{(Eq 1-14)}$$

Setting Eq 1-14 equal to zero leads to

$$2\left(\frac{d_1}{\eta}\right)^3 + 3\left(\frac{d_1}{\eta}\right)^2 - 1 = 0 \quad \text{(Eq 1-15)}$$

Equation 1-15 has one positive root given by $(d_1/\eta) = 0.5$. Hence, the ratio of microscale to particle diameter is

$$\left(\frac{\eta}{d_1}\right) = 2.0 \quad \text{(Eq 1-16)}$$

For $(\eta/d_1) = 2.0$, it can be shown that the second derivative of Eq 1-13 is positive. Hence, this value is a minimum. Under the viscous subrange conditions, the destabilization rate is a minimum when the microscale is twice the diameter of the colloid.

Inertial subrange. For the inertial subrange, consider Eq 1-11. With similar assumptions, the rate of destabilization is

$$\left(\frac{dD}{dt}\right)_i = C_4(d_1 + \eta)^{7/3}(\varepsilon)^{1/3} \quad \text{(Eq 1-17)}$$

Substituting Eq 1-7

$$\left(\frac{dD}{dt}\right)_i = C_i(d_1 + \eta)^{7/3}\left(\frac{1}{\eta}\right)^{4/3} \quad \text{(Eq 1-18)}$$

Where:

C_i = vC_4 = constant

Differentiating Eq 1-18

$$\frac{d}{d\eta}\left[\left(\frac{dD}{dt}\right)_i\right] = \frac{C_i}{3}\left[\left(\frac{d_1}{\eta}\right) + 1\right]^{4/3}\left\{7 - 4\left[\left(\frac{d_1}{\eta}\right) + 1\right]\right\} \quad \text{(Eq 1-19)}$$

Again for $(d/d\eta)[(dD/dt)_i] = 0$, $\{7 - 4[(d_1/\eta) + 1]\} = 0$, therefore $(d_1/\eta) = 0.75$. The ratio of microscale to particle diameter at the turning point for a minimum is

$$\left(\frac{\eta}{d_1}\right) = 1.33 \qquad \text{(Eq 1-20)}$$

Therefore, under the inertial subrange conditions, the destabilization rate is a minimum when the microscale is 1.33 times the colloid particle diameter.

The above result for minimum destabilization for the inertial subrange occurred at $(\eta/d_1) = 1.33$, whereas the initial assumption for the inertial subrange was $(\eta/d_1) \ll 1.0$. This seeming contradiction does not invalidate the theory. The reasons for the extensions of the results beyond their initial rigorous assumptions are discussed by Amirtharajah and Trusler (1986), and experimental data validating such extrapolations for flocculation are presented by Tambo and Watanabe (1979b).

The preceding analysis is reasonably complete and gives the beautifully simple result that the rapid mixing conditions for particle destabilization in the charge-neutralization mode under all conditions of the universal equilibrium range of turbulence should avoid the range of conditions in which η is approximately equal to $1.33d_1$ to $2.0d_1$. It is remarkable that analyses in both subranges of turbulence give the minimum conditions so close to one another.

Experimental Results for Inorganic Coagulants

Experimental investigation. Preliminary experimental data supporting the preceding theoretical analysis have been presented (Amirtharajah and Trusler 1986). The experiments were conducted on a direct-filtration pilot plant with a baffled tank backmix reactor as the rapid-mix unit and a dual media filter. Both alum and ferric chloride were used as coagulants to destabilize suspensions of silica (min-u-sil) used as turbidity slurries. The chemical dosage and pH conditions were controlled close to the charge-neutralization corona region of the coagulation diagrams. For example, the alum experiments at a dosage of 8 mg/L (as $Al_2(SO_4)_3 \cdot 14.3H_2O$) and pH = 6.9 are very close to the region of zero zeta potential, as can be seen in Figure 1-3. The intent was to produce small, destabilized floc that had a high filterability. The efficiency of destabilization was measured by the response of filtration and the electrophoretic mobility.

The mixing parameters in the rapid mixer were computed using standard expressions for power dissipation in terms of impeller speed and dimensions. Cutter's results (1966) indicated that the energy dissipation in an annular zone of thickness ½ in. (12 mm) around the impeller blades, which is the zone of maximum turbulence intensity, has an energy dissipation ε, that is 56.5 times the mean energy dissipation ε in the entire reactor. The destabilization theory was checked using the Kolmogoroff microscales calculated in this zone of maximum turbulence around the impeller. The computed parameters for impeller speed, mean velocity gradient, velocity gradient, and microscales in the zone of maximum turbulence are shown in Table 1-1. Recent research on the power dissipation at various locations within a stirred tank are reported by Laufhutte and Mersmann (1984, 1985a, 1985b).

The response of the filter to variations in the destabilization rate was determined using a filterability number that incorporated both the average effluent turbidity and the total head loss developed in the filter run. The best filtration efficiency corresponds to the lowest filterability number.

Table 1-1 Computed Parameters for Rapid-Mix Tank

Impeller Speed, rpm	Total Power Dissipation, ft-lb/s	Mean Power Dissipation per Unit Mass $\bar{\varepsilon}$, ft^2/s^3	Mean Velocity Gradient $\bar{G}$, s^{-1}	$\overline{GT}$	Velocity Gradient Around Impeller G_{imp}, s^{-1}	Power Dissipation per Unit Mass Around Impeller ε, ft^2/s^3	Turbulence Microscale in Zone Around Impeller η_i, μm
50	0.02	0.044	70	3640	520	2.5	40.8
100	0.18	0.40	210	10,900	1560	22.6	23.5
150	0.62	1.38	380	19,760	2900	78.0	17.3
250	2.74	6.09	810	42,100	6100	344.1	11.9
400	11.21	24.91	1640	85,300	12,300	1407.4	8.4
500	21.90	48.66	2290	119,100	17,200	2749.3	7.1
600	37.84	84.08	3000	156,000	22,600	4750.5	6.2
700	60.09	133.52	3800	197,600	28,500	7543.9	5.5

NOTE: 1 ft-lb/s = 1.36 W; 1 ft = 0.305 m.

Results and analysis. A comparison between theory and some experimental results is made in Figures 1-7 and 1-8. In the top sections of the figures, the experimental data of electrophoretic mobility and filterability number, which measure particle destabilization, are plotted versus mean velocity gradients. All of the electrophoretic mobility values shown are negative. The bottom sections of the figures show the corresponding theoretical computations of the destabilization rate curves (Eq 1-13 and Eq 1-18) as a function of the ratio of impeller zone microscale to particle diameter. The maximum turbulence occurs in this zone around the impeller.

The data in Figure 1-7 for ferric chloride treatment of particles with a mean size of 3 μm show a good match of experiment with theory. The minimum in destabilization rate at $(\eta_i/d_1) = 2.0$ for the viscous subrange almost exactly matches the maxima in electrophoretic mobility and filterability number. For best destabilization, the mean velocity gradient should be in the range of 700–1000 s^{-1} or greater than 3500 s^{-1}.

The experiments with ferric chloride were repeated with the higher mean size of particles of 6 μm. The filterability data in Figure 1-8 show small peaks at mean $\bar{G}$-values of 800–1000 s^{-1} and 3000 s^{-1} corresponding to (η_i/d_1) values of 2.0 and 1.0. The electrophoretic mobility data show a peak at a $\bar{G}$-value of 1700 s^{-1} corresponding to an $(\eta_i/d_1) = 1.33$ for the inertial subrange. The data are not incontrovertible. The major weaknesses of the experiments were not using single-sized particles and not using a mixing zone where the power dissipation and the microscale eddies were uniform in size.

Some practical evidence. In addition to the experimental results shown above, the theory seems to have indirect experimental verification by practical and empirical rules of thumb used for the design of rapid mixers over several decades. Most rapid mix units used in practice are designed as backmix reactors ($\bar{G} \sim$ 700–1000 s^{-1}) or inline blenders ($\bar{G} \sim$ 3000–5000 s^{-1}). A comparison of Figures 1-7 and 1-8 shows that these values of velocity gradient are in fact on either side of the region that have a minimum destabilization rate.

The theory and results presented above have application for particle destabilization with inorganic coagulants. The evidence in the literature for rapid mixing

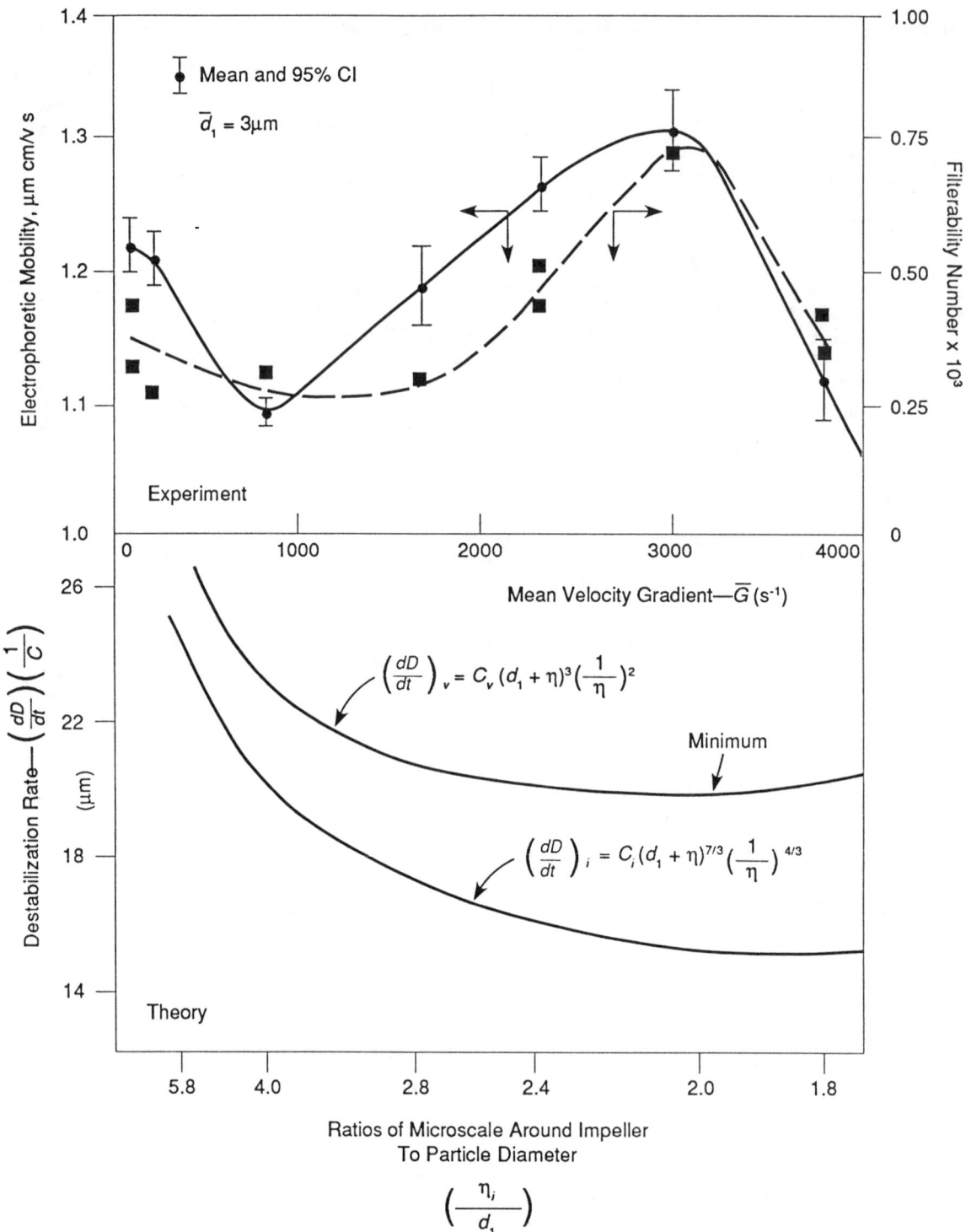

Source: Amirtharajah, A. & Trusler, S.L. 1986. *Destabilization of Particles by Turbulent Rapid Mixing.* Jour. Envir. Engrg., Div. ASCE, 112:6:1085, December 1986. Reprinted with permission of ASCE.

Figure 1-7 Comparison of theory and experiments: ferric chloride treatment of particles with mean size of 3 μm.

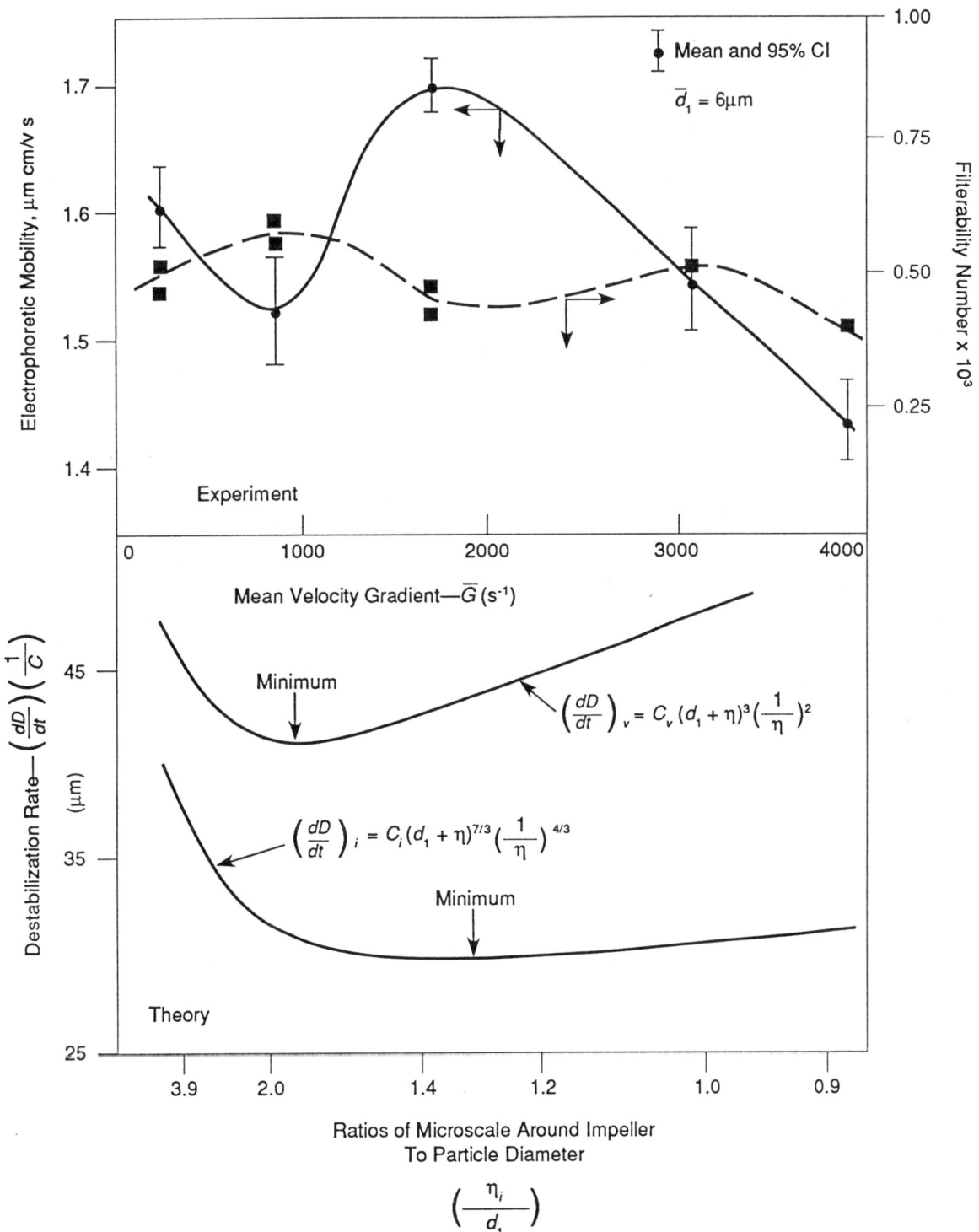

Source: Amirtharajah, A. & Trusler, S.L. 1986. Destabilization of Particles by Turbulent Rapid Mixing. Jour. Envir. Engrg., Div. ASCE, 112:6:1085, December 1986. Reprinted with permission of ASCE.

Figure 1-8 Comparison of theory and experiments: ferric chloride treatment of particles with mean size of 6 μm.

with polymers is partially different from that presented. The implications of the theory developed for organic polymers have not been assessed.

Rapid Mixing for Polymer Coagulants

The mechanisms of coagulation with organic polymers are charge neutralization and interparticle bridging. Since the competing reactions of adsorption onto the colloids and precipitation as a hydroxide, which occur simultaneously with inorganic coagulants during the rapid mix stage, do not occur with organic polymers, it may be assumed that the high intensities of mixing are also not imperative. The evidence in the literature (Amirtharajah and Kawamura 1983; Tomi and Bagster 1978; Keys and Hogg 1978) supports this view, and suggests that for polymers the interparticle bridging process causing floc aggregation and possible breakup of aggregated flocs due to the high intensities of turbulence are important controlling mechanisms.

Tomi and Bagster (1978) showed that at the initial stages of flocculation with polymers aggregates grow rapidly in size, reaching a critical maximum size (d_{max}). This maximum size is inversely related to the energy input during mixing ($d_{max} \propto \varepsilon^{-0.5}$). For any condition of mixing there is an optimum concentration of polymer flocculant, which decreases with impeller speed. This is consistent with the greater compression of polymer at higher speeds, leading to reduced flocculant dosage to maximize bridging. At the critical size, rupture occurs, followed by a gradual floc degradation leading to a diminution in size. This erosion mechanism is a function in time T and can be expressed empirically as

$$d_{max} \propto T^{-\beta} \qquad \text{(Eq 1-21)}$$

Where:

β = 0.5 to 0.2
$T > T^*$ = time for optimum flocculation or maximum growth in floc size

The critical times T^* were 1–3 min at an impeller speed of 500 rpm and 6–8 min at a speed of 250 rpm. These results suggest that time and intensity of mixing are both important parameters for mixing with polymers. These concepts are shown schematically in Figure 1-9.

Keys and Hogg (1978) studied mixing with polymers in a baffled tank reactor with a turbine impeller having six flat blades. The coagulant was a nonionic polyacrylamide polymer with a molecular weight of 15×10^6 daltons. They found that the optimum $\overline{G}$-values were approximately 250 s^{-1} to 750 s^{-1} and the $\overline{GT}$-values ranged from 16,000 to 28,000. The corresponding times for mixing were approximately 60 s at the lower $\overline{G}$-value to 30 s at the higher $\overline{G}$-value. In contrast to these low $\overline{GT}$ values, Leu and Ghosh (1988) found that optimum $\overline{GT}$-values were in the range of 200,000. The latter researchers also demonstrated that flocs formed with polymers ruptured at $\overline{G}$-values greater than 350 s^{-1}. It needs to be emphasized that the $\overline{GT}$-values noted are for polymers used alone, and the time periods include the flocculation process, too. While the $\overline{G}$- and $\overline{GT}$-values for optimum flocculation indicated mixing times for rapid mixing plus slow mixing of several minutes (4–10 min), the authors indicated that the most rapid growth of floc occurred during the first few seconds of rapid mixing. They concluded that rapid mixing (i.e., higher $\overline{G}$-values) and limitations on $\overline{GT}$-values were critical in flocculation with polymers.

When polymers are used as coagulant aids in combination with an inorganic coagulant, it is useful to consider rapid mixing in two stages, the first with higher

A. CHARGE NEUTRALIZATION AND BRIDGING

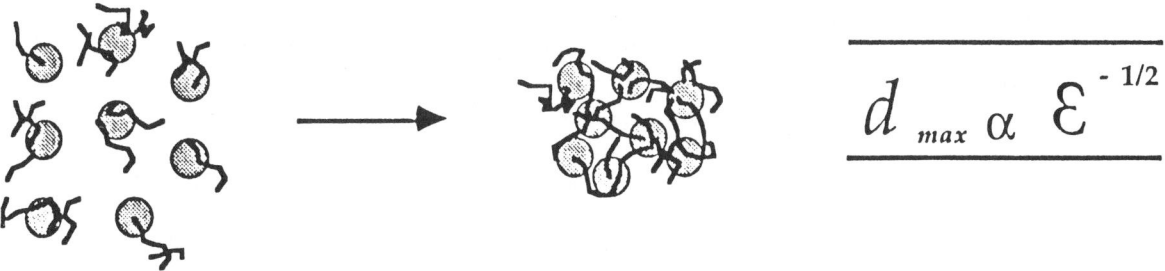

B. EROSION AND BREAKUP (Follows first stage, not simultaneous)

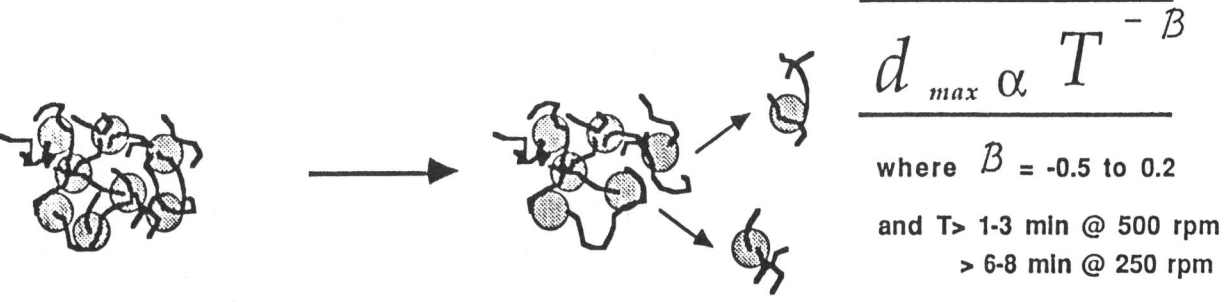

Figure 1-9 The two stages of polymer flocculation: (a) charge neutralization and bridging; (b) erosion and breakup (following first stage, not simultaneous).

$\overline{G}$-values (700–1000 s^{-1} for a few seconds) to destabilize the colloids with the inorganic coagulants and the second to enhance bridging and floc formation with the polymers with lower $\overline{G}$-values (200–400 s^{-1} for a few minutes). Very often when inorganics and polymers are added together for coagulation, engineering cost criteria would tend to give least-cost designs based on the mixing requirements for polymers ($\overline{G}$-values of 200–400 s^{-1} for a few minutes).

Floc Matrix of Clay and Color Colloids With Inorganic Coagulants

Floc is composed of particles that are to be coagulated and of precipitated metal coagulants. Most of the particles to be coagulated are clay particles and natural organic color colloids. Aquo aluminum complexes in soluble and precipitated forms constitute the majority of precipitated metal coagulants.

The very basic element of a floc is called the elementary particle. Flocs that have grown are agglomerated forms of the elementary particles. The characteristics of a grown floc are determined by the composition, or structure, of the elementary particle and the mode of the flocculation reaction, or agitation.

Characteristics of the elementary particles are determined by the kind of particles to be coagulated and their concentration, the kind of metal coagulants used and their concentration, pH, water temperature, intensity of flash mixing, and similar influences. The relationships between these influences and the resulting particle characteristics are very complicated and will be discussed as precisely as possible in other chapters of this book. This chapter discusses the basic patterns of the composition of the elementary particles. An overview of the mode of flocculation reaction and agitation is presented in the section on mixing for flocculation.

The most common particles to be coagulated are clay particles, which are the major component of suspended matter in a natural water. The size of an individual clay particle is approximately 1 μm to a few μm. The clay particles are dispersed in water by the surface negative charges, which cause the interacting electrical double layers to repel the particles from each other. The number of clay particles in a 1-mg/L kaolinite suspension is about 10^5–10^6/mL.

Another major constituent of natural water is organic color, which is composed primarily of humic substances. Humic substances in natural water can be separated into two major fractions: humic acid and fulvic acid.

Effect of constituent size on coagulation. The importance of the size of organic matter on coagulation and other processes has been recognized by many investigators. Gjessing and Lee (1967) suggested the fractionation of humic substances on Sephadex columns. Brodsky and Prochazka (1975) carried out Sephadex G-75 gel-chromatography analyses for several river raw waters and their chemically treated coagulated water. They revealed that the higher molecular weight fractions of humic substances were removed effectively by coagulation. Tambo and Kamei (1978) and Tambo et al. (1989) reported comprehensive, quantitative treatability evaluations with unpolluted and polluted natural waters, and wastewaters on the Sephadex G-15 gel chromatogram. They revealed that humic substances having molecular weights of less than 1500 are not removed effectively by alum coagulation, but that the higher molecular weight fractions are removed effectively at pH 5.0–6.0. Semmens and Ayers (1985) used ultrafiltration to fractionate the humic substances in Mississippi River water and its coagulated water. They reported that coagulation can remove intermediate and high molecular weight substances (1–10K and 10–100K) effectively, but low molecular weight substances (<1000) cannot be removed well. Nearly 25 years ago, Hall and Packham (1965) noted that there is no distinction between humic and fulvic acids in relation to the mechanism of coagulation from the viewpoint of complex formation, but fulvic acid (usually having smaller molecular weights, i.e., less than 1500) cannot be removed effectively by alum and iron coagulation.

From the above knowledge it is evident that only the higher molecular weight fraction of natural organic color, i.e., humic substances, may be removed through aluminum or iron coagulation. A total organic carbon concentration (TOC) of 1 mg/L (i.e., about 4 color units [cu]) is nearly equivalent to 2–3 mg/L of humic substances in some colored water. If the average molecular weight of removable higher molecular weight humic substances is about 3000, and they are about 60–80 percent of the total humics, then 1 mg/L of humic substances as TOC contains coagulable color colloids of the order of 10^{14} particles per millilitre.

The active aquo-aluminum complexes at pH 5.0–5.5 could include such species as $Al_8(OH)_{20}^{4+}$, $Al_2(OH)_2^{4+}$, and $Al_{13}(OH)_{24}^{7+}$ (Matijevic et al. 1961; Stumm and Morgan 1962; Dempsey et al. 1984). A polymer such as $Al_8(OH)_{20}^{4+}$, which has a molecular weight of about 600, is often postulated as an effective polymer for neutralizing the negative charge of colloidal particles and humic substances (Matijevic et al. 1961;

Stumm and Morgan 1962; Edwards and Amirtharajah 1985). At the same time, precipitated aluminum species, which are as large a 1 μm or more, exist at pH > 5.

Electrophoretic mobility. To identify the nature of the aluminum species, electrophoretic mobilities of aluminum species were measured using the electrophoretic mass-transfer method in a U tube (Kruyt 1952) with respect to various aluminum concentrations and hydrolysis reaction times given by Tambo and Itoh (1977) as shown in Figure 1-10. Figure 1-11 shows particle size distribution of hydrolyzed aluminum species obtained through ultrafiltration with respect to various pH's (Tambo and Itoh 1977). These measurements were made after 5-min reaction time with the aluminum concentration of 0.8 mg/L as Al^{3+}. These two

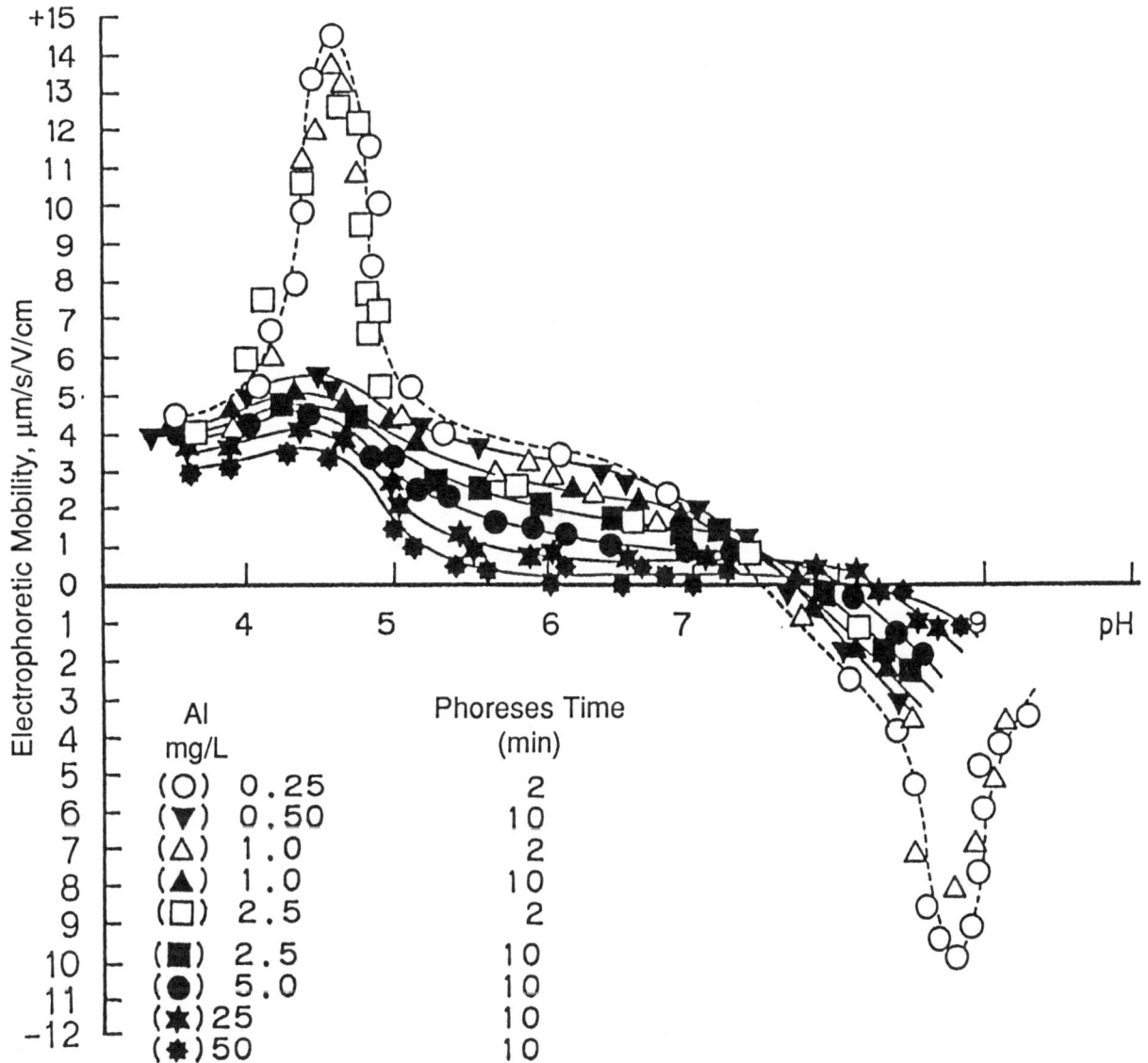

Source: Tambo, N. & Itoh, H. 1977. Electrophoretic Studies on Natural Color Colloid Coagulation by Alum. Jour. Jpn. Wtr. Wks. Assn., 508:38.

Figure 1-10 Change of electrophoretic mobility of aluminum with reaction time.

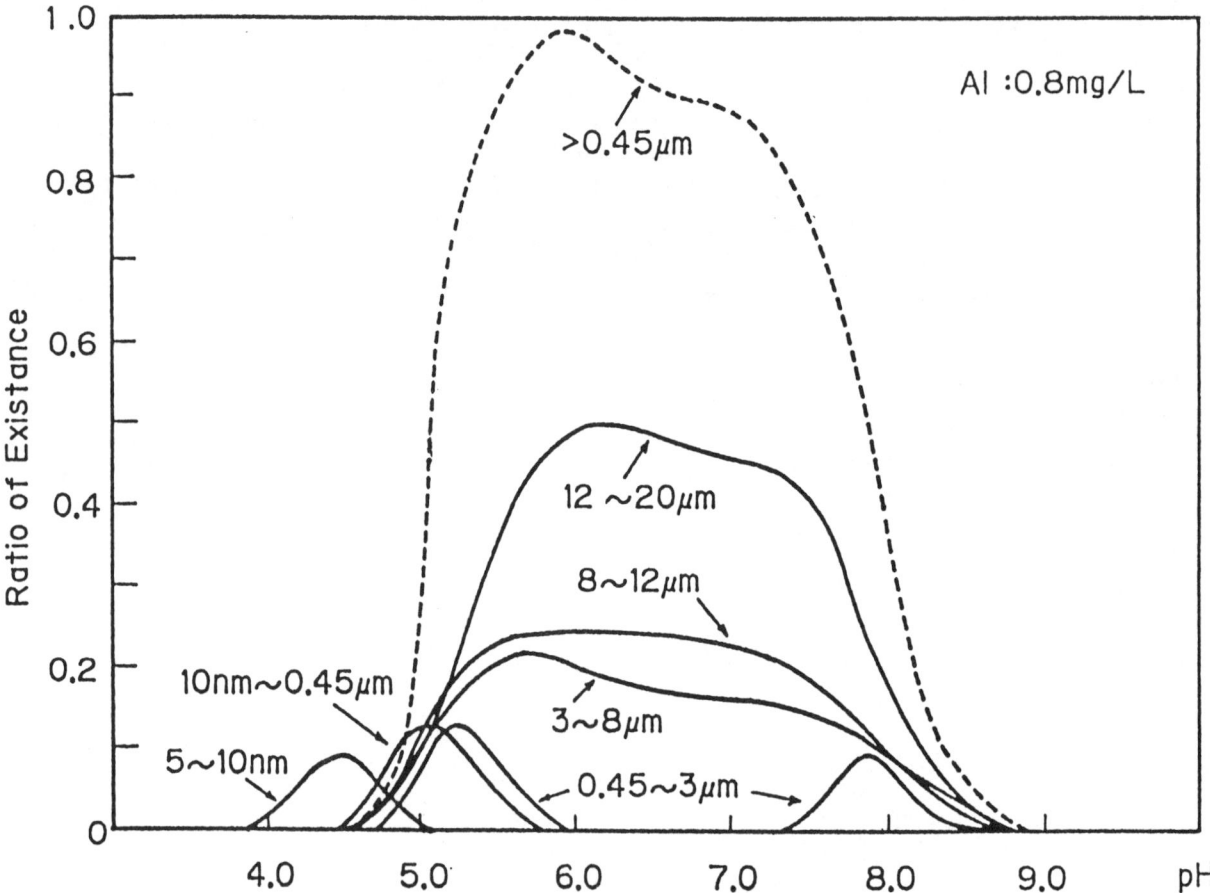

Source: Tambo, N. & Itoh, H. 1977. Electrophoretic Studies on Natural Color Colloid Coagulation by Alum. Jour. Jpn. Wtr. Wks. Assn., 508:38.

Figure 1-11 Particle size distribution of hydrolyzed aluminum.

figures reveal that the highest cationic aluminum polymers of small size exist at pH 4.5 and weakly cationic precipitated aluminum hydroxides exist from pH 5.0–8.5. The change of positive mobility with time at pH 5.0–5.5 was shown to be rather slow. Therefore, competitive reactions of aluminum with both OH^- ions in water and similar types of functional groups on the surfaces of clay particles or humic substances can occur.

The number of OH^- ions present in water is approximately 10^{14}/mL and 10^{12}/mL at neutral and weakly acidic regions, respectively. The number of aluminum ions are 2×10^{16}/mL in an acidic region where aluminum exists as Al^{3+} and 3×10^{15}/mL as $Al_8(OH)_{20}^{4+}$ in weakly acidic regions for a 1-mg/L Al solution. At neutral pH, aluminum hydroxide precipitates, and the volume of 1 mg of aluminum hydroxide is about 50 times as high as 1 mg of clay particles (Tambo and Matsui 1988) and the size of the precipitated particles is more than 1 μm.

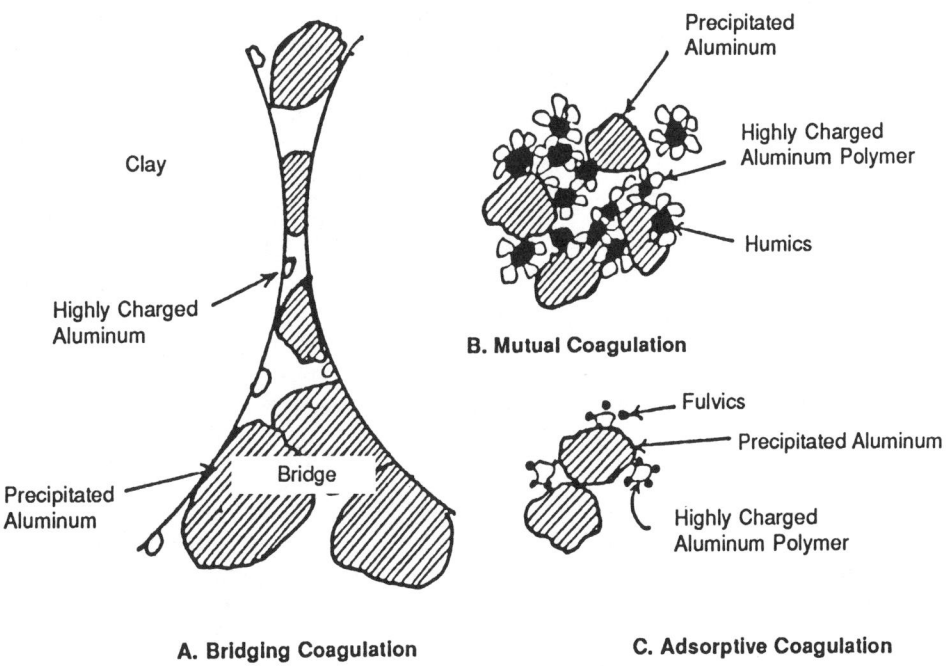

Source: Tambo, N. et al. 1989. Aquatic Humic Substances: Influence on Fate and Treatment of Pollutants. In Humic Substance Removal by Coagulation, *ed. by I.H. Suffet and P. McCarthy, chap. 28, Copyright 1989. American Chemical Society, Washington, D.C. Reprinted with permission.*

Figure 1-12 Matrix of clay or color colloids with hydrolyzed metal coagulant in a floc.

The combination of such cationic and anionic species (i.e., two anionic species as clay and/or humics, and cationic aluminum species) to form flocs is schematically illustrated in Figure 1-12 (Tambo et al. 1989).

Coagulation reaction. In the case of aluminum–clay flocs, relatively small numbers of clay particles, as compared to aluminum and OH⁻ ions, cause the reaction of aluminum hydroxide generation to predominate at the first stage of coagulation. Then the weakly positively charged aluminum hydroxide, which is precipitated, destabilizes and bridges clay particles. The aluminum ion or soluble aquo-aluminum complex that reacts directly with the functional groups on the clay particles is a small percentage. If the concentration of the clay suspension increases, the ratio of the aluminum ion or aquo-aluminum complex that reacts directly with functional groups on the clay becomes slightly higher. However, highly charged soluble aquo-aluminum complexes reverse the surface charge of clay with very small dosages and often cause the restabilization of clay suspensions. Hence, coagulation of clay suspensions is performed close to a neutral pH zone where requirements of both charge neutralization and bridging can be satisfied by weakly positively charged aluminum hydroxide precipitates as shown in Figure 1-12A (Tambo 1965a, 1965b). The restabilization at lower pH values and the effectiveness of coagulation under neutral pH conditions are also shown in the alum coagulation diagram of Figure 1-3.

In the case of coagulation of color colloids, competitive reactions of aluminum humate production and hydrolysis of aluminum during the initial stages of the coagulation operation can occur (Snodgrass et al. 1984; Dempsey et al. 1984). However, in reality, the hydrolysis of aluminum proceeds at a much quicker rate than direct humate generation at the optimum coagulation pH of about 5.5. Hence, the major association of aluminum species and humic substances may occur between soluble hydrolyzed aquo-aluminum complexes and humic substances (Tambo and Itoh 1977; Tambo et al. 1989). In this case, the relative size of the effective aluminum species and the humics to be coagulated may be important. If the size of humic substances to be coagulated is much smaller than the effective aquo-aluminum complex, normal coagulation or even mutual coagulation does not proceed effectively and an adsorption-like mechanism with poor coagulation occurs (Figure 1-12C). However, effective charge neutralization by the highly charged soluble polymers of aluminum with a molecular weight of 1000 or a little less results in the formation of aluminum–humic complexes when interaction occurs with the higher molecular weight (MW) humic substances (MW $\gg 10^3$). The charge-neutralized aluminum–humic complexes are bridged by precipitated aluminum hydroxide to form settlable flocs. Both of the above requirements are satisfied at about pH 5.0–5.5, with the formation of neutral aluminum–humic complexes and precipitated aluminum hydroxide at the same time. The color-aluminum floc matrix generated under these conditions is illustrated in Figure 1-12B.

When humic substances and clay colloids coexist, coagulation proceeds effectively in the weakly acidic region, similar to the coagulation of color colloids reported by many researchers (Tambo 1984; Edwards and Amirtharajah 1985; Semmens and Ayers 1985). In this case, the aluminum–humic-complex formation proceeds first, and the resulting products serve as bridging precipitates for the agglomeration of clay particles with the effect of charge neutralization of the clay. The floc matrix formed on the basis of the above mechanism is illustrated in Figure 1-13.

Floc Matrix With Organic Polymer

The use of synthetic organic polymers to improve coagulation has become common. The polymers used for this purpose can be classified into two categories on the basis of their use. The first category is cationic polymers, which are used as a charge neutralizer (destabilizer) and, at the same time, as a bridging agent of the particles. In the second category are anionic and nonionic polymers, which are usually used with metal coagulants and behave as a bridging or bridge-reinforcement agent.

Cationic polymers. The cationic polymers commonly used have molecular weights of less than 10^5. These polymers with relatively smaller molecular weight can be used economically to neutralize the negative surface charge of the particles to be coagulated. For clay particle coagulation, cationic polymers of molecular weight 10^5 are effectively used as the charge neutralizer and bridging agent at the same time. The number concentration of polymer of this size in a 1-mg/L polymer solution is on the order of 10^{13}/mL. This number is relatively smaller than that of color colloids to be coagulated. Hence, cationic polymers are not effectively used for colored-water coagulation during conventional treatment.

Anionic and nonionic polymers. Weakly charged anionic polymers or nonionic polymers with molecular weights of 10^6–10^7 are very often used to strengthen clay or color colloids being destabilized by aluminum or iron coagulants. These polymers behave as an intermicrofloc bridging agent. Hence, the minimal amounts of metal coagulants should be added to destabilize clay or color colloids prior to

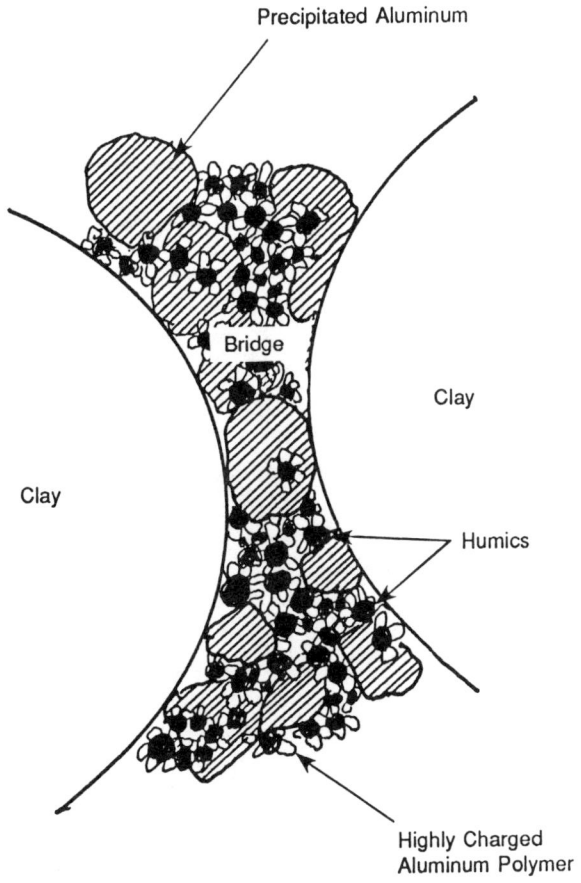

Figure 1-13 Matrix of clay and color colloids with hydrolyzed metal coagulant in a floc.

application of the bridging polymers. The numbers of the polymers with molecular weight of about 10^7 is of the order of 10^{10} particles per millilitre. Hence, the dosage is controlled by the number of microflocs to be bridged. Uniform dispersion of the polymers through rapid mixing without damaging the microfloc structure is essential to achieve optimum use of the polymers in good floc formation. The mixing conditions for such flocs have been indicated previously. As will be discussed in the next section and in chap. 7, the optimal use of polymers under a suitable agitation condition can cause the formation of pellet-like, dense aggregates. A schematic diagram of the floc structure with cationic polymers, and metal coagulant and anionic polymers is shown in Figure 1-14.

MIXING FOR FLOCCULATION

Chemical flocculation of suspended particles is achieved by two steps: coagulation, or destabilization, and floc growth. This section summarizes a few typical modes of floc growth, or agglomeration of the microflocs under characteristic mixing conditions.

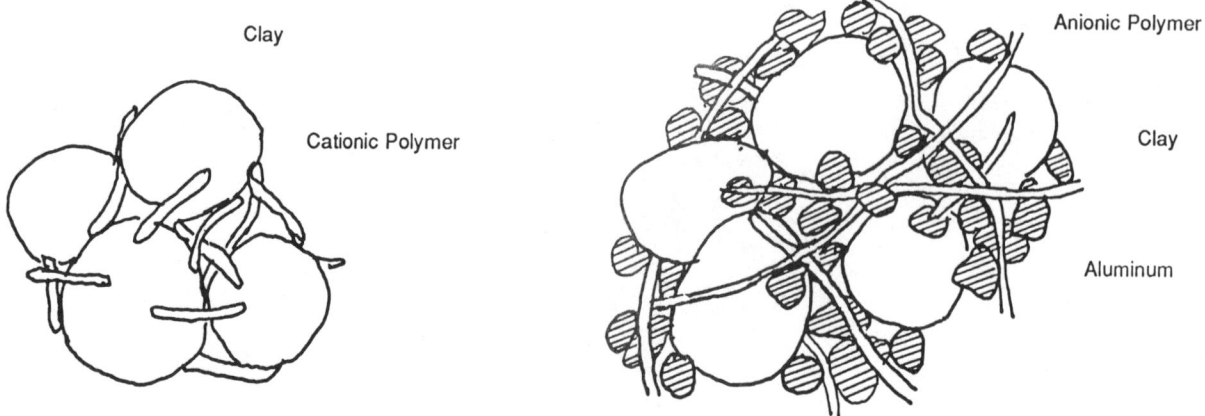

Figure 1-14 Matrix of clay with cationic polymer coagulant, and with metal coagulant and anionic flocculant in a floc.

The Collision Agglomeration Process

Most theoretical analyses of the flocculation step have focused on determining the frequency of collisions between suspended particles. The early theories of flocculation kinetics were developed by Smoluchowski (1916) under laminar flow conditions and with Brownian motion. As noted previously, Camp and Stein (1943) applied the Smoluchowski theory under laminar flow conditions to the flocculation of particles under water treatment conditions and proposed that the $\overline{G}$-value and the $\overline{GT}$-value be used as criteria for flocculation. In a later paper, Camp (1953) proposed the practical use of the $\overline{G}$-value and $\overline{GT}$-value theory for flocculator design.

However, almost all flocculators are operated under turbulent flow conditions. Hence, the flocculation theory developed for laminar flow conditions is usually not applicable to the quantitative evaluation of processes for water and wastewater treatment. Levich (1962) proposed a fundamental equation for the frequency of collisions between particles under turbulent flow conditions. The Levich equation is applicable to flocculation in water treatment under conditions of turbulent flow.

Similarities of Flocculation Under Laminar and Turbulent Flow Conditions

It is fortunate that under both laminar flow and the viscous subrange of isotropic turbulent flow, the main index of flocculation intensity can be described by the same parameter $\overline{G} = \sqrt{\varepsilon/\nu}$, in which ε is the total energy dissipated per unit time per unit mass of fluid, and ν is the kinematic viscosity. Because Camp's $\overline{G}$-value is identical with the parameter $\sqrt{\varepsilon/\nu}$, the $\overline{G}$-value is applicable as the flocculation intensity index, even for practical turbulent flocculator design.

Fair and Gemmell (1964), Swift and Friedlander (1964), Hudson (1965), Tambo (1965c), Cockerham and Himmelblau (1974), Tambo and Watanabe (1979b), and many others have developed equations and methods to describe the floc-growth process using the fundamental equations for different flow systems.

The following sections summarize three typical modes of flocculation associated with different mixing flow patterns: random flocculation, contact flocculation, and pellet flocculation.

Random flocculation. Random flocculation is the most common floc-growth pattern in a conventional flocculator used for water treatment. At the initial stages of flocculation, all individual particles are considered to be covered by or attached to coagulants and can be treated as elementary particles (see Figure 1-1). The elementary particles or aggregates of small numbers of elementary particles can be termed as primary particles from which flocculation starts.

Floc size. If these primary particles are introduced into a mixing basin, collisions between primary particles generate two fold particles. During the course of flocculation, floc particles of numerous sizes can appear as a result of collisions of primary particles and two fold particles, which produce three fold particles; collisions of two fold particles and three fold particles produce five fold particles, and so on. Suppose an i-fold particle is defined as the particle that contains i primary particles in a floc.

The number of collisions between i- and j-fold particles per unit volume per unit time N_{ij} can be calculated by Eq 1-22.

$$N_{ij} = K(d_i/2 + d_j/2)^3 n_i n_j$$
$$= K' d_1^3 (i^{1/3} + j^{1/3})^3 n_i n_j \qquad \text{(Eq 1-22)}$$

Where:

d_1 = diameter of primary particle, in centimetres
d_i = $d_1 i^{1/3}$ = diameter of i-fold particles, in centimetres
d_j = $d_1 j^{1/3}$ = diameter of j-fold particles, in centimetres
n_i = number of i-fold particles per unit volume (here, per cubic centimetre)
n_j = number of j-fold particles per unit volume (here, per cubic centimetre)
K = rate constant of floc collision (K is proportional to the $\overline{G}$-value s^{-1})
K' = the rate constant of floc agglomeration (K' is proportional to the $\overline{G}$-value s^{-1})

Many sizes of floc particles may be generated in random collisions among various sized floc particles under a given mixing condition. The maximum floc size attainable is limited to the S-fold particles on the basis of a balance between floc strength and intensity of agitation (see chap. 7). Hence, theoretically, S kinds of floc particles can exist during the course of flocculation with random collisions of two different sized floc particles. Simultaneously, the agitation causes floc breakup. A balance between floc growth due to random collisions and breakup due to agitation establishes an equilibrium floc size distribution that is usually attained after a certain duration of agitation (Tambo and Watanabe 1979b).

Floc density. During the course of floc growth under the random flocculation process, a large amount of water is encompassed in the void space of the agglomerated floc particles. Thus, the floc density decreases with an increase in the size or degree of agglomeration (Tambo and Watanabe 1967, 1979a; Lagvankar and Gemmell 1968). This is a very important feature of the random flocculation process.

As an *i*-fold particle is defined as a floc comprising *i* primary particles, the volume of an *i*-fold floc particle is

$$(\pi/6) d_i^3 = (\pi/6) d_1^3 i + V_w \quad \text{(Eq 1-23)}$$

Where:

V_w = volume of the internal water, in cubic centimetres

The effective density of an *i*-fold floc, ρ_{ei}, is defined by the following equation:

$$\rho_{ei} = \rho_i - \rho_w \quad \text{(Eq 1-24)}$$

Where:

ρ_i = density of an *i*-fold floc, in grams per cubic centimetre
ρ_w = density of water, in grams per cubic centimetre

Thus, the mass balance of an *i*-fold floc gives

$$(\pi/6) d_i^3 (\rho_{ei} + \rho_w) = (\pi/6) d_1^3 i (\rho_{e1} + \rho_w) + \rho_w V_w \quad \text{(Eq 1-25)}$$

From Eq 1-23 and Eq 1-25,

$$d_i^3 \rho_{ei} = i d_1^3 \rho_{e1} \quad \text{(Eq 1-26)}$$

A floc density function (Tambo and Watanabe 1967, 1979a) relates floc diameter and its corresponding effective density as shown in the following equation (see also chap. 7).

$$\rho_{ei} = \rho_i - \rho_w = a/(d_i/1)^{k\rho} \quad \text{(Eq 1-27)}$$

Where:

a = constant, in grams per cubic centimetre
$k\rho$ = dimensionless constant

The magnitudes of the constants depend upon the coagulation conditions. From Eq 1-26 and Eq 1-27,

$$d_i = i^{1/(3-k\rho)} d_1 \quad \text{(Eq 1-28)}$$

Accordingly, under the random flocculation regime, the numbers of collisions between *i*- and *j*-fold particles cannot be described by Eq 1-22, but can be correctly evaluated by Eq 1-29, which includes the modification due to floc density variation (Tambo and Watanabe 1979b).

$$N_{ij} = K' d_1^3 \left[i^{1/(3-k\rho)} + j^{1/(3-k\rho)} \right] n_i n_j \quad \text{(Eq 1-29)}$$

By an introduction of the relative number of *i*-fold particles as $N_i = n_i/n_0$ with respect to the initial number of primary particles n_0, Eq 1-29 may be rewritten as

Eq 1-30, which shows the relative numbers of collision–agglomeration N_{ij}, based upon the primary particle number.

$$N_{ij} = K' d_1^3 \left[i^{1/(3-k\rho)} + j^{1/(3-k\rho)} \right] N_i N_j \qquad \text{(Eq 1-30)}$$

From Eq 1-30, one can understand that the collision–agglomeration process of i- and j-fold particles proceeds in a similar pattern, with the only difference being the rate is evaluated by factors of $K' d_1^3 n_0$. The value of the rate constant K' can be obtained from the particle transfer rate and the collision agglomeration factor and is proportional to the $\overline{G}$-value. The product of $d_1^3 n_0$ is proportional to the initial volume concentration of primary floc C_0. Therefore, the progress of flocculation with time T can be determined by the nondimensional product of $\overline{G}C_0T$ (Tambo et al. 1970; Tambo and Watanabe 1979b). Camp (1953) denoted $\overline{G}(s^{-1})$, as the rate controlling factor and $\overline{G}T(-)$, as the factor to control the attainment of flocculation. However, in reality, the rate constant should be the $\overline{G}C_0$-value (s^{-1}), and the factor to show the attainment of flocculation should be $\overline{G}C_0T$-value $(-)$, as shown by Tambo et al. (1970) in the case of random flocculation.

On the basis of the rate constant and attainment factors, the kinetics of the process of flocculation can be derived on a population balance calculation of floc-size distribution.

Contact flocculation. Contact flocculation can be characterized as a mode of flocculation in which larger, well-grown floc particles adsorb incoming minute flocs onto the surface. For the sake of convenience, in this section, the former is designated *grown-floc* and the latter *microfloc*. Contact flocculation is typically seen in solids contact clarifiers (Tambo and Hozumi 1979) and partially seen in conventional flocculators with backmixing flow (Tambo and Watanabe 1984).

In solids contact clarifiers, well-grown larger flocs of high concentration circulate through a turbulent flow field or are suspended in an upflow stream. These flow fields cause collisions and agglomeration to occur between the grown-flocs and the incoming microflocs. Through the collisions between these two kinds of flocs, which have greatly different sizes, an adsorption-like flocculation process occurs. Therefore, this kind of flocculation can be analyzed as a sort of adsorption of microflocs onto the surfaces of grown-flocs.

Because of the presence of grown-flocs with high concentration C, resulting in high $\overline{GCT}$-values, the required duration of flocculation can be shortened remarkably. In addition, a very high concentration of grown-flocs usually yields the maximum floc size distribution (i.e., the final equilibrium floc size distribution) attainable under the given agitation intensity.

Under equilibrium conditions, the grown-floc sizes are distributed approximately between the maximum floc diameter attainable under the given agitation intensity (see chap. 7) and one-half of the maximum diameter (Tambo and Watanabe 1979b). In practice, the maximum floc diameter in a turbulent flocculation chamber is of the order of 10^{-1} to 10^{-2} cm. On the other hand, the incoming microflocs are in the range of 10^{-3} cm. Therefore, contact flocculation occurs between these floc groups with greatly different floc diameters.

The number of collisions between the grown-flocs and microflocs per unit time N_c is described in Eq 1-31, which has been derived by Tambo and Hozumi (1979).

$$N_c = K(D/2 + d/2)^3 Nn = K'' C_v n \qquad \text{(Eq 1-31)}$$

Where:

D = mean diameter of grown-flocs, in centimetres
d = mean diameter of microflocs, in centimetres
N = numbers of grown-floc per unit volume (here, per cubic centimetre)
n = numbers of microflocs per unit volume (here, per cubic centimetre)
C_v = $(\pi/6)\,ND^3$, volumetric concentration of grown-floc (dimensionless)

In the derivation of Eq 1-31, the following assumptions were used:

- The grown-flocs are in the final equilibrium (maximum) floc size distribution. Under these conditions, collisions between grown-flocs are not effective for flocculation.
- Collisions between microflocs occur with a collision diameter of d. The collision diameter d is much smaller than the collision diameter between grown-flocs and microflocs (i.e., $[D + d]/2$). Hence, in practice, for the progress of contact flocculation the collisions between microflocs are considered to be negligible in comparison with collisions between the grown-flocs and microflocs.
- Because of the great difference in diameter between grown-flocs and microflocs, the collision diameter between these flocs can be written as $(D + d)/2 \sim D/2$.
- The criteria $D \gg d$ and $ND^3 \gg nd^3$ are generally valid. Hence, the relative rate of floc-volume increase during a restricted duration of flocculation can be considered small. Therefore, the floc volume can be kept approximately constant by means of suitable intermittent adjustment of floc concentration (i.e., draw-off of the sludge). Therefore, the condition that C_v is proportional to ND^3 and is also constant can be assumed.

Equation 1-31 revealed that in contact flocculation, only the concentration of microflocs n declines with time as a first-order reaction with regard to n. Therefore, the rate constant is proportional to the product of $\overline{GC_v}$ and the extent of flocculation (i.e., decrease of n) is also defined by the $\overline{GC_vT}$-value.

The above-mentioned relationships (i.e., $\overline{GC_v}$- and $\overline{GC_vT}$-values) for contact flocculation are approximately applicable to flocculators in which backmixing is prominent. In a flocculator comprising a small number of tanks in series with complete mixing in each chamber, a type of contact flocculation of newly incoming microflocs with existing growing flocs can exist. Hudson proposed a first-order decrease of incoming microflocs in a flocculator as early as 1965. While his discussion is not strictly for contact flocculation, his proposal should be recognized as the first to mention the importance of floc-volume concentration. Tambo and Watanabe (1972, 1984) handled flocculation in a flocculator with backmixing as the combination of the modes of random flocculation and contact flocculation.

Another type of contact flocculation (different from the mixing basin flocculation mentioned previously) occurs in an upflow clarifier. In a floc blanket of an upflow clarifier, microflocs enter the floc blanket from the bottom, then contact and agglomerate with the suspended grown-flocs. The number of collisions and agglomeration N_b between the suspended grown-flocs and the incoming microflocs by contact flocculation in a floc blanket can be formulated as Eq 1-32 under the following assumptions made by Tambo and Hozumi (1979):

- Grown-flocs with a mean diameter D are suspended statically in an upflow velocity v_s.
- Microflocs with diameter d move upward with upflow velocity v_s.

- Collisions between two kinds of floc particles with diameters D and d in an upflow fluid field occur by a relative approaching velocity v_s and a collision diameter $(D + d)$.
- Collisions between microflocs are negligible for flocculation and those between grown-flocs are ineffective.

$$N_b = K^* v_s (D + d)^2 Nn = K^* v_s D^2 Nn = K^* v_s (C_v/D) n \qquad \text{(Eq 1-32)}$$

Where:

K^* = collision and agglomeration coefficient (dimensionless)

This equation is similar to Camp's equation for flocculant settling (Camp 1946).

From Eq 1-32, the progress of contact flocculation in a floc blanket is characterized as a first-order reaction with respect to microfloc concentration. The rate constant is proportional to the product of the volumetric concentration of the suspended grown-floc C_v and the inverse of its diameter $1/D$.

Pellet flocculation. In a conventional flocculation process, charge-neutralized elementary particles collide randomly with one another and form bulky agglomerates called *random floc*. During the course of floc growth through random flocculation, the density of floc decreases with increase of the floc size, since more void water is incorporated in the agglomerates. If floc particles with a much higher density and with a larger size could be produced by excluding much of the void water, which is inevitably included in random flocs, then the settling velocity of these particle would increase drastically. A process called pellet flocculation, or pelleting, is effective for this purpose.

The pelleting operation of a concentrated suspension was first proposed by Yusa and Gaudin (1964) as a promising process to dewater coagulated sludges discharged from rapid sand filtration systems, using a horizontal-drum dewatering apparatus with polymer addition. Later, Yusa (1977), Suzuki et al. (1980), and Kobayashi et al. (1981) proposed a fluidized-bed pellet contact clarifier to treat high-concentration suspensions with polymer addition. The apparatus was developed as an extension of the conventional solids-contact clarifier system, which had processes composed of a flash mixer, flocculator, and upflow floc blanket. The difference from a conventional solids contact clarifier is that the pellet contact clarifier has slow mixing blades in the floc blanket to control the irregular growth of the pellet floc. Tambo and Matsui (1989) showed that input of flocculated material to the fluidized bed did not generate good pelletization and resulted in poor effluents in experiments with higher concentrations of kaolinite suspensions of 10^2–10^3 g/m^3. On the basis of these results, Tambo and Matsui proposed that the pellet flocculation process is associated with the metastable state concept.

The metastable state can be explained generally as follows (Stumm and Morgan 1970). A supersaturated solution for a labile concentration forms a precipitate spontaneously. However, in a metastable solution, precipitation may not result over a relatively long period of time; the process needs an aging time to reach an equilibrium state. However, if a large amount of the same kind of solid phase is introduced into the metastable state, precipitation of the solute upon the surfaces of the introduced solid occurs very quickly, and very often dense solid aggregates are generated.

This kind of process can be used to improve contact flocculation by using a greater understanding of the metastable state. The metastable state products are those particles that already possess elementary structures of the final aggregates but do not have a sufficient size and/or concentration to form the desirable aggregates in a very short time period. In the case of clay suspensions, primary floc particles generated by the addition of a metal coagulant (i.e., destabilized clay particles) can be considered the elementary particles. If the elementary particles are introduced continuously into a suspension of large agglomerates of the same kind, a one-by-one attachment of the elementary particles onto the large agglomerates occurs. Introduction of polymer flocculant to reinforce interparticle bridging strength into the one-by-one attachment mode of flocculation at the bottom entrance of the fluidized bed may cause very tough and dense pellet particles under moderate agitation. The moderate agitation provided by rotating paddles prevents the random local growth of the adhered elementary particles on the surface of the pellet and thus causes the growth of a good spherical pellet without introducing unnecessary void volume. As regular layered deposition of the elementary particles occurs, high-density pellets are generated.

References

AMIRTHARAJAH, A. 1981. Initial Mixing. Proc. AWWA Seminar on Coagulation and Filtration: Back to the Basics. AWWA, St. Louis, Mo.

AMIRTHARAJAH, A. & KAWAMURA, S. 1983. System Design for Polymer Use. Proc. AWWA Seminar on Use of Organic Polyelectrolytes in Water Treatment. AWWA, Denver, Colo.

AMIRTHARAJAH, A. & MILLS, K.M. 1982. Rapid-Mix Design for Mechanisms of Alum Coagulation. *Jour. AWWA*, 74:4:210.

AMIRTHARAJAH, A. & TRUSLER, S.L. 1986. Destabilization of Particles by Turbulent Rapid Mixing. *Jour. Envir. Engrg. Div. ASCE*, 112:6:1085.

BRODSKY, A. & PROCHAZKA, J. 1975. Water-Treatment Analysis by Gel Chromatography. *Jour. AWWA*, 67:1:23.

CAMP, T.R. 1946. Sedimentation and the Design of Settling Tanks. *Trans. ASCE*, 3:895.

———. 1953. Flocculation and Flocculation Basins. *Proc. ASCE*, 79:9:1.

CAMP, T.R. & STEIN, P.C. 1943. Velocity Gradients and Internal Work in Fluid Motion. *Jour. Boston Soc. Civil Engrg.*, 30:10:217.

COCKERHAM, P.W. & HIMMELBLAU, M. 1974. Stochastic Analysis of Orthokinetic Flocculation. *Jour. Envir. Engrg. Div. ASCE*, 100:EE2:279.

CUTTER, L.A. 1966. Flow and Turbulence in a Stirred Tank. *AIChE Jour.*, 12:35.

DELICHATSIOS, M.A. & PROBSTEIN, R.F. 1975. Coagulation in Turbulent Flow: Theory and Experiment. *Jour. Colloid Inter. Sci.*, 51:394.

DEMPSEY, B.A. ET AL. 1984. The Coagulation of Humic Substances by Means of Aluminum Salts. *Jour. AWWA*, 76:4:141.

EDWARDS, J.K. & AMIRTHARAJAH, A. 1985. Removing Color Caused by Humic Acids. *Jour. AWWA*, 77:3:50.

FAIR, G.M. & GEMMELL, R.S. 1964. A Mathematical Model of Coagulation. *Jour. Colloid Sci.*, 19:4:360.

GJESSING, E. & LEE, F.G. 1967. Fractionation of Organic Matter in Natural Waters on Sephadex Columns. *Envir. Sci. Technol.*, 1:8:631.

HAHN, H.H. & STUMM, W. 1968. Kinetics of Coagulation With Hydrolyzed Al(III). *Jour. Colloid Inter. Sci.*, 28:1:134.

HALL, E.S. & PACKHAM, R.F. 1965. Coagulation of Organic Color With Hydrolyzing Coagulants. *Jour. AWWA*, 55:9:1149.

HUDSON, H.E. JR. 1965. Physical Aspects of Flocculation. *Jour. AWWA*, 57:885.

JOHNSON, P.N. & AMIRTHARAJAH, A. 1983. Ferric Chloride and Alum as Single and Dual Coagulants. *Jour. AWWA*, 75:5:232.

KEYS, R.O. & HOGG, R. 1978. Mixing Problems in Polymer Flocculation. *Water 1979*, AIChE Symp. Ser. 190, 75:63.

KOBAYASHI, ET AL. 1981. Application of Pelleting Flocculation Process to Solid–Liquid Separation. Proc. 2nd World Cong. of Chem. Engrg., Vol. 4.

KRUYT, H.R. 1952. *Colloid Science*, Vol. 1. Elsevier, Amsterdam. p. 214.

LAGVANKAR, A.L. & GEMMELL, R.S. 1968. A Size–Density Relationship for Flocs. *Jour. AWWA*, 60:9:1040.

LAUFHUTTE, H.D. & MERSMANN, A.B. 1984. *Chem. Ing. Tech.*, 56:11:862.

———. 1985a. Proc. 5th European Conf. on Mixing, Wurzburg, Germany, 331–340. BHRA, Cranfield, U.K.

———. 1985b. *Ger. Chem. Engrg.*, 8:371.

LETTERMAN, R.D.; QUON, J.E.; & GEMMELL, R.A. 1973. Influence of Rapid-Mix Parameters on Flocculation. *Jour. AWWA*, 65:716.

LEU, R.J. & GHOSH, M.M. 1988. Polyelectrolyte Characteristics and Flocculation. *Jour. AWWA*, 80:4:159.

LEVICH, V.G. 1962. *Physicochemical Hydrodynamics*. Prentice-Hall, Englewood Cliffs, N.J. p. 213.

MATIJEVIC, E. ET AL. 1961. Detection of Metal Ion Hydrolysis by Coagulation, III: Aluminum. *Jour. Phys. Chem.*, 65:5:826.

O'MELIA, C.R. 1972. Coagulation and Flocculation. Chap. 2 In *Physicochemical Processes for Water Quality Control*, ed. by W.J. Weber Jr. Wiley-Interscience, New York.

———. 1980. Aquasols: The Behavior of Small Particles in Aquatic Systems. *Envir. Sci. & Tech.*, 14:9:1052.

SAFFMAN, P.G. & TURNER, J.S. 1956. On the Collision of Drops in Turbulent Clouds. *Jour. Fluid Mechanics*, 1:16.

SEMMENS, M.J. & AYERS, K. 1985. Removal by Coagulation of Trace Organics From Mississippi River Water. *Jour. AWWA*, 77:5:79.

SMOLUCHOWSKI, M.V. 1916. Versuch einer Mathematischen Theorie der Koagulations Kinetik Kolloider Losungen. *Zeitschrift f. Physik. Chemie*, 92:126.

SNODGRASS, ET AL. 1984. Particle Formation and Growth in Dilute Aluminum (III) Solutions. *Wtr. Res.*, 18:4:479.

SPIELMAN, L.A. 1978. Hydrodynamic Aspects of Flocculation. *The Scientific Basis of Flocculation*, ed. by K.J. Ives. Sijthoff and Noordhoff, Alphen aan den Rijn, the Netherlands. pp. 63–88.

STUMM, W. & MORGAN, J. 1962. Chemical Aspects of Coagulation. *Jour. AWWA*, 54:8:971.

———. 1970. *Aquatic Chemistry*. Wiley-Interscience, New York.

SUZUKI, K. ET AL. 1980. New Clarifier Is Small, Economic and Energy Efficient. *Pulp & Paper, Can.*, 81:6:1.

SWIFT, D.L. & FRIEDLANDER, S.K. 1964. The Coagulation of Hydrosols by Brownian Motion and Laminar Shear Flow. *Jour. Colloid. Sci.*, 19:7:621.

TAMBO, N. 1965a. A Fundamental Investigation of Coagulation in Water Works I. *Memoirs of Faculty of Engrg., Hokkaido Univ.* (Japan), 11:581.

———. 1965b. A Fundamental Investigation of Coagulation (III). *Jour. Jpn. Wtr. Wks. Assn.*, 365:25.

———. 1965c. A Fundamental Investigation of Flocculation I. *Jour. Jpn. Wtr. Wks. Assn.*, 372:9:10.

———. 1984. Treatment of Raw Water With High Turbidity and/or Colour. Proc. IWSA Cong., Monastir, Tunisia. Special Subject No. 6. IWSA, London.

TAMBO, N. ET AL. 1970. Rational Design of Flocculators (I). *Jour. Jpn. Wtr. Wks. Assn.*, 431:8:20.

———. 1989. Aquatic Humic Substances: Influence on Fate and Treatment of Pollutants. In *Humic Substances Removal by Coagulation*, ed. by I.H. Suffet and P. McCarthy, chap. 28. American Chemical Society, Washington, D.C.

TAMBO, N. & HOZUMI, H. 1979. Physical Characteristics of Flocs, II. *Wtr. Res.*, 13:5:441.

TAMBO, N. & ITOH, H. 1977. Electrophoretic Studies on Natural Color Colloid Coagulation by Alum. *Jour. Jpn. Wtr. Wks. Assn.*, 508:38.

TAMBO, N. & KAMEI, T. 1978. Treatability Evaluation of General Organic Matter: Matrix Conception and Its Application for a Regional Water and Waste Water System. *Wtr. Res.*, 12:931.

TAMBO, N. & MATSUI, Y. 1988. Unpublished data.

———. 1989. Performance of Fluidized Pellet Bed Separator for High Concentration Suspension Removal. *Jour. Wtr. Supply Res. & Tech. (AQUA)*, 1:10.

TAMBO, N. & WATANABE, Y. 1967. A Study of the Aluminum Floc Density. *Jour. Jpn. Wtr. Wks. Assn.*, 397:10:2.

———. 1972. Rational Design of Flocculators (IV). *Jour. Jpn. Wtr. Wks. Assn.*, 454:7:27.

———. 1979a. Physical Characteristics of Flocs, I: The Floc Density Function and Aluminum Floc. *Wtr. Res.*, 13:5:409.

———. 1979b. Physical Aspect of Flocculation Process, I. *Wtr. Res.*, 13:5:409.

———. 1984. Physical Aspect of Flocculation, III: Flocculation Process in a Continuous Flow Flocculator With a Back-Mix Flow. *Wtr. Res.*, 18:6:695.

TOMI, D.T. & BAGSTER, D.F. 1978. The Behavior of Aggregates in Stirred Vessels. *Trans. Inst. Chem. Engrs.*, 56:1:1.

YUSA, M. 1977. Mechanism of Pelleting Flocculation. *Inst. Jour. of Min. Engrg.*, 4:293.

YUSA, M. & GAUDIN, A.H. 1964. Formation of Pellet-Like Flocs of Kaolinite by Polymer Chains. *Am. Cere. Bull.*, 43:5:402.

2

Mixing in Liquids

J.C. Godfrey, Ph.D., Department of Chemical Engineering, University of Bradford, Bradford, West Yorkshire, United Kingdom

Appiah Amirtharajah, Ph.D., P.E., Professor, School of Civil Engineering, Georgia Institute of Technology, Atlanta, Ga. USA

INTRODUCTION

Mixing Phenomena

Turbulent mixing. Because most mixing applications are concerned with low-viscosity liquids, turbulent mixing is commonly encountered. Its main advantage is cost; for low-viscosity fluids, turbulent flow conditions can be obtained with relatively inexpensive equipment and low operating costs. Turbulent flow is frequently the preferred mixing condition because it requires minimal mixing times and therefore equipment size. In this chapter, as in mixer usage and research, the main consideration will be the characteristics of agitated-tank mixers. Other mixers will be discussed—jet mixers, static mixers, dynamic in-line mixers—but the discussion in this section will focus mainly on the agitated tank, with some comparison to static mixers.

Turbulent flow brings about more rapid mixing than laminar flow. In addition to the circulating flow observed in agitated tanks, turbulent flow is characterized by additional, fluctuating velocities. Thus, there are two velocity terms that contribute to mixing: $\bar{u}$, the mean velocity, which principally contributes to large-scale circulation and blending, and u, the fluctuating velocity, which results in fine-scale mixing. The precise nature of the velocities in $\bar{u}$ and u, and their relationship to one another are very dependent on the type of equipment in which the mixing process is being conducted.

When the character of turbulent flow is observed at a single point in a mixer (the Eulerian view) the flow appears to consist of the mean flow $\bar{u}$, with a superimposed fluctuating velocity that changes both direction and magnitude over time. However, if the observer could follow the flow of a small "packet" of fluid as it circulates in the mixer (the Lagrangian view) the eddy characteristic of turbulent flow could be observed. This characteristic is analogous to that of the swirling eddy flows that can be observed in rivers. The turbulent eddy in mixing equipment can be assigned a characteristic size λ_e and a characteristic velocity $u_{\lambda e}$.

Thus, a simple view of turbulent mixing in an agitator is of rapid, circulating bulk flow on which is superimposed small-scale circulating flow patterns. The circulating eddies are generated at the impeller and, at this point in the mixer, are of similar size. As the eddies leave the impeller and circulate in the tank they are said to "decay." The result is that at positions farther from the impeller, the scale of the eddies is smaller and they are more numerous. A characteristic of the agitated tank therefore is that the turbulence condition is highly nonuniform, being most intense in the impeller region, with intensity decreasing away from the impeller.

Three separate mixing mechanisms can be envisaged in turbulent flow mixing: large-scale mixing associated with recirculating flows, small-scale mixing associated with flows within eddies, and diffusional mixing following the fine-scale mixing generated in eddies.

The nature of the bulk circulating flow is a function of the impeller type. When the rotational (tangential) velocities are large (e.g., turbine impellers) centrifugal effects lead to vortex formation. As this can lead to practical mixing problems—e.g., the ingestion of air, particularly in smaller mixers—vertical baffles are often fitted to the vessel wall to suppress vortex formation. The use of baffles changes the flow pattern and the energy-dissipation characteristics of the agitated tank, although mixing performance in terms of energy usage is not necessarily improved.

Another significant element of bulk flow is the residence time characteristics of continuously operated agitated tanks with feed and outlet streams. Under turbulent conditions the contents of the tank can be considered to be well-mixed (the concentration being the same at all points), and to have a "fully backmixed" characteristic. The concentration of the outlet stream is therefore the same as that in the mixer, and the mixer has the characteristic of damping out short-term fluctuations on the feed stream(s). This fully backmixed condition is characterized by a distribution of residence times, which indicates that the fluid elements leaving the mixer have not all spent the same time in the mixer; this is a disadvantage for some reactions.

Turbulent flow in pipes, and in static mixers, is in many ways much simpler than in agitated tanks. While there is some velocity distribution across the pipe diameter, the situation is less complex than the circulating flows in a tank. The turbulent pipe flow can be regarded as an axial velocity upon which is superimposed a turbulent fluctuating velocity. Relative to the agitated tank, turbulence is much more uniform in a pipe. The turbulent component gives good radial mixing, and at any axial position the contents of the pipe can be regarded as approaching uniformity in the radial direction. This flow pattern gives pipe flow a near plug-flow characteristic. Elements of fluid leaving the pipe all have similar residence times, which is advantageous for some chemical reactions but does not provide any smoothing of time variations arising in feed streams. The characteristic of turbulent flow in static mixers is very similar to that of pipe flow, but the turbulence level is higher. The fitting of mixing elements in a pipe increases pressure drop and turbulence levels, so the radial velocity distribution and residence times are more uniform and mixing rates are higher.

The theory of turbulence and mixing is still incomplete, but computing packages are available to allow some analysis of mixer performance (chap. 3). Similarly detailed work has been done on the influence of mixing and turbulence on fast chemical reactions (chap. 5). While both topics present practical and theoretical complexities, there are many simple solutions and parameters that allow a preliminary assessment of mixing problems. Similarly, a number of simple parameters can be used to describe the basic characteristics of turbulent mixing in single-phase systems (see "Mixer Performance," p. 38).

Multiphase mixing. As with single-phase mixing most multiphase mixing processes are conducted under turbulent flow conditions. However, it is the continuous phase that is turbulent, while the dispersed phase—be it solid, gas, or liquid—may have other characteristics that need to be considered. The phenomena of importance in multiphase mixing principally relate to the behavior of the dispersed phase: minimum conditions for dispersion; distribution of dispersed phase within the mixing tank; holdup of the dispersed phase in continuously operated tanks; and, for gas–liquid and liquid–liquid mixing, size of dispersed phase. Mass-transfer characteristics can be important for both phases.

In solid–liquid mixing the minimum criterion for dispersion generally used is the "just suspended" condition, defined by the impeller speed N_{JS}, where all the particles are considered to be in motion, with no particle remaining at the tank bottom for more than 1 or 2 s. However, at this level of agitation the particles are far from uniformly distributed throughout the tank, and a marked axial concentration profile, impeller-speed dependent, is present.

This phenomenon has been attracting study for some time and is relevant to the holdup characteristics of continuously operated tanks. When the contents of a continuous mixer are not homogeneous, it is likely that the contents of the tank will not correspond to the value determined from the flows to the tank. An additional consequence is that the residence time of the two phases is different in these circumstances. Experimentally, the holdup (volume fraction) of the solid phase is found to be a function of the impeller speed. The continuous-phase mass-transfer coefficient is an important but separate characteristic and is dependent on impeller speed.

The phenomena of gas–liquid mixing have something in common with the solid–liquid system. While there is no corresponding minimum mixing condition, the holdup of the dispersed gas phase is strongly impeller-speed dependent. Holdup is dependent on both the impeller speed and the gas throughput rate, with various gas flow patterns evident under different conditions. Bubble behavior is an important phenomenon in design and the interpretation of data, bubble size being dependent on impeller speed and system physical properties. Bubble-to-bubble coalescence and redispersion occur within the mixer and are very dependent on the gas–liquid system and any solutes present. Even small concentrations have a significant effect on the coalescence process.

Most of the characteristics of liquid–liquid mixing can be illustrated by analogy with other multiphase mixing processes. A concept of minimum agitation for complete dispersion is based on the—usually visual—determination that dispersion has been achieved in all parts of the mixing tank. The size of the dispersed-phase drops, as with bubbles, is a function of the level of agitation, as are mass-transfer coefficients. In most cases drops are small and therefore stagnant internally, while the continuous phase is turbulent; consequently, the mass-transfer coefficients for the two phases are very different.

Naturally, design calculations for batch and continuous multiphase systems differ in a number of ways. Some characteristics are regarded as sufficiently similar for the same approach to be used in both cases, e.g., minimum conditions for dispersion, bubble or drop size, mass-transfer coefficients. For continuous-flow tanks the holdup characteristic illustrates the need to consider the residence times of both phases. Furthermore, both phases have their own residence time distribution characteristics, which are a function of the level of agitation and possibly quite different from the fully backmixed condition frequently assumed for agitated tanks.

The discussion so far has been dominated by ideas of multiphase mixing in agitated tanks. However, other mixer geometries are used—increasingly, the static mixer. Some considerations of multiphase mixing in static mixers are similar to those for agitated tanks: minimum conditions for dispersion (a limiting velocity), size of dispersed phase, and mass-transfer coefficients. The static mixer is simpler in that, when it is correctly operated, holdup is identical to the flow ratio value and residence time characteristics are simpler, both phases having near plug-flow characteristics.

Laminar mixing. Laminar mixing is usually associated with the processing of high-viscosity products, and in agitated tanks the components of the mixture are homogenized by a combination of recirculating flow, laminar shear, and diffusion. Coarse-scale mixing is achieved by recirculating flows, which distribute the components throughout the tank, and the final reduction of scale at molecular level can only be achieved by diffusion.

In static mixers the coarse-scale mixing mechanism is somewhat different, depending on repeated subdivision of flow. Mixing at smaller scales is achieved, as in agitated tanks, by laminar shear and diffusion. The coarse-scale mixing mechanism of the static mixer results in a strong radial mixing component, which produces a near plug-flow residence time characteristic.

Not all laminar mixing processes are concerned with high-viscosity fluids. Blending processes where the mixing problem is small or where the volumes are very large (as in some water industry applications) are conducted under laminar flow conditions to economize on capital and running costs for the agitator.

Mixer Performance

Mixing time. The concept of mixing time is a batch-oriented concept and is based on experimental studies of the addition of a tracer to an agitated tank and monitoring concentration at one or more points. The typical measured response is illustrated in Figure 2-1. Initially large fluctuations of concentrations are observed, these die away as mixing proceeds, and a steady-state concentration is ultimately reached. The time taken to achieve this steady-state condition is referred to as the mixing time. Many experimental data are available from mixing-time studies of the effects of impeller speed, mixer geometry, physical properties, and other variables.

These data should be regarded as qualitative, since they were influenced by the experimental technique used and other variables such as detector or sample size, tracer volume, point of addition, and location of sample collection. Particularly influential is the scale at which the measurement is made, typically about 0.01 m. At this scale the measurement gives a good idea of the coarse-scale mixing achieved and some indication of the fine-scale mixing due to laminar shear.

These experiments have identified a simple relationship between impeller speed and mixing time. A certain number of impeller revolutions is required to

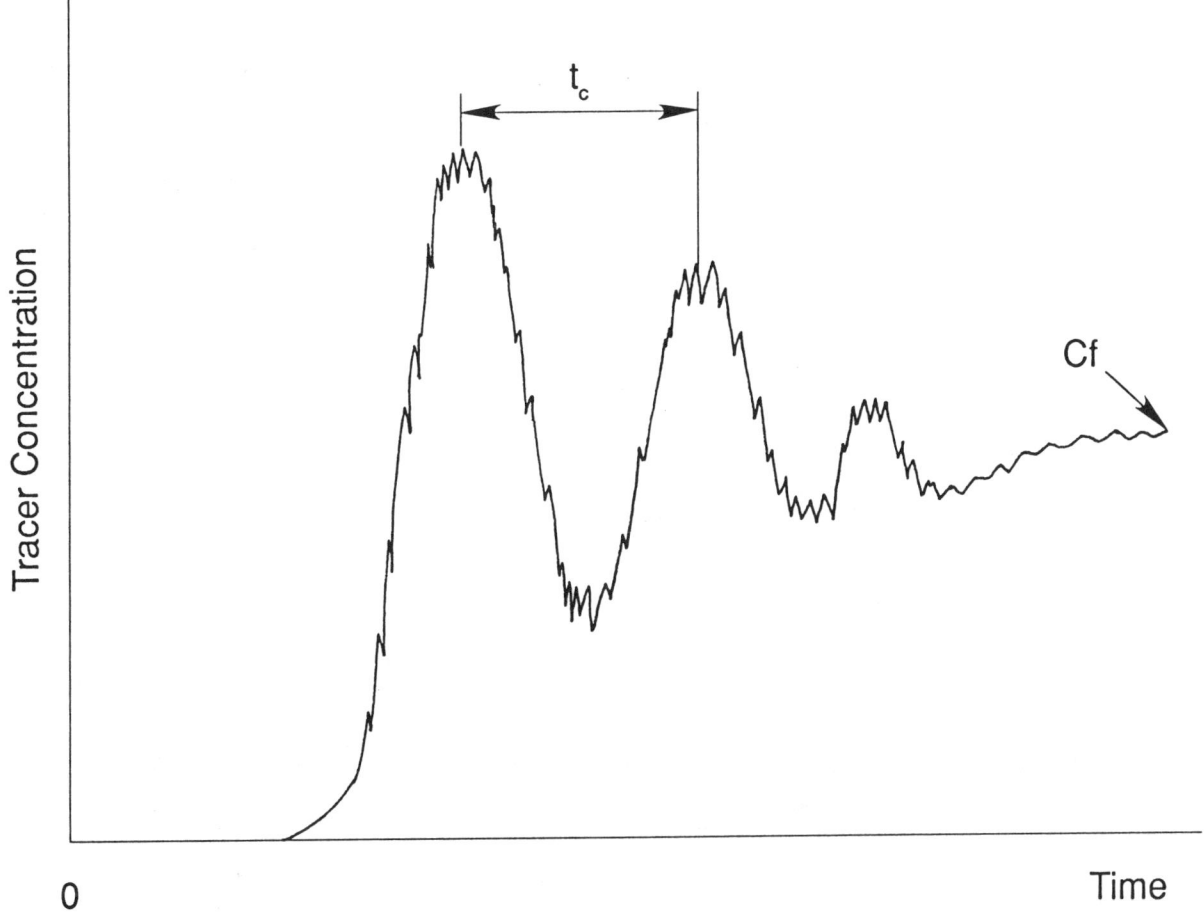

Figure 2-1 Experimental mixing time observation.

produce a "fully mixed" condition, and the product of impeller speed and mixing time is found to be constant for a particular mixing problem. The measurement of mixing time (θ) is also useful as a first estimate of order of magnitude of circulation time (t_c), as it has been shown that $\theta \sim \alpha t_c$, where α is in the range 3 to 5.

Other concepts of mixing time are necessary to deal with problems more complex than single-phase blending. In the consideration of competing fast chemical reactions where bi-product formation can cause problems, total mixing time includes both coarse-scale and diffusive mixing times. As a simple estimate, the two times can be regarded as additive. The coarse-scale mixing time reflects the time required for the distribution of components down to the smallest turbulent eddy scale and is approximated by the circulation time, as discussed above. The diffusive mixing time reflects the time required for diffusion within the eddy to produce the final state of homogenization.

When batch multiphase systems are studied, further complexities in the concept of mixing time are encountered. For the dissolution of solids, it may be that only the mass-transfer-rate characteristics of the liquid phase will determine mixing time if there is no chemical reaction. However, a range of possibilities exist, from mass-transfer-rate control to chemical-rate control if a chemical reaction is involved, the outcome depending on the relative rates of these two processes. Similar situations exist for gas–liquid and liquid–liquid mixing where simple solution or chemical reaction processes can occur.

Progress of mixing. There are many limitations to the simple concept of mixing time, and attention is increasingly being paid to more detailed and sometimes more relevant descriptions of the progress of mixing. These studies are related not only to the concept of mixing time but also to various hydrodynamic aspects— e.g, concentration profiles—that may be controlling. The advantage of these more detailed studies is that they lead to greater precision, ease of comparison of data from different sources, and an emphasis on those variables that actually control mixing performance.

For simple batch blending, various indices for nonuniform concentration have been used to relate mixture quality to time. Such an approach avoids the errors of judgment implicit in determining mixing time. Because the approach to a fully mixed condition is usually found to be asymptotic, the possibility of error is considerable. One procedure for establishing a more precise mixing time is the definition of a "fractional unmixedness" X:

$$X = \frac{C_f - C_\theta}{C_f - C_o} \qquad \text{(Eq 2-1)}$$

Where:

C_o = initial concentration
C_θ = concentration at time θ
C_f = final concentration

which can be related to the corresponding number of impeller revolutions:

$$N\theta = K_m G \log(2/X) \qquad \text{(Eq 2-2)}$$

Where:

N = impeller speed
G = geometry parameter

and K_m is determined by impeller type and G is a function of the impeller/tank diameter ratio; correlations of this type are available for turbulent mixing with a variety of impellers. Similarly, the progress of mixing with time (or impeller revolutions) can be followed in terms of the concentration variance of the tank contents determined from a number of sample positions:

$$\frac{d\alpha^2}{dt} = \beta\alpha^2 \qquad \text{(Eq 2-3)}$$

Where:

α^2 = variance
β = mixing rate constant

and α^2 is the mean square concentration fluctuation for the whole vessel and can be theoretically related to operating physical conditions and properties. For both turbulent and laminar blending it appears that the number of impeller revolutions fixes progress of mixing, analogous to observations of mixing time. These experimental findings are also in good agreement with theoretical and modeling studies on mixing rate.

These studies of blending rate illustrate a simple characteristic found in many studies of mixing rate processes—an asymptotic approach to the fully mixed condition. In general, it is not possible to select a mixing speed or mixing time that gives a fully mixed condition; rather, a required degree of mixing needs to be specified and the corresponding degree of agitation determined.

Hydrodynamics. All mixing devices are to some extent complicated, but there are a number of simple parameters that are widely used to describe mixing equipment. Perhaps the greatest concern has been the need to describe energy requirements at given operating conditions. For agitated tanks this can be achieved quite effectively by the long established power number (Po) and Reynolds number (Re) relationships. The power number can be regarded as the impeller drag coefficient. The Reynolds number is a ratio of inertial to viscous forces, low for laminar flow and high for turbulent. The two numbers are defined for impeller agitation as:

$$\text{Po} = \frac{P}{\rho N^3 D^5} \qquad \text{(Eq 2-4)}$$

$$\text{Re} = \frac{D^2 N \rho}{\mu} \qquad \text{(Eq 2-5)}$$

Where:

P = mixer power input
D = impeller diameter
ρ = liquid density
μ = liquid viscosity

A typical plot of the relationship between these two dimensionless groups is shown in Figure 2-2. In the laminar regime, power number is inversely proportional to Reynolds number:

$$\text{PoRe} = K_p \qquad \text{(Eq 2-6)}$$

Before turbulent flow is achieved at a higher Reynolds number, there is a transitional flow regime, covering a range of Reynolds number, where the system is only partially turbulent. In this regime no simple generalization can be made regarding the power number–Reynolds number relationship. When baffled tanks are used, many impellers give a constant value of power number when turbulence is fully established. The range of values of power number can vary quite widely, but generally $0.1 < \text{Po} < 10$. There is much published power number data for both laminar and turbulent flow.

The power number–Reynolds number relationship is also useful in deciding on suitable operating conditions for a particular mixing problem. For most low-viscosity mixing, turbulent flow is the logical operating condition, and the minimum Reynolds number for turbulence can easily be selected from the power number–Reynolds

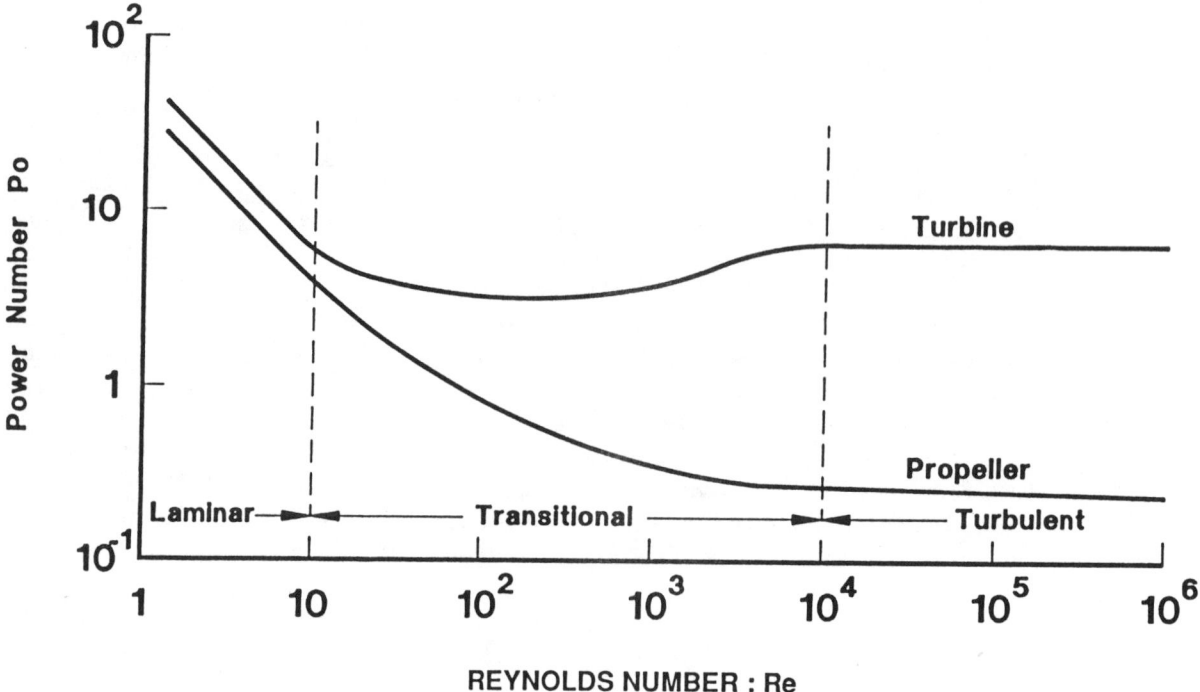

Figure 2-2 Typical power curves for baffled tanks.

number curve. Subsequent power-consumption calculations allow the first choice of Reynolds number to be optimized in terms of acceptable power consumption.

Some information is also available on the energy requirements of static mixers where the consideration of laminar, transitional, and turbulent flows is again required. The energy-consumption characteristics are described, as with pipe flow, by friction factor (C_D)–Reynolds number (Re) relationship:

$$C_D = \frac{2d_t \Delta p}{L_t \rho u^2} \qquad \text{(Eq 2-7)}$$

$$\text{Re} = \frac{u\, d_t\, \rho}{\mu} \qquad \text{(Eq 2-8)}$$

Where:

d_t = tube diameter
Δp = pressure drop
L_t = tube length
u = fluid mean velocity

This relationship, analogous to that for the agitated tank, is shown in Figure 2-3. In the laminar regime, friction factor is inversely proportional to Reynolds number; in the turbulent regime, friction factor is constant, with a transition regime existing between turbulence and laminar flow. In many cases the static-mixer friction factor can be regarded as a multiple of that for the empty pipe. The multiplier is generally 10 to 100 for laminar flow and in excess of 100 for turbulent flow. As with the agitated tank, a preliminary choice of operating Reynolds number can be made from the friction factor–Reynolds number curve and used for preliminary power calculations.

For agitated tanks, pumping and head characteristics may provide some indication of the suitability of the various impellers available. These are described by dimensionless groups: pumping coefficient (N_q) and head coefficient (N_h):

$$N_q = \frac{Q}{ND^3} \tag{Eq 2-9}$$

$$N_h = \frac{gH}{N^2D^2} \tag{Eq 2-10}$$

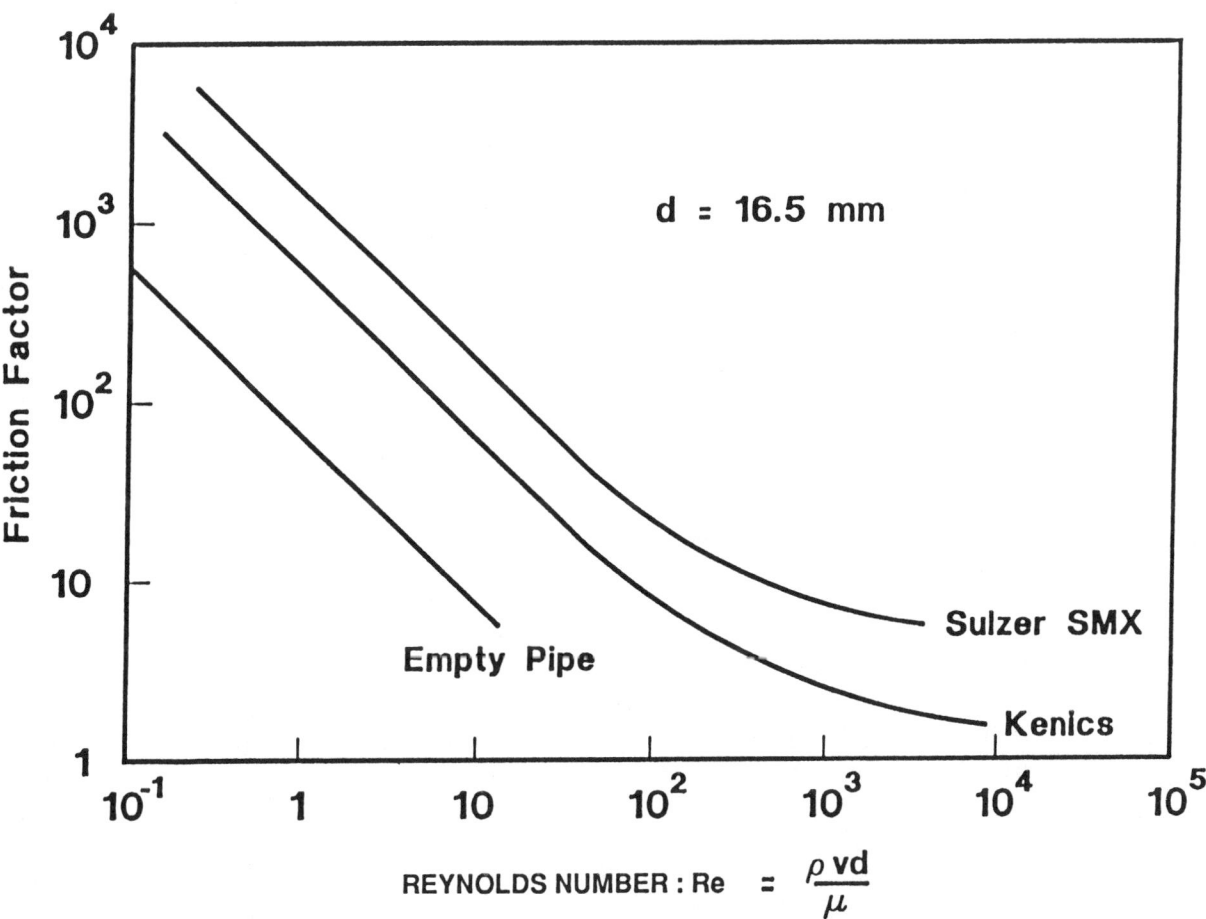

Source: Adapted from Novak, V. and Reiger, F. 1985 Homogenization Efficiency of Motionless Mixers. Proc. 5th European Conf. on Mixing. BHRA, Wurzburg, Germany.

Figure 2-3 Typical friction factor data for static mixers.

Where:

> Q = circulation rate
> H = head generated
> g = gravitational acceleration

Data for these two coefficients are less readily available than power number and friction factor, but N_q values have been appearing more often in recent publications. The head coefficient relates to the level of turbulence generated in the impeller region; high values indicate a good characteristic for dispersive mixing. Alternatively, a high pumping coefficient indicates a predominance of circulation. For any level of energy input, there is a balance to be considered between head and flow:

$$P = \eta Q \rho g H \qquad \text{(Eq 2-11)}$$

$$\text{Po} = \eta N_q N_h \qquad \text{(Eq 2-12)}$$

Where:

> η = impeller efficiency

At a particular value of power input per unit volume, different impellers will show a different balance of head and flow generation.

Measurements of flow have been used extensively to develop models of mixing in agitated tanks. Various modeling approaches have been developed, offering a range of details in their accuracy of description of mixer characteristics. Their applicability to mixing problems depends on the ability to simulate the controlling mechanisms of the problem under consideration. Understandably, the models are more or less complex, depending on the detail of the simulation. Simple models have been developed on the basis of a matrix of well-mixed cells with fluid interchange between cells and can be used to simulate concentration variations with time at various positions in an agitated tank. More complex models (coalescence–redispersion) are used to model the behavior of turbulent eddies and to describe both the exchange of material between eddies and the homogenization of the contents of an eddy. Computer packages have been available for some time for the analysis of flows in agitated tanks and, in principle, can be used for the prediction of velocity distribution in a specified geometry.

Mixing Equipment

Agitated tanks. The agitated tank is the most widely used item of mixing equipment. Consequently, there is more available information—data, theory, models—than for any other mixer. It is used in both laminar and turbulent installations and also in specific and general-purpose applications. Thus, there are many variations in design.

Of the agitated-tank applications, turbulent mixing is the most extensively used. Many varieties of mixer geometry are found. Typically, cylindrical tanks are used with depth approximately equal to diameter. Axial-flow (propeller type) and radial-flow (turbine type) impellers are both used extensively (Figure 2-4). Impeller diameters are usually one-third to one-half the tank diameter, but both larger and smaller impellers are used. Tank diameters range from 0.1 m for pilot and development work up to 10 m or greater for production use. In the mixing of low-viscosity

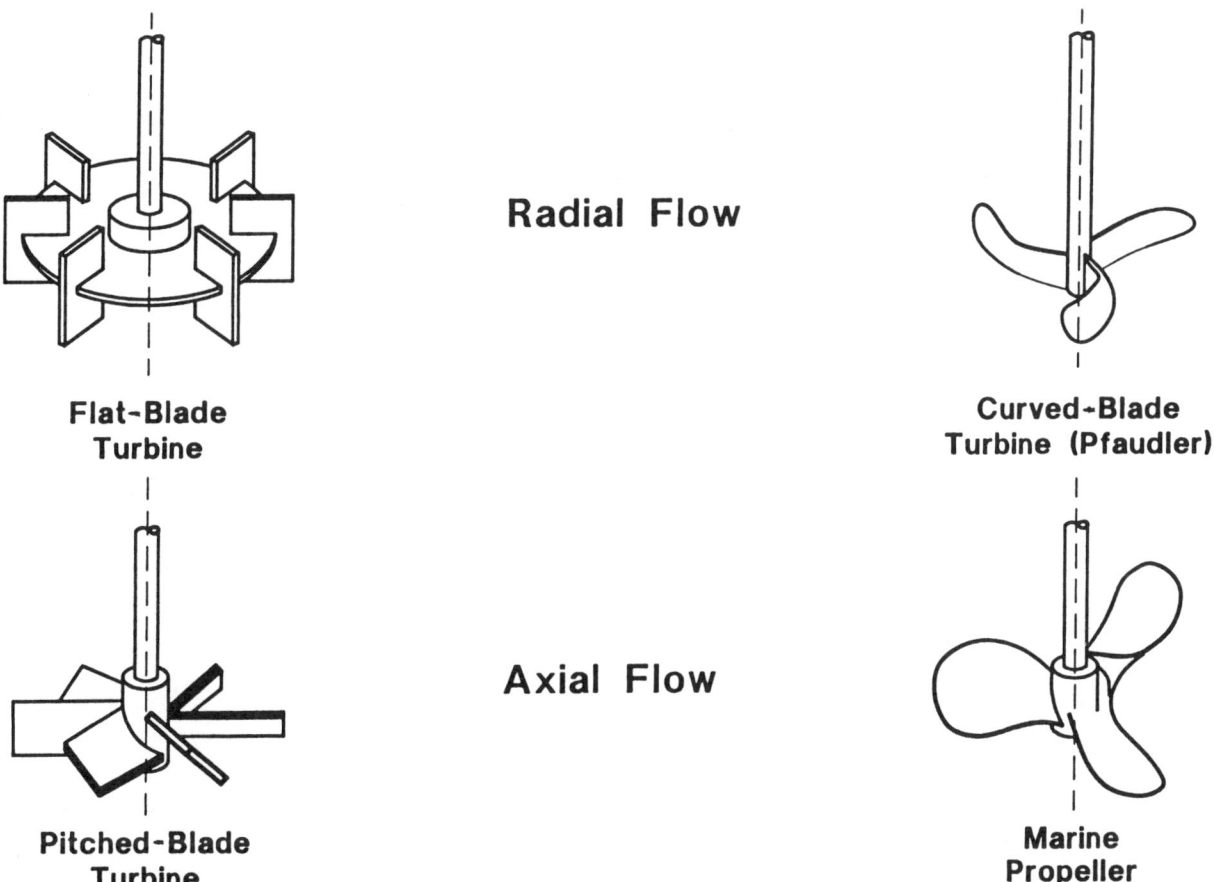

Figure 2-4 Mixing impellers.

fluids, vertical baffles are often used to suppress vortex formation. The most often used geometry is four equally spaced vertical baffles, one-tenth the width of the vessel diameter, fixed to the vessel wall. Although most agitated tanks have only one impeller, multiple impellers on one or more shafts are used in special applications.

Turbine-type impellers generate a radial flow and are classified by this characteristic. The radial discharge from the impeller leads to the formation of two circulation loops, one above and one below the impeller. Figure 2-5 gives an indication of the axial flow patterns, but there are also strong rotational flows and turbulent velocity fluctuations.

Axial-flow impellers produce a single-loop characteristic in their axial-flow pattern (Figure 2-6). There are also strong rotational flow components and turbulent fluctuations. With the axial-flow impeller there is the option of choosing the direction so as to give either "pumping down" or "pumping up" flow patterns, a useful facility in some mixing applications.

Although the best known and best documented of the impellers for turbulent mixing are the turbine and the marine propeller, many other impeller types are

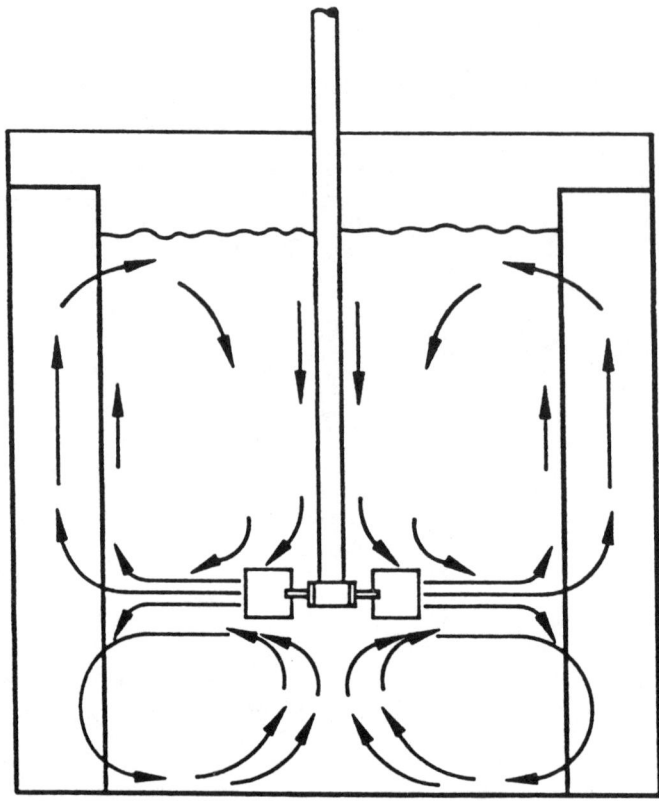

Figure 2-5 Flow pattern for axial-flow impeller—double loop.

being used. Among radial-flow impellers "diskless" turbines with a range of blade numbers—commonly between four and eight—are used. Radial-flow impellers are primarily high-shear impellers; this function is also achieved by sawtooth impeller types. Axial-flow impellers are mainly used where their recirculating pattern is advantageous—e.g., solid–liquid mixing—but are also being increasingly used as general-purpose impellers. In these applications pitched-blade turbines are used, combining aspects of both axial- and radial-flow mixing, as well as the relatively recent complex-profile "foil" impellers, which have been developed to enhance favored flow patterns (see chap. 9).

Both radial-flow and axial-flow impellers are capable of producing effective mixing with more viscous materials in the transitional and laminar-flow regimes. Larger impellers, like the anchor agitator, where the impeller sweeps much of the vessel volume are better, but better still are the more complex impellers like the helical ribbon or helical screw, which generate both rotational and axial flows (Figure 2-7).

A variety of other mixers fall into the category of agitated tanks but are less easily classified. The high-shear rotor-stator mixer is used for both low-viscosity and high-viscosity applications and is of value when high shear is advantageous, although power consumption is too high for simple applications. Blending in large storage tanks can perhaps also be regarded as another version of agitated-tank mixing. Side-entry and bottom-entry shafts are often used to minimize shaft length.

MIXING IN LIQUIDS 47

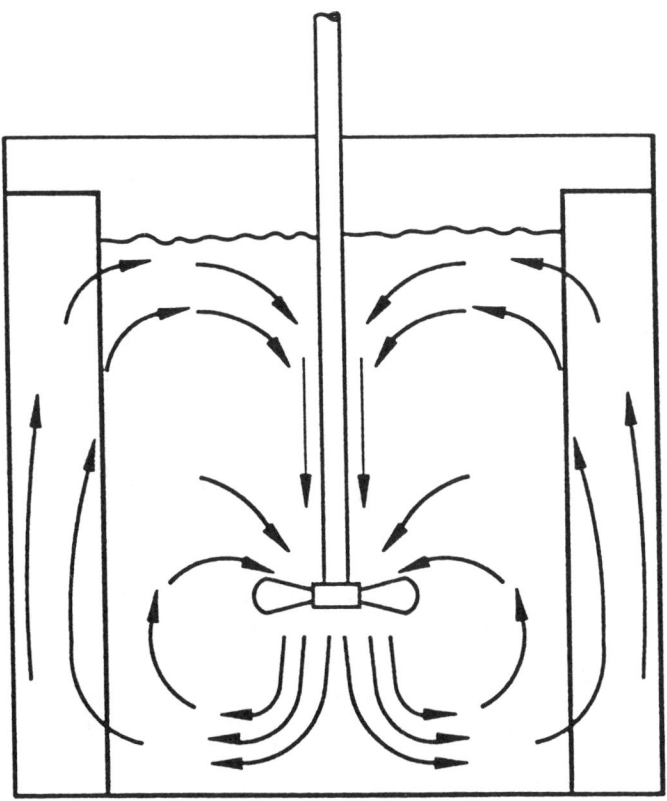

Figure 2-6 Flow pattern for axial-flow impeller—single loop.

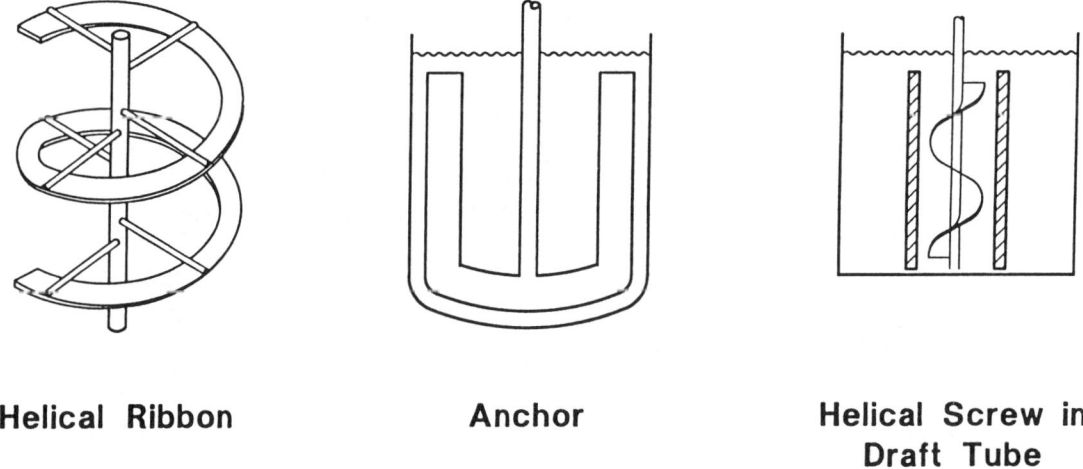

Helical Ribbon **Anchor** **Helical Screw in Draft Tube**

Figure 2-7 Close-clearance impellers for high-viscosity fluids.

The energy levels used are generally low, and axial-flow impellers are used to give a flow field that is closer to jet mixing than the conventional agitated tank.

Static mixers. The static mixer is a relatively new development. It has only been used extensively in the past 20 years. The first static mixers were mostly designed with in-line blending of viscous liquids in mind, but it soon became apparent that there were turbulent-flow applications for the static mixer if pressure losses were not too high. Like all mixers, the static mixer requires energy input, and energy costs need to be considered along with capital costs when different mixers are being compared. Static mixers and agitated tanks have similar cost requirements.

The static mixer is not just a possible replacement for the agitated tank. The static mixer is, in principle, a continuous-flow device. It can be adapted (by combining its use with recycling to storage tanks or agitated tanks) for batch processes, but it always retains a continuous-flow mixing characteristic—i.e., mixture quality generally depends on mixer length but is not very sensitive to flow rate. More importantly, the static mixer is a radial mixer, improving the degree of mixing or uniformity across the pipe section but not along the pipe length. Thus, the flow ratios of constituents to be mixed must be carefully controlled, as the static mixer has no ability to damp out fluctuations in time; this is in direct contrast to the agitated tank. The static mixer suggests itself for simple mixing duties like the blending of reagents into low-viscosity process streams. Often, simple blending can be achieved in turbulent pipe flow, but the pipe must be longer. Combinations of static mixer and turbulent pipe flow can also be envisaged, with the static mixer used to augment the mixing characteristic of the available pipe.

Turbulent flow offers many potential applications of the static mixer. Strong radial mixing and an almost complete absence of axial mixing create an advantageous condition for some difficult chemical reactions. For a given turbulence level the conditions provided for a reaction are more favorable in the near plug-flow regime provided by the static mixer than in the near backmixed flow provided by a continuous-flow agitated tank. Reaction rates and selectivity among competing reactions improve accordingly. There may also be some advantage in simpler computations, in that the distribution of turbulence is more uniform than in an agitated tank and the distribution of residence times is not so complex. Multiphase processes might also be conducted under turbulent flow conditions in static mixers, in particular gas–liquid and liquid–liquid dispersion.

Of the very large number of static-mixer designs, only relatively few are well documented, particularly with respect to the manufacturer's literature. One of the first mixers to have success in the field is the Kenics mixer, now marketed by the Chemineer* mixing company. Also long established are the Sulzer mixers, marketed by Ross.† More recently a static mixer design has been developed by Lightnin‡ (Oldshue 1985). Many manufacturers offer a choice of design for laminar or turbulent operations. The main requirement is acceptable pressure drop in the flow regime required.

Jet mixers. A variety of jet mixing techniques are available for a range of applications in low-viscosity mixing. The two main areas of application are pipeline mixing and tank mixing. The pipeline mixing applications are similar to the

---

*Chemineer, Inc., Dayton, Ohio.

†Charles Ross & Son Company, Hauppauge, N.Y.

‡Lightnin, Rochester, N.Y.

short-mixing-time processes using static mixers, and the two technologies are often competitive. Tank mixing applications are mostly concerned with simple mixing requirements involving long mixing times and simple equipment.

Pipeline jet mixing techniques are used to improve on the mixing characteristics of simple, turbulent pipe flow. The characteristics of pipeline jet mixing are similar to the corresponding static-mixer application: turbulent flow and short mixing times, plug flow and a requirement for flow ratio control. In-line jet mixers come in two major categories: coaxial jet and mixing tees. The coaxial jet brings the two flows to be mixed together via one or more inlet tubes located in a pipe (Figure 2-8). The mixing tee brings the two liquids together via a simple tee junction in a pipe (Figure 2-9). The latter is simpler to construct and has better characteristics for blending unequal flow rates.

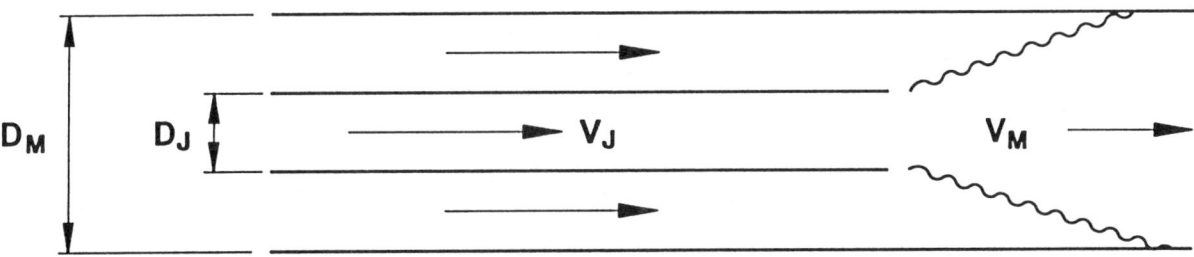

Figure 2-8 Coaxial jet mixer.

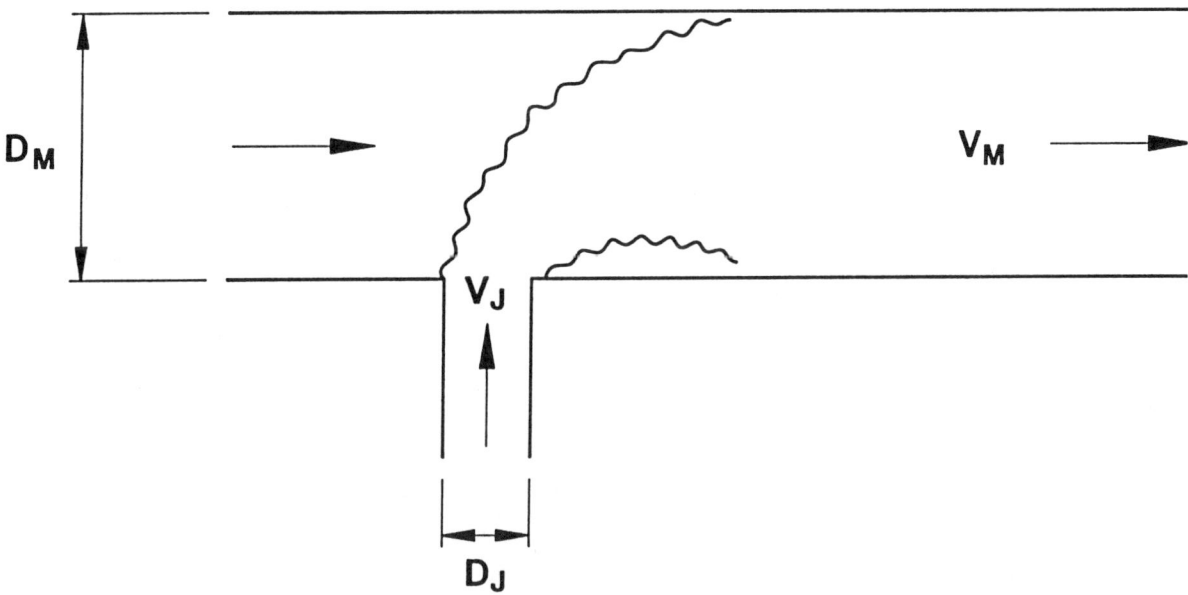

Figure 2-9 Side entry or tee jet mixer.

Jet mixing in tanks primarily involves simple blending of low-viscosity liquids in storage tanks. A wide range of tank and jet geometries have been used to suit different applications. The process is often operated batchwise, with a recycle stream and pump used to provide mixing energy. Energy requirements are probably higher than for the corresponding agitated tank, but jet mixing is preferred for some applications for its simplicity and low capital and maintenance costs. The design and application of jet and pipe mixers in water treatment applications are discussed in detail in chap. 6.

Dynamic in-line mixers. Dynamic in-line mixers offer another alternative to the static mixer or jet mixer. Most designs can be regarded as a modification of the agitated tank (Figure 2-10). The mixers are significantly much smaller than the corresponding agitated tank, but the power per unit volume levels are significantly greater. The levels of power per unit volume are also greater than can be guaranteed in static mixers, so the dynamic mixer is suitable for the production of emulsions or dispersions where large interfacial areas are required. As with static mixers, the dynamic design has been used for both laminar and turbulent processes.

Dynamic-mixer design promises a wide range of applicability, although it has not yet been achieved. In principle, dynamic mixers could replace many larger installations, but the absence of much design or research data is probably responsible for their comparatively low usage at present. The equipment may also be more expensive than static-mixer installations.

Dynamic mixers have been used for a range of applications including liquid blending; and solid–liquid, gas–liquid, and liquid–liquid dispersion. Claims have also been made for the capacity of the dynamic mixer to smooth time variations. The dynamic mixer offers more potential than static or jet mixers for smoothing, but the

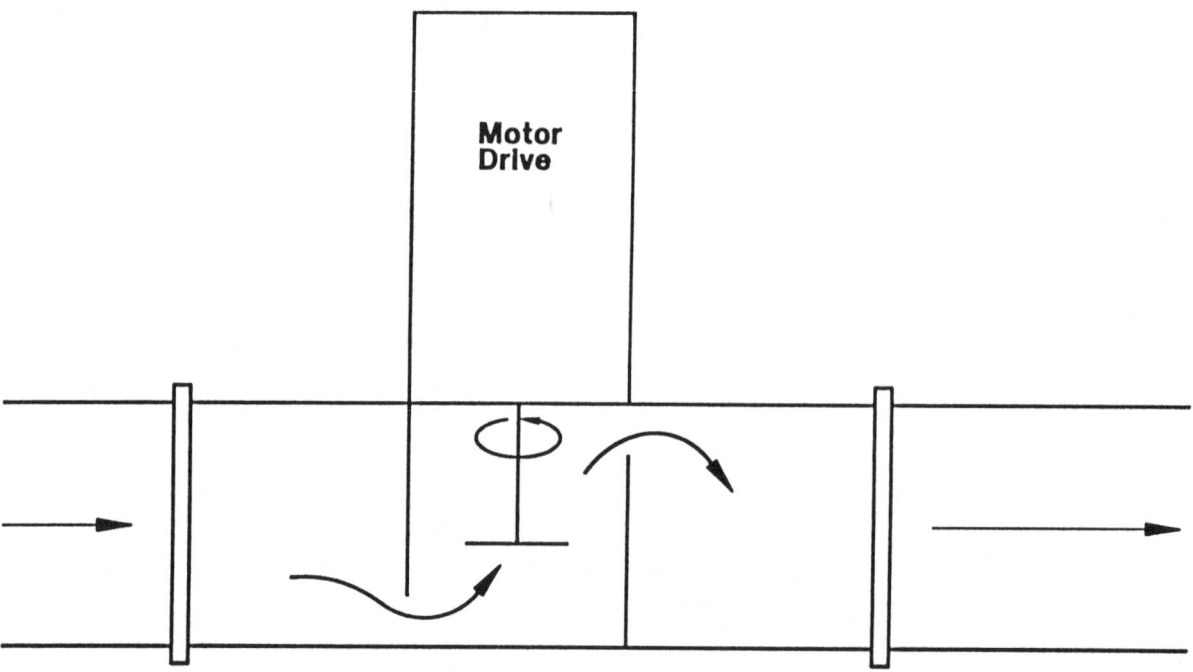

Figure 2-10 Impeller-agitated in-line mixer.

small volume of the mixing chamber still subordinates it to the conventional agitated tank.

Design parameters of the dynamic mixer are similar to those of any in-line agitated tank: mixing time (and hence mixer volume) and energy input (power per unit volume). The dynamic mixer is therefore a very small, very high-specific-power device and is mainly suitable only for applications requiring such characteristics. As with any other mixer selection, comparison on the basis of capital and operating costs is desirable.

In water treatment practice, the dynamic in-line mixer offers significant advantages when used with inorganic coagulants such as alum in the charge neutralization mechanism of coagulation (see chap. 1 and Amirtharajah 1987). For these applications it is used with average velocity gradients ($\overline{G}$-values) of 3000 to 5000 s^{-1} with mixing times less than 1 s.

STIRRED-TANK REACTORS

The main characteristics of stirred-tank reactors have been discussed earlier in the introduction. In the sections that follow, the nature of these characteristics is discussed in greater detail together with available design or performance correlations.

Single-Phase Turbulent Mixing

Power consumption. Of all of the parameters of mixer performance, power consumption is perhaps the simplest to measure and correlate. The power number–Reynolds number relationship, Eq 2-5 and 2-6, is easy to evaluate and can include the influence of a number of variables, including scale, if used carefully. However, many variables that affect the value of power number—e.g., impeller type, use of baffles, multiple impellers, off-center impellers—cannot be described by a simple correlation. In addition, although the power number is often a constant value for a particular impeller–tank combination at high Reynolds number (Re > 2 × 10^4), it must be remembered that power number is a function of Reynolds number for some geometries. The exact dimensions of an impeller can also be important, and attention must be paid to the influence of impeller blade thickness, and other geometrical parameters in some cases. Thus, while power-consumption correlations are very helpful and simple to use, correlations must be selected carefully.

The relationship between power number and Reynolds number increasingly includes additional parameter groups to describe the influence of mixer geometry and other variables. Occasionally there is some discussion of the influence of the Froude number, but it seems generally that any effect is small (Novak, Ditl, and Rieger 1982). This conclusion applies to both baffled and unbaffled vessels, and, in the latter, vortex formation is more marked and is related to Froude number but does not influence power consumption. In the turbulent regime many impellers show only a very small dependence of power number on Reynolds number. This has lead to the practice of using power numbers to characterize particular impeller types and to make power calculations. However, it cannot be assumed that the power number is constant in the turbulent regime for all mixer geometries. From the earliest detailed studies—e.g., Rushton, Costich, and Everett (1950)—the influence of baffles on the power number has been evident, particularly with the characteristic of a constant power number for a particular impeller. Baffles also minimize the effect of the impeller diameter/tank diameter ratio for some impellers, notably the Rushton turbine. The influence of baffles has been reviewed in detail (Nagata 1975), with much data collected. Varying degrees of baffling are possible, interpreted in terms of the ratio of power number with baffles to power number without. The traditional four-baffle arrangement with pitched-blade turbines also produces a substantially constant power number for a range of Reynolds numbers above Re = 2×10^4, but some dependence of impeller size (i.e., D/T where D = diameter of impeller and T = diameter of vessel) is seen, e.g., Bujalski et al. (1987).

It is sometimes assumed that if power number does not vary with Reynolds number, the system is fully baffled and not susceptible to the influence of the impeller to vessel diameter ratio, D/T. The error of this assumption is illustrated by Godfrey and Grilc (1988), where tanks of square cross section are shown to give a constant power number in the absence of baffles, with Po = 3.5 for a Rushton turbine of D/T = 1/3. Other values of D/T give different values of power number. Recent surveys have emphasized the lack of agreement between power consumption data generated by different laboratories. Bujalski et al. (1987) illustrate a number of aspects of this problem, evaluating the influence of disk thickness (x_1) and vessel diameter (D_v) for Rushton turbines:

$$\text{Po} = 222.5 \, (x_1/D)^{-0.2} \, D_v^{0.065} \qquad \text{(Eq 2-13)}$$

Other variables of practical interest have been presented in recent work: the influence of off-center impeller locations and vessel shapes (Novak, Ditl, and Rieger 1982), the use of multiple impellers where power numbers seem additive for axial-flow impellers but not for radial-flow impellers (Kuboi and Nienow 1982), the effect of liquid depth (Greaves and Loh 1969), the influence of pitched-blade impeller geometry (Fort and Medek 1988). There is also a vast review of older data (Nagata 1975).

Discharge rate. Because the performance of most mixers in use is flow controlled, it is useful to have some understanding of the impeller discharge characteristics. In general, impellers can be regarded as pumps, either axial or radial, and their performance can be interpreted accordingly (see Figure 2-11). Although the flow pattern is very complicated, a simple characterization of many agitated systems can be provided by the dimensionless flow number N_q: Q/ND^3 (see Eq 2-9).

For baffled tanks N_q is essentially constant for a particular impeller type and, with careful design, independent of scale. The increasingly reliable values of N_q now available are useful in impeller selection, especially when used in conjunction with corresponding Po values. In addition to impeller selection, Q values are increasingly being used in the interpretation of mixing-time data and offer the prospect of a basis for mixing-time calculations in processes that are mainly flow controlled.

The flow number, N_q, is a dimensionless number and, confusingly, is also known as the pumping number, the flow, pumping, or discharge coefficient. The discharge rate, Q, is the discharge rate from the impeller only. This rigid definition has been shown to be important (Revill 1982), in ensuring the compatibility of data from different sources. Extensive reexamination of data for the disk turbine has revealed that virtually all of the available data for baffled tanks can be described by $N_q = 0.75 \pm 0.15$ for the range of impeller diameters $0.20 < D/D_v < 0.50$. Constant values of N_q are reported in recent studies of a very wide range of impeller types (Weetman and Oldshue 1988). Thus, the discharge rate characteristics of many impeller-agitated tanks with baffles are described by a single parameter that only depends on impeller type and is independent of fluid properties and mixer size. Obviously, some caution must be exercised, since a constant value of N_q is only a characteristic of baffled tanks and requires that all impeller dimensions (including

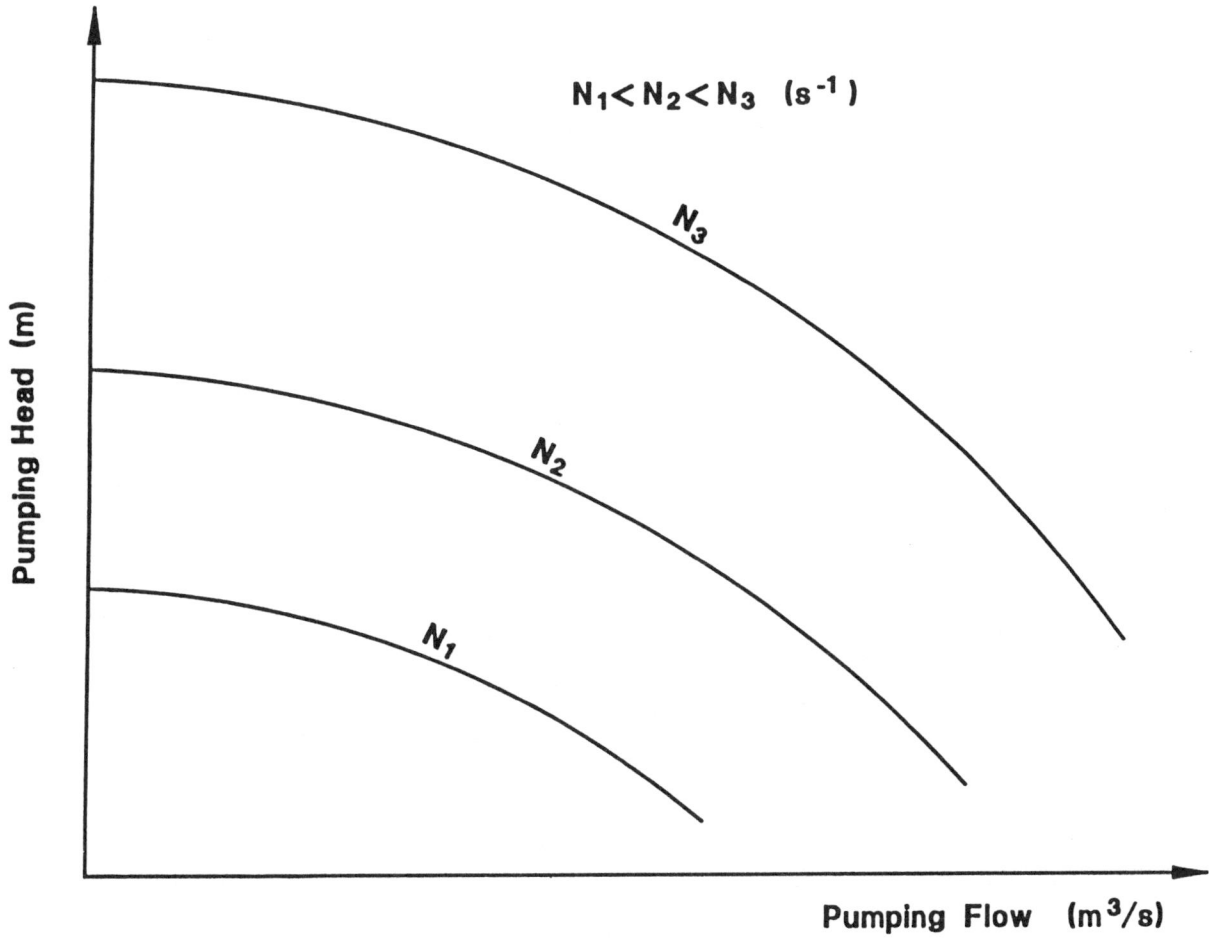

Figure 2-11 Effect of impeller speed on pump-mix head and flow.

blade thickness) be accurately scaled. A constant value also requires turbulent flow, transitional flow values being lower.

Most recent data have been generated by laser–Doppler techniques, which have the advantages of high precision in velocity measurement, no influence on the flow field, the ability to measure steady and fluctuating velocity with high-frequency response, and the ability to measure two velocity components. This improves the accuracy of the determination of values Q, e.g., for the disk turbine it is the radial velocity that is of interest:

$$Q = \pi D \int_{-w/2}^{+w/2} v_r \, dz \qquad \text{(Eq 2-14)}$$

With axial-flow impellers, v_z is the required value:

$$Q = 2\pi \int_0^r r v_z \, dr \qquad \text{(Eq 2-15)}$$

Where:

$$r = D/2$$

The laser–Doppler technique is expensive and also requires very good "visibility" for the transmission and detection of laser light. Consequently, it is a technique best suited to studies of optically clean liquids in transparent tanks in a clean laboratory. It is interesting to note that useful data have been recently published for an old technique—the "flow follower" (Faraday 1987). A large (visible) particle was used to determine a mean circulation time from which the circulation rate was estimated:

$$Q = V/t_c \qquad \text{(Eq 2-16)}$$

This technique is clearly limited with regard to the accuracy of the value of Q and requires good visibility, but the data obtained are worth consulting as an example of a cheap and relatively quick investigation.

Where circulation is the most important mixing mechanism, impeller selection may be based on pumping capacity. Axial-flow impellers generally give greater flow for a given power input than radial-flow impellers. Power number and flow number values can be used to calculate flow per unit power. While circulation is the major consideration for some processes—blending, solids suspension, heat transfer—other applications require the consideration of shear parameters, as discussed below.

Head characteristics. To improve the quantitative description of impeller performance and provide more detailed parameters for impeller comparison and selection, the impeller–pump analogy can be extended to include the concept of head in Eq 2-10 to 2-12. The variations of N_h versus N_q for three different impellers are shown in Figure 2-12.

The dimensionless formulations for P and Po are the basis of comparison of impeller performance in terms of Po and N_q and various concepts of efficiency. In addition, mixing applications can be considered in terms of whether head or flow is more important to the success of the process. Processes that require high head are frequently those that require high shear rate, e.g., deagglomeration.

Thus, head, efficiency, and shear rate can be viewed as related concepts for impeller or mixing process evaluation. A range of impeller applications can be envisaged, extending from one extreme, high flow requirement, to the other, high shear requirement. High-flow processes such as blending, heat transfer, and suspension of solids require large circulating flows and little shear or head and usually employ axial-flow impellers. High-shear processes such as the breaking of agglomerated particle assemblies and the creation of liquid–liquid dispersions or emulsions require more shear and usually employ radial-flow impellers. Between the two extremes, more and more mixing processes are being identified that would benefit from a balance of flow and shear, and more detailed examination of impeller performance is beginning to establish a more precise relationship between process and impeller type. Comparisons of power requirement and efficiency at constant flows can also be made.

Head characteristics are usually discussed in terms of the dimensionless head, N_h (see Eq 2-10). However, measurements of head are not generally made except where the information is specific to the application as with pump-mix (or lifter) impellers or with impeller draught-tube combinations (Lol, Baird, and Hanson 1983). Although the head characteristics of general-purpose impellers are not usually measured, they are increasingly discussed. One interpretation being used by

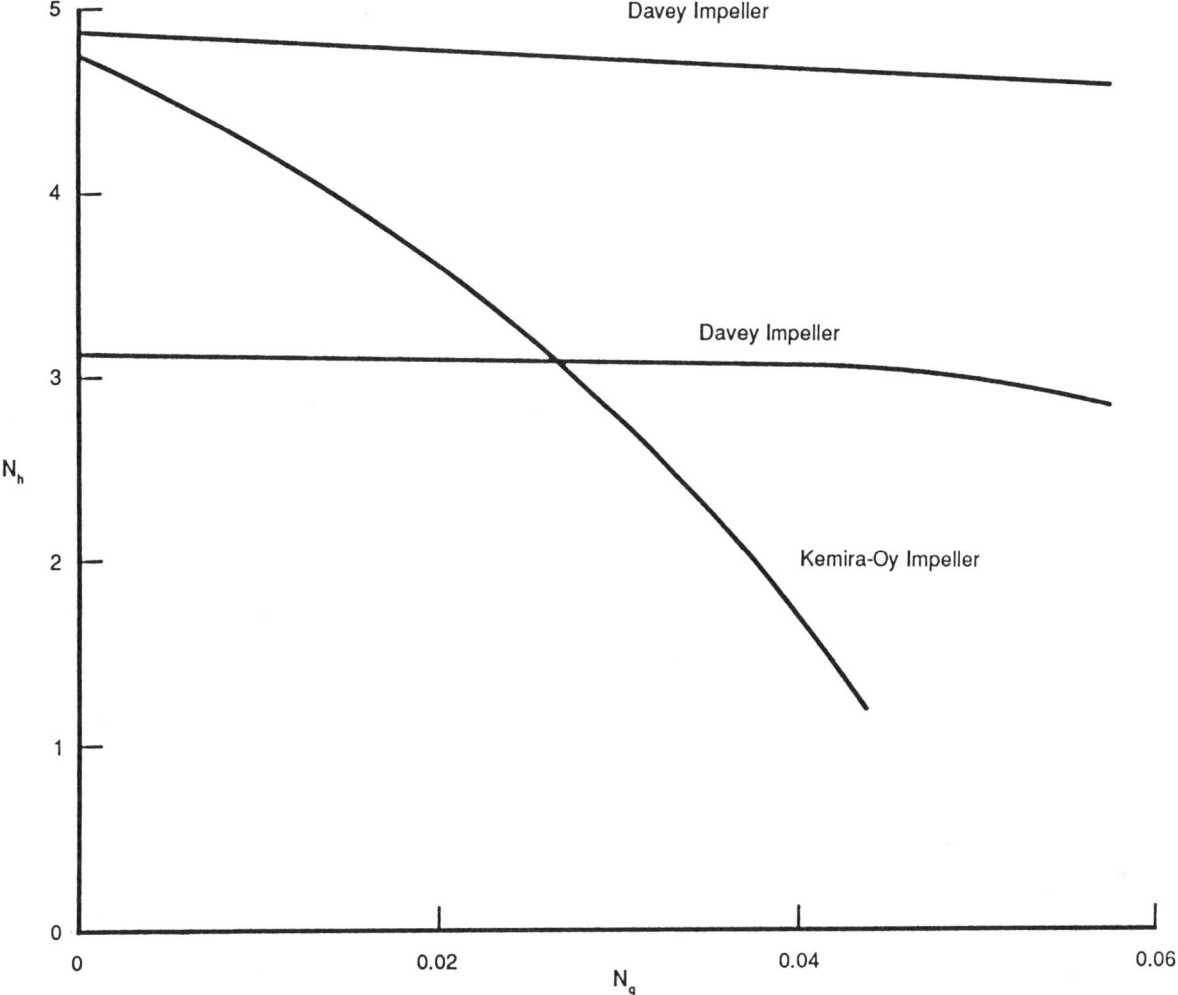

Figure 2-12 Dimensionless head-flow relationships for pump-mix impellers.

impeller manufacturers is the consideration of head as the sum of static head (or pressure) and viscous shear terms (Weetman and Oldshue 1988). This provides a distinction between different types of head requirement: head when substantial density differences exist between the materials being mixed, shear when breakage or dispersion is required. Alternatively, relationships for head and efficiency have been formulated in terms of pressure difference, head difference, and discharge coefficients leading to comparisons of flow number and efficiency (Fort and Medek 1988). The relationship derived between efficiency and flow takes the form

$$\frac{\eta_1}{\eta_2} = \frac{\left[N_h^3/Po\right]_1}{\left[N_h^3/Po\right]_2} \quad \text{(Eq 2-17)}$$

for the comparison of two impellers. Thus, while N_h and η are indeterminate for most impellers, efficiencies can be compared on the above basis. A similar comparison is achieved on the basis of circulation per unit power, using the ratio Q/P as the comparison parameter for fixed values of Q and D. It is interesting that the ratios of the Rushton turbine at the low end of the scale and the "fluidfoil" axial-flow impeller at the high end vary by a factor of more than 10. It is important to note that the comparison will vary according to the value of Q chosen as the basis of comparison, so comparisons based on Q/P are only relevant for particular value of Q. Equations 2-10 and 2-11 can be combined to give

$$\frac{Q}{P} = \frac{1}{\rho N^2 D^2 \eta N_h} \qquad \text{(Eq 2-18)}$$

where the product ηN_h is a constant; and from Eq 2-12

$$\eta N_h = \frac{Po}{N_q} \qquad \text{(Eq 2-19)}$$

These equations allow the Q/P ratio to be determined for known values of Po and N_q, and a specified value of Q.

Although head and efficiency of most impellers cannot be measured directly, the related parameter shear rate can be measured. Increasing use of laser measurements for the collection of velocity (for N_q) and turbulence (for modeling) data has also yielded velocity profile data that can be used to compute shear rates. Data have recently been presented to describe the shear rate characteristics for the impeller discharge region of the Rushton turbine, the pitched turbine, and the fluidfoil impeller. Average shear rate values for the three impellers correlate directly with impeller speed for a range of impeller sizes (Weetman and Oldshue 1988). At first, it appears that shear rates will be much higher for the Rushton turbine than for the other impellers. It is true that operating shear rates will generally be higher, but operating speeds are generally lower for the same power input per unit volume.

Thus, while measurements of head can be difficult and are rarely published, the concept can be easily related as dimensionless head to dimensionless flow and power number. This does not solve the problem of evaluation, because of the need to consider efficiency. However, the concept is being increasingly used to improve the evaluation of aspects of the mixing process that are dependent on head (mixing of materials of different density) and shear (breakage of agglomerates).

Turbulence. A number of the characteristics of turbulence—velocities, eddy scale, turbulent stress—can be estimated from a knowledge of power input and system properties. Perhaps of most interest is the size of the smallest eddy, because of its relationship to the smallest scale of nonhomogeneity achieved in turbulent mixing processes:

$$\lambda_e = (\mu^3 V / P\rho^2)^{1/4} \qquad \text{(Eq 2-20)}$$

Not surprisingly, it is generally believed that the smallest nonhomogeneity is smaller than the smallest eddy—i.e., nonhomogeneities can be regarded as distributed within the eddy. The size of eddy calculated above, the Kolmogoroff length scale, is the smallest an eddy can be before disappearing by viscous dissipation.

Other turbulence parameters that can be estimated from energy input are the root mean square velocity

$$\bar{u}^2 = C_{RMS} (Pd/\rho V)^{2/3} \qquad \text{(Eq 2-21)}$$

which refers to variations over the distance d, where d may be the dimension of a particle, bubble, or drop and C_{RMS} is a coefficient. This root mean square velocity is a measure of the influence of turbulence velocities on a particle of size d and can be used as the basis for estimating breakage forces due to turbulence. For eddies larger than the smallest—i.e., λ_e calculated from Eq 2-20 above—a turbulence shear stress π can be estimated from

$$\pi = C_\pi (P\rho/dV)^{2/3} \qquad \text{(Eq 2-22)}$$

in which C_π is a coefficient.

For some time it has been possible to make measurements of velocity in agitated tanks, showing both large-scale circulating velocities and small-scale fluctuating velocities. Although both collection and processing of such data are very time consuming, considerable work is being done in this area.

Turbulence energy-dissipation rates, ε, cannot be measured directly but can be estimated from measurements of steady and fluctuating velocities. Data available for the Rushton turbine suggest that 60 percent of all energy dissipation occurs in the 3 percent of tank volume in the impeller discharge region. Two impeller-blade widths into the discharge stream, the value of ε has fallen almost to zero; beyond this point the system appears to be isotropic. The characteristics of the impeller discharge stream are fairly simple, in that the ratio ε/N^3D^2 is not much affected by impeller speed, so some economy of measurement is possible. Values of turbulence energy, k, can also be determined from three-dimensional fluctuating velocity measurements. The variation of the two turbulence parameters k and ε is essential for some models of turbulent mixing (e.g., FLUENT) and it appears that a knowledge of the variation of the turbulence dissipation rate ε with position in the vessel can be used to calculate yield for mixing-controlled reactions, one of the most demanding of mixing conditions.

Mixing time. Many mixing processes are simple to achieve, e.g., the turbulent blending of two low-viscosity miscible fluids. For simple mixing tasks the concept of mixing time is very helpful and probably adequate. A variety of concepts are used in the description of mixing time: mixing time itself, the product of mixing time and impeller speed (often regarded as a constant), and circulation time. However, some of the concepts lack precision, and more demanding aspects of mixing require more accurate descriptive techniques. There are three main aspects to the question of accuracy: the measurement technique, the definition of mixedness, and the nature of the influence of time. There are also some interrelationships between these three elements.

When accuracy is important, it is necessary to consider the relationship between the assessment of quality used to describe the product and the technique used to follow the progress of mixing. Conducting tracers are often used in the study of mixing but are not necessarily relevant to the description of product quality. Further, the characteristic size of each measuring technique will affect the perceived mixture quality. In most cases, it is important that the scale of the measuring device is not changed during scale-up exercises, as the measuring scale relates to product quality. It is also important to specify how measured concentration fluctuations are assessed. For example, it is very difficult to be sure exactly when a fully mixed condition is achieved since the approach to the final concentration steady state is asymptotic. A study of the relationship between mixture quality and mixing time is useful in this respect, as a rate constant can then be used to describe the process and the problem of specifying an arbitrary end point is removed.

Other difficulties remain, even when it is possible to precisely define the progress of mixing as a function of time. Most methods of measurement can only follow mixing at a relatively coarse scale and cannot detect variations in mixture quality at the scale of turbulent eddies. This will present problems in mixing processes where control lies with small-scale turbulent characteristics. There are also a range of problems associated with scale-up. In addition to the influence of the scale of the measuring device, the influence of mixer size is complex and differs according to the controlling mechanism.

The concept of mixing time usually relates to the monitoring of the blending of a quantity of tracer into a mix (see Figure 2-1). As mentioned above, a problem exists in identifying an exact end point. This is avoided in some work by specifying a mixing condition, e.g., within a percentage of the steady-state concentration.

When an attempt is made to compare mixing-time data from different sources it is important to remember the arbitrary nature of most mixing end points. The location of the detector used for tracer measurement is another major factor in comparing data from different sources. Examination of these effects, using several detectors in mixing measurements, has been made by Rielly and Britter (1985). Various tracing techniques have been used, including reactions that cause coloring or decoloring of the mixture. Hiby (1985) discusses the results to be expected from different techniques and concludes that the simple decoloring techniques can indicate dead zones and that mixing time obtained from fast-reaction experiments is also relevant to nonreactive blending. The topic of the suitability of different mixing indices is also considered, and the coefficient of variation (σ/C_f) is preferred to intensity of segregation (σ^2/σ_o^2) when there is a large difference in volume fraction between the two liquids to be mixed.

Although there are obviously a number of difficulties associated with the use of mixing time as a description of performance, the concept is widely used and is an effective basis for the comparison of different mixer types where a precise description of mixture quality is not important and the results are obtained in a consistent manner. There is also a simple relationship between impeller speed and mixing time that can be helpful in planning laboratory and pilot-scale test programs. Many mixing studies have shown that, for a particular mixer under turbulent flow conditions, the product of impeller speed and mixing time is a constant. While there has been much discussion of the dependence of the product, Nt_m, on the various possible influencing variables, e.g., Reynolds number, the dependence in general has been small (Rielly and Britter 1985).

Other time parameters have also been used to describe turbulent mixing, e.g., circulation time (t_c). Circulation time has been estimated from the concentration fluctuations observed in mixing-time experiments (see Figure 2-1). Results for disk turbine impellers of various diameters are described in a manner analogous to mixing time (Holmes, Voncken, and Dekker 1964):

$$Nt_c = 0.85 \, (D_v/D)^2 \qquad \text{(Eq 2-23)}$$

Measurements have also been made by monitoring the position of a transmitting "radio pill," by Middleton (1979), giving a similar relationship:

$$Nt_c = 0.5 \, V^{0.3} \, (D_v/D)^3 \qquad \text{(Eq 2-24)}$$

More recent laser measurements of velocities in agitated tanks could be used to further explore this relationship. It is frequently assumed that there is a direct relationship between circulation time and mixing time, of the order:

$$t_M \sim 3t_c \qquad \text{(Eq 2-25)}$$

for the turbine impeller in a baffled tank located at $c = H/3$, above the base of the tank. Recent studies for two impeller systems in deep tanks, $H \sim 2 \times T$, have given the approximation:

$$t_M \sim 2t_c \qquad \text{(Eq 2-26)}$$

In addition to the use of circulation time as an indication of mixing time, many commercially important competing reactions are considered to be controlled by bulk circulation and therefore circulation-time dependent.

Other mixing times of relevance to reacting systems can be estimated from the equations of turbulence. An estimate of the diffusive mixing time associated with a turbulent eddy is discussed by Bourne (1985):

$$t_{MD} = 0.5 \lambda_e^2 / 4D \qquad \text{(Eq 2-27)}$$

where the turbulent microscale λ_e is for the smallest eddies (Eq 2-20).

The scale of λ_e is approximately 10 to 30 μm for turbulent aqueous liquids, and the corresponding value of t_{MD} is expressed in terms of milliseconds. To estimate the time required to achieve molecular homogeneity, Bourne (1985) suggests that the diffusive mixing time (t_{MD}) and bulk mixing time (t_M) can be added.

Progress of mixing. More precise estimates of mixing time can be made if the mixing end point specified is also precise. This approach is long established, Prochazka and Landau (1961) having related a fractional unmixedness (Eq 2-1) to the mixing performance of various impellers:

Propeller $\qquad N\theta = 3.48 \left[\dfrac{T}{D}\right]^{2.05} \log\left[\dfrac{2}{X}\right] \qquad$ (Eq 2-28)

Pitched-blade turbine $\qquad N\theta = 2.02 \left[\dfrac{T}{D}\right]^{2.2} \log\left[\dfrac{2}{X}\right] \qquad$ (Eq 2-29)

Disk turbine $\qquad N\theta = 0.905 \left[\dfrac{T}{D}\right]^{2.56} \log\left[\dfrac{2}{X}\right] \qquad$ (Eq 2-30)

A very similar relationship has been more recently proposed (Khang and Levenspiel 1976) to relate the amplitude of concentration fluctuations during mixing, A, to mixing time:

$$A = 2 \exp(-Kt) \qquad \text{(Eq 2-31)}$$

The rate constant, K, is related to impeller speed and tank geometry for two impeller types:

Propeller (pitch = 1.5) $$\frac{N}{K}\left[\frac{D}{T}\right]^2 = 0.9 \qquad \text{(Eq 2-32)}$$

Disk turbine $$\frac{N}{K}\left[\frac{D}{T}\right]^{2.3} = 0.5 \qquad \text{(Eq 2-33)}$$

This is essentially the same relationship as that of Prochazka and Laudau (Eq 2-28 and 2-30) and when rearranged gives

Propeller $$N\theta = 0.9\left[\frac{T}{D}\right]^2 \ln\left[\frac{2}{A}\right] \qquad \text{(Eq 2-34)}$$

Turbine $$N\theta = 0.5\left[\frac{T}{D}\right]^{2.3} \ln\left[\frac{2}{A}\right] \qquad \text{(Eq 2-35)}$$

Direct comparison is not possible because of the difference between X and A. However, the influence of the impeller diameter expressed as T/D is similar. The choice of mixing index should be regarded as potentially important but unresolved. Certainly the choice of different indices will lead to different mixing times or different characteristics of the progress of mixing. As mentioned above, Hiby prefers the coefficient of variation, σ/c_f, to account for the relative volume fractions of the two liquids being blended. However, Rielly and Britter (1985) argue that the volume fraction will not influence measured mixing rate unless the length scale is less than the turbulent microscale, λ_e. As the microscale is generally very small, their suggestion is that, for practical mixing processes, the volume fraction will not have any influence.

Present experimental evidence suggests that the long-standing representation of mixing times in terms of the simple correlation

$$Nt_M = \text{constant} \qquad \text{(Eq 2-36)}$$

is still a good approximation. The influence of impeller diameter is frequently expressed by the ratio $(T/D)^2$, although the exponent (2) is seen to vary in the results of various workers (see Eq 2-28 to 2-35).

In recent work it has become accepted that a mixed end point must be carefully defined if results from different sources are to be comparable. The influence of tracer volume, and perhaps volume ratio of blended components, seems unresolved, with arguments for and against. Likewise, the influence of mixer volume is unresolved, but the problem of scale-up is now generally accepted and examined when opportunities arise. These considerations are important and similar for mixing time and progress of mixing studies. Both measurement methods provide data relating to bulk mixing performance, as distinct from mixing at turbulent eddy scale. Recent comparisons of data obtained using inert or reactive tracers have shown that the results are, in many respects, very similar and, thus, both appear to measure the same characteristic, i.e., bulk mixing.

Recent experimental work in mixing of two liquids has also begun to explore the influence of differences in density or viscosity (Rielly and Pandit 1988). As the difference in density increases so the value of Nt_M increases markedly and can be correlated with a dimensionless "Richardson" number N_R, which gives a ratio of buoyancy and inertia forces:

$$N_R = \frac{\Delta \rho g H}{\rho_L N^2 D^2} \qquad \text{(Eq 2-37)}$$

For large viscosity ratios ($\mu_2 > \mu_1$) mixing time increases at constant impeller speed, but no quantitative relationship for the dependence is available.

Turbulent Solid–Liquid Mixing

The main consideration in most solid–liquid mixing applications is the determination of the minimum impeller speed required to ensure that particles do not settle on the bottom of the tank. The just-suspended condition provides adequate mixing for many applications, including mass transfer or chemical reaction, but in some cases—e.g., continuous-flow tanks—a more uniform dispersion of particles is required. There has also been considerable recent study of gas–liquid–solid systems, and procedures for determining the three-phase just-suspended condition have been developed.

Minimum mixing condition. A just-suspended criterion was proposed by Zweitering (1958) to define a minimum mixing condition:

$$N_{JS} = \frac{S \nu^{0.1} d_p^{0.2} \left[\frac{g \Delta \rho}{\rho}\right]^{0.45} X_p^{0.13}}{D^{0.85}} \qquad \text{(Eq 2-38)}$$

Where:

N_{JS} = impeller speed for a just-suspended condition
X_p = the weight fraction of solids in the mixture
d_p = the particle diameter
S = a mixer geometry parameter. This parameter is dependent on impeller type, clearance between tank bottom and impeller clearance, and the ratio of tank diameter to impeller diameter.
ν = kinematic viscosity (m^2/s)

Values of S were determined for a range of impellers and geometry variations. A summary of this data and additional values are given by Nienow (1985a). Different geometric combinations lead to quite different power requirements for the just-suspended condition.

Generally, axial-flow impellers have lower energy requirements than radial-flow impellers. However, the shape of the bottom of the vessel can have a significant influence and the magnitude of this effect can be very different for the two impeller types. In a recent analysis of available data, Gray and Oldshue (1986) concluded that the combination of axial-flow impeller and dished tank bottom was preferred for low power requirements, with fluidfoil axial-flow impellers able to provide even lower-energy mixing. The use of axial-flow impellers with draught tubes has also been discussed for low energy requirements (Nienow 1985a). Nienow concludes that the Zweitering correlation remains the most useful for practical purposes.

In terms of equipment selection, it appears that many impeller–tank combinations provide good performance as judged by the just-suspended criterion. Evidence discussed in recent reviews (Nienow 1985a; Gray and Oldshue 1986) suggests that some combinations are best avoided, e.g., radial-flow impellers and flat-bottomed vessels. Also, as far as can be estimated from the available data, it is likely that radial-flow impellers require significantly more power than axial-flow impellers. As the power requirements for simple solid–liquid mixing processes are not high, approximately 0.5 W/kg (mixer contents), many applications will be satisfactorily served by a downward-pumping axial-flow impeller in the usual diameter range $1/3 < D/D_v < 1/2$. As there is some lack of agreement about the exact influence of the vessel bottom, a dished bottom may be an acceptable compromise. The conventional arrangement using form vertical baffles with width $W = 0.1\ D_v$ should be adequate. The impeller should be positioned closer to the tank than for most conventional mixing processes, i.e., $1/6 < C/T < 1/4$.

There is a useful discussion on shape and the possibility of much reduced energy requirements by the use of a profiled bottom (Aesbach and Bourne 1972). There is also the question of possible advantageous effects of pipework or steady bearings located in the center of the tank bottom, by way of eliminating dead zones in that region. However, the reviews do show that available data do not provide a clear picture of the influence of variables on solid–liquid mixing. This can be examined from the point of view of a relationship for N_{JS}, e.g., Zweitering (1958):

$$N_{JS} = \frac{S_v^\alpha d_p^\beta \left[\frac{g\Delta\rho}{\rho}\right]^\gamma X_p^\delta}{D^\varepsilon} \qquad \text{(Eq 2-39)}$$

The mixer-geometry term is dependent on impeller type, vessel type, baffling, impeller diameter, and impeller clearance. Different values for the same geometry have been reported. Similarly the values of the exponents α, β, γ .... have been given different values by different researchers. This variability reflects the simple form of the correlation and the simple definition of the just-suspended condition. The consequences of this lack of precision influence the prediction of N_{JS}. Where the precision of the calculation is important, more than one correlation should be examined. Following the discussions of Nienow (1985a), it is sensible to make a first prediction on the basis of the Zweitering equation (Eq 2-38). A second estimate would be best made using a correlation determined under conditions that are relevant to the project in question. In analyzing the results of the calculation, there are two points worthy of consideration:

1. For many processes the energy requirements for the mixer are likely to be small (< 0.5 W/kg); therefore, the costs of overdesign may not be significant compared to the possible failure of the process. Further, excess power is unlikely to have any deleterious effects and may have some small advantages.

2. If N_{JS} is too low and particles accumulate at the bottom of the tank, the savings in power could be significant (Bourne and Sharma 1974). If a small accumulation of particles is acceptable, this option can be considered.

The nature of the process dictates which of the two options is more suitable. Consideration of practical applications and design leads immediately to the question of scale-up. The two options outlined above are also relevant to the choice of the scale-up procedure.

It seems certain that the conventional scale-up parameter of constant power per unit volume is conservative for the just-suspended particle mixing condition. On the other hand, direct use of the Zweitering correlation, leading to P/V proportional to $D_v^{-0.55}$, seems unduly optimistic. As there is reasonable agreement between the criteria of the manufacturer, Gates, and a research study (Chapman, Nienow, and Middleton 1981), the proposed condition—i.e., P/V proportional to $D_v^{-0.28}$—is worth consideration. Thus, a design exercise should be based on the estimation of a value of N_{JS} for the size of mixer for which the correlation was established, followed by scale-up on the basis of P/V proportional to $D_v^{-0.28}$. Alternatively, a value of N_{JS} may be established by pilot studies and then the scale-up rule applied. As with all scale-up studies, the bigger the pilot equipment, the better. The question of scale-up methods is not closed. Voit and Mersmann (1988) discriminate between scale-up to avoid settling and scale-up for off-bottom suspension. In this example the Froude number is used to define the scale-up criterion; earlier work by Kneule and Weinspach (1967) has used a two-regime analysis and the Froude number scale-up criterion.

Mass transfer. It is appropriate to consider mass transfer and chemical reaction at this point, since most studies of the subject assume that the just-suspended condition is adequate for good mass transfer. High levels of agitation are considered to provide relatively small increases in mass-transfer rate for large increases in power consumption. However, if good mass-transfer rates are essential, the power levels are unlikely to be a deterrent in terms of total plant energy requirement.

For mass-transfer calculations based on the just-suspended condition, Nienow (1985a) proposes a simplified calculation procedure. Mass transfer is described by the Froessling equation:

$$\text{Sh} = 2 + 0.72\,\text{Re}_p^{1/2}\,\text{Sc}^{1/3} \qquad \text{(Eq 2-40)}$$

requiring a particle velocity that is approximated to the terminal velocity. For $d_p < 500\ \mu\text{m}$

$$u_T = \frac{0.152 g^{0.71} d_p^{1.14} \Delta\rho^{0.71}}{\rho^{0.29} \mu^{0.43}} \qquad \text{(Eq 2-41)}$$

while for $d_p > 1500\ \mu\text{m}$

$$u_T = \left[\frac{4 g d_p \Delta\rho}{3\rho}\right]^{1/2} \qquad \text{(Eq 2-42)}$$

where a value of $C_D = 1.0$ has been used in the Newton's Law region because of the higher turbulent drag coefficients characteristic of agitated tanks. Using an appropriate value of velocity, a mass-transfer coefficient can be calculated from Eq 2-40. For particles larger than 40 μm some mass-transfer enhancement can be expected and this is described by

$$\frac{k_{js}}{K_T} = \left[\frac{d_p^1}{40}\right]^{-0.08} \qquad \text{(Eq 2-43)}$$

where d_p^1 is expressed in micrometres. Further enhancement of the mass-transfer coefficient by using impeller speeds higher than N_{JS} is based on the relationship

$$K_s \propto N_s^{0.4} \qquad \text{(Eq 2-44)}$$

where the exponent of 0.4 is thought to be conservative. The enhancement of the mass-transfer coefficient by raising the impeller speed is expensive in terms of the additional power required—i.e., $P \propto N^3$—and will only be justified in special cases such as reaction rates that are mass-transfer-rate controlled for selectivity.

Homogeneous suspension. For many solid–liquid mixing applications the just-suspended condition provides an adequate level of agitation. However, some applications (Ford, Etchells, and Short 1985) require a near homogeneous distribution of solid particles. There is interest in processes involving high-value products and in both batch and continuous processing. Concern centers on both the uniformity of the suspension within the tank and the ability to withdraw material from the tank that is representative of the tank contents as a whole. Various practical procedures have been explored in the search for better uniformity, e.g., multiple impellers, partial baffles not reaching the vessel bottom, and, for more effective product withdrawal such as isokinetic sampling, choice of sampling position.

In many cases of product withdrawal the material removed must be representative of the mixer contents. The effectiveness of the withdrawal procedure is described by a separation coefficient K:

$$K = \frac{C_w}{C_m} \qquad \text{(Eq 2-45)}$$

Where:

C_w = the concentration of solids withdrawn
C_m = the mean concentration in the mixer

Some experimental studies have found that K is substantially independent of time. For the continuous withdrawal of suspension from a batch agitated tank (washout) the relationship between C_w and time can be described by

$$\frac{C_{wt}}{C_{mo}} = K e^{-t/t_L K} \qquad \text{(Eq 2-46)}$$

and

$$\frac{C_{wt}}{C_{wo}} = e^{-t/t_s} \qquad \text{(Eq 2-47)}$$

Where:

C_{wt} = the withdrawal concentration at time t
C_{mo} = the mean mixer concentration at the commencement of the experiment
C_{wo} = the concentration in the withdrawal stream at the commencement of the experiment

Using these relations, the mean residence time of the liquid, t_L, the mean residence time of the solid, t_s, and K can be determined by a simple graphical procedure (Nienow 1985b). The value of K reflects the vessel geometry, the withdrawal procedure, and the system physical properties and can be used in mixer design (Bourne and Zabelka 1980). A number of procedures have been used to obtain good withdrawal performance such as isokinetic sampling, modified mixer geometry, and withdrawal pipe location (Nienow 1985b). It may be possible to obtain good withdrawal characteristics in the absence of a homogeneous suspension if a suitable withdrawal point, at the mean mixer concentration, can be located.

As impeller speed is increased above the just-suspended value, so the distribution of solids in the mixer becomes more uniform. However, the distribution is complex for settling solids, because there are generally higher concentrations at the bottom of the tank than at the top. Attempts have been made to model the concentration profile in terms of particle settling velocity and mixer dispersion coefficient, but the complexity of the profile and the highly nonuniform turbulence conditions in the mixer create difficulties.

Some recent studies have described the condition of the suspension in terms of the dependence of the coefficient of variation on the impeller speed. Barresi and Baldi (1986) have examined the relationship between standard deviation σ and a mixer Peclet number Pe,

$$\frac{1}{\text{Pe}} = \frac{\text{Po}^{1/3} N D}{\mu_t} \quad \text{(Eq 2-48)}$$

for several impellers and a range of particle concentrations and present their data as plots of σ/c and $1/(c^{0.13} \text{Pe})$. The impellers used, both axial- and radial-flow types, show similar characteristics but differ in detail. Rieger, Ditl, and Havelkova (1988) also used the variation coefficient to examine the performance of pitched-blade turbine impellers of different diameters, with a range of vessel sizes. Performance was assessed in terms of power input and impeller tip speed, but no unambiguous scale-up criterion emerged. For scale-up Voit and Mersmann (1986) recommend that the two requirements of the "avoidance of settling" and "off-bottom suspension" be considered separately with respect to the ratio of particle size to tank size. Scale-up recommendations are based on Froude number but are dependent on particle and tank size.

Floating solids. The mixing of solid particles with a more dense liquid can be considered in terms similar to those used for settling solids. Briefly, a certain minimum impeller speed is necessary to ensure that most particles are entrained in the recirculating liquid flow, but considerably higher impeller speeds are necessary if the dispersion of particles in the mixer is required to approach homogeneity. Joosten, Schilder, and Broere (1977) made a study of energy requirements and found that the combination of a single, short baffle with a four-blade pitched turbine positioned near the bottom of a round-bottom tank gave the lowest power requirements for mixing of the systems tested. The mixer produced an off-center vortex, which seemed advantageous for mixing as judged visually from the disappearance of stagnant zones of particles on the liquid surface. For the optimum geometry the disappearance of stagnant zones could be correlated by a minimum Froude number,

$$\text{Fr}_{\min} = 0.036 \left[\frac{D}{T}\right]^{-3.65} \left[\frac{\Delta\rho}{P}\right]^{-0.42} \quad \text{(Eq 2-49)}$$

A recent study (Hemrajani et al. 1988) appears to confirm the importance of the vortex in the mixing of floating solids. A variety of mixer geometries were tested for both surface agitation and for particle concentration in a withdrawal stream. Location of the pitched-blade turbine impeller off center and use of narrow baffles with $w = T/50$ were advantageous. For floating solids, another possibility is to monitor the quantity of particles on the surface at any time (Ellis, Godfrey, and

Majidian 1987). Such monitoring reveals a lower limit below which the quantity of particles on the surface cannot be reduced, corresponding to a percentage of the total particles present and independent of total quantity.

Turbulent Multiphase Mixing

In addition to solid–solid mixing, two other areas of multiphase mixing technology may be of relevance to the water industry: gas–liquid and liquid–liquid mixing. Both processes are relatively expensive in water industry terms but may have application in more specialized processes. For both gas–liquid and liquid–liquid mixing, the emphasis in this section will be on agitated contacting equipment. Some of the applications discussed may be appropriate for the static mixer, as it is an effective device for both systems and possibly very attractive if contact times are short.

Liquid–liquid mixing. By comparison with solid–liquid and gas–liquid mixing, liquid–liquid technology attracts little interest in the literature of mixing technology and research. However, a moderate amount of information can be found in the literature of solvent extraction. Liquid–liquid mixing is a complex process and is most easily understood, like gas–liquid mixing, as a collection of phenomena and a range of flow regimes. The phenomena of concern are power consumption, drop size, holdup, preferred dispersed phase, entrainment, coalescence, coalescence-redispersion, mass transfer, and chemical reaction. The performance of the two main types of equipment, mixer settler and column contactor, is quite different. For the most part the two types of equipment also serve quite different purposes. To attempt to generalize, mixer settlers are used for large flows or slow extraction rates; and for columns, vice versa.

The mixer settler is a two-function device, with the mixer dispersing one liquid in another and providing some measure of circulation and blending of the resultant dispersion. The dispersion then overflows into the settler, where the drops are allowed to coalesce and the two separate phases reform. The drops formed in the mixer of a mixer settler are small ($d_{vs} < 1$ mm) and this dictates the need for a settler. Because such small drops are usually slow to coalesce, the settler should really be thought of as a "coalescer."

Column contactors do not usually require a settler because the drops produced are larger ($d_{vs} > 1$ mm). Columns are gravity-flow devices, with the heavier liquid introduced at the top and the lighter at the bottom. As with mixer settlers either phase may be chosen as the dispersed phase, but a suitable flow ratio and start-up procedure are necessary to ensure that the required dispersion is produced. Of potential interest to the water industry is the development of column contactors for biochemical processes and perhaps the pump-mix design of mixers where the impeller is used for both pumping and mixing duties, offering considerable capital savings in a large plant.

Gas–liquid mixing. Gas–liquid contacting is effected in many ways, but the agitated tank is probably the most widely used. In a typical agitated-tank geometry, the gas stream is introduced at the bottom of the tank and dispersed, and recirculated to some degree, by the impeller. Impeller design and power input influence the degree of dispersion, which is reflected by the holdup of the gas phase in the column and by the bubble size. Because the density difference between the gas and liquid phases is significant, a large gas holdup can only be achieved by the input of considerable impeller power. The power characteristics of the system are complex, the presence of the gas reducing power input in comparison to single-phase operation at the same impeller speed. Thus, many aspects of gas–liquid mixing would benefit from quantitative analysis and, in addition to the topics above, there are the questions of mass transfer and chemical reaction characteristics, scale-up, and, increasingly, gas–liquid–solid systems.

However, although the subject is complex, there is much recent literature and research of considerable assistance in evaluating gas–liquid processes. Mann (1983) gives an extensive review of the subject up to that time, and in later work Mann (1988) presents modeling procedures to give a detailed description of gas hydrodynamics and reactions in agitated vessels. Oldshue (1985) has reviewed the characteristics of a wide range of gas–liquid contacting devices and gives guidance regarding equipment selection for particular contacting requirements. In recent work, Weetman and Oldshue (1988) reassess the use of the axial-flow impeller, showing modified designs that are capable of good performance characteristics and are perhaps particularly suitable for high-uniformity applications as in some biochemical processes. Impeller characteristics, including multiple impeller systems, continue to attract a great deal of study in recent publications and conferences on mixing. The influence of impeller type on the characteristics of a closed-loop, open-channel, oxidation ditch is perhaps of some interest to the wastewater industry (Roustan, Heziane, and Faup 1988). In another review, Middleton (1985) has outlined the characteristics of a variety of gas–liquid contactors. The main emphasis is on the performance of agitated tanks and the selection procedures for design and scale-up, including consideration of mass transfer, chemical reaction, and their interaction. The complexity of gas–liquid contacting can be greatly clarified by considering the various operating regimes of gas–liquid mixing (Smith and Warmoeskerken 1985). This review describes the very extensive studies on gas-bubble behavior. It also includes the choice of correlations for predicting the hydrodynamic and mass-transfer performance of industrial equipment, many of which are also discussed by Middleton (1985).

Further clarification of gas–liquid behavior on the basis of interpretation of low regimes appears in more recent work (Warmoeskerken and Smith 1986). Five regimes are discussed, together with quantitative relationships, to define the onset of each regime. The regimes are defined mainly with respect to the behavior of the gas in the vicinity of the impeller blades and the character of the gas "cavities" that exist behind the blades.

Laminar Mixing

The main reason for conducting a mixing process in the laminar flow regime, when the Reynolds number is 10 or less, is to avoid excessive power consumption in blending of high-viscosity fluids (Godfrey 1985). For most mixing equipment operating in the laminar region, power consumption can be described in the form

$$\text{Po} = K_p/\text{Re} \qquad \text{(Eq 2-50)}$$

where K_p is a constant of proportionality, which is frequently independent of mixer scale but very dependent on mixer geometry. When non-Newtonian fluids are involved the definition of Re needs to be examined with care. As with turbulent mixing processes, mixing-rate data for many laminar processes can be summarized by a relationship similar to Eq 2-2:

$$N\theta = K \qquad \text{(Eq 2-51)}$$

where the constant K is a function of mixer type but not viscosity; for non-Newtonian fluids, mixing-rate characteristics are more complex and there is some difference of opinion regarding the influence of non-Newtonian properties.

The influence of non-Newtonian properties on power consumption has been reasonably well accommodated for shear-thinning fluids by using an average mixer shear rate ($\dot{\gamma}_{av}$) to define the appropriate apparent viscosity for the Reynolds number. In the simplest analysis the mixer shear rate is assumed proportional to impeller rotational speed:

$$\dot{\gamma}_{av} = K_s N \quad \text{(Eq 2-52)}$$

where K_s is the mixer shear rate proportionality constant and is a function of impeller type, depending, to a lesser degree, on details of impeller design (Edwards and Ayazi-Shamlou 1983).

The influence of shear-thinning properties on mixing rate is much more conjectural. Some data show mixing rate for the helical impeller decreasing as non-Newtonian properties become more marked (Hall and Godfrey 1971). However, other data for the helical ribbon show no influence of non-Newtonian properties (Nagata 1975). There are many potential complications for mixing of shear-thinning fluids, because they are often elastic as well. It seems likely that fluids that have little or no elastic character but are shear-thinning show mixing-rate characteristics similar to those for Newtonian fluids. This assumption is supported by the circulation time measurements for the helical ribbon, which show only a small influence of shear-thinning properties for fluids believed to have little elastic character (Takahashi et al. 1988).

IN-LINE MIXERS

Jet Mixing

When a high-velocity stream of liquid is introduced into a large volume of substantially stagnant liquid the characteristics of the jet used in mixing can be monitored. The jet Reynolds number is

$$\text{Re}_J = \frac{\sigma_J V_J D_J}{\mu_J} \quad \text{(Eq 2-53)}$$

Where:

σ_J = density of jet liquid
μ_J = viscosity of jet liquid
V_J = velocity of jet liquid
D_J = diameter of jet

In most applications the jet liquid is the same as the bulk liquid. For jet Reynolds numbers above 2000 the jet is turbulent; this is the flow condition of interest for mixing applications. As the jet enters the stagnant liquid it entrains liquid from its surroundings and expands in size (Figure 2-13). The jet consists of two regions: a flow development, or core, region and a fully developed region, beginning at approximately $10\,D_J$ from the jet inlet. In the fully developed region the centerline velocity may be approximated as

$$V_c \sim V_J \frac{6 D_J}{Z} \quad \text{(Eq 2-54)}$$

where Z = distance from jet inlet

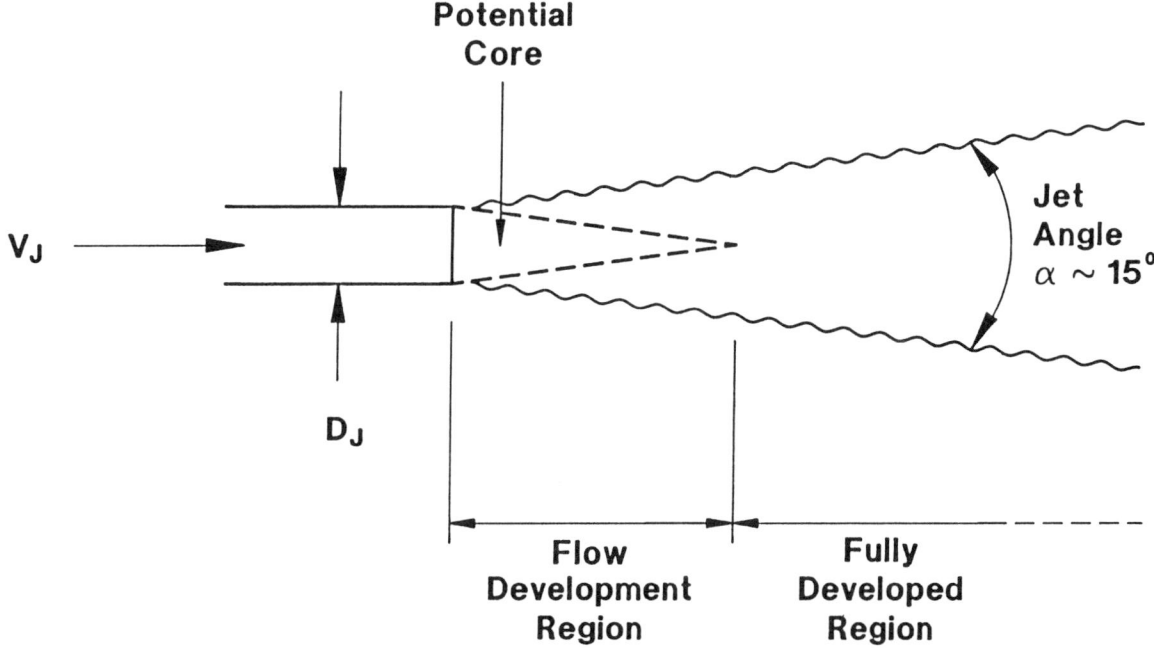

Figure 2-13 Characteristics of a turbulent jet.

The centerline concentration of material in a jet may be approximated as

$$C_c \sim C_J \frac{4.5 D_J}{Z} \qquad \text{(Eq 2-55)}$$

The other characteristics of the jet are more difficult to estimate. The jet expands by entrainment of liquid from the immediate surroundings. For a liquid jet, the jet angle α is approximately $15°$. The volumetric entrainment rate is related to the jet flow:

$$Q_e \sim Q_J \frac{0.3 Z}{D_J} \qquad \text{(Eq 2-56)}$$

The exact forms of the relationships for velocity, concentration, and entrainment are likely to show some small influence of Reynolds number.

These relationships can be used to estimate jet characteristics up to about 500 jet diameters from the inlet. Farther from the inlet the mixing action of the jet is minimal.

Mixing in tanks. A typical, simple mixed tank geometry with a jet is shown in Figure 2-14, but a number of variations on jet arrangements are possible. The jet shown is directed diagonally from bottom to top and is typical in that the jet is directed along one of the longer axes of the tank. It would be possible to introduce the jet at the top of the tank directed diagonally downward, or to use centrally located jets pointing either up or down. Multiple jets might also be used in large vessels or when there are geometrical problems, e.g., shallow tanks. In most

70 MIXING

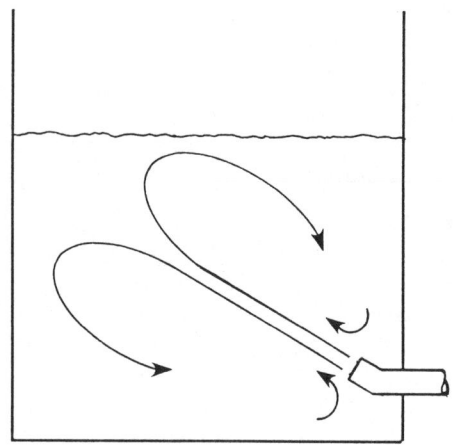

Side-Entry Jet

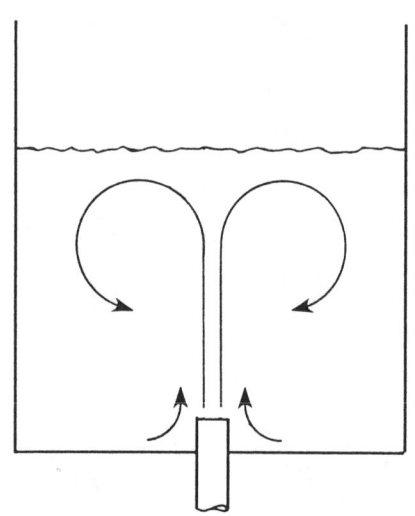
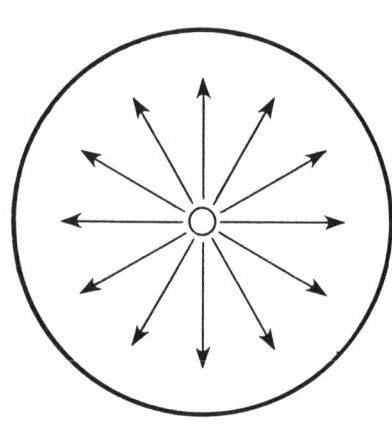

Axial Jet

Figure 2-14 Jet mixing in tanks.

applications it is usual to direct the jet to produce the greatest uninterrupted flow path, which minimizes energy losses by avoiding wall effects.

The choice of jet/tank geometry is commonly between axial and diagonal jets, axial jets usually being chosen for

$$0.75 < \frac{H}{T} < 3.0 \qquad \text{(Eq 2-57)}$$

where H = liquid height in the tank of diameter T

Diagonal (side entry) jets are chosen for

$$0.25 < \frac{H}{T} < 1.25 \qquad (Eq\ 2\text{-}58)$$

Multiple diagonal jets are used when $H/T > 3.00$; multiple jets of either type are used when $H/T < 0.25$. Obviously, the exact values of the H/T parameters must be regarded as somewhat arbitrary. As the jet will be oriented along the longest tank dimension L, this leads to an estimate of suitable jet diameter D_J, i.e., $D_J > L/500$. (If the jet discharge is oriented along the diagonal of a tank, then L will be the diagonal length.) The upper limit of jet diameter is approximated by $D_J < L/50$ on the basis of minimizing energy wastage. When there is a significant density difference between the liquids being mixed, i.e. $\Delta\rho/\rho > 0.05$, where ρ is the heavier liquid the jet should point upward for $V_c/V > 0.5$ or downward for $V_c/V < 0.5$ where V_c and V are the light liquid and tank volumes, respectively.

The jet velocity and therefore the flow rate and pumping requirements can be determined on the basis of the required blending time:

$$\frac{\theta V_J}{D_J} = A \left[\frac{T}{D_J}\right]^{1.5} \left[\frac{H}{D_J}\right]^{-0.5} \qquad (Eq\ 2\text{-}59)$$

Where

$$A = \frac{30{,}000}{Re_J}, \quad Re_J \leq 5000$$

$$A = 6, \quad Re_J \geq 5000$$

From the velocity obtained for a specific mixing time, the pressure losses in the jet and associated pipework can be calculated and hence the energy requirements. It should be remembered that the energy requirement could be a large value and unacceptable if the combination of mixing time and jet size is unfavorable. A range of values for the jet diameter ($L/500 < D_J < L/50$) will indicate the range for the power requirement. If the values of power are too large, a longer mixer time may be necessary.

In general, the energy requirements are higher for jet mixing than the corresponding agitated tank, but the jet mixer has the advantages of lower capital and maintenance costs. Some compromise between jet mixing and the agitated tank may be reached by using impellers to create jet-like flows while avoiding recirculating pressure losses in ancillary pipework (Merzouk, Ramcl, and Bertrand 1988; Roustan and Heziane 1986).

Jet mixers in pipes. For many applications jet mixing in pipes offers mixing conditions somewhere between an empty pipe and a static mixer in terms of mixing time, energy, and cost. The equipment is suitable for low-viscosity fluids and like all pipeline mixers has the usual advantages and disadvantages of a near plug-flow residence time distribution. The conditions it offers can be particularly favorable for some chemical reactions, e.g., fast consecutive reactions (Tebel and May 1988). A recent review describes the comparative performance of pipeline, jet, and static mixers for turbulent-flow applications (Etchells and Short 1988). There are two principal geometries for jet mixing in pipes: coaxial and side entry (or tee). Of the two, it seems that the side-entry geometry is preferred in many cases because of its simplicity and superior mixing ability for mixing unequal flow rates.

Revill (1982) recommends a design procedure based on mixing time and jet velocity for the side-entry jet mixer. Depending on the relationship between jet velocity, V_J, and the second fluid velocity, V_s, two alternatives are possible. First, if

$$V_J < 1.2\ V_s \qquad \text{(Eq 2-60)}$$

and the jet does not penetrate to the pipe wall opposite the inlet, then a mixing condition is produced that is equivalent to simple pipeline mixing:

$$\theta \sim 150\frac{D_m}{V_m} \qquad \text{(Eq 2-61)}$$

Where:

D_m = mix pipe diameter
V_m = mix velocity

Under the second condition,

$$V_J > 1.2\ V_s \qquad \text{(Eq 2-62)}$$

the jet expands to meet the opposite pipe wall at some point before it loses its identity. This condition is more relevant to rapid mixing, but mixing time is more difficult to estimate.

For a side-entry jet mixer the penetration distance χ of the jet into the secondary flow is related to the mixing distance L by

$$\frac{\chi}{D_J} = \left[\frac{\rho_J V_J^2}{\rho_s V_s^2}\right]^{0.4}\left[\frac{L}{D_J}\right]^{0.3} \qquad \text{(Eq 2-63)}$$

As with all mixing processes the result is dependent on operating conditions, and a range of calculations should be made to explore the potential of the technique, including the associated energy requirements. There appears to be considerable potential for the side-entry jet mixer (Revill 1982; Etchells and Short 1988).

Design estimates for the coaxial jet mixer proceed along similar lines. When the jet velocity is similar to the secondary-flow velocity, the behavior of the mixer is similar to that of a simple pipe mixer and, as before, $\theta \sim 100\ D_m/V_m$. For jet velocities at least 50 percent higher than the secondary velocity, true jet mixing can be achieved and mixing time is a function of the distance between the inlet and the point at which the jet contacts the wall. For a coaxial jet mixer a relationship between mixing time and the point at which the jet reaches the pipe wall is as follows:

$$\theta = \left(6.6 + 1.7\log\frac{Q_s}{Q_j}\right)\frac{L}{V_m} \qquad \text{(Eq 2-64)}$$

where L = distance from jet outlet to point at which the expanding jet meets the opposite wall. In this case the length parameter is given by

$$L = \frac{(D_m - D_J)}{2\tan\alpha/2} \qquad \text{(Eq 2-65)}$$

where α is the jet angle. Equations 2-64 and 2-65 can be used to estimate the mixing time.

Jet-mixing technology appears to have much to offer the water industry, particularly if installed pumps can be used alternately for both pumping and mixing duties. Energy costs range from high for tank mixing to competitive for pipeline mixing, but capital and maintenance costs are low. Jet mixing with application for particle destabilization and flocculation in water treatment is considered in more detail in chap. 6.

Static Mixers

Static mixers offer performance characteristics similar to those of the in-line jet mixers, with regard to both mixing rate and energy requirements. In general, the balance will be toward faster mixing rates, higher energy consumption, and higher capital cost for the static mixer.

Mixer types. A large number of static mixers are available commercially. It is important to note, as in the Koch/Sulzer range, that some mixers are designed for a specific flow regime, laminar or turbulent, or with specific applications in mind, e.g., heat transfer. Other designs—e.g., Kenics/Chemineer—are multipurpose designs. For most applications, laminar or turbulent, a large free area is necessary if pressure losses, and therefore energy requirements, are not to be prohibitive. A large number of static mixers and aspects of their performance have been described by Pahl and Muschelknautz (1979). In addition to the well-known Sulzer/Koch and Kenics/Chemineer designs, the Ross and Lightning geometries are well known in North America.

Laminar mixing. Mixing rate is often assessed in terms of the number of pipe diameters necessary to achieve a specified mixture quality, all of the specified length being fitted with static-mixer elements. The number of pipe diameters, i.e., L_m/D_m, is regarded as independent of flow rate and Reynolds number in the laminar regime, also as independent of scale as a basis for design. These assumptions probably go too far but are the basis of many current design and research procedures. The mixture-quality parameter most frequently used is the variation coefficient (σ/C), which is thought to reflect the influence of flow ratio on the difficulty of the mixing process. A variation coefficient of $\sigma/C = 0.05$ is considered an indication of a generally well-mixed condition. Thus, a preliminary comparison of static-mixer types can be based on the value of L_m/D_m required to produce a variation coefficient of $\sigma/C = 0.05$.

However, both mixing rate and pressure drop are important considerations. Pressure-drop characteristics are often described in terms of $\Delta P_m/\Delta P_o$, the ratio of pressure loss in a static mixer to that in an empty tube of the same length and diameter. The performance of different mixer types can be compared by examining the product of these two terms:

$$\left[\frac{L_m \Delta P_m}{D_m \Delta P_o}\right] \quad \text{(Eq 2-66)}$$

the lower values corresponding to greater efficiency (Godfrey 1985).

Turbulent mixing. The application of static mixers to turbulent mixing is likely to be more important to the water treatment industry. At an introductory level the main design concepts are similar: the number of pipe diameters required to give adequate mixing, mixture quality expressed in terms of a variation coefficient of 0.05 as a general-purpose mixing criterion, and the pressure drop ratio ($\Delta P_m/\Delta P_o$) as the basis of energy characteristics.

There is relatively little information available on the relationship between mixture quality and mixer length, the basic design question. Some data are available in commercial literature (Sulzer, e 23.27.06.04 – V.87 – 100) in the form of log variation coefficient against number of elements for various flow ratios [$x = Q_1/(Q_1 + Q_2)$]. It may be important to note that Sulzer recommends a mixer-element length equal to pipe diameter for diameters less than 100 mm, and length one-half the diameter for larger diameters. This guideline may relate to changes in packing geometry for different diameters.

An alternative relationship for the Sulzer SMV (Etchells and Short 1988) is

$$\frac{\sigma}{\sigma_o} = 2 \exp(-1.5\, L/D) \tag{Eq 2-67}$$

where σ/σ_o is the relative standard deviation. For this estimation the relationship between variation coefficient and (σ/σ_o) is given by

$$\frac{\sigma}{C} = \frac{\sigma}{\sigma_o} \sqrt{\left[\frac{1}{C} - 1\right]} \tag{Eq 2-68}$$

With the general lack of data it is interesting to compare the Sulzer plot data with Eq 2-67 for typical operating conditions. For example, with flow ratios of $x = 0.1$ and 0.01 and a quality specification of $\sigma/C = 0.05$, the Sulzer plots give L/D values of 3.2 and 4.6, and Eq 2-67 gives 3.2 and 4.0. This seems fairly good agreement for mixing data, but the agreement deteriorates for conditions away from this "typical" range of operation. However, these approximate values are in the range generally quoted when estimates are made giving $L/D \sim 4$ for the Sulzer SMV and $L/D \sim 8$ for the Kenics mixer (Revill 1985). A similar value for the Kenics mixer is quoted (Novak and Rieger 1985) at $L/D \sim 9$. In the absence of more extensive data it is significant that the correlation given by Eq 2-67 has been transformed by the authors to include a friction factor term:

$$\frac{\sigma}{\sigma_o} = 2 \exp(-1.54\, f^{0.5}\, L/D) \tag{Eq 2-69}$$

and is in a very similar form to their derivation for pipeline mixing:

$$\frac{\sigma}{\sigma_o} = 2 \exp(-1.5\, f^{0.5}\, L/D) \tag{Eq 2-70}$$

Obviously, this suggests that mixing times can be estimated from friction factors for different mixers and, using the values for σ/C of 0.1 and 0.01 as before, values of L/D of 5.5 and 6.9 are obtained. Clearly, these are of the correct order of magnitude and somewhat smaller than expected, as was the case with similar calculations for the Sulzer SMV. Some caution is necessary in the choice of friction factors since there are several definitions in use, as discussed below.

The energy characteristics of static mixers in turbulent flow can be reasonably well-described by the friction factor concept used for pipe flow. The friction factor used in Eq 2-69 is defined as

$$f = \frac{D\Delta P}{2\rho v^2 L} \qquad \text{(Eq 2-71)}$$

However, some available data (Godfrey 1985; Novak and Rieger 1985) use the definition

$$f' = \frac{2D\Delta P}{\rho v^2 L} \qquad \text{(Eq 2-72)}$$

so care is required when using (mostly European) literature data. A further complication arises in the European use of the Newton number

$$\text{Ne} = \frac{D\Delta P}{\rho v^2 L} \qquad \text{(Eq 2-73)}$$

as used by Sulzer to describe pressure drop data. For the Sulzer mixer SMV the Newton number, Ne, is approximately 2; hence f is approximately 1. This value is quoted for $\text{Re} > 6 \times 10^3$, above which Ne is constant for the SMV mixer. However, below this level Ne and, therefore, f are a function of Reynolds number. Approximate values for friction factor (Pahl and Muschelknautz 1982) are useful: Kenics, $f \sim 0.75$; Sulzer SMV, $f \sim 1.0$; Sulzer SMX, $f \sim 3$.

For precise calculations the influence of Reynolds number should be considered. This can be accomplished by using available friction factor Reynolds number plots or correlations for pressure drop ratio ($\Delta P_m/\Delta P_o$), which are available for some mixers. Equipment suppliers should be able to provide these data.

Under turbulent flow conditions the static mixer can produce an increase in heat-transfer coefficient of approximately a factor of three, but with a possible pressure-loss increase of a hundredfold or more. Thus, the energy costs of potential applications need to be carefully examined.

The basis for the static-mixer design is reasonably well-established. Estimates of mixer length and pressure drop can be made from available correlations. For both laminar and turbulent flow, mixer length or number of elements required is independent of flow rate as far as can be judged from the correlations available. Corrections are sometimes required for the flow ratio when the criterion of a specified value of the variation coefficient is used. Corrections for the viscosity ratio are recommended by some equipment suppliers, e.g., Sulzer for laminar flow and the SMX mixer.

Dynamic Mixers

The dynamic mixer offers an alternative to the static mixer where higher energy or shear rate levels are required or where some degree of backmixing is an advantage. Characteristically, the dynamic mixer offers a continuous flow, high energy, small volume, and short residence time. The dynamic mixer may be able to replace agitated-tank mixers if mixing requirements are precisely known. Dynamic mixers have not attracted much research or review study. Oldshue (1985) has discussed some equipment types and design procedures.

Mixer types. Dynamic mixers are available for both high-viscosity and low-viscosity applications. Many are very similar to small agitated tanks and use impellers that are familiar from agitated-tank practice. Other specialist rotor-stator designs are used for multiphase, high-viscosity applications; ultrasonic mixers also fall into this class.

For the blending of high-viscosity fluids the helical-ribbon impeller has been used, while rotor-stator or ultrasonic devices are used when dispersion is required. Dispersions of the solids in liquids or stable emulsions of one liquid in another are produced in this way. Gas–liquid and liquid–liquid dispersions are also produced with turbine impellers when the product is of low viscosity. Turbine-agitated dynamic mixers may have two impellers on the same shaft, with the geometry of the mixing chamber arranged to approach the condition of two agitated tanks in series. A variety of different impellers are available for the agitated-tank designs of dynamic in-line mixers.

Mixing time. In all applications mixing must be accomplished in a short time. Typical residence times are a few seconds, and typical mixer volumes range from 5 to 100 L. As most applications are for relatively high flow rates, it is difficult to collect laboratory data. However, useful batch studies might be conducted by using mixers of the same scale intended for the final application. In a number of mixer processes the reduction of variance (or standard deviation) can be approximated to a first-order process in which case the batch data may be described by

$$\frac{\sigma^2}{\sigma_o^2} = 1 - e^{-k't} \qquad \text{(Eq 2-74)}$$

The rate constant, k', can then be used to estimate mixture quality for continuous flow:

$$\frac{\sigma^2}{\sigma_o^2} = 1 - \frac{k't}{1 + k't} \qquad \text{(Eq 2-75)}$$

Because dynamic mixers, which are impeller agitated, have some degree of backmixing they smooth time fluctuations to some degree. This smoothing function is less significant than that of agitated tanks but much more significant than can be achieved by a static mixer and may be an advantage for some in-line applications. Some indication of damping characteristics is given by Oldshue (1985).

Mixing power. Many dynamic-mixer applications have high power-per-unit-volume requirements: solid–liquid dispersions, liquid–liquid dispersions. Further, as residence times are short, power per unit volume needs to be high to impart the necessary mixing energy even for less-demanding processes. Energy inputs range from 20 to 100 kW/m^3, usually higher in smaller mixers. When it comes to estimating the power characteristics of particular impellers, little data are available. However, the basic power number–Reynolds number relationships will not be greatly changed from the batch conditions if the geometry is not too different, and estimates may be made on this basis.

References

AESBACH, S. & BOURNE, J.R. 1972. Attainment of Homogeneous Suspensions in a Continuous Stirred Tank. *Chem. Engrg. Jour.*, 4:3:234.

AMIRTHARAJAH, A. 1987. Rapid Mixing and the Coagulation Process. Proc. AWWA Seminar on Influence of Coagulation on the Selection, Operation, and Performance of Water Treatment Facilities. AWWA, Denver, Colo.

BARRESI, A. & BALDI, G. 1986. Solid Dispersion in an Agitated Vessel. Act. Colloque Agitation Mecanique (ENSIGC), Toulouse, France.

BOURNE, J.R. 1985. *Mixing in the Process Industries*, ed. by N. Harnby, M.F. Edwards, and A.W. Nienow. Butterworths, London.

BOURNE, J.R. & SHARMA, R.N. 1974. Suspension Characteristics of Solid Particles in Propeller-Agitated Tanks. Proc. 1st European Conf. on Mixing (B3). BHRA, Cambridge, England.

BOURNE, J.R. & ZABELKA, M. 1980. The Influence of Graduation Classification on Continuous Crystallization. *Chem. Engrg. Sci.*, 35:3:533.

BUJALSKI, W. ET AL. 1987. The Dependence on Scale of Power Numbers of Rushton Turbines. *Chem. Engrg. Sci.*, 42:4:317.

CARREAU, P.J.; PATTERSON, I.; & YAP, C.Y. 1976. Mixing of Viscoelastic Fluids with Helical Ribbon Agitators. *Can. Jour. Chem. Engrg.*, 54:2:135.

CHAPMAN, C.M.; NIENOW, A.W.; & MIDDLETON, J.C. 1981. Particle Suspension in a Gas Sparged Rushton-Turbine Agitated Vessel. *Trans. Inst. Chem. Engrs.*, 59:2:134.

EDWARDS, M.F. & AYAZI-SHAMLOU, P. 1983. *Low Reynolds Number Flow Heat Exchangers*. Hemisphere Publishing Corp., New York.

ELLIS, D.I.; GODFREY, J.C.; & MAJIDIAN, N. 1987. A Study of the Influence of Impeller Speed on the Mixing of Floating Solids in a Liquid. *Inst. Chem. Engrs. Symp. Ser.*, 108:181.

ETCHELLS, A.W. & SHORT, D.G.R. 1988. Pipeline Mixing—A User's View, I: Turbulent Blending. 6th European Conf. on Mixing. BHRA, Pavia, Italy.

FARADAY, A.G. 1987. A Computerized Flow-Follower Technique for Assessing Agitation. *Inst. Chem. Engrs. Symp. Ser.*, 108:63.

FORD, W.N.; ETCHELLS, A.W.; & SHORT, D.G.R. 1985. An Industrial Mixing Laboratory. Proc. 5th European Conf. on Mixing. BHRA, Wurzburg, Germany.

FORT, I. & MEDEK, J. 1988. Hydraulic and Energetic Efficiency of Impellers With Inclined Blades. Proc. 4th European Conf. on Mixing. BHRA, Pavia, Italy.

GODFREY, J.C. 1985. *Mixing in the Process Industries*, ed. by N. Harnby, M.F. Edwards, and A.W. Nienow. Butterworths, London.

GODFREY, J.C. & GRILC, V. 1988. The Agitation of Liquid-Liquid Dispersions in Square Cross-Section Tanks: Power Consumption Characteristics. 6th European Conf. on Mixing. BHRA, Pavia, Italy.

GRAY, J.B. & OLDSHUE, J.Y. 1986. *Mixing: Theory and Practice*, Vol. 3, ed. by V.W. Uhl & J.B. Gray. Academic Press, London.

GREAVES, M. & LOH, V.Y. 1969. Power Consumption Effect in Three Phase Mixing. *Inst. Chem. Engrs. Symp. Ser.*, 89:69.

HALL, K.R. & GODFREY, J.C. 1971. The Effect of Fluid Properties and Impeller Design on the Performance of Helical Ribbon Impellers. Proc. Chemical 1970, III. Butterworths, London.

HEMRAJANI, R.R. ET AL. 1988. Suspending Floating Solids in Stirred Tanks: Mixer Design, Scale-up and Optimization. Proc. 6th European Conf. on Mixing. BHRA, Pavia, Italy.

HIBY, J.W. 1985. Problems With Measuring Mixing Times. 5th European Conf. on Mixing. BHRA, Wurzburg, Germany.

HOLMES, D.B.; VONCKEN, R.M.; & DEKKER, J.A. 1964. Fluid Flow in Turbine-Stirred Baffle Tanks, I: Circulation Time. *Chem. Engrg. Sci.*, 19:3:201.

JOOSTEN, G.E.H.; SCHILDER, J.G.M.; & BROERE, A.M. 1977. The Suspension of Floating Solids in Stirred Vessels. *Trans. Inst. Chem. Engrs.*, 55:3:220.

KHANG, S.J. & LEVENSPIEL, O. 1976. The Mixing Rate Number for Agitator-Stirred Tanks. *Chem. Engrg.*, 83:21:141.

KNEULE, F. & WEINSPACH, P.M. 1967. Suspendieren von Feststoff Partikeln im Ruhreegefass. *Verfahrenstechnik*, 1:12:531.

KUBOI, R. & NIENOW, A.W. 1982. The Power Drawn by Dual Impeller Systems Under Gassed and Ungassed Conditions. Proc. 4th European Conf. on Mixing. BHRA, Noordwijkerhout, the Netherlands.

LOL, T.; BAIRD, M.H.I.; & HANSON, C. 1983 *Handbook of Solvent Extraction.* John Wiley and Sons, New York.

LOTT, J.B.; WARWICK, G.C.I.; & SCUFFNAM, J.B. 1972. Design of Large Scale Mixer Settlers. *Trans. Soc. Min. Engrg.* (AIME), 252:27.

MANN, R. 1983. Gas–Liquid Mixing. *Industrial Research Fellow Report.* Inst. of Chem. Engrs., Rugby, England.

———. 1988. Fundamentals of Gas–Liquid Mixing in a Stirred Vessel: An Analysis Using Networks of Backmixed Zones. Proc. 4th European Conf. on Mixing. BHRA, Pavia, Italy.

MATILLA, T.K. 1974. The Kemira Mixer-Settler Extractor. Proc. Intl. Solvent Extraction Conf. (Society of Chemical Industry, Lyon, France), 1:169.

MERZOUK, K.; RAMEL, C.; & BERTRAND, J. 1988. A First Study of the Submersible Agitators Flygt. 6th European Conf. on Mixing. BHRA, Pavia, Italy.

MIDDLETON, J.C. 1979. Measurements of Circulation Within Large Mixing Vessels. Proc. 3rd European Conf. on Mixing. BHRA, Cranfield, U.K.

———. 1985. *Mixing in the Process Industries,* ed. by N. Harnby, M.F. Edwards, and A.W. Nienow. Butterworths, London.

NAGATA, S. 1975. *Mixing: Principles and Applications.* Halsted Press, Tokyo.

NIENOW, A.W. 1985a. *Mixing in the Process Industries,* ed. by N. Harnby, M.F. Edwards, and A.W. Nienow. Butterworths, London.

———. 1985b. *Mixing of Liquids by Mechanical Agitation,* ed. by J.J. Ulbrecht and G.K. Patterson. Gordon and Breach, New York.

NOVAK, V.; DITL, P.; & RIEGER, F. 1982. Mixing in Unbaffled Vessels: The Influence of an Eccentric Impeller Position on Power Consumption and Surface Aeration. 4th European Conf. on Mixing. BHRA, Noordwijkerhout, the Netherlands.

NOVAK, V. & RIEGER, F. 1985. Homogenization Efficiency of Motionless Mixers. Proc. 5th European Conf. on Mixing. BHRA, Wurzburg, Germany.

OLDSHUE, J.Y. 1985. Scale-up of Unique Industrial Fluid Mixing Process. Proc. 5th European Conf. on Mixing. BHRA, Wurzburg, Germany.

PAHL, M.H. & MUSCHELKNAUTZ, E. 1979. Einsatz und Auslegung Statischer Mischer. *Chem. Ing. Tech.,* 51:5:347.

———. 1982. Static Mixers and Their Applications. *Intl. Chem. Engrg.,* 22:2:197.

PORTNER, R.; LANGER, G.; & WERNER, U. 1988. Homogenization of Non-Newtonian Fluids in Stirred Tanks. Proc. 6th European Conf. on Mixing. BHRA, Pavia, Italy.

PROCHAZKA, J. & LANDAU, J. 1961. Studies on Mixing, XII: Homogenization of Miscible Liquids in the Turbulent Region. *Coll. Czech. Comm.,* 26:12:2961.

REVILL, B.K. 1982. Pumping Capacity of Disk Turbine Agitators: A Literature Review. Proc. 4th European Conf. on Mixing. BHRA, Noordwijkerhout, the Netherlands.

———. 1985. *Mixing in the Process Industries,* ed. by N. Harnby, M.F. Edwards, and A.W. Nienow. Butterworths, London.

RIEGER, D.H.; DITL, P.; & HAVELKOVA, O. 1988. Suspension of Solid Particles: Concentration Profiles and Particle Layer on the Vessel Bottom. Proc. 6th European Conf. on Mixing. BHRA, Pavia, Italy.

RIELLY, C.D. & BRITTER, R.E. 1985. Mixing Times for Passive Tracers in Stirred Tanks. Proc. 5th European Conf. on Mixing. BHRA, Wurzburg, Germany.

RIELLY, C.D. & PANDIT, A.B. 1988. The Mixing of Newtonian Liquids With Large Density and Viscosity Differences in Mechanically Agitated Contactors. Proc. 6th European Conf. on Mixing. BHRA, Pavia, Italy.

ROUSTAN, M. & HEZIANE, A. 1986. Circulation Engendree par des Mobiles dAgitations a Axes Horizontaux Dans un Bassin Type Chenal. Act. Colloque Agitation Mecanique (ENSIGC), Toulouse, France.

ROUSTAN, M.; HEZIANE, A.; & FAUP, G. 1988. Circulation and Dispersion of Gas–Liquid Induced by Agitators With Horizontal Axes in a Closed Loop Open Channel. Proc. 6th European Conf. on Mixing. BHRA, Pavia, Italy.

RUSHTON, J.H.; COSTICH, E.W.; & EVERETT, H.J. 1950. Power Characteristics of Mixing Impellers. *Chem. Engrg. Prog.,* 46:8:395; 46:9:467.

SMITH, J.M. 1985. *Mixing of Liquids by Mechanical Agitation*, ed. by J.J. Ulbrecht and G.K. Patterson. Gordon and Breach, New York.

SMITH, J.M. & WARMOESKERKEN, M.M.C.G. 1985. The Dispersion of Gases in Liquids With Turbines. Proc. 5th European Conf. on Mixing. BHRA, Wurzburg, Germany.

TAKAHASHI, K. ET AL. 1988. A Study of Circulation Time for Pseudoplastic Liquids in a Vessel Agitated by Helical Ribbon Impellers. Proc. 6th European Conf. on Mixing. BHRA, Pavia, Italy.

TEBEL, K.H. & MAY, H.O. 1988. Confined Jets in Tubular Reactors: An Effective Mixing Device for Suppressing Fast Consecutive Reactors. Proc. 6th European Conf. on Mixing. BHRA, Pavia, Italy.

ULBRECHT, J.J. & CARREAU, P.J. 1985. *Mixing of Liquids by Mechanical Agitation*, ed. by J.J. Ulbrecht, and G.K. Patterson. Gordon and Breach, New York.

VOIT, H. & MERSMANN, A.B. 1986. Scale-up of Agitated Vessels for Suspending. Act. Colloque Agitation Mecanique (ENSICG), Toulouse, France.

———. 1988. Calculation of Mixing Times for Gas–Liquid Mixing Vessels. Proc. 6th European Conf. on Mixing. BHRA, Pavia, Italy.

WARMOESKERKEN, M.M.G.C. & SMITH, J.M. 1986. Flow Maps for Gas–Liquid Mixing Vessels. Act. Colloque Agitation Mecanique (ENSICG), Toulouse, France.

WEETMAN, R.J. & OLDSHUE, J.Y. 1988. Power, Flow and Shear Characteristics of Mixing Impellers. Proc. 6th European Conf. on Mixing. BHRA, Pavia, Italy.

ZWEITERING, T.N. 1958. Suspending Solid Particles in Liquids by Agitations. *Chem. Engrg. Sci.*, 8:3/4:244.

3

Turbulence and Mixing: Modeling Effects on Chemical Reactions

Gary K. Patterson, Ph.D., P.E., Associate Dean,
  School of Engineering, University of Missouri—Rolla,
  Rolla, Mo. USA

Robert P. Zipp, Ph.D., Department of Chemical Engineering,
  University of Michigan, Ann Arbor, Mich. USA

INTRODUCTION

Coagulation, which seems to be most effective with very fast mixing, must include a very fast precipitation reaction. The rate of this reaction is limited by diffusion of the reactants. Modeling of the turbulent mixing process involved in coagulation would, therefore, be profitable when developing optimum configurations and better scale-up. Flocculation requires much longer residence time and a gentler mixing regime, but once the process is adequately understood, modeling of the flow patterns and turbulence distribution should also prove useful.

Modeling for Mixing and Chemical Reactions

Models for mixing and chemical reaction may be classified as Lagrangian and Eulerian (see chap. 2). Lagrangian modeling of mixed chemical reactors, in which fluid elements are followed as batch reactors, has generally not led to successful results for complex flow systems. Eulerian approaches, which make use of balances of mass, momentum, turbulence energy, and components over a stationary volume element, have been more successful. Both two- and three-dimensional spatial profiles of such variables as velocity, turbulence energy, rate of energy dissipation, concentrations of all chemical components, and temperature have been computed, and some compare well with experimental data. Such profiles have been made possible by the development of several versions of computer codes, which solve the parabolic or elliptic balance equations for the relevant variables in steady flows and unsteady flows.

One Lagrangian modeling approach, random coalescence–dispersion, has led to good results when compared with experiments and has valuable capabilities. Random coalescence–dispersion has the advantage over other methods of handling complex reaction kinetics, but it is computationally intensive.

Further progress in this area will depend on better flow models that are based on more extensive use of three-dimensional modeling and on the development and testing of new and innovative closures for the chemical reaction terms.

Mechanically Stirred Vessel Design

Mechanically stirred vessels of many types and scales are used in process industries and in water treatment for a variety of purposes, such as mixing liquid reactants to bring about a chemical reaction, dispersing immiscible phases into a liquid, controlling crystallization, precipitating and agglomerating chemicals, coagulating and flocculating, or providing a controlled environment for microorganism activity. The correct design of such vessels can be crucial to the profitability of the process by virtue of its influence on reaction yield, slurry uniformity, crystal size or microorganism growth rate and production, and ultimately on separation and purification costs.

Current design methods rely on correlations of overall vessel-average parameters, which are obtained from laboratory and pilot-scale experimental data. Such experimental programs are expensive and time-consuming, and have not covered all the relevant parameters. The scale-up of the information established is difficult, especially since most vessels have to perform several functions simultaneously (for example, dispersion, reaction, and heat transfer), which do not scale-up equally. The semiempirical nature of the correlations limits the design to geometric similarity with the small-scale laboratory apparatus, which may not lead to the optimum configuration for the particular process. In water treatment, for example, the laboratory unit is rarely geometrically similar to the final plant.

The greatest shortcoming of such methods is that "local" effects cannot be determined. Consequently, processes in which inhomogeneities in concentration are important present great problems in industrial design. Current scale-up methods based on vessel-average parameters are inapplicable in such cases, and serious errors can result from attempts to use them.

These shortcomings have stimulated a considerable body of research in the development of detailed, numerically based modeling of mixed reactors. The objectives have been to be able to simulate the mixing behavior in every region of a stirred reactor and the resulting transport and chemical reactions, leading to geometrical distributions of concentrations, conversion, yield, and temperature.

REVIEW OF TURBULENCE AND MIXING THEORY

The turbulent mixing process takes place in three stages. First, bulk diffusion, the result of the convective transport of the flow, carries the fluid into the turbulent field. In the second stage of mixing, turbulence breaks large lumps of fluid into smaller lumps. This turbulent dispersion reduces the size (scale of segregation) and increases the available diffusion cross section between different fluids. In the third stage, molecular diffusion causes the final homogenization on a molecular scale; this is a reduction of the intensity of segregation. The effect of the turbulence, therefore, is to speed up the molecular diffusion process by increasing the available surface area and by reducing the diffusion distances.

A prerequisite to understanding fluid mixing is a thorough knowledge of the fluid dynamics within a mixing system. Such knowledge requires the study of velocities and turbulence parameters. These studies, however, do not fully quantify the behavior of mixing fluids. Therefore, numerous studies on the mixing of reacting and nonreacting fluids have been made. Both areas of study have guided in the selection of models that would best predict mixing behavior. However, application of these models was feasible only for very restricted conditions. With the advent of high-speed computer technology, it is now possible to simulate mixed chemical reactors.

Velocity and Turbulence

Before a study of turbulent mixing can be attempted, the motion and turbulence characteristics of the fluid must be understood. General review papers can be found in Uhl and Gray (1966), Brodkey (1975), Nagata (1975), and Ulbrecht and Patterson (1985).

Early investigations were based on the observation that the power consumption per unit volume should remain constant during scale-up in order to keep the mixing rate constant. Rushton, Costich, and Everett (1950) obtained correlations between impeller speed and power consumption in various geometries. They defined the power number as

$$N_P = \frac{P}{\rho N^3 D^5} = f\left(\frac{\rho N D^2}{\mu}\right) \qquad \text{(Eq 3-1)}$$

Where:

P = power consumption
D = impeller diameter
N = rotation rate
ρ = fluid density
μ = viscosity

They discovered that the power number could be correlated to the Reynolds number for a given impeller type and tank geometry (see chap. 2).

Further understanding of the mixing process required an understanding of the fluid behavior within the vessel (in most cases, a tank). Velocity measurements were first obtained by photographing tracer particles suspended in the fluid, the length of the particle streak being proportional to the local velocity. This technique was first used by Ruston and Oldshue (1953). They discovered that impeller discharge rates were proportional to the impeller speed and that a significant amount of fluid was entrained by the impeller jet. Numerous studies have used this technique, including Cutter (1966).

The development of hot-wire and hot-film anemometry made it possible to measure both the velocity and the turbulence levels. The first study using this technique in stirred tanks was made by Bowers (1965). Turbulence levels and jet velocities were found to be proportional to the impeller tip speed. Again, numerous studies have been made using this technique.

Both the hot-film anemometry and photographic methods suffer from serious limitations. The hot-film probe causes interference with the flow, and the photographic methods have poor spatial resolution. A more recent device, the laser-Doppler anemometer, overcomes these problems. The first application of this method to stirred tanks was made by Reed, Princz, and Hartland (1977). Turbulence levels of several hundred percent were observed.

Hot-wire, hot-film, and laser-Doppler anemometry can provide additional relevant information about the turbulence. For example, the ability to measure correlation functions gives spectral information about the turbulence. Such studies were made by Kim and Manning (1964), Sato et al. (1967) Sato, Kamiwano, and Yamamoto (1970), and Fort et al. (1974). The results yielded energy spectra very much like those proposed by theory.

One recent study of particular relevance is that of Wu (1988). Laser-Doppler anemometry was used to provide velocities, turbulence levels, and energy dissipation rates in the impeller stream and throughout the vessel. The turbulence levels were corrected to account for periodic fluctuations due to the impeller blades. The data from this study can be used as a basis for simulation of flow patterns and turbulence.

Concentration and Concentration Fluctuations

Many distinct approaches have been taken to quantify the effects of imperfect mixing in stirred tank reactors. Initial attempts treated the stirred tank as a "black box" and attempted to relate some relevant parameter to the state of the fluid at the exit. Other studies defined a mixing time and related it to the Reynolds number, in a manner analogous to the correlation of the power consumption with the Reynolds number. The most relevant studies distributed measurements of concentration and unmixedness at locations within the tank.

The most relevant parameter in studying mixing in chemical reactors is the extent of reaction. Worrell and Eagleton (1964) examined the simple second-order reaction of sodium thiosulfate with hydrogen peroxide. The resulting effluent conversion was correlated with the power input. Paul and Treybal (1971) determined that second-order, competitive–consecutive (series–parallel) reactions were more sensitive to micromixing. They examined the reaction between iodine and l-tyrosine. The resulting yield of one of the two products, which varied between 50 and 72 percent, was correlated to the average turbulence levels in the tank. Truong and Methot (1976) used the saponification of ethylene glycol diacetate by sodium hydroxide to study the mixing. They concluded that segregation is not a property of the fluid, but is a result of the interaction of hydrodynamics, and that the chemical reaction invalidated the assumptions upon which many models, including residence time distribution models, were based. Bourne, Moergeli, and Rys (1977) also studied the effect of the mixing process upon the yield and product distribution of series–parallel reactions. The reaction between l-naphthol and diazotized sulfanilic acid produces two dyes with concentrations that can be measured spectrophotometrically. Extensive study of this system has been made by Bourne, Moergeli, and Rys (1977), Angst, Bourne, and Dell'Ava (1984), and Bourne and Dell'Ava (1987), and the results have been used to determine parameters in their model. The alkaline hydrolysis of nitromethane was used by Klein, David, and Villermaux (1980), and the results were used to determine the parameters in their model. Spencer and Lunt (1980) combined the features of residence time distribution experiments with those of conversion experiments by measuring the conversion of p-nitrophenylate in response to square wave inputs of the reactants. However, interpretation of the results was inconclusive. An interesting competitive–consecutive reaction was investigated by Barthole, David, and Villermaux (1982), in which the optical density of the precipitated product, barium sulfate, was related to the segregation within the reactor.

In all of the above studies, only the effluent or final values of conversion or yield were measured. No study related the distribution of these values throughout the tank to the mixing and hydrodynamics. The parameters obtained are dependent on the gross behavior of the tank, not the specific local behavior of the fluid within the tank.

The most suitable method for determining the behavior of stirred tanks is to directly measure the relevant mixing parameters at various points within the tank. Measurement of the average velocities determines the flow patterns in the tank, and measurement of the mean square of the concentration fluctuations determines the pattern of mixedness. However, since the flow is highly three-dimensional, the interpretation of this data is difficult.

The first study of this nature was made by Manning and Wilhelm (1963). To avoid entrance effects, water was introduced at the bottom of the tank, and a conductive solution was injected just below the impeller; the mixed solution was allowed to overflow the top of the tank. The temporal-mean and root-mean-square (RMS) concentration, and the spectral distribution of the fluctuations was measured using a conductivity probe at various locations within the tank. The average tracer concentration was uniform throughout the tank, but the RMS fluctuations were maximum near the impeller tip and decreased as radial distance increased. This is to be expected, as the turbulent radial jets of fluid emitted by the impeller decrease the gradients of concentration. Reith (1965) made studies in a similar geometry, but the entrance solutions were both injected into the impeller stream, one above and one below. Higher RMS concentrations were obtained, but this is a result of using lower impeller speeds.

One limitation of these conductivity studies is the interference caused by introducing a probe into the domain. This difficulty was overcome by Zipp (1981), who used a light-scattering technique to measure mean and fluctuating tracer concentrations within the tank. A high-intensity light beam was passed through the window in the wall of the tank, and a photomultiplier was focused on the beam through another window. The measuring volume was the intersection of the two. Local concentration of particles introduced in one of two feed streams was proportional to the voltage across the photomultiplier. This technique was modified by Bockelman (1983) and Quigley (1984) to allow measurements in reacting flows. Concentration and segregation were measured by detecting the intensity of fluorescence of the reaction product. The reaction between cerium (IV) and sodium azide was used. Cerium (III) was the fluorescent product of the reaction. The results of passive mixing studies, in which cerium (III) was in one feed and not the other, closely agreed with the conductivity and light-scattering measurements.

Theory of Turbulent Mixing

The term "mixing" has been ambiguously defined in the literature, resulting in some confusion. In this book, the concern is with turbulent mixing of low-viscosity liquids. This process occurs in three stages, all of which occur simultaneously at different scales. The first stage, bulk diffusion, is the overall convection of fluid within the domain. Bulk diffusion can be easily understood, since it is the result of the overall motion of the fluid, which can be represented by an average velocity. If the domain is turbulent, large groups of molecules move about randomly; these turbulent eddies cause a fine-scale blending. The second stage is eddy diffusion, which can only reduce the scale to the smallest scale of the turbulence, which is usually considered to be the Kolmogoroff microscale. The final stage, molecular diffusion, occurs at all scales, and is the result of relative molecular motion. Molecular diffusion causes the components in a fluid mixture to intermingle at a molecular level. Since eddy diffusion causes a significant increase in the diffusion cross section available for molecular diffusion between species, and since the intermingling at the molecular level is predominant at small scales, mixing to the molecular level in a turbulent flow occurs much more rapidly than would be possible in a laminar flow.

Two criteria that describe the level of species mixedness in a fluid are used: the degree to which the material has been redistributed by the turbulence (decrease of scale of segregation) and the approach to uniformity by the action of molecular diffusion (decrease of intensity of segregation). In defining the significant parameters of the turbulence, the length scale and kinetic energy of the turbulence are determined in terms of measurable statistical quantities of the velocity field. A similar analysis of the statistical quantities of the concentration field was first made by Danckwerts (1953). The scale of segregation at point ξ is defined as

$$L_c = \int_0^\infty q_c(r)\, dr \qquad \text{(Eq 3-2)}$$

Where:

$q_c(r)$ = the Eulerian correlation coefficient at separation distance r

$$q_c(r) = \frac{\overline{c_A(\xi)c_A(\xi+r)}}{\overline{c_{AO}^2}} \qquad \text{(Eq 3-3)}$$

Where:

c_A = the concentration fluctuation of species A

$\overline{c_{AO}^2}$ = the initial segregation of species A

Since the scale of segregation is an average of the spatial correlation function, it is a measure of the large-scale diffusion process.

The intensity of segregation is defined as

$$I_s = \frac{\overline{c_A^2}}{\overline{c_{AO}^2}} \qquad \text{(Eq 3-4)}$$

It is equivalent to the correlation coefficient $q_c(r)$ at zero separation distance. By normalizing by the initial value, the intensity of segregation varies from unity for total unmixedness to zero for complete mixedness.

Turbulence is a multiscale phenomenon; energy cascades from large-scale eddies to small-scale eddies, where it is dissipated. Methods have been developed for examining the energy spectrum associated with the turbulence. In a similar manner, an analysis of the scalar spectrum can provide insight into the mixing process.

In order to perform a spectral analysis, it is necessary to assume that eddies of a given size are associated with a fluctuation frequency. The wave number k can then be related to the frequency n

$$k = \frac{2\pi n}{U_c} \tag{Eq 3-5}$$

Where:

U_c = the convection velocity

The wave number is inversely proportional to the eddy size. Information about the spatial structure of the turbulence is contained in the correlation function

$$Q_{ij}(r) = \overline{(u_i)_\alpha (u_j)_\beta} \tag{Eq 3-6}$$

Where:

$(u_i)_\alpha$ = the velocity fluctuation in direction i at location α
$(u_j)_\beta$ = the fluctuation velocity in direction j at location β
r = the separation distance between α and β

The Fourier transform of the correlation function is more useful, as it expresses the spectrum of the velocity fluctuations among various sized eddies. For example, the one-dimensional turbulent energy spectrum for direction 1 is

$$E_1(k) = \frac{2}{\pi} \int_o^\infty Q_{11}(r) \, kr \sin(kr) dr \tag{Eq 3-7}$$

The turbulence intensity in direction 1, $\overline{u_1^2}$, can be obtained from the spectrum by integrating over all wave numbers

$$\overline{u_1^2} = \int_o^\infty E_1(k) \, dk \tag{Eq 3-8}$$

A similar analysis can be performed on the concentration correlation $Q_c(r)$,

$$Q_c(r) = \overline{(c_i)_\alpha (c_i)_\beta} \tag{Eq 3-9}$$

Where:

$(c_i)_\alpha$ = the fluctuation concentration of species i at location α
$(c_i)_\beta$ = the fluctuation concentration of species i at location β separated by distance r

By integration over all r, one obtains a concentration energy spectrum function $E_c(k)$,

$$E_c(k) = \frac{2}{\pi} \int_0^\infty Q_c(r)\, kr \sin(kr)\, dr \qquad \text{(Eq 3-10)}$$

from which one can obtain the segregation $\overline{c_i^2}$,

$$\overline{c_i^2} = \int_0^\infty E_c(k)\, dk \qquad \text{(Eq 3-11)}$$

A typical three-dimensional turbulence energy spectrum $E(k)$ is shown in Figure 3-1. This curve represents the energy contained in eddies of wave number k, and the curve may be divided into two distinct regions. In the inertial-convective subrange, energy is transferred from large eddies to small eddies, and dissipation of energy to heat is insignificant compared to the amount of energy transferred. Furthermore, the amount of energy transferred is large compared to the rate of

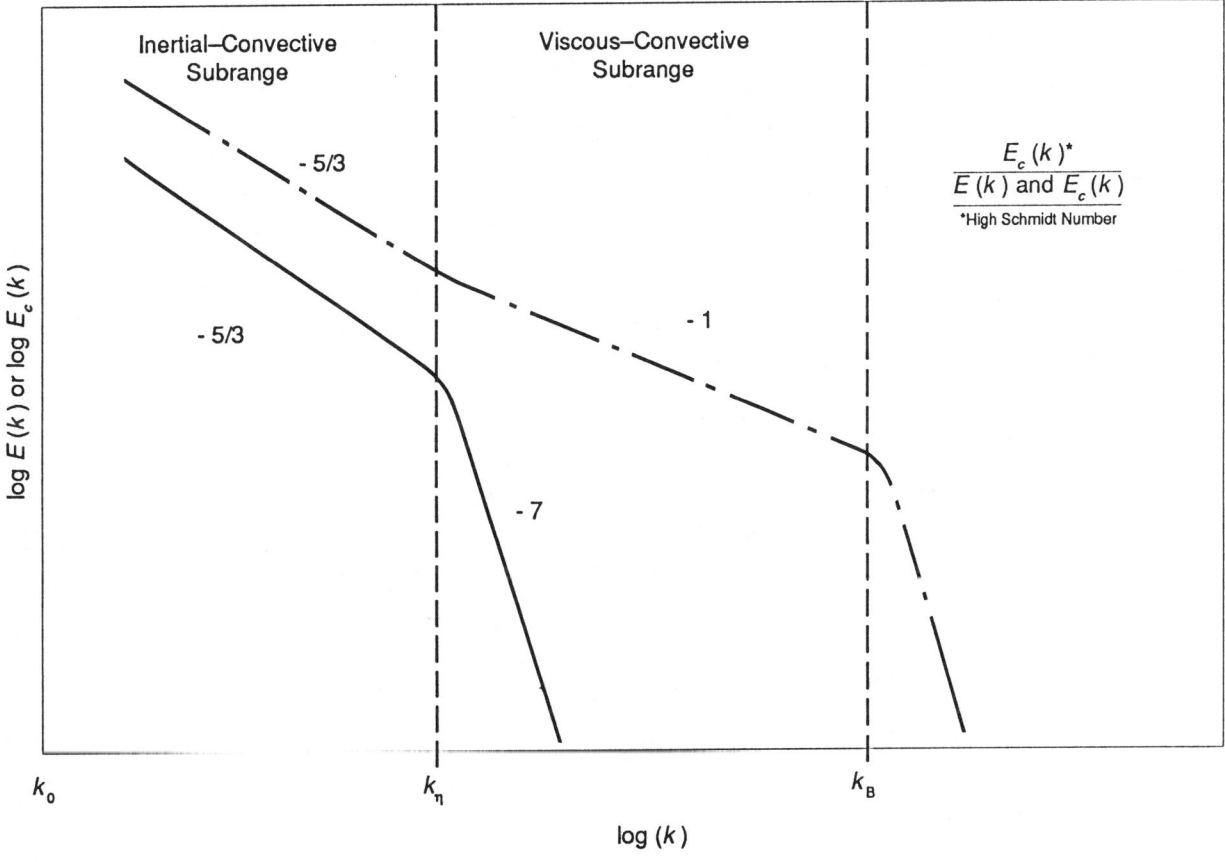

Source: Brodkey, R.S. 1967. *Phenomena of Fluid Motions. Addison-Wesley Co., Reading, Mass.*

Figure 3-1 Velocity and concentration spectra.

change of the energy of the eddies so that they can be considered to be in statistical equilibrium with one another. This theory was first proposed by Kolmogoroff (1941):

> At sufficiently high Reynolds numbers there is a range of high wave numbers where the turbulence is in equilibrium and uniquely determined by the parameters ν (the kinematic viscosity) and ε (the turbulence energy dissipation). This state of equilibrium is universal.

This has become known as the universal equilibrium theory.

At a slightly higher wave number, transfer of energy from lower to higher wave numbers occurs. In this inertial–convective subrange, the energy spectrum function is represented by

$$E(k) = A_1 \varepsilon^{-2/3} k^{-5/3} \qquad \text{(Eq 3-12)}$$

The scalar spectrum function $E_c(k)$ behaves in a similar manner in this range

$$E_c(k) = A_2 \varepsilon^{-1/3} \varepsilon_c k^{-5/3} \qquad \text{(Eq 3-13)}$$

Where:

ε_c = the rate of destruction of $\overline{c_i^2}$

At higher wave numbers, dissipation becomes the dominant mechanism, and energy transport becomes minimal. Kolmogoroff hypothesized that when the eddy Reynolds number approaches unit, viscous effects will become dominant. The length scale at which this occurs is the Kolmogoroff length scale

$$\eta = \left(\frac{\nu^3}{\varepsilon}\right)^{1/4} \qquad \text{(Eq 3-14)}$$

The dissipation wave number k_η is of the order $1/\eta$. Above this wave number, Heisenberg obtained

$$E(k) = A_3 (\varepsilon \nu^5)^{1/4} k^{-7} \qquad \text{(Eq 3-15)}$$

which implies a significant loss of energy at high wave numbers greater than k_η. This is the dissipative subrange for $E(k)$. However, in liquids, the ratio of the kinematic viscosity ν to the molecular diffusivity $\mathcal{D}_c$ is much greater than one, which implies that mixing on the molecular level is incomplete until significantly higher wave numbers are reached. Based on dimensional reasoning, Batchelor (1953) deduced that the spectral density in this range should be

$$E_c(k) = A_4 \frac{\varepsilon_c}{\gamma} k^{-1} \qquad \text{(Eq 3-16)}$$

Where:

ε_c = the dissipation of segregation
γ = a strain rate parameter

In this viscous–convective subrange, $E_c(k)$ is proportional to k^{-1}. Beyond the Batchelor wave number $k_B = \left(\varepsilon/\nu \mathcal{D}_c^2\right)^{1/4}$, the mixing nears completion and the scalar spectrum function decays rapidly. Figure 3-1 illustrates these model spectra.

Corrsin (1964) analyzed the decrease of segregation in a domain of decaying turbulence. This analysis was based on the similarity between the velocity field and the scalar field. The rate of mixing was expressed as

$$\frac{\overline{-dc_i^2}}{dt} = \frac{\overline{c_i^2}}{\tau_c} \qquad \text{(Eq 3-17)}$$

Where:

τ_c = the mixing time constant, which can be related to the Taylor scalar microscale λ_c as

$$\tau_c = \frac{\lambda_c^2}{12\mathcal{D}_c} \qquad \text{(Eq 3-18)}$$

By definition, the Taylor scalar microscale can be obtained from the scalar energy spectrum

$$\lambda_c = \frac{\int_0^\infty E_c(k)\,dk}{\int_0^\infty k^2 E_c(k)\,dk} \qquad \text{(Eq 3-19)}$$

By integrating Eq 3-13 over the inertial–convective subrange and Eq 3-16 over the viscous–convective subrange, Corrsin obtained

$$\tau_c = \frac{1}{2}\left[4\left(\frac{L_c^2}{\varepsilon}\right)^{1/3} + (\nu/\varepsilon)^{1/2} \ln Sc \right] \qquad \text{(Eq 3-20)}$$

Where:

L_c = the integral length scale for the concentration field

This derivation applies only for Schmidt numbers $Sc = \nu/\mathcal{D}_c$ much greater than 1.0, as is the case for liquids.

PERSPECTIVE

When reactants within a chemical reactor are incompletely mixed with each other and when entering reactants are diluted with product fluid, the reaction rates are slowed significantly and yield is affected. Ideally, mixing effects on chemical reactor performance would be nonexistent. The prospect of a stirred chemical reactor not being well mixed has always vexed designers. How fluids move and how turbulent and diffusive mixing are distributed in a reactor vessel have, until very recently, been considered beyond computation. Difficulties in solving the equations of motion for fluid mechanics are evidence enough of the impossibility of developing realistic, geometrically distributed models for the fluid motion in mixed vessels of various geometries.

Geometrically distributed modeling is an old concept in process engineering. When a design or modeling problem required a distributed approach, it was developed. Without one-dimensional distributed models for shell-and-tube heat exchangers, packed distillation, extraction absorption, and stripping columns, process engineering would not exist (Bennett and Meyers 1962; McCabe and Smith 1967). Two-dimensional models also find extensive use for velocity distributions in pipe flows, temperature distributions in flow and no-flow situations, distributions in boundary layers, and so on. All of this has, of course, been explained in books on transport phenomena since about 1960 (Bird, Stewart, and Lightfoot 1960; Welty, Wicks, and Wilson 1976; Fahien 1983).

Even when fluid flow patterns were too complex to model in detail, discrete one-dimensional models became useful approximations. Such models include staged unit models for mass-transfer operations, in which each stage is considered to be a uniformly mixed mass-transfer unit with no gradients of important concentrations or temperatures, except in the microscale (Faust et al. 1980). Equilibrium is assumed to prevail in each stage. It is important to assume the correct number of well-mixed, equilibrium stages in such models in order to result in the experimentally observed separation. With some correction factors for lack of equilibrium, the staged unit approach is a close approximation of true staged processes, such as those using plates or trays to separate stages.

For mixed chemical reactors, an equivalent approximation of reality has generally been avoided. Certainly, the discrete unit approach has been used as a numerical approximation for the tubular reactor, particularly where the plug flow approximation was valid (Levenspiel 1972; Smith 1982), and to simulate circulation pattern effects in stirred tanks (Middleton 1979; Wen and Fan 1975). However, these reactors had reactants that were always well mixed, or assumed to be, even before injection at the feed point. The problem of interest here is that flow and mixing patterns of a two- or three-dimensional nature affect the performance of the chemical reactor. Turbulent mixing effects are of primary interest in this review; laminar mixing, treated by a different approach, is related to turbulent mixing only at the last, or smallest, stage of the process (Ottino, Ranz, and Mocosko 1979; Ottino 1981; Palpepu, Adler, and Edwards 1981).

LAGRANGIAN MODELING

The term "Lagrangian" to describe some of the modeling approaches discussed in this section may be objectionable to some readers. The term is used here for the concept of characterizing the fluid element (and its chemical components in a reaction mixture) by its history of locations within the reactor and by its history of concentrations, hence its history of chemical reaction and mass and heat transfer rates. This type of modeling approach requires approximations for the histories of a representative number of fluid elements in the reactor. These approximations are necessary for computing and averaging fluid conditions (concentrations, temperature, conversion, yield) at the reactor effluent.

Residence Time Distribution

While formulating simple and easily understood Lagrangian models of mixed reactors, the concept of the residence time distribution emerged (Danckwerts 1953). Two ideal residence time distributions are the perfectly mixed tank (decaying exponential) and plug flow (delta function) (see Figure 3-2). The dispersion model produces residence time distributions that indicate the degree of deviation from plug-flow behavior. Frequently, measured distributions are used (Levenspiel 1972; Smith 1982). Turner (1983) has reviewed the history and uses of residence time distributions.

A complete history for the many fluid elements flowing through a mixed reactor has never been compiled or modeled. Simpler models have, however, been developed. The first level of improvement over the perfectly mixed reactor model is one in which perfectly mixed elements of fluid are fed to the reactor; their concentrations change within the reactor according to batch kinetics; they are the same as their neighbors; and they exit at various times according to a residence time distribution. Such models have a distribution of effluent concentrations, the weighted average of which is the model effluent concentration. For the perfectly mixed tank residence time distribution, the average effluent concentration equals the average concentration within the reactor. This is the *micromixed* or maximum

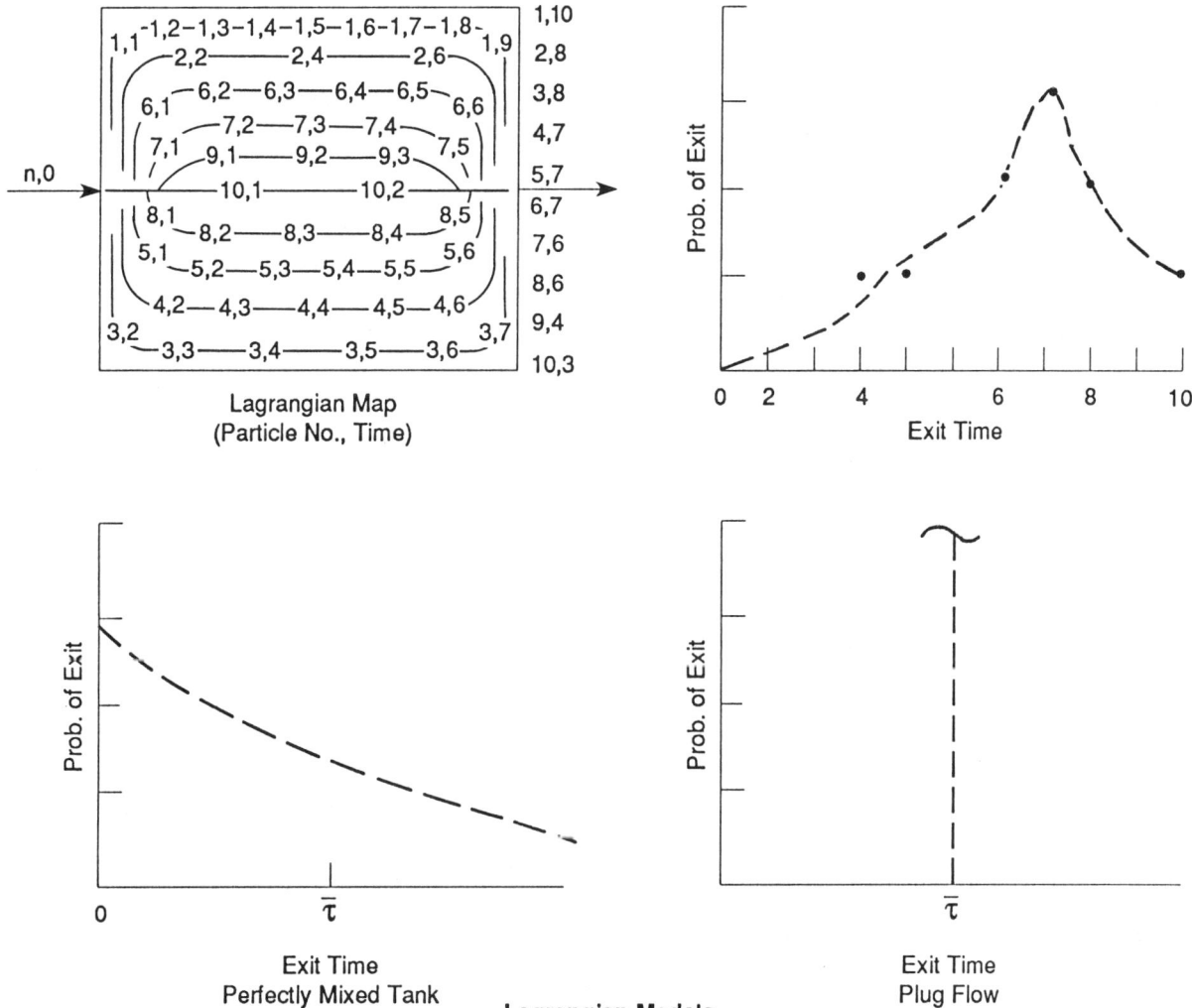

Figure 3-2 Examples of exit time distributions resulting from Lagrangian viewpoint.

mixedness model (Zwietering 1959). It can be shown that for first-order reactions the residence time distribution of the fluid elements has no effect on effluent concentrations, hence no effect on conversion. For other reaction orders there is an effect. A real reactor must be very well-mixed for this model to be applicable.

In order to approximate less well-mixed reactors, the *segregated* reactor model was conceived (Danckwerts 1953). In that model perfectly mixed elements of fluid are fed to the reactor; their concentrations change within the reactor according to batch kinetics; they are *not* the same as their neighbors; no mass transfer occurs; and they exit at various times according to a residence time distribution. A good way to visualize the difference between the micromixed and segregated reactors is as follows (see Figure 3-3).

Micromixed. Fluid is injected at points in the reactor. This fluid instantly mixes with local fluid such that the resulting residence times correspond to those required for the modeled residence time distribution.

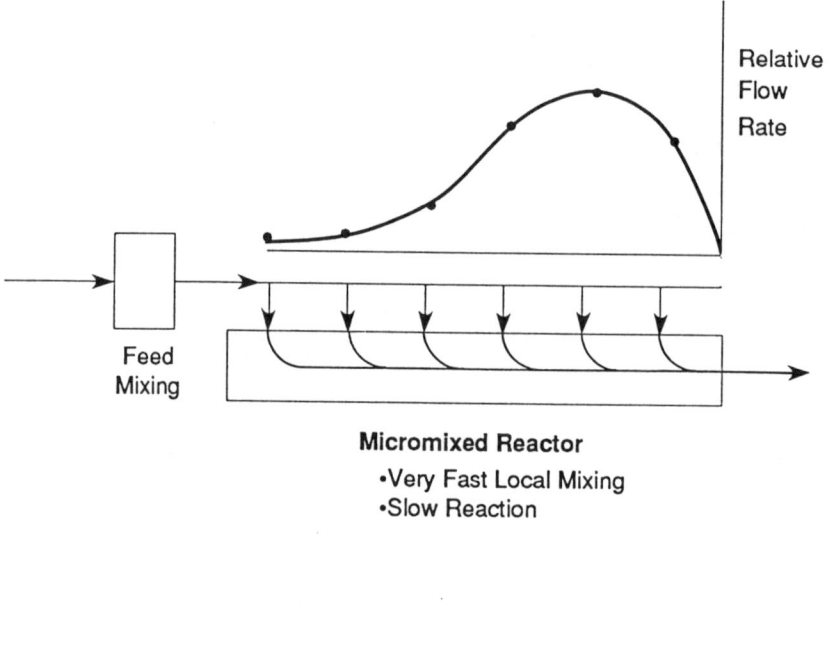

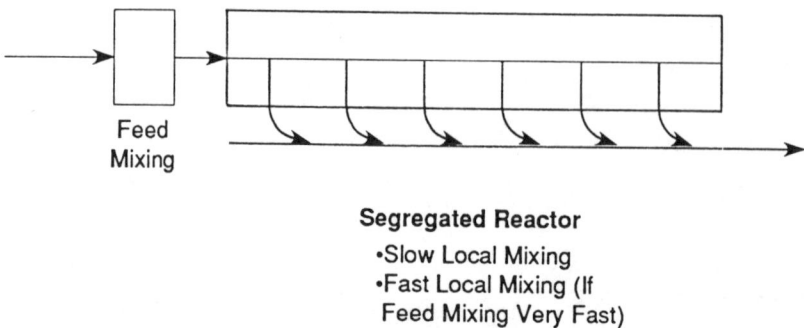

Figure 3-3 Illustrations of the micromixed and segregated reactors.

Segregated. Fluid is removed at points in the reactor. This fluid does *not* mix with surrounding fluid, such that the resulting residence times correspond to those required for the modeled residence time distribution.

No real reactor is either micromixed or segregated. If chemical reactions are slow enough, however, the reactor may be considered to be micromixed. Of course, under most circumstances with such slow reactions, the residence time distribution has little effect. Consequently, the *perfectly mixed* reactor model is used. If the chemical reactions are very fast relative to the mixing rate, the reactor may be considered to be segregated, *if* the feed is *premixed*. It is, however, impossible to premix feeds for a fast chemical reaction. Thus, for single-phase systems, the segregated model is useful in few circumstances. If the feed is not premixed, the segregated model is not applicable.

Levels of segregation. Considerations of real reactor behavior led to concern about intermediate levels of segregation. The degree of *residence time (age) segregation* was defined by Danckwerts (1958) as

$$J = \frac{1}{V} \int_v (\alpha_p - \overline{\alpha})^2 \, dv \Big/ \int_o^\infty (\alpha - \overline{\alpha})^2 \, I(\alpha) \, d\alpha \qquad \text{(Eq 3-21)}$$

Where:

V = reactor volume
α = fluid age in the reactor
$I(\alpha)$ = the age distribution in the reactor
subscript p means at a point
overbar indicates local average

Experimental measurements of segregation based on Eq 3-21 may be made, but little has been done to relate measured values to mixer conditions (see Wen and Fan 1975, p. 460).

In order to account for reactor conditions under which complete segregation of fluid elements does not occur, various combinations of the micromixed and segregated models have been proposed and demonstrated. These are generally called *multi-environment* models (Ng and Rippin 1965; Weinstein and Adler 1967; Villermaux and Zoulalian 1969; Richie and Togby 1979; Mehta and Tarbell 1983; also see chap. 5). Each reactor segment, whether micromixed or segregated, is an environment. Various combinations of flow between environments have been proposed, the relative flow rates in some cases depending on the relative sizes of environments. Such models give rise to arbitrary parameters to describe the relative sizes of environments and the relative flow rates between them. Table 3-1 shows some typical multi-environment models with their parameter requirements.

The main weakness of the multi-environment approach is the difficulty of relating parameter values to measured flow rates, velocities, levels of turbulence, or other physical characteristics of mixed reactors. Mehta and Tarbell (1983) have recently attempted to relate the parameters of a multi-environment model to measurable mixing characteristics. More is discussed about this aspect in chap. 5.

Summary

It should be noted that the Lagrangian methods outlined above are only vaguely related to mixed reactor geometry or flow patterns. Two- and three-dimensional flow patterns and circulation are not explicitly accounted for in the models. Many

Table 3-1 Typical Multi-Environmental Models

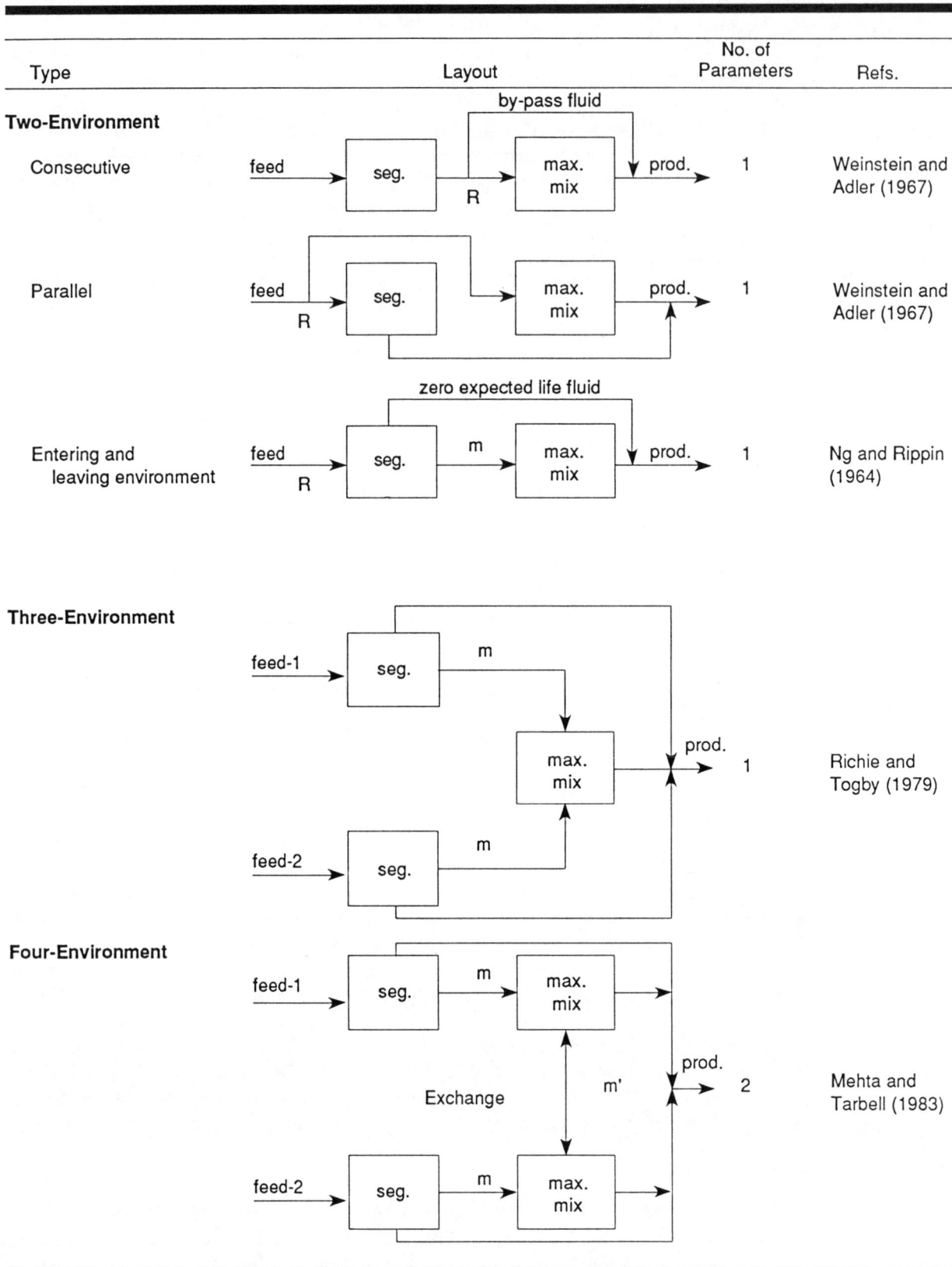

different flow patterns may produce the same residence time distribution, and many different multi-environment combinations may produce the same residence time distribution and reaction conversion, so no uniqueness of models results.

If the Lagrangian approach is extended to incorporate information on flow patterns, such that the location histories of many fluid elements are used in the simulation, a more physically realistic model is generated. Such a model can contain more local mixing information. The random coalescence–dispersion model discussed below is one way to accomplish such detailed Lagrangian modeling.

EULERIAN MODELING

In contrast to Lagrangian concepts, which are based on the histories of fluid elements, Eulerian modeling focuses on variations of local (laboratory coordinate) conditions and rates of change of those conditions as influenced by convection (flow) and diffusion toward and away from the locality. Concentration on reaction fluid history has tended to divert efforts away from the Eulerian approach.

The tremendous respect shown by reaction engineers for the work of Danckwerts on residence time distributions and segregation has had a controlling influence on the models developed. As has been pointed out by Brodkey (1967), use of the term "segregation" by Danckwerts for both species mixing and age mixing has created much confusion. The definition of species segregation is the mean-square fluctuation about the mean for the component concentration in question. Brodkey has shown that total segregation of two components is given as follows:

$$\overline{c_o^2} = \overline{C}_A \overline{C}_B \qquad \text{(Eq 3-22)}$$

Where:

$\overline{C}_A$ and $\overline{C}_B$ = average concentrations after mixing

Turner (1983) and Bourne (1983) made it quite clear that much work has been done to solve nonproblems because of applications of residence-time-distribution-based models to unrealistic or impossible reactor situations, such as the fast reaction with premixed feed.

In order to formulate more detailed and more realistic descriptions of turbulent mixing effects in chemical reactors, many have turned to the Eulerian approach. Rather than determine changes to elements of fluid as they flow through a mixed reactor, changes in, or steady-state conditions of, stationary control volumes are modeled. The fullest expression of such a model consists of partial differential equations describing the Reynolds-averaged balances of mass, momentum, turbulence energy, turbulence energy dissipation rate, possibly turbulent shear stresses, thermal energy, thermal energy fluctuation intensity, concentrations of each component, and the species segregation of each component from the other components. These balances are generally of the form

$$D\overline{\xi}/Dt = \frac{\partial}{\partial x_i}\left(\frac{\nu_t + \nu}{\sigma_\xi} \times \frac{\partial \overline{\xi}}{\partial x_i}\right) + \text{PROD}_\xi - \text{DISS}_\xi \qquad \text{(Eq 3-23)}$$

ξ can be any of the balanced variables, and correlations (i.e., $\overline{\xi_1 \xi_2}$) may occur in the production and dissipation terms, giving rise to a closure problem. The result of a solution of such a model is a two- or three-dimensional spatial distribution of each of the balanced components within a prescribed set of boundary conditions (see Figure 3-4).

The problem that arises in the formulation of the above balance equation is the generation of the higher moments (correlations) of the balanced variables. The higher moments must, at some point, be expressed in terms of lower ones in order to close the set of equations—the so-called closure problem. For the mixing–reacting part, the higher moments are products of fluctuating concentrations and, possibly, temperature.

Turbulent Transport Equations

A full discussion of the equations necessary for a complete turbulent mixing model is given by Patterson (1981). A summary of the approximate equations that must be solved are shown below. (Equations 3-24 through 3-27 are the $k - \varepsilon$ model for turbulent flow.)

Mass:

$$\frac{D\bar{\rho}}{Dt} = -\bar{\rho}\left(\frac{\partial \overline{U}_i}{\rho x_i}\right) \qquad \text{(Eq 3-24)}$$

Momentum:

$$\frac{D\overline{U}_i}{Dt} = -\frac{1}{\rho}\frac{\partial \bar{p}}{\partial x_i} + \frac{\partial}{\partial x_j}\left(\frac{v_t}{1}\frac{\partial \overline{U}_i}{\partial x_j}\right) + \bar{g}_i \qquad \text{(Eq 3-25)}$$

$$\phantom{\frac{D\overline{U}_i}{Dt} = -\frac{1}{\rho}} \text{II} \phantom{\frac{\partial \bar{p}}{\partial x_i} +} \text{I} \phantom{\frac{\partial}{\partial x_j}\left(\frac{v_t}{1}\frac{\partial \overline{U}_i}{\partial x_j}\right)} \text{II}$$

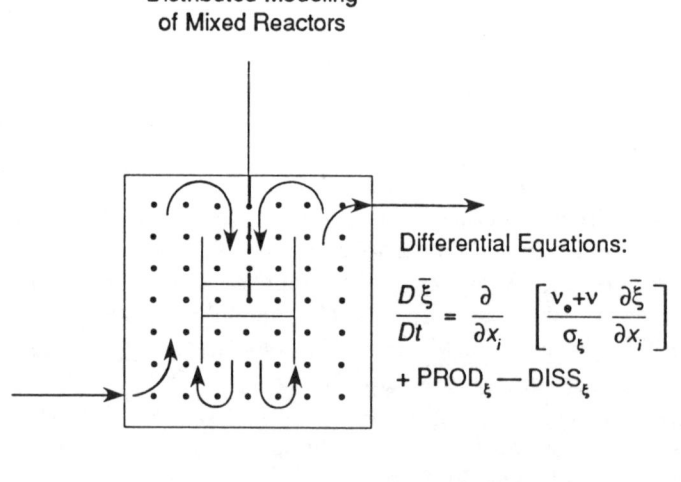

Figure 3-4 Representation of the distributed modeling method.

Turbulence energy ($k_t = \frac{1}{2} \sum_i \overline{u_i^2}$):

$$\frac{Dk_t}{Dt} = -\underbrace{\frac{\partial}{\partial x_j}\left(\frac{\nu_t}{Sc_k}\frac{\partial k_t}{\partial x_j}\right)}_{\text{I}} + \underbrace{\nu_t \left\{\left(\frac{\partial \overline{U}_1}{\partial x_2} + \frac{\partial \overline{U}_2}{\partial x_1}\right)^2 + 2\left(\frac{\partial \overline{U}_i}{\partial x_i}\right)^2\right\}_1}_{\text{II}} - \underbrace{\varepsilon}_{\text{III}} \quad \text{(Eq 3-26)}$$

Turbulence energy dissipation rate:

$$\frac{D\varepsilon}{Dt} = -\underbrace{\frac{\partial}{\partial x_j}\left(\frac{\nu_t}{Sc_\varepsilon}\frac{\partial \varepsilon}{\partial x_j}\right)}_{\text{I}} + \underbrace{C_1 \nu_t \left\{\cdot\right\}_1 \varepsilon/k_t}_{\text{II}} - \underbrace{C_2 \varepsilon^2/k_t}_{\text{III}} \quad \text{(Eq 3-27)}$$

Component mass*:

$$\frac{D\overline{C}_i}{Dt} = \underbrace{\frac{\partial}{\partial x_j}\left(\frac{\nu_t}{Sc_c}\frac{\partial \overline{C}_i}{\partial x_j}\right)}_{\text{I}} - \underbrace{K_R\left(\overline{C}_i\overline{C}_j + \overline{c_i c_j}\right)}_{\text{III}} \quad \text{(Eq 3-28)}$$

Segregation ($s_i = \overline{c_i^2}$):

$$\frac{Ds_i}{Dt} = \underbrace{\frac{\partial}{\partial x_j}\left(\frac{\nu_t}{Sc_s}\frac{\partial s_i}{\partial x_j}\right)}_{\text{I}} + \underbrace{C_{g_1}\nu_t\left(\frac{\partial \overline{C}_i}{\partial x_j}\right)^2}_{\text{II}} - \underbrace{\frac{2s_i}{C_M(k_t\varepsilon)}^\dagger}_{\text{III}}$$

$$- \underbrace{K_R\left(\overline{C}_i\,\overline{c_i c_j} + \overline{C}_j\,\overline{c_i^2} + \overline{c_i^2 c_j}\right)_2}_{\text{III}} \quad \text{(Eq 3-29)}$$

In the above equations, terms I are diffusion, terms II are production, and terms III are dissipation.

The equations for turbulent transport and mixing may be dimensionally analyzed in order to generate important groups for correlation. These groups may be used to present the results of experiment and modeling (see Patterson 1985).

For Eq 3-24 through 3-27, let

$$U_i = VU_i^*\,;\, x_i = Lx_i^*\,;\, k_t = Kk_i^*\,;\, \varepsilon = E\varepsilon^*\,;\, \rho = \rho_R\rho^*\,;\, p = \pi p^*\,;\, t = Lt^*/V_v$$

$$\frac{Dp^*}{Dt^*} = -\rho^* \frac{\partial U_i^*}{\partial x_i^*} \quad \text{(Eq 3-30)}$$

---

*Second-order reaction.

†More complete form: $\dfrac{2s_i}{C_M(k_t/\varepsilon) + (\nu/\varepsilon)^{1/2} \ln(N_{SC})}$ (Corrsin 1964).

$$\frac{DU_i^*}{Dt^*} = \left[\frac{\pi}{V_v^2}\right]\left[\frac{\partial p^*}{\partial x_i^*}\right] + \left[\frac{\nu_t}{LV_v}\right]\frac{\partial^2 U_i^*}{\partial x_j^{*2}} + \left[\frac{Lg_i}{V_v^2}\right] \quad \text{(Eq 3-31)}$$

In Eq 3-31, $\left[\dfrac{\nu_t}{LV_v}\right] = \dfrac{1}{\text{Re}_t}$ and $\left[\dfrac{Lg_i}{V_v^2}\right] = \dfrac{1}{Fr}$

$$\frac{Dk_t^*}{Dt^*} = -\left[\frac{\nu_t}{Sc_k LV_v}\right]\frac{\partial^2 k_t^*}{\partial x_j^{*2}} + \left[\frac{\nu_t V_v}{KL}\right]\{-\}_1^* - \left[\frac{LE}{V_v K}\right]\varepsilon^* \quad \text{(Eq 3-32)}$$

In Eq 3-32 $\left[\dfrac{\nu_t}{Sc_k LV_v}\right] = \dfrac{1}{Sc_k \text{Re}}$; the combination of $\left[\dfrac{\nu_t V_v}{KL}\right]$ and $\left[\dfrac{LE}{V_v K}\right]$ with

$E = P/(\rho L^3)$ and $\nu_t \propto LV_v$ leads to $\left[\dfrac{P}{\rho L^2 V_v^3}\right] = $ Po, the power number.

Equation 3-27 leads to the same dimensionless groups because it is the same form as Eq 3-26.

For Eq 3-28 and 3-29, let $\overline{C_i} = CC_i^*$; $s_i = Ss_i^*$; $\overline{c_j c_i} = SFs_i^*$;

$$\frac{DC_i^*}{Dt^*} = \left[\frac{\nu_t}{Sc_c LV_v}\right]\frac{\partial^2 C_i^*}{\partial x^{*2}} + \left[\frac{K_R CL}{V_v}\right]C_i^* C_j^* - \left[\frac{K_R SL}{CV_v}\right]Fs_i^* \quad \text{(Eq 3-33)}$$

$$\frac{Ds_i^*}{Dt^*} = \left[\frac{\nu_t}{Sc_s LV_v}\right]\frac{\partial^2 C_i^*}{\partial x_j^{*2}} + \left[\frac{Cg_1 \nu_t C^2}{SLV_v}\right]\left(\frac{\partial C_i^*}{\partial x_j^*}\right)^2$$

$$- \left[\frac{PoV^2}{K}\right]\frac{S_i^* \varepsilon^*}{2k_t^*} - \left[\frac{K_R CL}{V_v}\right]\{-\}_2 \quad \text{(Eq 3-34)}$$

In Eq 3-32, 3-33, and 3-34, the $\left[\dfrac{\nu_t}{ScLV_v}\right]$ terms are reciprocals of products of Schmidt and turbulent Reynolds numbers, $\left[\dfrac{K_R CL}{V_v}\right]$ is the Damköhler number Da, and $\left[\dfrac{K_R SL}{CV_v}\right]$ is a mixing intensity number. $\left[\dfrac{PoV_v^2}{K}\right]$ is a turbulence power number and the significance of $\left[\dfrac{Cg_1 \nu_t C^2}{SLV_v}\right]$ is difficult to assess.

The most important aspect of the above dimensionless groups is that they may be applied on a local basis in a distributed model or on the average to the entire process unit. When a group is applied to the entire process unit, the significant variables must be the dominant ones in the process over the range of scale involved.

Eulerian Models for Combustion Engineering

The area of reactor engineering where greatest use has been made of the Eulerian mixing model is combustion engineering (Lockwood and Naquib 1975; Borghi 1975; Elghobashi, Pun, and Spalding 1977; Ellail et al. 1978). Most applications were for two-dimensional geometries in which the flow patterns could be reliably modeled without using the third dimension. In some cases for axisymmetric, two-dimensional burners with swirl in the third dimension (tangential), the swirl characteristics were modeled without using a three-dimensional model. The variables modeled (about which balance equations were solved) in most cases have been mass (continuity), momentum, turbulence characteristics affecting momentum transport, average temperature (thermal energy), and component concentrations. In general, quite reasonable solutions have been obtained for many combustion geometries and many fuel mixtures. The chemistry for combustion can, in most cases, be simplified since the energy produced, as shown by the temperature distribution in the flame, is the most important aspect unless product pollutants are of interest.

Numerous comparisons between the results of complete or nearly complete mixing models for combustion and experimental measurements of distributions of concentration and temperature have been made, some of which are described by Ellail et al. (1978), Edelman and Harsha (1978), Spalding (1978), and Varma, Fishburne, and Donaldson (1978). A variety of source and sink terms were used in the models, the most variation being in the terms for component concentrations and segregations. The closures of those balance equations are the main problem in modeling reactions influenced by mixing. These comparisons show a relatively high degree of success in modeling complex combustion systems.

The source–sink terms for component and segregation balances are the most important aspect of mixed reaction modeling in general, as is the case for combustion in particular. If the temperature variations in the mixing vessel are small enough to ignore viscosity variations, the fluid mechanics (flow and turbulence) part of the model is independent of the mixing–reacting part. That means that any method may be used to obtain the needed distributions of velocities, turbulence energy, and other needed turbulence variables, such as rate of turbulence energy dissipation in the $k - \varepsilon$ model (Patterson 1985; Lockwood and Naquib 1975; Jones and Launder 1973; Toor 1962; Vassilatos and Toor 1965; Bilger 1979; Gosman et al. 1969). Because of this independence for all but the strongest endo- or exothermic-mixed reactions, the remainder of this chapter concentrates on the source–sink terms (closure terms) for the component and segregation balances. Completely satisfactory interaction models for temperature fluctuation effects (that is, the degree of temperature–concentration correlations) have not yet been achieved (Borghi 1975; Donaldson 1975).

CLOSURES FOR COMPONENT AND SEGREGATION BALANCES

The primary problem involved in modeling reactions between mixing components is shown in the following equation for reaction rates:

$$\text{DISS}_{C_A} = - \partial \overline{C}_A / \partial t = \overline{K_R C_A C_B} = K_R [\overline{C}_A \overline{C}_B + \overline{c_A c_B}] \qquad \text{(Eq 3-35)}$$

Where:

$$C_A = \overline{C}_A + c_A$$
$$C_B = \overline{C}_B + c_B$$

and c_A and c_B are fluctuations about the average concentration, K_R is assumed to be constant. The value of $\overline{c_A c_B}/\overline{C}_A \overline{C}_B$ ranges from –1 to +1 depending on conditions. Modeling the values of $\overline{c_A c_B}$ in terms of other variables depends on the closure method chosen.

Equilibrium Assumption Closure

One of the simplest closures for mixed reactions is the *equilibrium assumption*. If the chemical reaction $(dC_A/dt)_R$ is assumed to be so fast that it reaches immediate equilibrium under each condition encountered, then no reaction rate equations are involved and the component closure terms are simplified. The rate of reaction becomes the rate of mixing and the segregation between reacting components stays total. That means that $\overline{c_A c_B}$ is always equal to $(-\overline{c}_A \overline{c}_B)$ and closure is simply based on computation of the distribution of segregation level. This approach is adequate for reactions where

$$(d\overline{C}/dt)_R \gg \left(\frac{d\sqrt{\overline{c_A^2}}}{dt}\right) \qquad \text{(Eq 3-36)}$$

An early model, which is very closely related to the equilibrium model, is the hypothesis by Toor (1962) that states that segregation level is independent of reaction. This results in the relationship

$$\overline{C}/C_o = \overline{c^2}/\overline{c_o^2} \qquad \text{(Eq 3-37)}$$

for reactant ratios $(\overline{C}_A/\overline{C}_B)$ of 1 and is exactly valid for very fast reactions. For $\overline{C}_A/\overline{C}_B \neq 1$, the relationship is more complex. This relationship is shown for plug-flow mixing in Figure 3-5.

Conserved Scalar Approach

Another closure for infinitely fast reaction is the *conserved scalar* approach. In contrast to the equilibrium assumption closure, this approach is based on the assumption of zero segregation between reacting components that have diffused into the same region. This means that this model is applicable only when large-scale diffusion of reactants is the controlling step in the reaction. The method has been used by Bilger (1979) and Gosman et al. (1969). Closure is based on computation of turbulent diffusion of components. No balance equation for segregation is involved.

Models have also been developed for relatively slow reactions in which mixing rate still has an effect, as is the case in large industrial mixers with very slow circulation. It is assumed that local mixing is rapid so that molecular reaction kinetics may be applied at each computational mode in the mixer. An early example of this is given by Mann, Middleton, and Parker (1977) and Middleton (1979) in which a competitive–consecutive five-reaction scheme (related to a real industrial case) is modeled in a stirred vessel. Loops of interconnected fully backmixed zones are used to model the liquid circulation and mixing within the vessel. This method requires a knowledge of the mean and distribution of circulation times in order to decide how many of these zones should be used. This information was obtained by measurement using a flow follower. Passage of the flow follower, which contains a

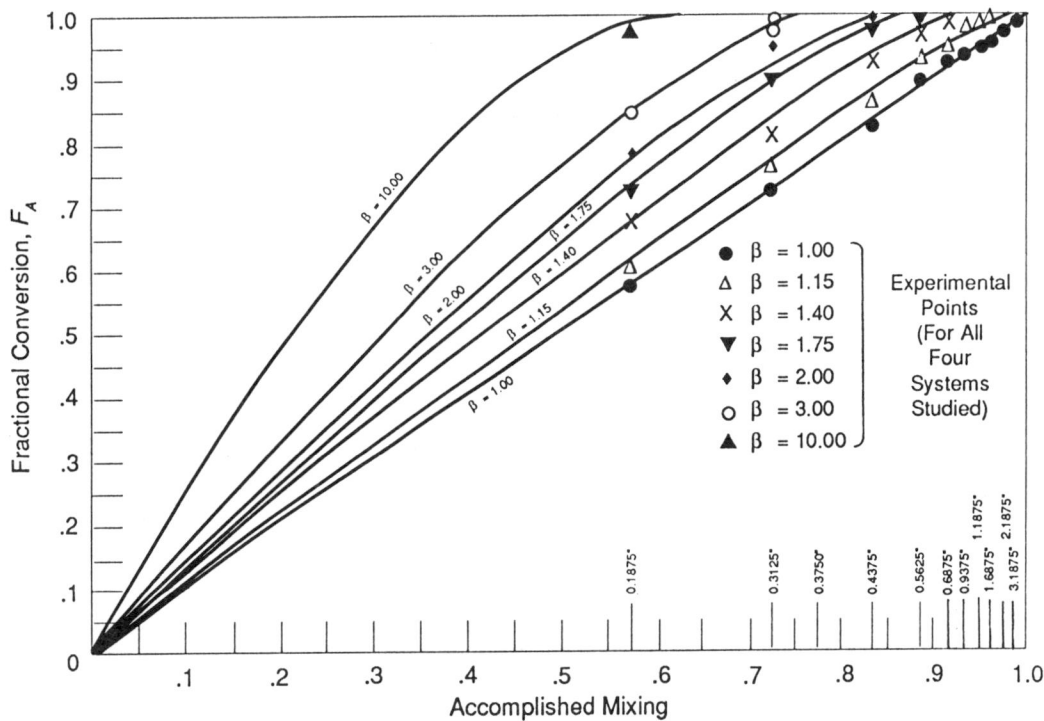

NOTE: The small-sized numbers inside the horizontal axis represent distance downstream.

Source: Vassilatos, G. & Toor, H.L. 1965. Second-Order Chemical Reactions in a Nonhomogeneous Turbulent Fluid. AIChE Jour., 11:4:666. *Reproduced by permission of the American Institute of Chemical Engineers.*

Figure 3-5 Fractional conversion versus accomplished mixing, very rapid reactions.

radio transmitter, is detected by antennae in the vessel. This method was used in research and production-sized vessels. One problem with this modeling method is that circulation time measurements must be made on the vessel in question or on many other vessel configurations and sizes to obtain working correlations, before it can be applied. A *predictive* modeling method was not achieved. Another problem is that this method cannot deal with very rapid reaction/mixing interactions in which very small-scale mixing is important.

More recently, the same research group has developed a computer-based model to solve the three-dimensional flow and turbulence equations for baffled, stirred vessels. The model uses a finite-difference method that incorporates the $k - \varepsilon$ turbulence model. It also successfully predicts single-phase velocities and turbulence as compared with measurements taken using laser velocimetry. It has been used successfully to predict the final yield of a two-reaction competitive–consecutive scheme, which is controlled by the large-scale convective mixing patterns (Middleton, Pierce, and Lynch 1984).

The above model has been applied where local mixedness could be assumed complete (zero local segregation). The program accounted for local feed and circulation effects (blending) where high reactant concentration plumes were produced. Local small-scale mixing effects were not in the program.

Figure 3-6 illustrates this application for one such industrial reaction in which local concentration levels (but not local mixedness) control the yield of the desired intermediate product in a competitive–consecutive reaction (A + B → R; R + B → S). The reactor was operated in the semibatch mode with reactant B suddenly injected over a very short time. Tables 3-2 and 3-3 show the final values of concentrations and yields ($X = 2C_s/(C_R + 2C_s)$) for two vessel sizes and three impeller sizes in each case. The impeller in each case was a standardized disk turbine of diameter equal to one-third the tank diameter. Figure 3-7 shows that yield is not constant with power per unit volume ($\sim N^3 D^2$), tip speed ($\sim ND$), or blending time ($t_M = V/ND^3$). The good correspondence with experimental values shown in Tables 3-2 and 3-3 illustrates the power of using realistic velocity patterns in scaling up such processes. Figure 3-8 shows typical velocity patterns computed by the fluid dynamics program.

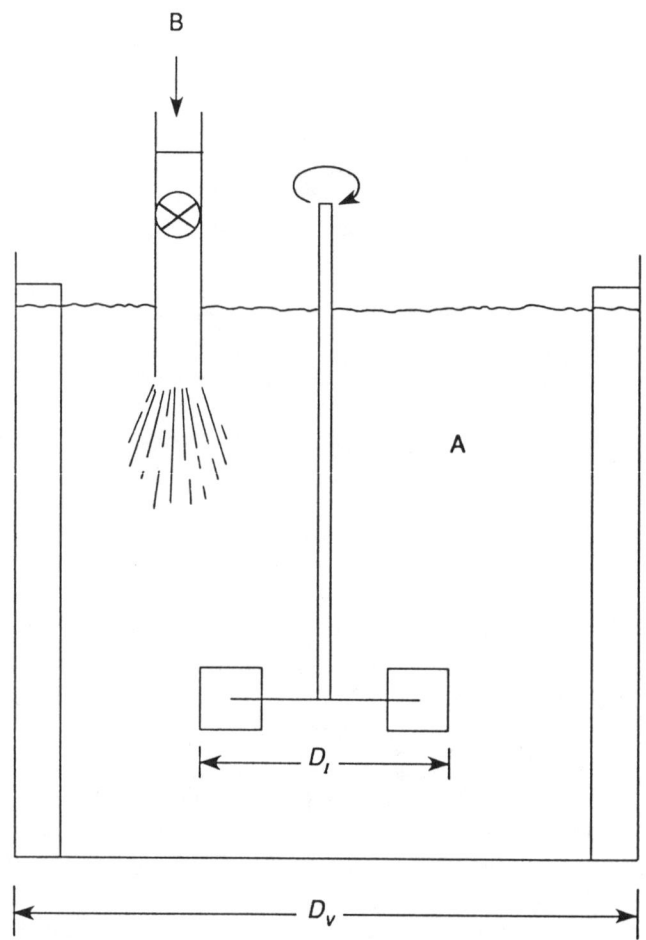

Source: Middleton, J.C.; Pierce, F.; & Lynch, P.M. 1984. Computations of Flow Fields and Complex Reaction Yield in Turbulent Reactors and Comparisons With Experimental Data. AIChE Symp. Series, 87:239. Reproduced by permission of the American Institute of Chemical Engineers.

Figure 3-6 Vessel configuration.

Table 3-2 Impeller Rotation Rate Effects on a Stirred Reactor: Results: 30-L Vessel

Speed, rpm	Final Concentrations (molar × 10^3)					
	Predicted			Measured		
	R	S	X	R	S	X
600	4.5	0.24	0.095	4.6	0.22	0.088
100	3.6	0.68	0.27	4.0	0.56	0.22
				3.7	0.62	0.25
50	3.2	0.92	0.37	3.2	1.0	0.39
				3.1	1.1	0.42

Source: Middleton, J.C.; Pierce, F.; & Lynch, P.M. 1984. Computations of Flow Fields and Complex Reaction Yield in Turbulent Reactors and Comparisons With Experimental Data. AIChE Symp. Series, 87:239. Reproduced by permission of the American Institute of Chemical Engineers.

NOTE: $X = \dfrac{2[S]}{[R] + 2[S]}$ for $A + B \rightarrow R; R + B \rightarrow S$

Table 3-3 Impeller Rotation Rate Effects on a Stirred Reactor: Results: 600-L Vessel

Speed, rpm	Final Concentrations (molar × 10^3)					
	Predicted			Measured		
	R	S	X	R	S	X
200	3.4	0.8	0.32	3.5	0.72	0.29
				3.8	0.53	0.22
				3.7	0.63	0.25
150	3.2	0.9	0.36	3.3	0.85	0.34
				3.3	0.77	0.32
				3.5	0.73	0.37
100	2.9	1.0	0.41	3.4	0.87	0.34
				3.0	1.00	0.40
				3.4	0.76	0.32
				2.8	1.05	0.43
				3.0	0.98	0.40

Source: Middleton, J.C.; Pierce, F.; & Lynch, P.M. 1984. Computations of Flow Fields and Complex Reaction Yield in Turbulent Reactors and Comparisons With Experimental Data. AIChE Symp. Series, 87:239. Reproduced by permission of the American Institute of Chemical Engineers.

NOTE: $X = \dfrac{2[S]}{[R] + 2[S]}$ for $A + B \rightarrow R; R + B \rightarrow S$

Higher-Order Moment Closure

The balance equations for the components and segregation always result in correlations (moments of fluctuating quantities; for example, $\overline{c_1 c_2}$) of order two or higher. The *higher-order moment closure* is a modeling approach that has been used for hydrodynamic equations (momentum and turbulence). In this approach, the highest-order moments are expressed as functions of the lower-order terms, for example, $\overline{c^2 c_2} = f(\overline{c_1 c_2}, \overline{c_1^2}, \overline{c_2^2})$.

Borghi (1975), among others, has applied this closure to mixing with chemical reaction. His complete model for a mixing second-order reaction required 13 balance equations. That seems to be an impractically large problem, but much larger ones will soon be practical on the new generation of computers that are now available.

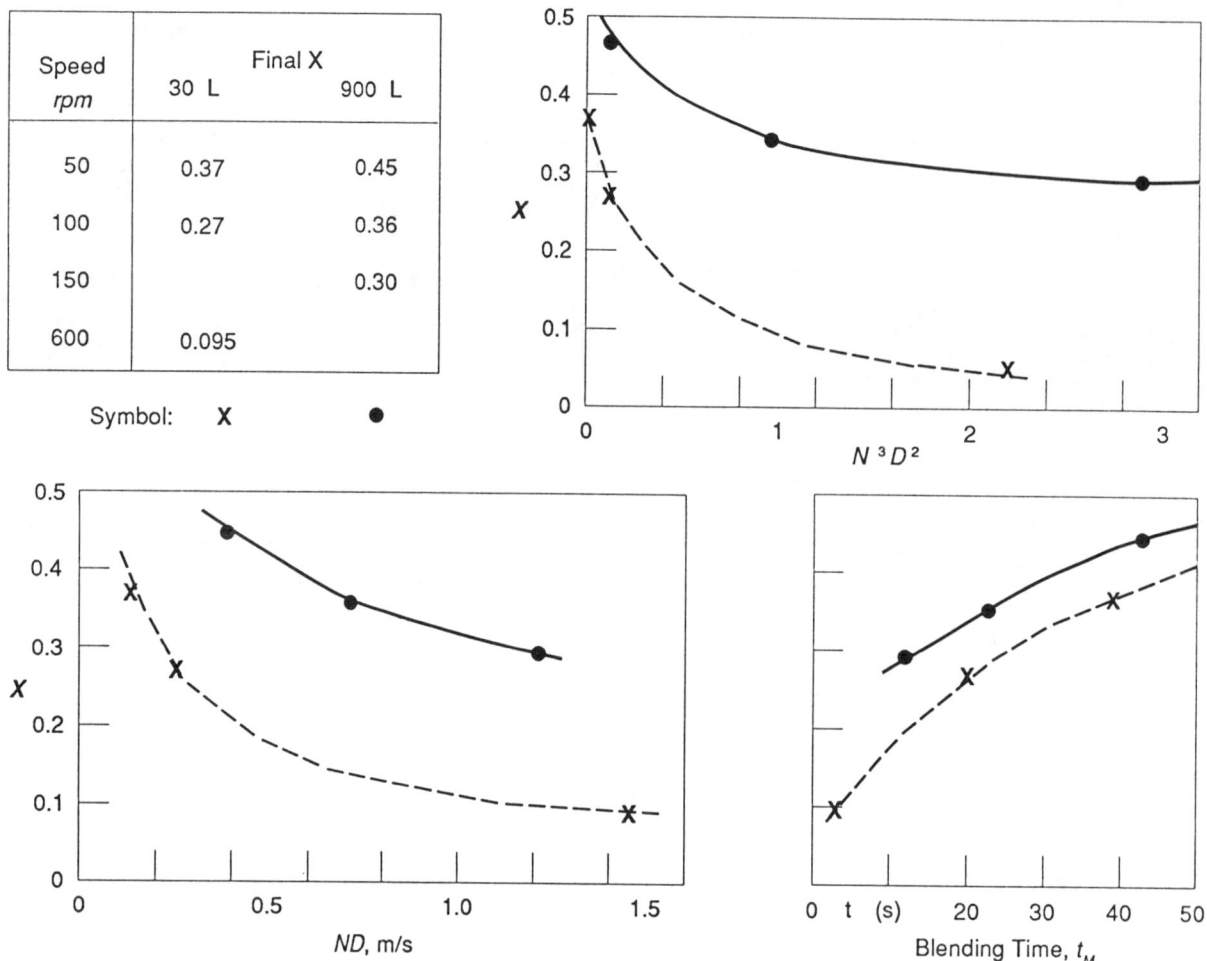

Source: Middleton, J.C.; Pierce, F.; & Lynch, P.M. 1984. Computations of Flow Fields and Complex Reaction Yield in Turbulent Reactors and Comparisons With Experimental Data. AIChE Symp. Series, 87:239. Reproduced by permission of the American Institute of Chemical Engineers.

Figure 3-7 Scale effects on yield in a stirred vessel.

Probability Distribution Functions

A way to simulate the distribution of values of the higher-order moments is by use of proposed distribution functions for the probability of concentration levels. The proposed functions then reduce the equations to a solvable set. Among those who have made use of *probability distribution functions* (pdf) of concentration to generate higher moments are Lockwood and Naquib (1975). A function such as $\overline{c_1 c_2} = \int_0^1 c_1(\phi) c_2(\phi) P(\phi) d\phi$ was used to relate $\overline{c_1 c_2}$ to distributions of c_1 and c_2 values. They used a clipped-normal pdf to simulate combustion reactions during diffusive mixing. Figures 3-9 and 3-10 compare a typical "real" distribution with the clipped normal. Their model-experiment comparisons showed good correspondence of reactant concentration profiles. The flexibility of their closure is limited, however,

TURBULENCE AND MIXING 105

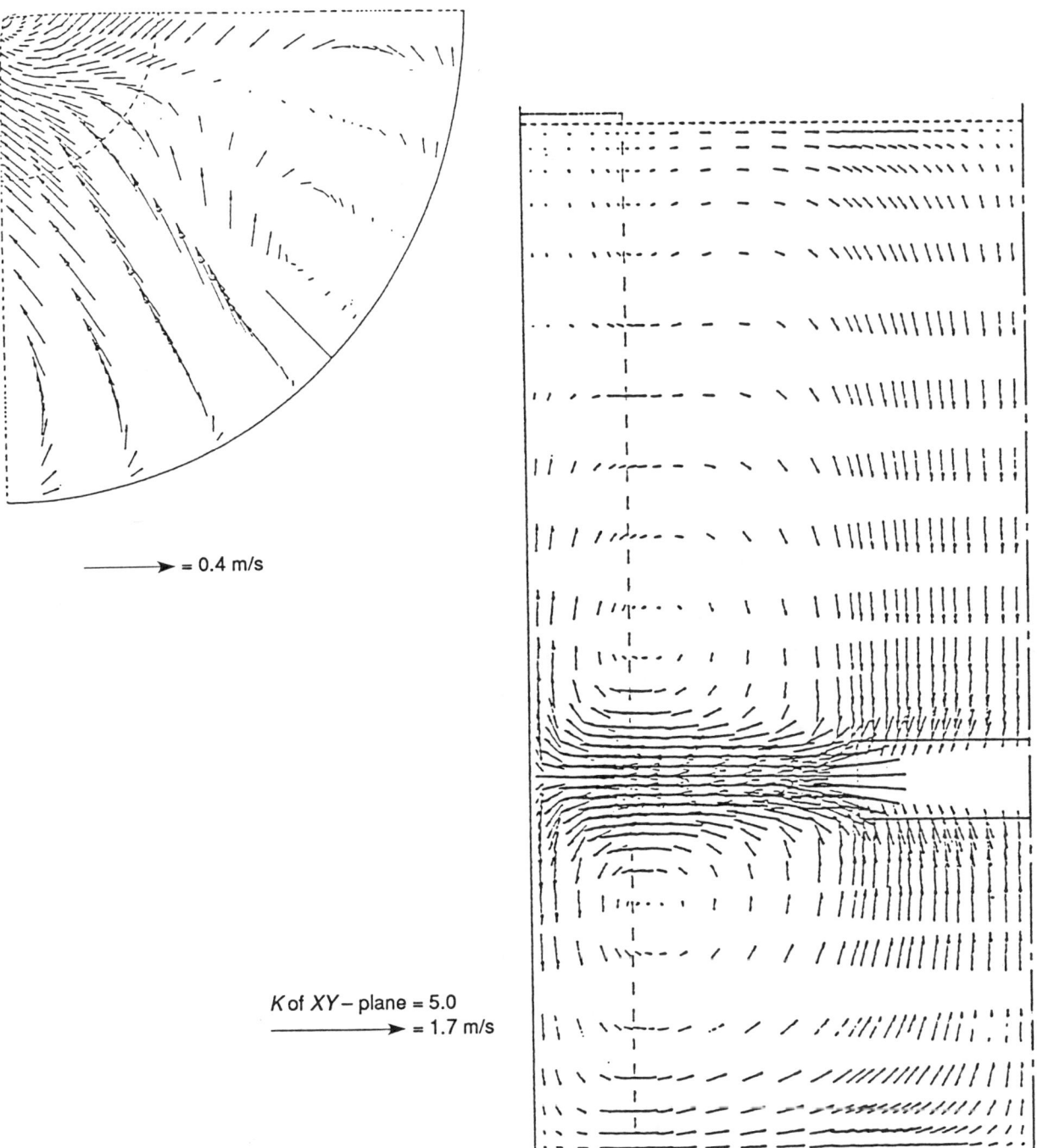

Source: Middleton, J.C.; Pierce, F.; & Lynch, P.M. 1984. Computations of Flow Fields and Complex Reaction Yield in Turbulent Reactors and Comparisons With Experimental Data. *AIChE Symp. Series,* 87:239. Reproduced by permission of the American Institute of Chemical Engineers.

Figure 3-8 Typical computed velocity vectors.

because the distribution does not adjust for changing reaction conversion and mixedness.

A simplification of the pdf approach is the *spiked pdf*, which has been used by Donaldson (1975). He called his approach the "most typical" eddy model. The spiked pdf is designed to simulate the real pdf of the moment in question (see Figure 3-11). A quasi-equilibrium set of relative spike heights is assumed in the model. Donaldson (1975) and Edelman and Harsha (1978) have tested this approach against experimental data with good success.

The simplest version of the pdf approach is the *interdiffusion* model (Patterson, Lee, and Calvin 1977; Patterson 1975a), previously called the hydrodynamic mixing

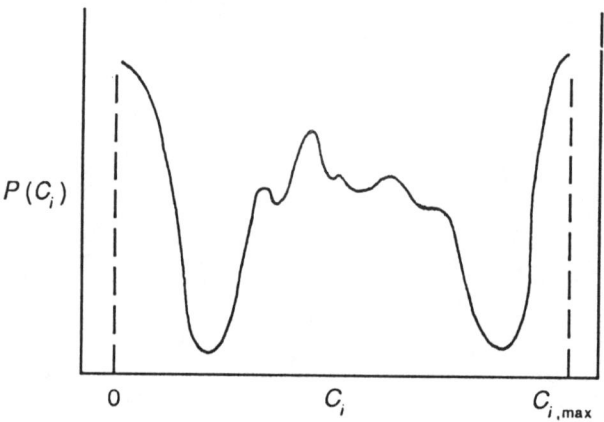

Source: Patterson, G.K. 1981. Application of Turbulence Fundamentals to Reactor Modeling and Scaleup. Chem. Engrg. Comm., *8:25.*

Figure 3-9 Typical real distribution function for component-i in a mixing system.

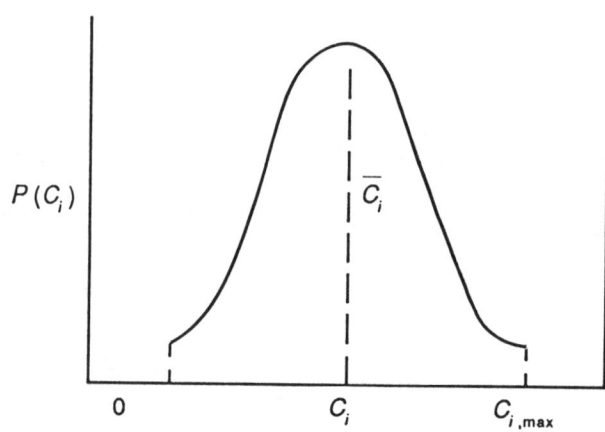

Source: Patterson, G.K. 1981. Application of Turbulence Fundamentals to Reactor Modeling and Scaleup. Chem. Engrg. Comm., *8:25.*

Figure 3-10 Clipped-normal distribution function for component-i.

model (HDM). Three spikes, as shown in Figure 3-12, are involved: one spike each appears at the maximum and minimum values of the concentration modeled and one at the final mixed value. The maximum and minimum spikes shrink and the mixed-value spike grows as mixing occurs.

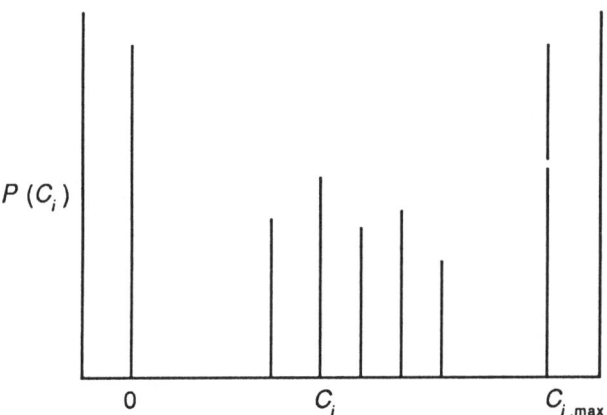

Source: Patterson, G.K. 1981. Application of Turbulence Fundamentals to Reactor Modeling and Scaleup. Chem. Engrg. Comm., 8:25.

Figure 3-11 Spiked distribution function for component-i.

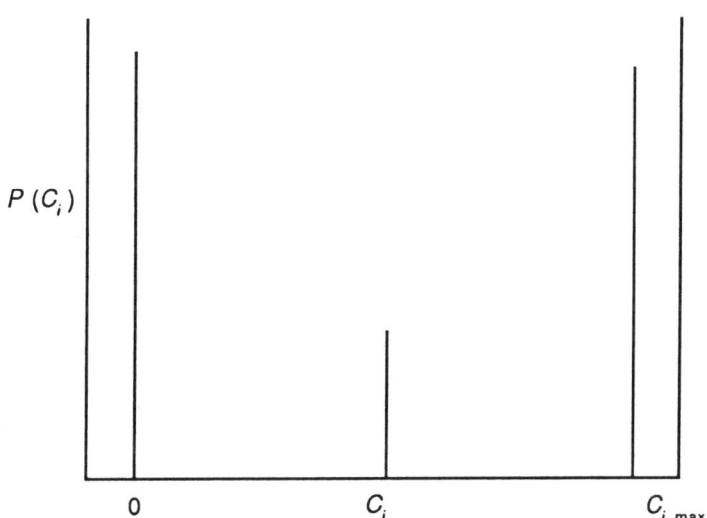

Source: Patterson, G.K. 1981. Application of Turbulence Fundamentals to Reactor Modeling and Scaleup. Chem. Engrg. Comm., 8:25.

Figure 3-12 Spiked distribution function for component-i with only one intermediate concentration.

The resulting equation for $\overline{c_1 c_2}$ is as follows:

$$\overline{c_1 c_2} = - \overline{c_1^2} (1 - \gamma) / [(\overline{C}_2/\overline{C}_1)(1 + \gamma)] \qquad \text{(Eq 3-38)}$$

Where

$\gamma = (\overline{C_1^2} - \overline{c_1^2})/(\overline{C_1^2} + \overline{c_1^2})$ (symmetry is preserved if C_1 and C_2 are switched).

Figure 3-13 shows the concentration profiles simulated by this model. Substitution of γ into Eq 3-38 results in $\overline{c_1 c_2} = - (\overline{c_1^2})^2/(\overline{C}_1 \overline{C}_2)$.

The results of the interdiffusion model have been extensively tested against experimental data for turbulent tubular flow mixing (Patterson 1973) (Figure 3-14), annular jet mixing (Patterson, Lee, and Calvin 1977) (Figure 3-15), and stirred tank mixing (Patterson 1982; Patterson, Bockelman, and Quigley 1982) (Figures 3-16, 3-17, and 3-18). The value of this model is that it is simple enough to solve with little computational difficulty, but it retains some features of a distribution function modeling approach.

An extension of the interdiffusion model, which was conceived for a single second-order reaction, is the *paired-interaction* model (Patterson 1980, 1981). It is based on the assumption that the higher moments between each pair of reactants are unaffected by the other moments; in other words, moments of more than two concentrations are zero. This model is aimed at multiple reaction cases in which relative yields of products are important. The equations for the closure model are

$$\text{DISS}_{\overline{C}_1} = \sum_k \sum_j \overline{K}_{ijk} (\overline{C}_k \overline{C}_j - \overline{c_k^2}\, \overline{c_j^2}/\overline{C}_k \overline{C}_j) \qquad \text{(Eq 3-39)}$$

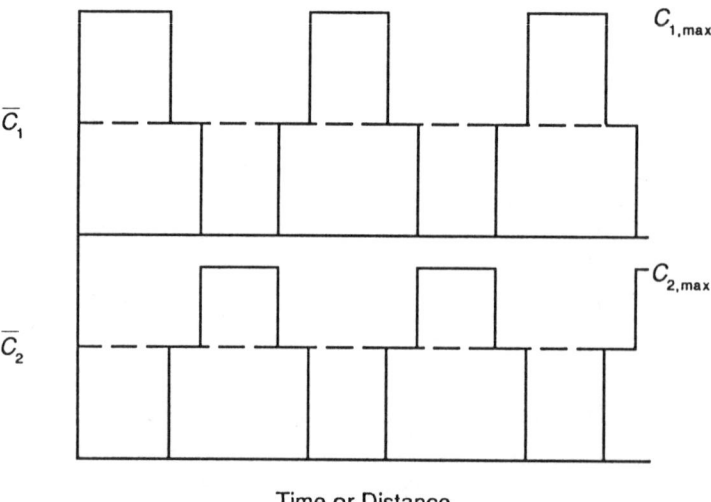

Source: Patterson, G.K. 1981. Application of Turbulence Fundamentals to Reactor Modeling and Scaleup. Chem. Engrg. Comm., 8:25.

Figure 3-13 Concentration profile corresponding to the distribution in Figure 3-11—the interdiffusion or HDM model.

$$\text{DISS}_{\overline{c_i^2}} = \sum_k \sum_j \overline{K}_{ijk} \, (\overline{c_j^2} \, \overline{C}_k - \overline{c_k^2} \, \overline{c_j^2}/\overline{C}_k) + \text{DISS}_{\text{mix}} \qquad \text{(Eq 3-40)}$$

A production term for $\overline{c_i^2}$ may also arise from large-scale diffusion. In the above, $\overline{K}_{ijk}$ are positive or negative rate constants, depending on whether C_i is a product or a reactant. The $\overline{K}_{ijk}$ are for Reynolds average temperature and do not account for $\overline{c_i T}$ correlations. So far, only cursory checks between literature data and model results have been achieved. Figure 3-19 shows a comparison of paired-interaction model predictions for mixing effects on series–parallel reactions of the form $A + B \xrightarrow{k_1} R, \, R + B \xrightarrow{k_2} S$ (Patterson 1980).

The yield of R (as moles R formed per mole B reacted) is plotted versus Damköhler number referred to component A. The other variables are relative

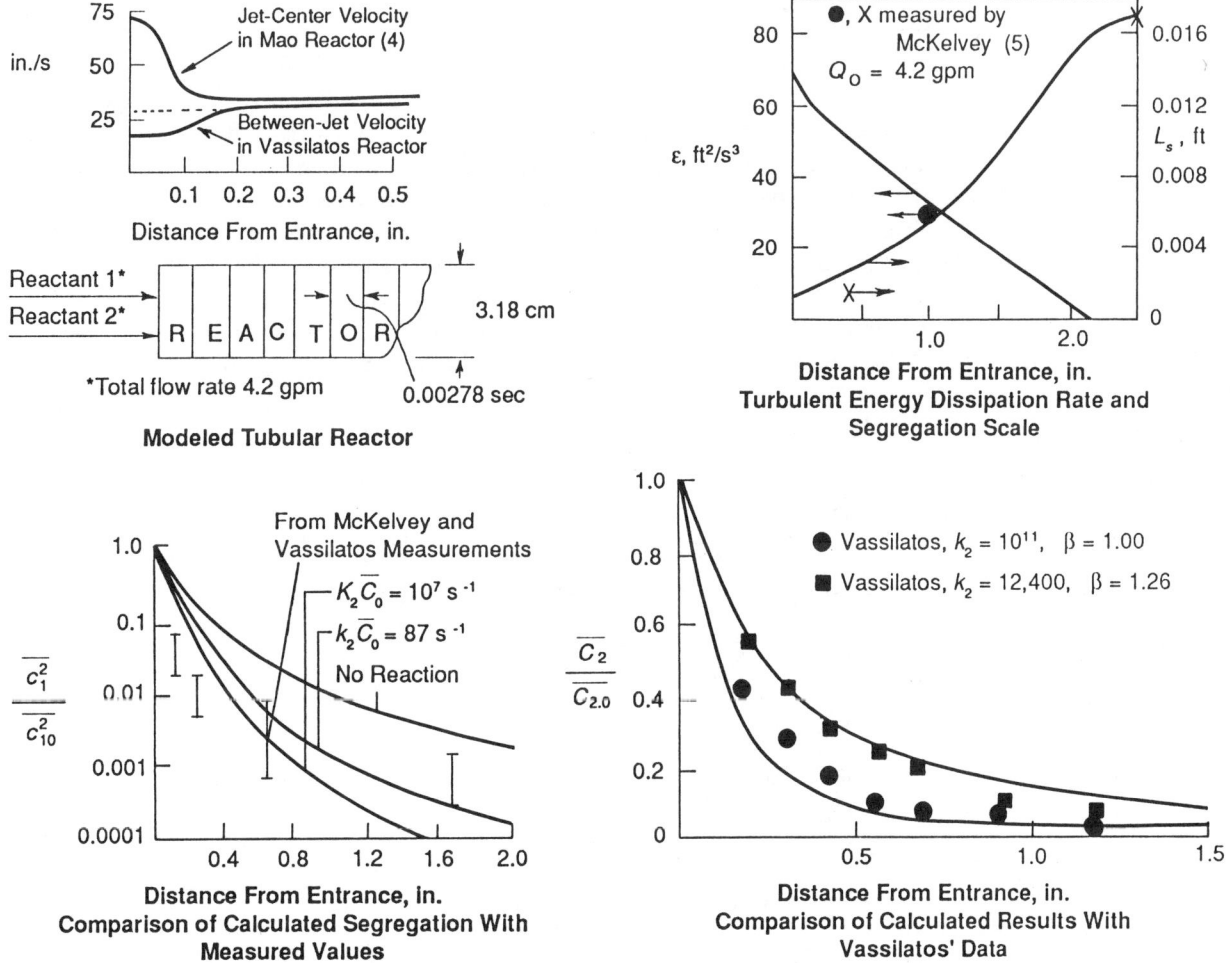

Source: Patterson, G.K. 1983. Turbulent Mixing and Its Measurement. In Handbook of Fluids in Motion, ed. by N.P Cheremisinoff and R. Gupta. Ann Arbor Science Publishers, Ann Arbor, Mich.

Figure 3-14 Comparison of results from the HDM model and experimental data (Vassilatos and Toor 1965) for second-order chemical reaction in a turbulently mixed tubular reactor.

pumping rate (ND^3/Q_O), feed molar ratio (C_{BO}/C_{AO}), and kinetic rate constant ratio (K_{R_1}/K_{R_2}). Scattered about the diagram are experimental yield results from various sources, most with some degree of experimental condition mismatch to the modeled system (baffled tank, disk turbine, continuous feed). The experimental results are, in general, in agreement with the model, but much more and better-suited data are needed. Recently Brodkey and Lewalle (1985) obtained yield curves of about the same shape as those in Figure 3-19 using a novel multi-environment approach (see Figure 3-20).

APPROXIMATE MODELS FOR LOCAL MIXING

In order to avoid full modeling of the mixing–reacting system, many chemical engineers working in this field have proposed and tested more approximate models for mixed reactors. "More approximate" models are models that were not, in general, intended for geometrically distributed modeling leading to solutions that give distributions of all relevant properties.

Coalescence–Dispersion Model

One of the first models used to approximate local mixing effects on chemical reactions was the random *coalescence–dispersion* (c–d) model (Curl 1963; Evangelista, Shinnar, and Katz 1969; Spielman and Levenspiel 1965; Kattan and Adler 1967). In the c–d model, mixing and reseparation of nearby fluid elements are used to simulate the mixing process. Most early applications applied the model as if mixing degree were uniform in the entire vessel, even when flow from section to section was simulated, as for a turbulent tubular reactor (Rao and Dunn 1970; Rao and Edwards 1971). In order to simulate effects of varying mixing rates from place to place within the mixed reactor, Canon et al. (1977) varied the c–d rate proportional to the mixing rate. A relationship between c–d rate, I_r, and turbulence energy dissipation rate, ε,

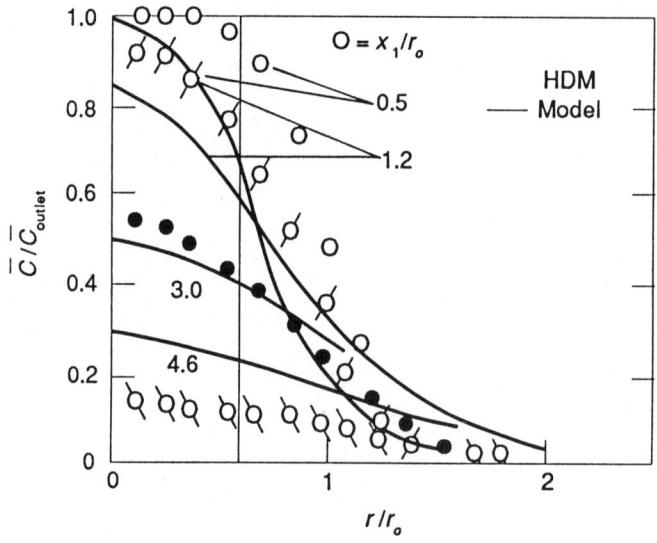

Figure 3-15 Concentration profiles in a coaxial jet without a chemical reaction.

and turbulence energy, k_t, was found. This relationship gave good quantitative comparisons between predicted and experimental reaction conversion distributions within the reactor (see Figure 3-21). The relationship used was

$$I_r = 1333 \, (\varepsilon/k_t)(\bar{\tau}/N_e) \qquad \text{(Eq 3-41)}$$

Where:

$\bar{\tau}$ = fluid residence time in a reactor section
N_e = the number of fluid elements in that section

Equation 3-41 is based on use of the Corrsin (1964) relationship for isotropic mixing, which is a weak function of molecular diffusivity. Evangelista and co-workers (1969) proposed a relationship $I_r = 24D\bar{t}/\lambda_s^2$, where λ_s is a fluid element thickness. This expression is a strong function of diffusivity and would, therefore, apply only to very low Schmidt number cases. Both Pratt (1976) and Patterson

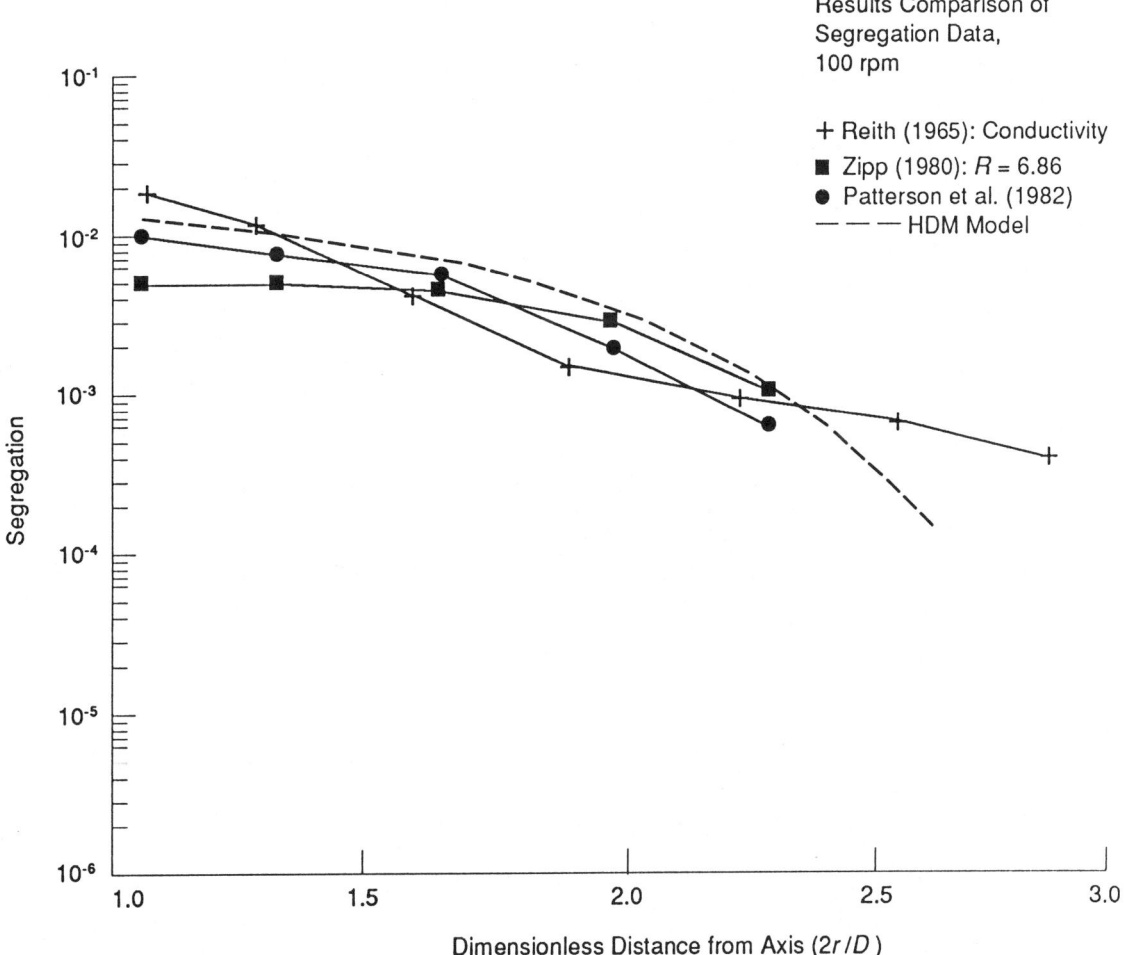

Figure 3-16 Comparison of segregation measurement results in the impeller stream.

(1977) have imbedded the c–d procedure within geometrically distributed models to model segregation and reaction conversion. The model yields results remarkably close to experiment, even though the fine details of the mixing process are ignored. Figure 3-22 shows comparisons with data in a plug-flow mixed reactor, and Figure 3-23 shows comparisons with yield data in a stirred tank.

Local Diffusion Model

Another approximate type of model, which includes formulations by several different investigators, is the *local diffusion* model. Included are the exchange-with-the-mean (IEM) model of David and Villermaux (1975) and others; the diffusing strand model of Truong and Methot (1976); the related lamellar mixing model of Ottino, Ranz,

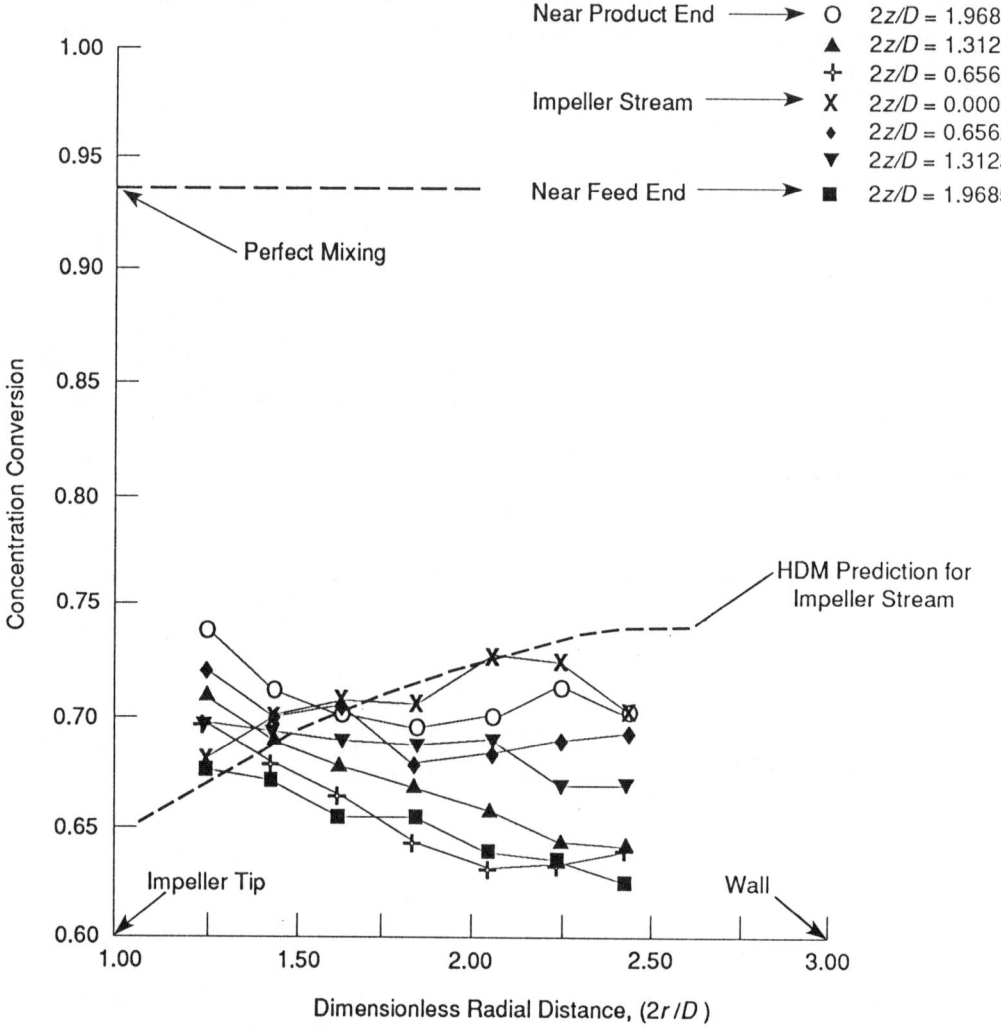

Source: Bockelman, W.D. 1983. Measurements of Conversion in a Stirred Tank Reactor by Fluorescence. Master's thesis, University of Missouri–Rolla, Rolla, Mo.

Figure 3-17 Measured and predicted (HDM) profiles for a feed concentration ratio of 2.5. $Z/R(0) = 0.0$ is the impeller stream. $N = 30$ rpm.

TURBULENCE AND MIXING   113

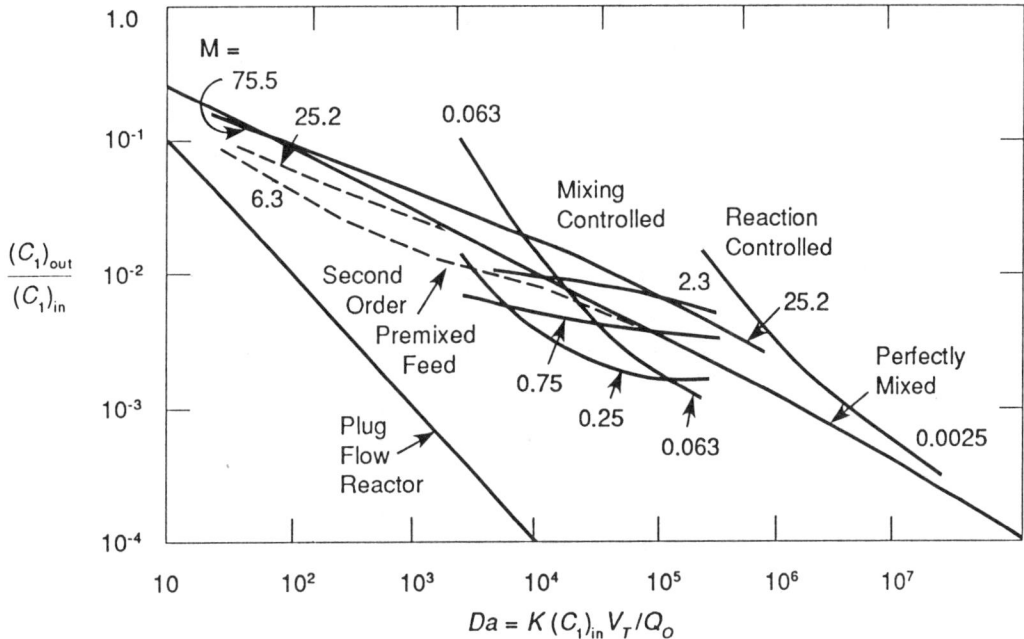

NOTE: Turbine centered in tank, $D = T/3$, feed at tank end near axis. M (mixing intensity) $= (\varepsilon/L^2)^{1/3}/K(C_1)_{in}$.

Source: Bockelman, W.D. 1983. *Measurements of Conversion in a Stirred Tank Reactor by Fluorescence.* Master's thesis, University of Missouri–Rolla, Rolla, Mo.

Figure 3-18 Conversion in the effluent for second-order reactions in a Rushton turbine stirred tank.

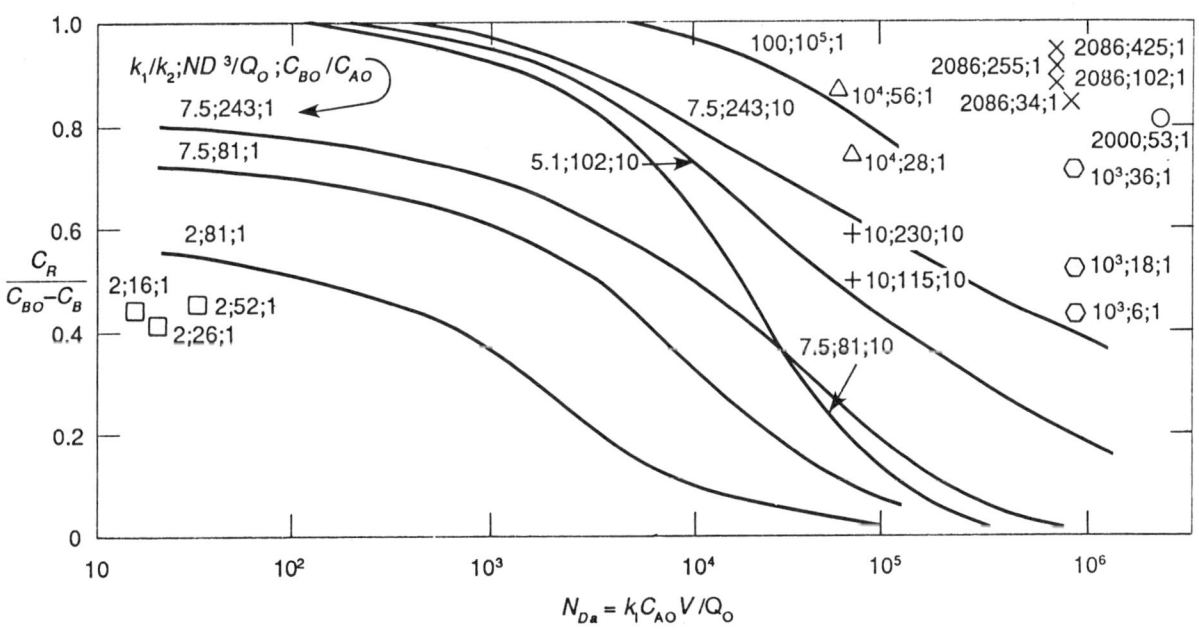

NOTE: The "paired-interaction" version of the HDM model was used. Feed was away from impeller near tank top.

Data: △ Bourne, Moergeli, and Rys (1977); □ Truong and Methot (1976); + Paul and Treybal (1971); X Bourne et al. (1981); O Angst, Bourne, and Kozicki (1979); ○ Nabholz, Ott, and Rys (1977).

Figure 3-19 Prediction of the yield of R in the reactions $A + B \xrightarrow{k_1} R$, $R + B \xrightarrow{k_2} S$ in a Rushton turbine stirred tank.

and Mocosko (1979); and the extensional flow model of Palpepu, Adler, and Edwards (1981); the diffusing slab model of Mao and Toor (1970); and the spherical eddy model of Bourne, Moergeli, and Rys (1977), which was developed further by Belevi, Bourne, and Rys (1981). For the most part, these models treat the molecular diffusion stage of mixing based on concentration gradients generated by the turbulent or otherwise mixing flow. Typically, the size of a quasi-equilibrium unmixed strand, lamellae, or sphere is determined based upon comparison with experiment or considerations of the hydrodynamics. In most cases, the models are not applied on a geometrically distributed basis, although in principal all of them could be. Ottino (1981) has applied the *lamellar model* to mixed reaction in a turbulent tubular reactor on a one-dimensional distributed basis, obtaining conversion profiles as a function of reactor length (see Figure 3-24).

Some controversy has developed regarding the difference in approach of component mixing models. These models are based on macroscale or Taylor microscale as the characteristic length scale for mixing rate and on the Kolmogoroff or smaller scales where molecular diffusion is controlling. The Corrsin (1964) isotropic mixer models mixing rate using the segregation macroscale as a reference upon which to base the rate of scale reduction and ultimate mixing. Although it is generally recognized that the smallest scales are responsible for the final molecular diffusion leading to complete mixing, the cascade theory for the rate of reduction of scales seems

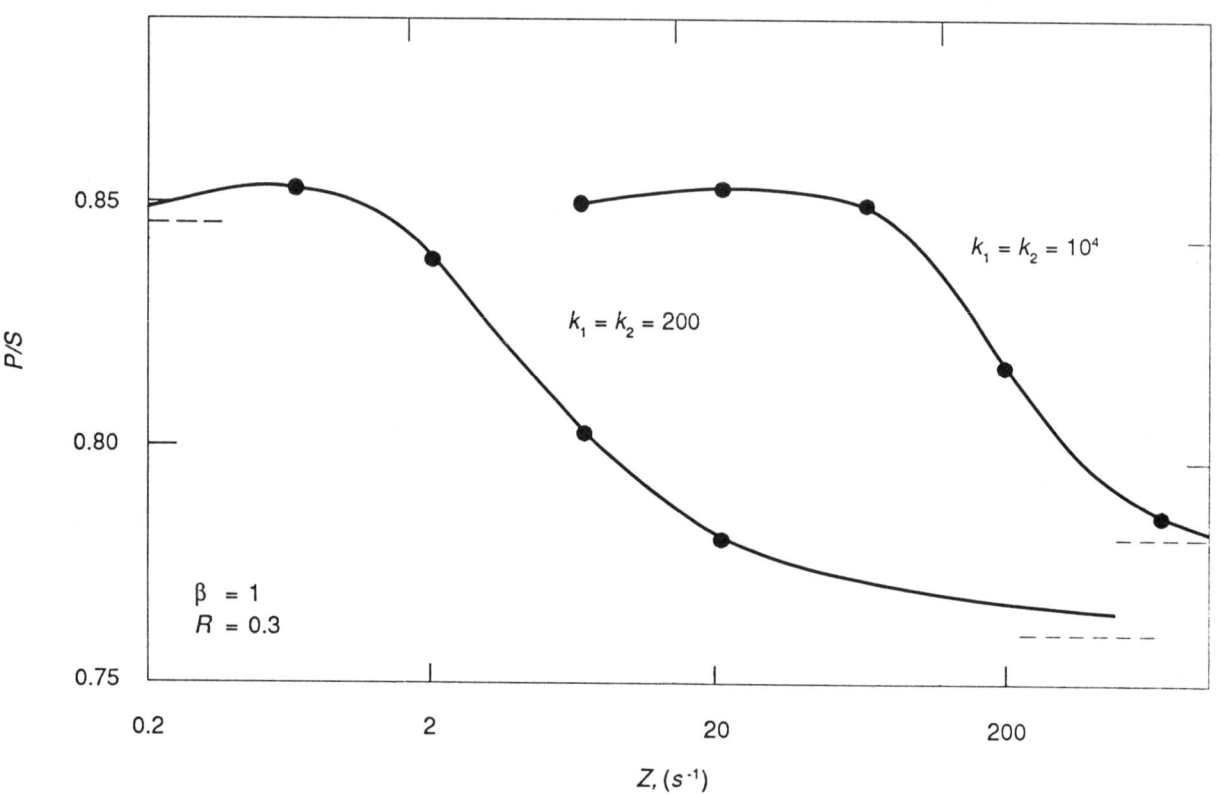

NOTE: P/S is the primary-to-secondary-product ratio and Z is a mixing rate parameter.

Source: Brodkey, R.S. & Lewalle, J. 1985. *Reactor Selectivity Based on First-Order Closures of the Turbulent Concentration Equations.* AIChE Jour., 31:111. *Reproduced by permission of the American Institute of Chemical Engineers.*

Figure 3-20 Selectivity as a function of mixing intensity from the Brodkey and Lewalle model.

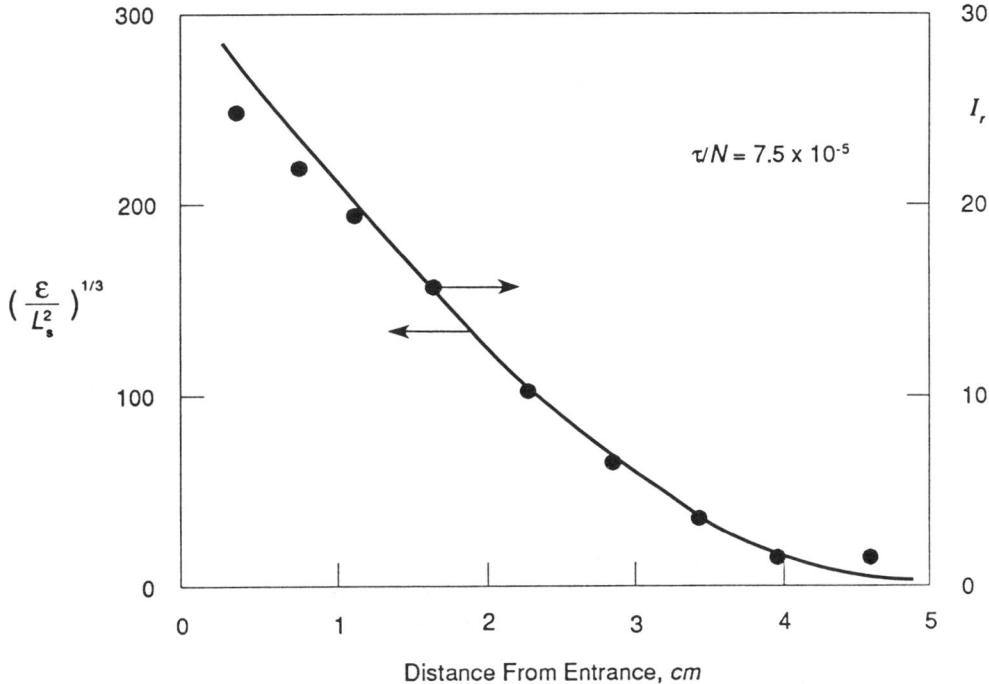

Source: Canon, R.M. et al. 1977. The Turbulence Level Significance of the Coalescence–Dispersion Interaction Parameters. Chem. Engrg. Sci., 32:1349.

Figure 3-21 Plots of hydrodynamic mixing parameter $(\varepsilon/L_s^2)^{1/3}$ and the coalescence–dispersion interaction parameter I versus distance from the reactor entrance.

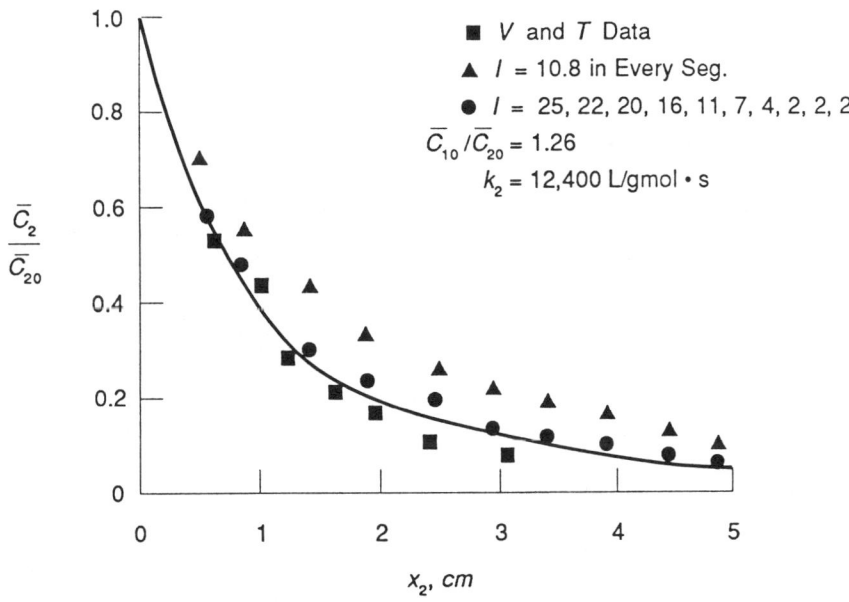

Source: Vassilatos, G. & Toor, H.L. 1965. Second-Order Chemical Reactions in a Non-homogeneous Turbulent Fluid. AIChE Jour., 11:666.; and Canon, R.M. et al. 1977. The Turbulence Level Significance of the Coalescence–Dispersion Interaction Parameters. Chem. Engrg. Sci., 32:1349.

Figure 3-22 Comparisons of coalescence–dispersion results with various coalescence–dispersion rates (I) with experimental data (Vassilatos and Toor 1965) in a tubular reactor.

to hold, making knowledge of the scale at the upper end of the spectrum adequate for modeling. Use of either reference scale would be adequate, but the large scale is much more convenient because it is very closely related to the scale given by the relationship $\ell = k\ell^{3/2}/\varepsilon$, which is involved in one k–ε turbulence model.

It seems clear that the models that will develop the most overall usefulness are those that are generalized to the geometrically distributed basis. These models may be used for any mixing–reacting geometry with proper boundary conditions. Any such model may still be used on a zero-dimensional (lumped) basis as an approximate model for spatially averaged processes.

MODELING COMPLEX CHEMICAL REACTIONS

The industrial mixed reactor problem of real significance is not the determination of the conversion of a single chemical reaction, but the determination of yield of a particular product when multiple reactions occur. For complex chemistry in which each chemical reaction is slow enough that the mixing rate should affect it only slightly, the interactions involved make mixing rate an important parameter. This was illustrated earlier by Paul and Treybal (1971) (Figure 3-23) and more recently by Bourne and Rohani (1983) (Figure 3-25), David and Villermaux (1975), and Zoulalian and Villermaux (1974) (Figure 3-26).

Most of the above-mentioned models may be used to attempt to predict mixing effects on complex reactions. In only a few cases have the predictions been compared with experiments for coalescence–dispersion (Canon et al. 1977), for the paired-interaction closure (Patterson 1981), for extensional flow modeling (Palpepu, Adler, and Edwards 1981), for the extension of Toor's hypothesis by McKelvey et al. (1975),

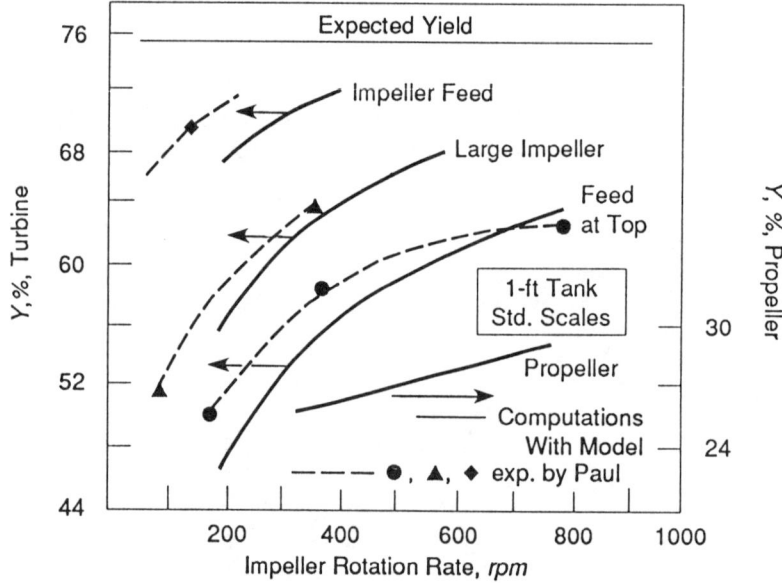

Source: Paul, E.L. & Treybal, R.E. 1971. *Mixing and Product Distribution for a Liquid-Phase, Second-Order, Competitive–Consecutive Reaction.* AIChE Jour., 17:718; and Canon, R.M. et al. 1977. *The Turbulence Level Significance of the Coalescence–Dispersion Interaction Parameters.* Chem. Engrg. Sci., 32:1349.

Figure 3-23 Semi-batch reactor—experimental (Paul and Treybal 1971) and model results (Canon et al. 1977).

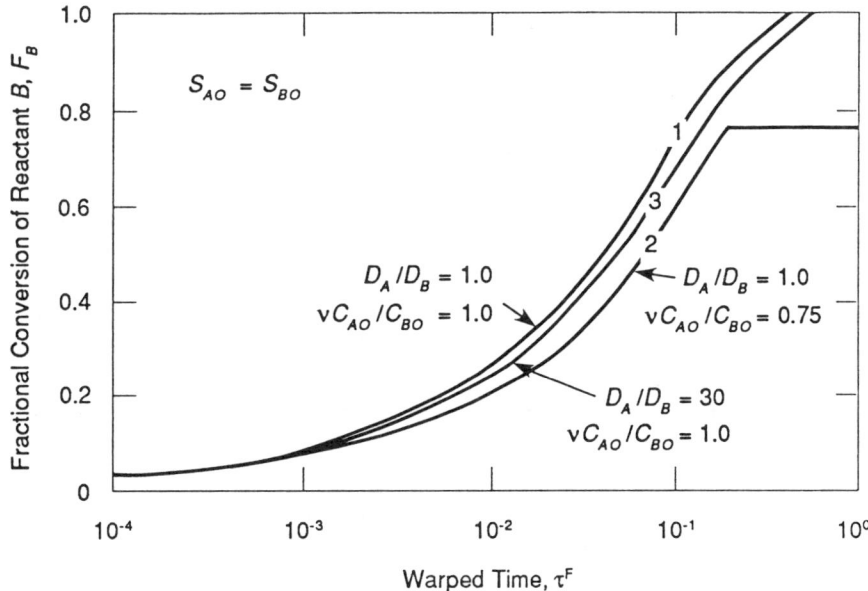

A. Extent of reaction F_B vs. τ^F (warped time) for equal striation thicknesses showing effects of different diffusion coefficients and stoichiometry (adapted from Fisher 1974). Fractional conversion is measured with respect to the excess reactant at S_x-scales.

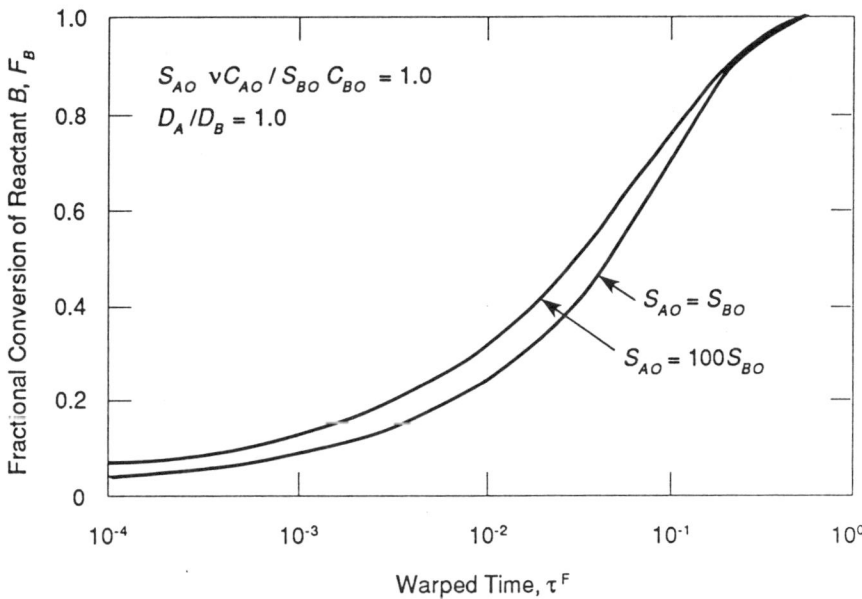

B. Extent of reaction F_A vs. τ^F (warped time) for equal diffusivities and stoichiometric amounts showing effect of unequal laminar thicknesses (adapted from Fisher, 1974).

Source: Ottino, J.M. 1981. Efficiency of Mixing From Data on Fast Reactions in Multi-Jet Reactors and Stirred Tanks. AIChE Jour., 27:184. Reproduced by permission of the American Institute of Chemical Engineers.

Figure 3-24 Effect of mixing computed using the lamellar model.

for the spherical eddy model (Bourne et al. 1981), and for a four-environment age-mixing model (Lagrangian basis) with an exchange rate parameter between the separate leaving environments for the two feed streams (Bourne et al. 1981).

Among those models, only the c–d and paired-interaction (see Eq 3-38 and 3-39 and Figure 3-25) models have been developed to the point that geometrically distributed results are obtained. The models may be solved with hydrodynamic model equations so that parameters based directly on experiment for the mixing rate are unnecessary. The c–d method involves a parameter, the constant 1333 of Eq 3-41, but it is relatively insensitive to geometry. The value of that constant might prove to be dependent on the Schmidt number. The paired-interaction method requires no parameters. Both methods are based on the distributed modeling approaches developed for combustion.

The combustion models have not been included as examples of models confirmed by experiment, because, in all cases known to the author, when complex chemistry was involved, the reaction kinetics were applied as if no local mixing effects were important, i.e., only large-scale turbulent diffusion of reactants was considered.

EFFECT OF AVAILABLE EXPERIMENTAL TECHNIQUES

Experimental techniques available for model confirmation have always exercised a discernible effect on the models that were most useful in mixed-reactor modeling. Methods for measuring effluent concentrations with fast response to concentration change made application of residence time distribution analysis and the associated age-mixing models possible. The advent of hot-wire and hot-film anemometry and laser–Doppler anemometry made detailed measurement of velocities and turbulence within a mixing vessel or other mixed reactor possible. Those techniques were embraced wholeheartedly by aerodynamics and combustion engineers. More recently, methods for measuring concentration and concentration fluctuations within mixed reactors are being developed, including conductivity (in use for some time but

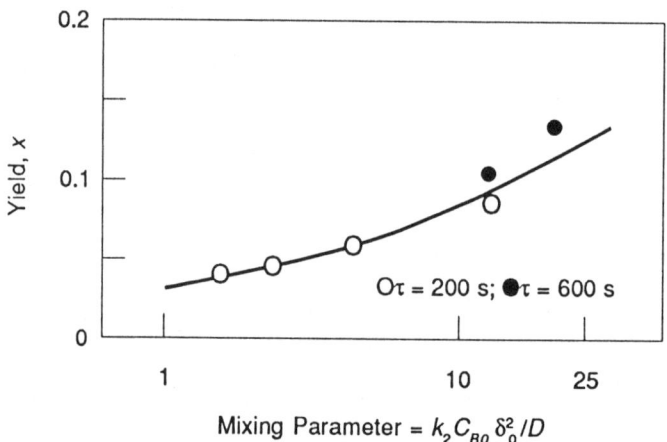

Reprinted with permission from Chemical Engineering Science, *Volume 38, J.R. Bourne and S. Rohani, "Mixing and Fast Chemical Reaction—VII: Determining Reaction Zone Model for the CSTR," Copyright 1983, Pergamon Press plc.*

Figure 3-25 Computed and experimental product distributions.

difficult to use), species absorption of light, and species fluorescence (first applied to gases, now to liquids). Most techniques require a probe within the fluid, but fluorescence can be done remotely if absorption of the light is not too strong. The techniques presently available have been reviewed by the author (Patterson 1983).

Now that methods for measuring distributions of concentration in reacting fluids are becoming available, more effort should be expended in attempting to model the chemical reaction phenomena on a distributed basis.

SCALE-UP OF MIXED REACTORS

The scale-up and scale-down of mixed reactors in which there are significant mixing effects on conversion and yield can only be successfully performed through use of models, which account for all the important effects. Since the hydrodynamic phenomena are nonlinear and, in turn, interact with mixing and reaction phenomena in a highly nonlinear way, only models that include the relevant relationships will prove generally adequate. Since only geometrically distributed models may incorporate these interactions without gross simplification, only these

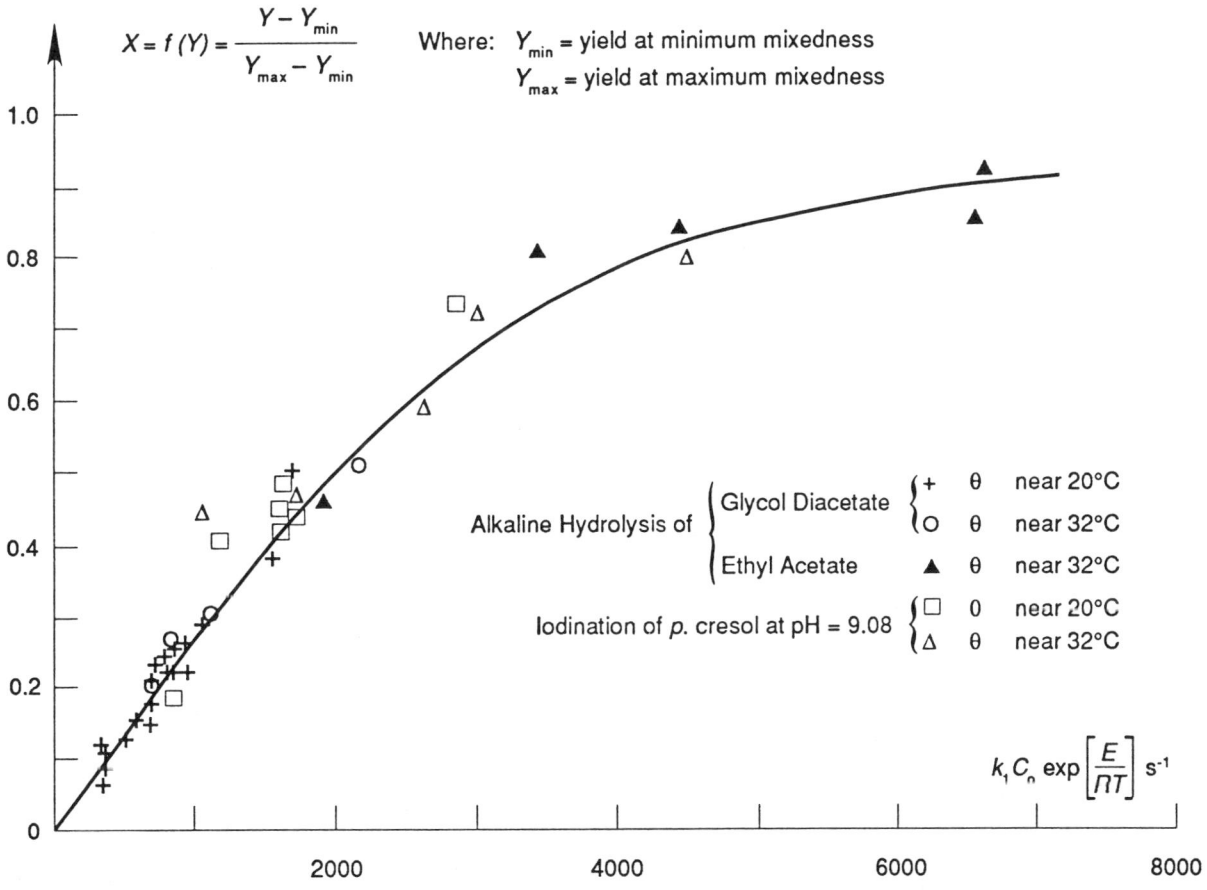

Reprinted with permission from Advances in Chemistry Series, *133, A. Zoulalian and J. Villermaux, "Influence of Chemical Parameters on Micromixing in a Continuous Stirred Tank Reactor," Copyright 1974, American Chemical Society.*

Figure 3-26 Effect of reaction rate on yield at constant mixing rate.

models will be generally successful in scale-up and scale-down where absolute dynamic similarity is not maintained. As is well known, dynamic similarity cannot usually be maintained for at least two reasons: (1) when more than one dimensionless parameter is involved, they cannot usually *both* be satisfied; and (2) the power requirements for mixing at large scale are too great under the same mixing rate conditions necessary at small scale; local mixing rates and even their geometric distribution are changed upon scale-up. Even when distributed models must be adjusted because of poorly understood boundary conditions or for other reasons that lead to lack of adequate fit to laboratory data, scale-up based on the adjusted model can be relied on with more confidence than on a lumped (volume-averaged) model, because more of the controlling physical variables are accounted for.

The problem of computer capability and cost for performing complete modeling of hydrodynamic phenomena is being solved by the advent of inexpensive large-scale computer memory and faster computer processors. The cost and inconvenience of performing a computer simulation that requires a large amount of memory have been greatly reduced and are not now significant obstacles, even for three-dimensional simulations. Three-dimensional simulation of furnace mixing and combustion has been done for some time (Ellail et al. 1978).

As in the case of staged and packed countercurrent separation (contacting) devices, most efficient scale-up may be achieved for mixed reactors when well-developed codes to simulate mixing with a variety of boundary conditions become generally available. Because of the complexity of all but the most special geometries, it will be necessary to ultimately make use of three-dimensional models.

RELATIVE COSTS OF MODELING METHODS

Assuming that the value of a distributed modeling method for scale-up and/or design of mixed reactors is accepted, the relative costs of the methods as well as the relative efficacy are of concern.

Finite Difference Method

Most distributed modeling has been done using a finite difference solution of the governing partial differential balance equations. Experience indicates that the computing cost to solve a system of five hydrodynamic and eight concentration and segregation (four components) equations is less than $100 at current CPU prices, if the two-dimensional region is represented by 48×48 (= 2304) grid points and the system is at steady state. A three-dimensional space of 20,000 grid points would cost proportionately more. An unsteady-state system (batch or semibatch or boundary disturbance) would also cost more. In general, it may be expected that the cost for a simulation is approximately proportional to the number of balance equations solved times the number of grid points, unless the additional equations seriously affect the approach to convergence. That can occur if source term equations, which generally contain the closure assumptions, are highly nonlinear or interactive.

Coalescence–Dispersion Method

No good comparisons exist between finite difference methods and other methods except for the random c–d approach. The random c–d method can be implemented for a mixed reactor in which the fluid dynamics (velocities, turbulence, energy dissipation rate) are known either from experiment or from a finite difference simulation. A disadvantage of c–d is that the fluid dynamics and mixer–reaction simulations must be done in two stages, whereas the finite difference approach can solve all equations simultaneously and account for the reaction effects of hydrodynamics, such as density and viscosity effects.

The advantage of the c–d method is that even though the spatial resolution is not as good for a given cost, it can handle any reaction rate equation for any number of chemical reactions and has been proven effective for at least one type of multiple reaction system—the competitive–consecutive reaction producing two products. For the two-dimensional finite difference case mentioned above with a 48 × 48 point grid, the equivalent c–d case costing about the same had a spatial resolution of 5 volume elements by 6 volume elements (30 in all) with a total of 3000 c–d fluid elements (100 per volume element). In addition to producing a final steady-state result, the c–d method has the benefit of producing unsteady-state information, since its convergence method is a timewise approach to steady-state.

k-ε Method

The $k-\varepsilon$ turbulence modeling method is ideal for generating the flow and turbulence information for the c–d method. This is due to the fact that Eq 3-41, relating c–d rate on-line to mixing rate developed above, has k_t and ε as its main parameters. Beyond that a volumetric flow rate between volume elements must be determined by integration of the distributed flow velocities computed by the $k-\varepsilon$ method.

Summary

In general, the finite difference solution for mixing reaction is preferred where closures for the chemical reactions are available and are known to be reliable. Where that is not true (which presently is the majority of the time), the c–d method is preferable. If reliable small-scale data exist for the distribution of conversion and other variables needed to determine the parameter or parameters in a diffusion-based model (such as the extensional flow, lamellar element, spherical eddy, or diffusing slab model), then distributed computation of reaction conversions for different reactor sizes could be done. Care must be exercised, however, to make sure that the parameters established do not change with reactor size, shape, or agitator speed, since they are not based on fundamental turbulence principles. No experience presently exists for guiding such approaches.

SIMILARITY APPROXIMATIONS FOR DISTRIBUTED VARIABLES

Dimensionless groups generated from Eq 3-24 through 3-29 and from other sources may sometimes be applied both locally and globally to the entire reactor. If geometric similarity is maintained in scale-up, it is often true that distributions of certain dimensionless variables also remain similar. An illustration is the power number $P/(\rho N^3 D^5)$, a global group for tanks, which leads to P/V proportional to $P/(\rho N^3 D^2)$ for geometric similarity at high Reynolds number. If P/V is proportional to $\rho \varepsilon$, at any constant location for all impeller speeds and sizes, then ε at any constant location should be proportional to $N^3 D^2$. That means that if ε distribution remains relatively similar, then information on the variation of P/V can be used to approximate variations of ε in stirred tanks during scale-up. If the shape of the vessel or the impeller type is changed, then constant ratios will not result.

Another similarity approximation, which is kinematic in nature, is the use of correlations of the discharge number, Q/ND^3, with other stirrer variables. If local velocity U_i is deemed proportional to ND and if discharge number is relatively constant for constant density and geometric similarity at high Reynolds number, then discharge number correlations may be used to approximate local velocity distributions in scale-up. Such kinematic similarity approximations are similar to those commonly made for jet flow (Patterson 1981).

Similarity approximations are very useful for partial modeling of turbulent mixing effects. If velocity and turbulence data at pilot or bench scale are used, they may be scaled to larger sizes in that way. Such similarity approximations may also be used to establish distributions of variables at larger scale for computing c–d rates if that type of modeling is used to compute distributed mixing and reaction rates.

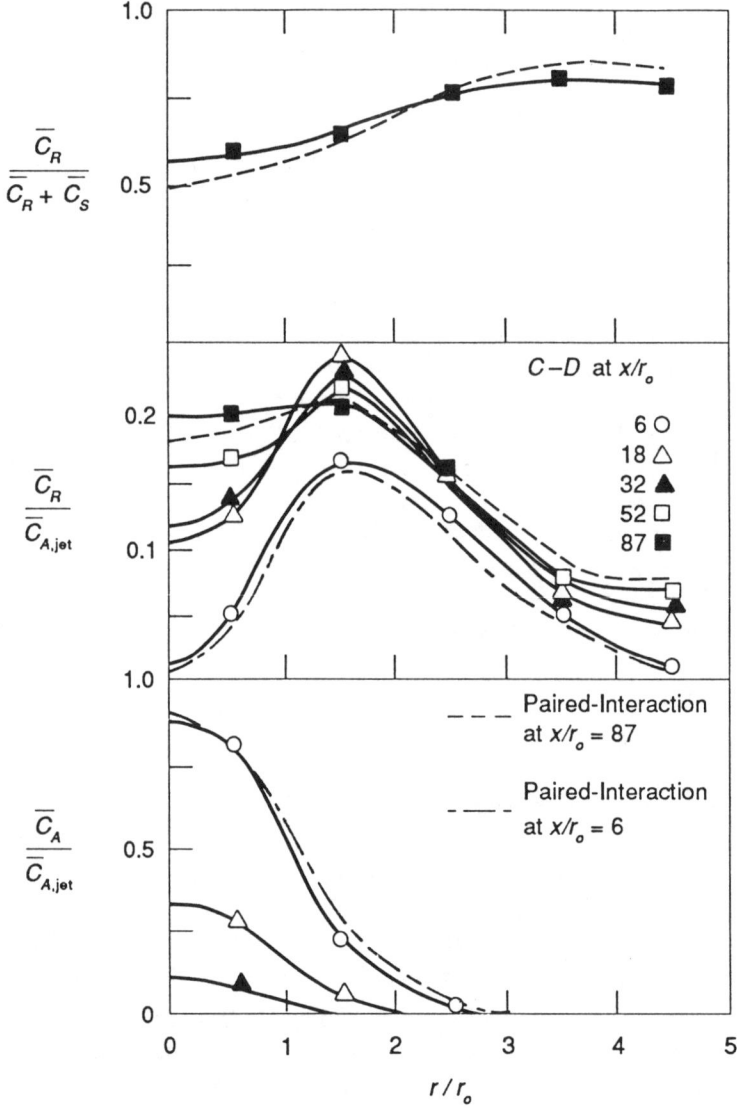

NOTE: Jet radius, r_o = 1.0 cam; v_{jet} = 10 cm/s; k_{123} = 10 cc/gmole/s; k_{234} = 1.0 cc/gmole/s.

Figure 3-27 Reactant-A conversion, product-R, and yield of product-R for a typical complex reaction.

That approach was used to compute distributions of yield for a competing reaction ($A + B \rightarrow R; R + B \rightarrow S$) in an annular jet mixer, as shown in Figure 3-27. The computations were done using Eq 3-41; similarity approximations to obtain distributions of ε, U_i, and k_t; and batch kinetics equations for the reactions in each element between coalescence–dispersion events. The same approach has been used for stirred-tank reactor computations (Patterson 1981).

GLOBAL VARIABLE CORRELATIONS FROM DISTRIBUTED MODELING

Distributed modeling may be used to generate large sets of data covering broad ranges of operating variables. Dimensional analysis, such as that presented above, may then lead to useful correlations of global variables, such as feed rate, impeller speed, reactant ratio, kinetics parameters, and output variables such as conversion and yield. Such a correlation, obtained from a partial study of mixing effects on yield and conversion in a tank stirred by a Rushton turbine, is shown in Figure 3-19, where yield is correlated with Damköhler number with feed rate/discharge rate ratio and reaction rate constant ratio as parameters. The correlated results were generated by a computer simulation that uses laboratory data for hydrodynamics combined with Eq 3-28 and 3-29 with a new closure method for multiple second-order reactions (Eq 3-39 and 3-40).

Results for the simple second-order reaction in a stirred-tank reactor are correlated in Figure 3-18 for the Rushton turbine stirred reactor. Conversion is correlated with Damköhler numbers for mixing and reaction, respectively.

CONCLUSION

In the not-too-distant future, most "experimental" data on transport-dominated processes, such as mixed chemical reactors, will be generated by computer simulations. The computations will be backed up by only a few experimental data, saving much time and money. The progress reviewed here is a small step toward that end.

References

ANGST, W.; BOURNE, J.; & DELL'AVA, P. 1984. Mixing and Fast Chemical Reaction IX, Comparison Between Models and Experiments. *Chem. Eng. Sci.*, 39:335.

ANGST, W.; BOURNE, J.R.; & KOZICKI, F. 1979. Some Measurements of Micromixing in Commercial-Scale Stirred Reactors. Proc. 3rd European Conf. on Mixing. York, England.

BARTHOLE, J.; DAVID, R.; & VILLERMAUX, J. 1982. A New Chemical Method for the Study of Local Micromixing Conditions in Industrial Stirred Tanks. *Chem. Reaction Eng.*—Boston, 545.

BATCHELOR, G.K. 1953. *The Theory of Homogeneous Turbulence*. Cambridge University Press, Cambridge, England.

BELEVI, H.; BOURNE, J.R.; & RYS, P. 1981. Mixing and Fast Chemical Reaction, II: Diffusion–Reaction Model for the CSTR. *Chem. Engrg. Sci.*, 36:10:1649.

BENNETT, C.O. & MEYERS, J.E. 1962. *Momentum, Heat, and Mass Transfer*. McGraw-Hill Publishers, New York.

BILGER, R.W. 1979. Turbulent Flows With Nonpremixed Reactants. In *Turbulent Reacting Flows*, ed. by P.A. Libby and F.A. Williams. Springer Verlag, Heidelberg, Germany.

BIRD, R.B.; STEWART, W.E.; & LIGHTFOOT, E.N. 1960. *Transport Phenomena*. John Wiley and Sons, New York.

BOCKELMAN, W.D. 1983. Measurements of Conversion in a Stirred Tank Reactor by Fluorescence. Master's thesis, University of Missouri-Rolla, Rolla, Mo.

BORGHI, R. 1975. Computational Studies of Turbulent Flows With Chemical Reaction. In *Turbulent Mixing in Non-reactive and Reactive Flows*, ed. by S.N.B. Murthy. Plenum Press, New York.

BOURNE, J.R. 1983. *Chem. Engrg. Sci.*, 38:5.

BOURNE, J.R. ET AL. 1981. Mixing and Fast Chemical Reaction, III: Model Experiment Comparisons. *Chem. Engrg. Sci.*, 36:1655.

BOURNE, J.R. & DELL'AVA, P. 1987. Micro- and Macro-mixing in Stirred Tank Reactors of Different Sizes. *Chem. Engrg. Res. Des.*, 65:180

BOURNE, J.R.; MOERGELI, U.; & RYS, P. 1977. Mixing and Fast Chemical Reaction: Influence of Viscosity on Product Distribution. Proc. 2nd European Conf. on Mixing. Cambridge, England.

BOURNE, J.R. & ROHANI, S. 1983. Mixing and Fast Chemical Reaction, VII: Determining Reaction Zone Model for the CSTR. *Chem. Engrg. Sci.*, 38:911.

BOWERS, R.H. 1965. An Investigation of Flow Phenomena in Stirred Liquids. *AIChE-IChE Symp. Ser.*, 10:8.

BRODKEY, R.S. 1967. *Phenomena of Fluid Motions*. Addison-Wesley, Reading, Mass.

BRODKEY, R.S., ed. 1975. *Turbulence in Mixing Operations*. Academic Press, New York.

BRODKEY, R.S. & LEWALLE, J. 1985. Reactor Selectivity Based on First-Order Closures of the Turbulent Concentration Equations. *AIChE Jour.*, 31:111.

CANON, R.M. ET AL. 1977. Turbulence Level Significance of the Coalescence–Dispersion Interaction Parameters. *Chem. Engrg. Sci.*, 32:1349.

CORRSIN, S. 1964. The Isotropic Turbulent Mixer: Part II, Arbitrary Schmidt Number. *AIChE Jour.*, 10:870.

CURL, R.L. 1963. Dispersed Phase Mixing: Theory and Effects in Simple Reactors. *AIChE Jour.*, 9:175.

CUTTER, L.A. 1966. Flow and Turbulence in a Stirred Tank. *AIChE Jour.*, 12:35.

DANCKWERTS, P.V. 1953. Continuous Flow Systems. *Chem. Engrg. Sci.*, 2:1.

———. 1958. The Effect of Incomplete Mixing on Homogeneous Systems. *Chem. Engrg. Sci.*, 8:93.

DAVID, R. & VILLERMAUX, J. 1975. Micromixing Effects on Complex Reactions in a CSTR. *Chem. Engrg. Sci.*, 30:1309.

DONALDSON, C.D. 1975. In *Turbulent Mixing in Non-reactive and Reactive Flows*, ed. by S.N.B. Murthy. Plenum Press, New York.

EDELMAN, R.B. & HARSHA, P.T. 1978. Some Observations on Turbulent Mixing With Chemical Reactions. In *Turbulent Combustion*, ed. by L.A. Kennedy. Amer. Inst. Aeron. Astron.

ELGHOBASHI, S.E.; PUN, W.M.; & SPALDING, D.B. 1977. Concentration Fluctuations in Isothermal Turbulent Confined Coaxial Jets. *Chem. Engrg. Sci.*, 32:161.

ELLAIL, M.M.M.A. ET AL. 1978. Description and Validation of a Three-Dimensional Procedure for Combustion Chamber Flow. In *Turbulent Combustion*, ed. by L.A. Kennedy. Amer. Inst. Aeron. Astron.

EVANGELISTA, J.J.; SHINNAR, R.; & KATZ, S. 1969. 12th Symp. on Combustion. The Combustion Institute.

FAHIEN, R.W. 1983. *Fundamentals of Transport Phenomena*. McGraw-Hill Publishers, New York.

FAUST, A.S. ET AL. 1980. *Principles of Unit Operations*. John Wiley and Sons, New York.

FORT, I. ET AL. 1974. Turbulent Characteristics of the Velocity Field in a System With Turbine Impeller and Radial Baffles. *Coll. Czech. Chem. Comm.*, 39:1810.

GOSMAN, A.D. ET AL. 1969. *Heat and Mass Transfer in Recirculating Flows*. Academic Press, London.

JONES, W.P. & LAUNDER, B.D. 1973. *Intl. Jour. Heat Mass Transfer*, 16:119.

KATTAN, A. & ADLER, R.J. 1967. A Stochastic Mixing Model for Homogeneous, Turbulent, Tubular Reactors. *AIChE Jour.*, 13:580.

KIM, W.J. & MANNING, F.S. 1964. Turbulence Energy and Intensity Spectra in a Baffled Stirred Vessel. *AIChE Jour.*, 10:747.

KLEIN, J.; DAVID, R.; & VILLERMAUX, J. 1980. Interpretation of Experimental Liquid Phase Micromixing Phenomena in Continuous Stirred Reactor With Short Residence Times. *Ind. Eng. Chem. Fund.*, 19:373.

KOLMOGOROFF, A.V. 1941. *Copt. Rend. Acad. Sci. U.R.S.S.*, 32:16.

LEVENSPIEL, O. 1972. *Chemical Reaction Engineering*, 2nd ed. John Wiley and Sons, New York.

LOCKWOOD, F.C.; & NAQUIB, A.S. 1975. The Prediction of the Fluctuations in the Properties of Free, Round Jet, Turbulent Diffusion Flames. *Combustion and Flame*, 24:109.

MANN, R.; MIDDLETON, J.C.; & PARKER, I.B. 1977. Proc. 2nd European Conf. on Mixing. Cambridge, England.

MANNING, F. & WILHELM, R. 1963. Concentration Fluctuations in a Stirred Baffled Vessel. *AIChE Jour.*, 9:1.

MAO, K.W. & TOOR, H.L. 1970. A Diffusion Model for Reactions With Turbulent Mixing. *AIChE Jour.*, 16:49.

MCCABE, W.L. & SMITH, J.C. 1967. *Unit Operations of Chemical Engineering*. McGraw-Hill Publishers, New York.

MCKELVEY, K.N. ET AL. 1975. Turbulent Motion, Mixing and Kinetics in a Chemical Reactor Configuration. *AIChE Jour.*, 21:1165.

MEHTA, R.V. & TARBELL, J.M. 1983. Four Environment Model of Mixing and Chemical Reaction: Parts I and II. *AIChE Jour.*, 29:320, 329.

MIDDLETON, J.C. 1979. Measurements of Circulation Within Large Mixing Vessels. Proc. 3rd Conf. on Mixing. BHRA, Cranfield, U.K.

MIDDLETON, J.C.; PIERCE, F.; & LYNCH, P.M. 1984. Computations of Flow Fields and Complex Reaction Yield in Turbulent Reactors and Comparisons With Experimental Data. Proc. 8th Intl. Symp. on Chemical Reaction Engineering, AIChE Symp. Ser., 87:239.

NABHOLZ, F.; OTT, R.J.; & RYS, P. 1977. Mixing Disguised Chemical Selectivities. Proc. 2nd European Conf. on Mixing. Cambridge, England.

NAGATA, S. 1975. *Mixing, Principles and Applications*. Halsted Press, Tokyo.

NG, D.Y.C. & RIPPIN, D.W.T. 1965. Effect of Incomplete Mixing on Conversion in Homogeneous Reactions. Proc. 3rd Symp. Chemical Reaction Engineering, Amsterdam.

OTTINO, J.M. 1981. Efficiency of Mixing From Data on Fast Reactions in Multi-Jet Reactors and Stirred Tanks. *AIChE Jour.*, 27:184.

OTTINO, J.M.; RANZ, W.E.; & MOCOSKO, C.W. 1979. A Lame Hour Model for Analysis of Liquid–Liquid Mixing. *Chem. Engrg. Sci.*, 34:877.

PALPEPU, P.T.; ADLER, R.J.; & EDWARDS, R.V. 1981. 74th Annual AIChE Meeting, New Orleans, La.

PATTERSON, G.K. 1973. Model With No Arbitrary Parameters for Mixing Effects on Second-Order Reaction With Unmixed Feed Reactants. Proc. Symp. on Fluid Mech. of Mixing, ASME, Atlanta, June.

———. 1975a. Simulating Turbulent-Field Mixers and Reactors. In *Application of Turbulence Theory to Mixing Operations*, ed. by R.S. Brodkey. Academic Press, New York.

———. 1975b. Average Molecular Weight Distributions in Stirred-Tank Reactors by Random Coalescence–Dispersion Simulation. 78th Annual AIChE Meeting. Houston, Texas.

———. 1977. Made King Complex Chemical Reactions in Flows With Turbulent, Diffusive Mixing. 70th Annual AIChE Meeting. New York.

———. 1980. Closure Approximations for Complex Multiple Reactions. 2nd Symp. on Turbulent Shear Flows. London.

———. 1981. Application of Turbulence Fundamentals to Reactor Modeling and Scaleup. *Chem. Engrg. Comm.*, 8:25.

———. 1982. Mixed Reactor Scaleup Through Geometrically Distributed Modeling. 75th Annual AIChE Meeting, Los Angeles, Calif.

———. 1983. Turbulent Mixing and Its Measurement. In *Handbook of Fluids in Motion*, ed. by N.P. Cheremisinoff and R. Gupta. Ann Arbor Science Publishers, Ann Arbor, Mich.

———. 1985. Modeling of Turbulent Reactors. In *Mixing of Liquids by Mechanical Agitation*, ed. by L.L. Ulbrecht and G.K. Patterson. Gordon and Breach, New York.

PATTERSON, G.K.; LEE, W.C.; & CALVIN, S.J. 1977. Measurement and Numerical Modeling of Turbulent Scalar Mixing and Reaction of Coaxial Jets. Symp. Turbulent Shear Flows, Pennsylvania State University, University Park, Pa.

PATTERSON, G.K.; BOCKELMAN, W.D.; & QUIGLEY, J.J. 1982. Measurements of Mixing Effects on Local Reaction–Conversion in Stirred Tanks. Proc. 4th European Conf. on Mixing. Noordwijkerhout, The Netherlands.

PAUL, E.L. & TREYBAL, R.E. 1971. Mixing and Product Distribution for a Liquid-Phase, Second-Order, Competitive–Consecutive Reaction. *AIChE Jour.*, 17:718.

PRATT, D.R. 1976. *Progress in Energy Combustion Science*, 1:73.

QUIGLEY, J.J. 1984. Fluorescence Based Measurement of Concentration in Stirred Tanks. Master's thesis, University of Missouri-Rolla, Rolla, Mo.

RAO, D.P. & DUNN, I.J. 1970. A Monte Carlo Coalescence Model for Reaction With Pospercion in a Tubular Reactor. *Chem. Engrg. Sci.*, 25:1275.

RAO, D.P. & EDWARDS, L.L. 1971. Conversion in Turbulent Tubular Reactors With Unmixed Feed. *Ind. Engrg. Chem. Fund.*, 10:798.

REITH, I.R. 1965. Generation and Decay of Concentration Fluctuations in a Stirred, Baffled Vessel. *AIChE-IChE Symp. Ser.*, 10:14.

RICHIE, B.W. & TOGBY, A.H. 1979. A Three Environment Micromixing Model for Chemical Reactors With Arbitrary Separate Feedstreams. *Chem. Engrg. Jour.*, 17:173.

RUSHTON, J.H.; COSTICH, E.W.; & EVERETT, H.J. 1950. Power Characteristics of Mixing Impellers. *Chem. Engrg. Prog.*, 46:395, 467.

SATO, Y. ET AL. 1967. Turbulent Flow in a Stirred Vessel. *Kagaku Kogaku*, 31:275.

SATO, Y.; KAMIWANO, M.; & YAMAMOTO, K. 1970. Turbulent Flow in a Stirred Tank; Effects of Impeller Types. *Kagaku Kogaku*, 34:104.

SMITH, J.M. 1982. *Chemical Engineering Kinetics*. McGraw-Hill Publishers, New York.

SPALDING, D.B. 1978. A Simple Model for the Rate of Turbulent Combustion. In *Turbulent Combustion*, ed. by L.A. Kennedy. Amer. Inst. Aeron. and Astron.

SPENCER, J. & LUNT, R. 1980. Experimental Characterization of Mixing Mechanisms in Flow Reactors Using Reactive Tracers. *Ind. Engrg. Chem. Fund.*, 19:142.

SPIELMAN, L.A. & LEVENSPIEL, O. 1965. A Monte Carlo Treatment for Reacting and Coalescing Dispersed Phase Systems. *Chem. Engrg. Sci.*, 20:247.

TOOR, H.L. 1962. Mass Transfer in Dilute Turbulent and Nonturbulent Systems With Rapid Irreversible Reactions and Equal Diffusivities. *AIChE Jour.*, 8:70.

TRUONG, K.T. & METHOT, J.C. 1976. *Can. Jour. Chem. Engrg.*, 54:572.

TURNER, J.C.R. 1983. Perspective in Residence-Time Distributions. *Chem. Engrg. Sci.*, 38:1.

UHL, V.W. & GRAY, J.B., eds. 1966. *Mixing: Theory and Practice*, Vol. 1 and 2. Academic Press, New York.

ULBRECHT, J.J. & PATTERSON, G.K., eds. 1985. *Mixing of Liquids by Mechanical Agitation*. Gordon and Breach, New York.

VARMA, A.K.; FISHBURNE, E.S.; & DONALDSON, D.C. 1978. In *Turbulent Combustion*, ed. by L.A. Kennedy. American Institute Aeron, and Astron.

VASSILATOS, G. & TOOR, H.L. 1965. Second-Order Chemical Reactions in a Nonhomogeneous Turbulent Fluid. *AIChE Jour.*, 11:666.

VILLERMAUX, J. & ZOULALIAN, Z. 1969. Etat de Melange du Fluide dans un Reacteur Continu. A Propos dun Model de Weinstein et Adler. *Chem. Engrg. Sci.*, 24:1513.

WEINSTEIN, H. & ADLER, R.J. 1967. Micromixing Effects in Continuous Chemical Reactors. *Chem. Engrg. Sci.*, 22:65.

WELTY, J.R.; WICKS, C.E.; & WILSON, R.E. 1976. *Fundamentals of Momentum, Heat and Mass Transfer*. John Wiley and Sons, New York.

WEN, C.Y. & FAN, K.T. 1975. *Models for Flow Systems and Chemical Reactors*. Marcel Dekker, New York. pp. 251–280.

WORRELL, G. & EAGLETON, L. 1964. An Experimental Study of Mixing and Segregation in a Stirred Tank Reactor. *Can. Jour. Chem. Engrg.*

WU, H. 1988. LDV Measurements and Numerical Modelling of the Turbulent Flow in a Stirred Mixer. Ph.D. dissertation, University of Arizona, Tucson, Ariz.

ZIPP, R.P. 1981. Light-Scattering Measurements of Turbulence and Concentration in a Stirred Tank. Master's thesis, University of Missouri-Rolla, Rolla, Mo.

ZOULALIAN, A. & VILLERMAUX, J. 1974. Influence of Chemical Parameters on Micromixing in a Continuous Stirred Tank Reactor. In *Advances in Chemistry Series*. American Chemical Society, Washington, D.C.

ZWIETERING, T.N. 1959. The Degree of Mixing in Continuous Flow Systems. *Chem. Engrg. Sci.*, 11:1.

4

Residence Time Distribution

E. Bruce Nauman, Ph.D., Professor, Department of Chemical
  Engineering, Rensselaer Polytechnic Institute,
  Troy, N.Y. USA

Mark M. Clark, Ph.D., Assistant Professor, Department of Civil
  Engineering, University of Illinois at Urbana–Champaign,
  Urbana, Ill. USA

Residence time distribution is the distribution of ages of particles leaving a mixing system or, equivalently, the distribution of life expectancies for particles entering the system. This chapter decribes an important theoretical and experimental tool for understanding flow and mixing processes in real systems. Most practical flow devices are too complex to allow a rigorous solution to the equations of motion, but residence time measurements using inert tracers often provide enough information for analysis, design, and troubleshooting. The first two sections of this chapter describe the mathematical properties of residence time distributions. Following sections discuss models that may be used to approximate real systems, as well as experimental techniques and the analysis of data, including example calculations for prototype flocculation tanks. The theory of residence time distributions is extended to include the concepts of segregation and micromixing in flow systems and the effects of these phenomena on chemical reactions. Finally, design and scale-up techniques based on residence time concepts are discussed.

THE SIMPLE CLOSED SYSTEM

Residence time theory deals with the transport of conserved particles through a system, described in the time domain. These particles could be molecules, atoms, or any small particles for which gravity forces are a negligible transport mechanism. Each particle flowing through the system has some first entrance into the system and some final exit from it. The time the particle spends inside the system is called the residence time t.

Residence time theory allows treatment of systems with many connections to their environments, with multidirectional flow in the connections, with time-dependent flow rates, and even with time-dependent system boundaries. Most of this chapter is devoted to the simple closed system illustrated in Figure 4-1.* A simple system has only one inlet and one outlet. In a closed system, particles cannot make temporary excursions outside of the system.

MATHEMATICAL PROPERTIES

In order to understand the residence time distribution, one must first understand its underlying mathematical properties: distribution functions, means and moments, and mean residence times and normalized distributions.

Distribution Functions

Not all particles flowing through a system have the same residence time. If a number of particles were tagged and their individual residence times in the system were recorded, a histogram could be constructed (see Figure 4-2). The histogram could be converted to a continuous function $f(t)$ if the number of recorded particles approached infinity while simultaneously the class width Δt approached zero (Figure 4-3). Then the infinitesimal area $f(t)dt$ would be equal to the probability that any randomly chosen particle would have a residence time between t and $t + dt$. The term $f(t)$ is a function that depends on the nature of the system; it is known by several names. It is called the residence time density function (in analogy to the probability density function), the differential distribution function of residence times, or the residence time frequency function. It has the important property that

$$\int_0^\infty f(t)\,dt = 1 \tag{Eq 4-1}$$

The only other restriction on $f(t)$ is that it be nonnegative:

$$f(t) \geq 0,\ t \geq 0 \tag{Eq 4-2}$$

Sometimes it is more convenient to use functions describing the probability of a residence time less than or greater than a specific value. These two functions are called respectively the residence time distribution function

$$F(t) = \int_0^t f(t')\,dt' \tag{Eq 4-3}$$

---

*For a more detailed treatment of possible complications, refer to the specialized literature (Nauman and Buffham 1983).

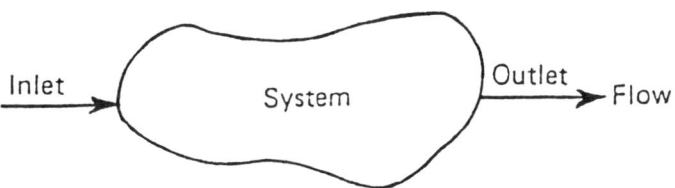

Source: Nauman, E.B. & Buffham, B.A. 1983. Mixing in Continuous Flow Systems. Copyright 1983, John Wiley and Sons, New York. Reprinted by permission of John Wiley and Sons, Inc.

Figure 4-1 A simple flow system.

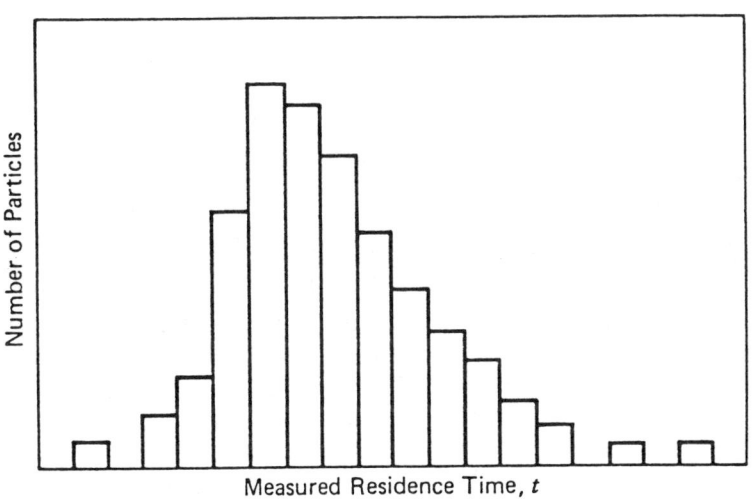

Source: Nauman, E.B. & Buffham, B.A. 1983. Mixing in Continuous Flow Systems. Copyright 1983, John Wiley and Sons, New York. Reprinted by permission of John Wiley and Sons, Inc.

Figure 4-2 Histogram for residence time measurements on individual particles.

and the washout function

$$W(t) = \int_t^\infty f(t')\, dt' \qquad \text{(Eq 4-4)}$$

The two functions are clearly related to each other by

$$F(t) + W(t) = 1 \qquad \text{(Eq 4-5)}$$

They are related to the density function by

$$f(t) = \frac{dF}{dt} = \frac{-dW}{dt} \qquad \text{(Eq 4-6)}$$

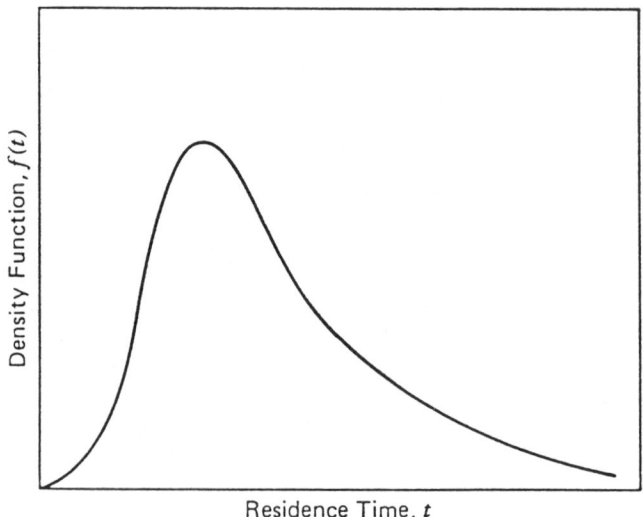

Source: Nauman, E.B. & Buffham, B.A. 1983. Mixing in Continuous Flow Systems. Copyright 1983, John Wiley and Sons, New York. Reprinted by permission of John Wiley and Sons, Inc.

Figure 4-3 Typical residence time density or differential distribution function.

Figure 4-4 shows a typical residence time washout function. By definition, $F(t)$ is a nondecreasing function over the interval $0 < t < \infty$, while $W(t)$ is nonincreasing over the same interval. In most real systems, $F(t)$ and $W(t)$ would be expected to be continuous functions. Yet pathological behavior, such as "bypassing," is theoretically possible; this behavior would result in a discontinuity in $F(t)$ at $t = 0$. A more common pathological situation is stagnancy, which is indicated by $F(\infty) < 1$.

Perfect Mixing and Piston-Flow Models

Probably the two most important residence time models in process engineering and environmental modeling are perfect mixing and piston (or plug) flow.

Perfect mixing. The model for perfect mixing can be derived by assuming compositional uniformity throughout the system, that is, concentration interior to the system is identical to concentration in fluid packets leaving the system. A perfect mixer always has an exponential distribution of residence times. The washout function is

$$W(t) = \exp\left(\frac{-t}{\bar{t}}\right) \quad \text{(Eq 4-7)}$$

Using Eq 4-6, the corresponding density function is

$$f(t) = \left(\frac{1}{\bar{t}}\right)\exp\left(\frac{t}{\bar{t}}\right) \quad \text{(Eq 4-8)}$$

An exponential distribution usually results from high circulation rates within a system, but this does not necessarily imply perfect mixing. In this discussion,

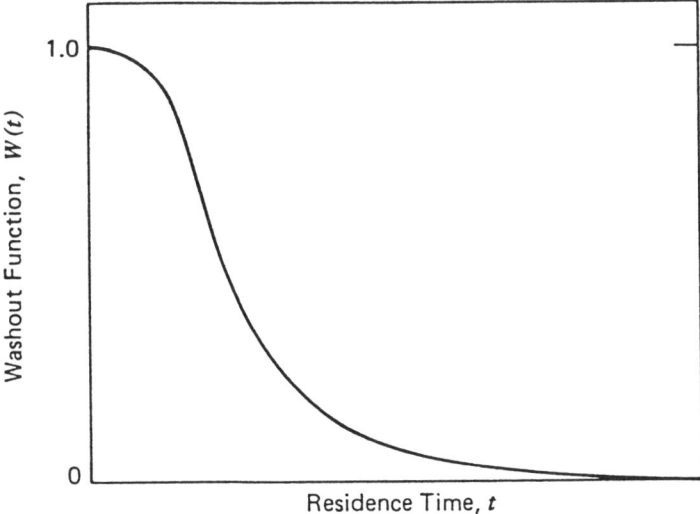

Source: Nauman, E.B. & Buffham, B.A. 1983. Mixing in Continuous Flow Systems. *Copyright 1983, John Wiley and Sons, New York. Reprinted by permission of John Wiley and Sons, Inc.*

Figure 4-4 Residence time without washout function.

a stirred-tank reactor* is a system having an exponential distribution of residence times.

Piston- or plug-flow model. Another ideal residence time distribution results from considering gravity or pressure flow in a pipe or other conduit in which a uniform velocity distribution across the flow section and minimal molecular diffusion are imagined; hence, all particles entering the system at the same time have the same residence time in the system. This is called the piston- or plug-flow model, for which the distribution function is

$$F(t) = 0, t < \bar{t} \quad \text{(Eq 4-9)}$$
$$= 1, t > \bar{t}$$

and the density function is

$$f(t) = \delta(t - \bar{t}) \quad \text{(Eq 4-10)}$$

Where:

δ = the Dirac delta function

---

*The stirred-tank reactor goes by several different names. For example, Levenspiel (1972) uses the "constant flow stirred-tank reactor" (CFSTR), while Weber (1972) uses the "completely mixed flow reactor" (CMF) and the "continuous stirred-tank reactor" (CSTR). All of these titles probably mean the same thing: an exponential distribution of residence times. As with the "stirred-tank reactor," it is important to remember that an exponential distribution of residence times does not imply perfect mixing.

The Dirac function has a value of zero everywhere except at $t = \bar{t}$, and also has the properties

$$\int_0^\infty \delta(t)\, dt = 1 \qquad \text{(Eq 4-11)}$$

and, for any continuous function $C(t)$,

$$\int_0^\infty C(t)\, \delta(t - t_o)\, dt = C(t_o) \qquad \text{(Eq 4-12)}$$

Means and Moments

As in the case of the probability density function, any $f(t)$ can, in principle, be characterized completely by its moments,

$$\mu_n = \int_0^\infty t^n f(t)\, dt \qquad \text{(Eq 4-13)}$$

Where:

μ_n = moment about origin of order n
n = 0, 1, 2, ...

The case $n = 1$ yields the *mean residence time*

$$\bar{t} = \mu_1 = \int_0^\infty t f(t)\, dt \qquad \text{(Eq 4-14)}$$

Special importance is attached to $\bar{t}$ because it is the average age of particles leaving the system and because it can often be estimated knowing only the ratio of the system volume to flow rate (see following section for a discussion of estimating mean residence times).

Occasionally, a useful alternative to Eq 4-13 is

$$\mu_n = n \int_0^\infty t^{n-1} W(t)\, dt \qquad \text{(Eq 4-15)}$$

$W(t)$ is often easily measured with a negative step-change tracer experiment (discussed in this chapter under the heading "Measurement Techniques"). Equation 4-15 then yields more accurate estimates of the moments because there is less weighing by the extreme values of the distribution (in which statistically there is always less confidence). From Eq 4-15, note that the mean residence time is simply the area under the $W(t)$ curve,

$$\bar{t} = \int_0^\infty W(t)\, dt \qquad \text{(Eq 4-16)}$$

Moments about the mean are defined as

$$\mu_n' = \int_0^\infty (t - \bar{t})^n f(t)\, dt \qquad \text{(Eq 4-17)}$$

which are also called the central moments of the distribution. These may be related to the ordinary moments merely by multiplying out $(t-\bar{t})^n$ in Eq 4-17. The results are

$$\mu'_0 = \mu_0 = 1 \qquad \text{(Eq 4-18)}$$

$$\mu'_1 = 0 \qquad \text{(Eq 4-19)}$$

$$\mu'_2 = \mu_2 - (\bar{t})^2 \qquad \text{(Eq 4-20)}$$

$$\mu'_3 = \mu_3 - 3\bar{t}\mu_2 + 2(\bar{t})^3 \qquad \text{(Eq 4-21)}$$

$$\mu'_4 = \mu_4 - 4\mu_3\bar{t} + 6\mu_2(\bar{t})^2 - 3(\bar{t})^4 \qquad \text{(Eq 4-22)}$$

and so on. The important special case of Eq 4-17 is for $n = 2$, because this gives the variance of the distribution:

$$\sigma^2 = \mu'_2 = \int_0^\infty (t-\bar{t})^2 f(t)\,dt = \mu_2 - (\bar{t})^2 \qquad \text{(Eq 4-23)}$$

The third central moment μ'_3 is known as the skewness of the distribution, and the fourth central moment μ'_4 is called the kurtosis. The inverse relationships of Eq 4-18 through Eq 4-22 can be found by successive substitutions

$$\mu_2 = \mu'_2 + (\bar{t})^2 \qquad \text{(Eq 4-24)}$$

$$\mu_3 = \mu'_3 + 3\mu'_2\bar{t} + (\bar{t})^3 \qquad \text{(Eq 4-25)}$$

$$\mu_4 = \mu'_4 + 4\mu'_3\bar{t} + 6\mu'_2(\bar{t})^2 + (\bar{t})^4 \qquad \text{(Eq 4-26)}$$

Mean Residence Times and Normalized Distributions

A great deal of attention has been devoted to the determination of $\bar{t}$ from physical considerations. The principal result for incompressible fluids in closed systems is

$$\bar{t} = \frac{V}{Q} \qquad \text{(Eq 4-27)}$$

Where:

V = the volume of the system
Q = volumetric flow rate

This can be generalized to

$$\bar{t} = \frac{I_p}{Q_p} \qquad \text{(Eq 4-28)}$$

Where:

I_p = the total inventory or holdup of particles in the system
Q_p = total outflow of dead particles

Dead particles are those that have left the system, never to return. A simple proof of Eq 4-28 results from considering conservation of age in the system at steady state and noting that age is created spontaneously within the system (Nauman and Buffham 1983).

Once $\bar{t}$ is known, either from Eq 4-28 or from measured residence time data, it can be used to generate a normalized, or dimensionless, residence time distribution. The area under all residence time density functions is unity, and all normalized washout functions also cover unit area

$$\int_0^\infty \mathcal{W}(\tau)\, d\tau = 1 \qquad \text{(Eq 4-29)}$$

Where:

$\mathcal{W}(\tau)$ = the normalized or dimensionless washout function
τ = the dimensionless residence time, which is defined as

$$\tau = \frac{t}{\bar{t}} \qquad \text{(Eq 4-30)}$$

The moments of the normalized distribution are obtained from the corresponding unnormalized moments by dividing by $(\bar{t})^n$

$$\nu_n = \int_0^\infty \tau^n f(\tau)\, d\tau = \frac{\mu_n}{(\bar{t})^n} \qquad \text{(Eq 4-31)}$$

and

$$\nu'_n = \int_0^\infty (\tau - 1)^n f(\tau)\, dt = \frac{\mu'_n}{(\bar{t})^n} \qquad \text{(Eq 4-32)}$$

Where:

ν_n = dimensionless moments about the origin
ν'_n = dimensionless moments about the mean
$f(\tau)$ = normalized density function

By normalizing distributions, different distributions in the same (nondimensional) time domain can be compared. Note that in dimensionless form, all piston-flow system distributions collapse to the same curve, $f(\tau) = \delta(\tau - 1)$, while all stirred systems collapse to the same curve, $f(\tau) = \mathcal{W}(\tau) = \exp(-\tau)$. The practical matter of transforming from dimensional to dimensionless distributions is accomplished by considering conservation of probability, and leads to the following simple transformations (Nauman and Buffham 1983)

$$f(\tau) = \bar{t} f(t) \qquad \text{(Eq 4-33)}$$

$$\mathcal{F}(\tau) = F(t) \qquad \text{(Eq 4-34)}$$

$$\mathcal{W}(\tau) = W(t) \qquad \text{(Eq 4-35)}$$

All normalized distributions have $v_0 = v_1 = 1$, so that differences between distributions appear only in the second and higher moments. The dimensionless variance v_2' (also represented by σ_τ^2) has a theoretical range of zero to infinity. However, the range $0 \leq \sigma_\tau^2 \leq 1$ covers most practical situations, since $\sigma_\tau^2 = 0$ for piston-flow systems and $\sigma_\tau^2 = 1$ for stirred systems.

MODELS FOR RESIDENCE TIME DISTRIBUTIONS

This section develops several models for residence time distributions. These models are either fundamental, that is, derived from first principles, or empirical, that is, containing adjustable parameters. Real residence time data probably will fit one or more of the models better than the others. Real data may fit a model that does not seem to correspond to the actual process configuration. For example, the tanks-in-series model may fit data from a field installation consisting of a single tank. In such cases, the underlying physical interpretation of the model may shed light on the actual mixing.

Formulation Requirements

All models considered in this chapter are well formulated with regard to residence time theory. This means they satisfy the requirements on distribution functions given in Eq 4-1 and Eq 4-2. Restrictions on $F(t)$ and $W(t)$ follow from these, for example, $F(\infty) = 1$.

Empirical models have one or more adjustable parameters, discounting the mean residence time $\bar{t}$ since this is often known independently of any residence time experiment. Any discrepancy between values calculated using Eq 4-27 or Eq 4-28 and those determined from experimental residence time measurements represents a material balance error and/or nonideal tracer injection or detection methods. It is often preferable to compare candidate models in terms of their dimensionless distributions. Because all first moments are equal to unity, the lowest-order moment available for use as an adjustable parameter is the dimensionless variance $\sigma_\tau^2 = \sigma^2/(\bar{t})^2$. As was previously pointed out, $\sigma_\tau^2 = 0$ for piston-flow systems and $\sigma_\tau^2 = 1$ for the exponential distribution of a stirred tank. Most turbulent-flow systems have intermediate values for σ_τ^2, that is, $0 < \sigma_\tau^2 < 1$. The following sections discuss various models suitable for fitting data in this range.

Fractional Tubularity Model

The simplest model capable of $0 < \sigma_\tau^2 < 1$ is a stirred tank in series with a piston-flow element as illustrated in Figure 4-5. This is known as the fractional tubularity model. Figure 4-5A gives one physical representation of the fractional tubularity model. One can easily demonstrate that the distribution is the same regardless of the order of the stirred-tank and piston-flow elements. It is assumed that the piston-flow and stirred-tank elements are statistically independent. The washout function for the composite system is

$$W(t) = \exp\left[\frac{-(t - t_p)}{t_s}\right], t > t_p \qquad \text{(Eq 4-36)}$$

136 MIXING

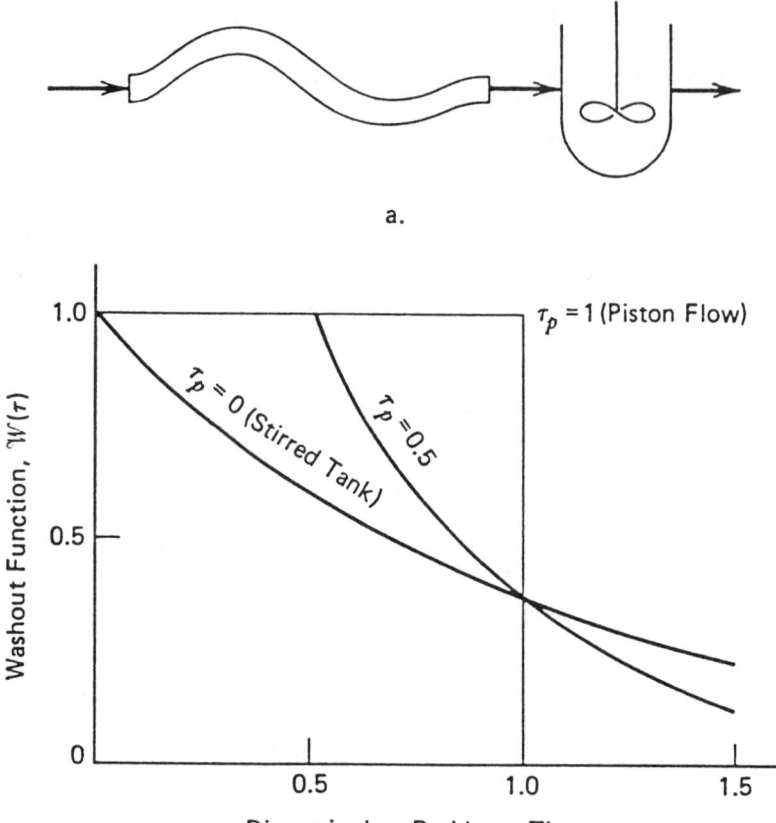

Source: Nauman, E.B. & Buffham, B.A. 1983. Mixing in Continuous Flow Systems. Copyright 1983, John Wiley and Sons, New York. Reprinted by permission of John Wiley and Sons, Inc.

Figure 4-5 The fractional tubularity model: (A) physical representation; (B) normalized washout function.

Where:

$t_p = V_p/Q$ = the mean residence time for the piston-flow element
$t_s = V_s/Q$ = the mean residence time for the stirred tank

Where:

V_p = volume of piston-flow element
V_s = volume of stirred tank

In dimensionless form,

$$W(T) = \exp\left[\frac{-(\tau - \tau_p)}{(1 - \tau_p)}\right], \tau > \tau_p \qquad \text{(Eq 4-37)}$$

Where:

τ_p = $V_p/(V_p + V_s)$, which is a single adjustable parameter called the fractional tubularity

Figure 4-5B depicts the response of the model for several values of τ_p. The dimensionless variance is given by

$$\sigma_\tau^2 = (1 - \tau_p)^2 = \tau_s^2 \qquad \text{(Eq 4-38)}$$

The model is well formulated only for $0 \le \tau_p \le 1$ and thus for $0 \le \sigma_\tau^2 \le 1$.

The fractional tubularity model is one of the simplest and most useful models in residence time theory. As shown in Figure 4-5B, it is characterized by a sharp first-appearance time $t_{min} = t_p = \tau_p \bar{t}$, where t_{min} = minimum residence time in system and by an exponential tail. The single parameter may be chosen from this first appearance time, from a measured variance, or by regression analysis in time domain (see discussion of data analysis later in this chapter, p. 152).

Tanks in Series

Process engineers often have been concerned with stirred-tank reactors, and the oldest residence time model (apart from the single stirred tank itself) is for a series arrangement of N stirred tanks of equal volume (see Figure 4-6). The residence time distribution for the tanks-in-series model is given by

$$f(t) = \frac{N^N t^{N-1}}{(\bar{t})^N (N-1)!} \exp\left(\frac{-Nt}{\bar{t}}\right) \qquad \text{(Eq 4-39)}$$

The normalized washout function is

$$\mathcal{W}(\tau) = \exp^{-N\tau} \sum_{i=0}^{N-1} \frac{N^i \tau^i}{i!} \qquad \text{(Eq 4-40)}$$

This function is illustrated in Figure 4-6B for various values of N. The normalized moments for the tanks-in-series model are

$$\nu_n = \frac{(N+n-1)!}{N^{n-1} N!} \qquad \text{(Eq 4-41)}$$

The dimensionless variance is

$$\sigma_\tau^2 = \frac{1}{N} \qquad \text{(Eq 4-42)}$$

If $N = 1$, all of the foregoing results reduce to those for a single stirred tank. If $N \to \infty$, the tanks-in-series model approaches piston flow. Corresponding to values of $N = 1, 2, 3, \ldots$ are discrete values of the dimensionless variance $\sigma_\tau^2 = 1, \frac{1}{2}, \frac{1}{3}, \ldots$. The tanks-in-series model can provide good representation of a wide range of mixing phenomena: Small values of N are appropriate for agitated

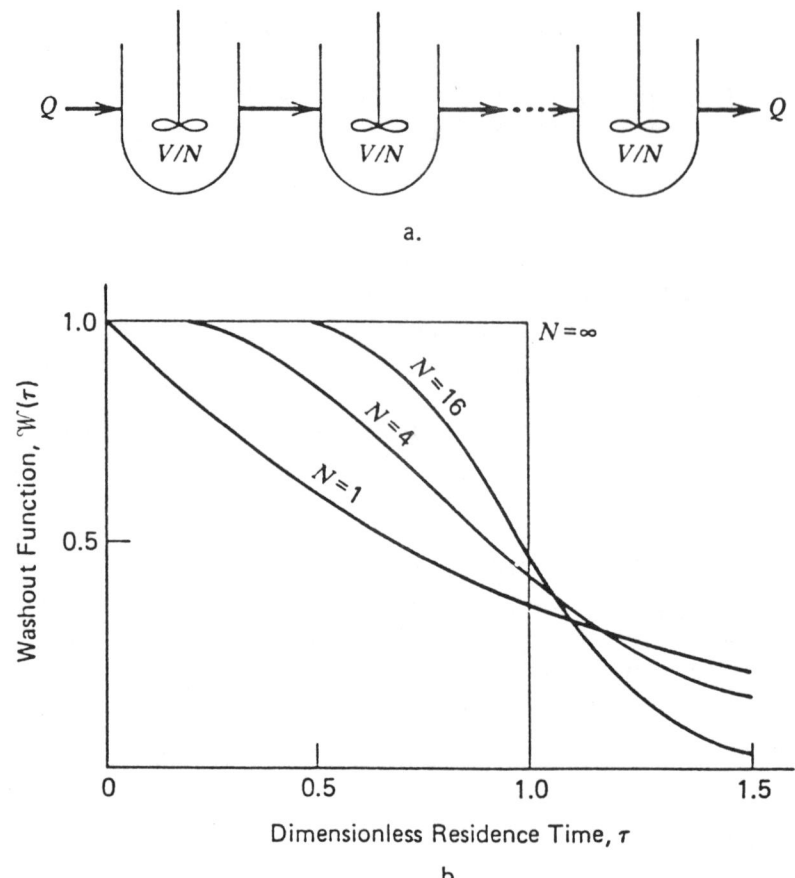

Source: Nauman, E.B. & Buffham, B.A. 1983. Mixing in Continuous Flow Systems. Copyright 1983, John Wiley and Sons, New York. Reprinted by permission of John Wiley and Sons, Inc.

Figure 4-6 The tanks-in-series model: (A) physical representation; (B) normalized washout function.

tanks, and larger values are appropriate for columns and long and narrow channels or tanks.

In this discussion, the tanks-in-series model has been defined only for integral values $N = 1, 2, \ldots$. This is no handicap when N is large, for small adjustments may be made through small changes in N. However, when N is small, a change in N causes relatively large differences in the residence time distribution function. Thus, there is a clear need for a method of interpolating between small values of N, particularly if the tanks-in-series model is to be used to model small deviations from the exponential distribution of the single tank. Two such extensions have been proposed: the gamma-function model and the fractional-tank model.

Gamma-function extension. The simplest extension of the tanks-in-series model is the gamma-function model. N is now treated as an adjustable index of mixing performance. When N is a continuous variable, the dimensionless density function is

$$f(\tau) = \frac{N^N \tau^{N-1} \exp^{-N\tau}}{\Gamma(N)} \qquad \text{(Eq 4-43)}$$

Where:

$\Gamma(N)$ = the gamma function, a tabulated function usually defined as

$$\Gamma(N) = \int_0^\infty \exp^{-x} x^{N-1} dx \qquad \text{(Eq 4-44)}$$

The gamma function can be thought of as a generalized factorial function because $\Gamma(N) = (N-1)!$ when N is a positive integer. Integration of Eq 4-43 gives

$$\mathcal{W}(\tau) = 1 - \frac{1}{\Gamma(N)} \int_0^{N\tau} \exp^{-x} x^{N-1} dx \qquad \text{(Eq 4-45)}$$

The definite integral is an incomplete gamma function, because the upper limit is finite rather than infinite, as in Eq 4-44. It plays the same role as the finite sum in Eq 4-40.

The simplicity of the gamma-function extension is not due entirely to the functional representation for $f(\tau)$; it is also due to the fact that Eq 4-42 still holds exactly, even when N is not an integer. This version of the tanks-in-series model is well formulated for all $N > 0$ and allows any σ_τ^2 in the range $0 < \sigma_\tau^2 < \infty$.

The major use of the gamma-function model is to fit small departures from the exponential distribution of a single stirred tank. Two sorts of departure are possible, depending on whether $\sigma_\tau^2 < 1$, or $\sigma_\tau^2 > 1$ for the experimental data. When $\sigma_\tau^2 < 1$, the model gives $N > 1$, and the outlet response to an impulse input rises somewhat more slowly than would be expected for a single stirred tank. When $\sigma_\tau^2 > 1$, the model gives $N < 1$, and the outlet response to an impulse rises more than expected. Physically, this corresponds to bypassing or short-circuiting of some entering fluid.

Short-circuiting and bypassing. The early chemical engineering literature uses the term *short-circuiting* for any kind of flow where some of the fluid experiences residence times much shorter than $\bar{t}$. The exponential distribution itself gives short-circuiting in this sense. On the other hand, the term *bypassing* often has been used to mean zero residence time for some portion of the fluid. An intermediate and more precise meaning for the term *bypassing* is needed. Bypassing is defined here as any flow for which $\mathcal{W}(\tau) < \exp(-\tau)$ for some suitably small values of τ. This means that bypassing exists when the washout function initially decreases more rapidly than would be expected for a perfect mixer. Since the normalized washout function must have a unit integral, any bypassing situation will have $\mathcal{W}(\tau) > \exp(-\tau)$ for some larger values of τ. In complete bypassing, some fraction of the entering fluid has zero residence time, that is, $\mathcal{W}(0+) < 1$. Bypassing and complete bypassing usually, but not always, give dimensionless variances greater than one.

The gamma-function extension to the tanks-in-series model shows bypassing when $N < 1$. The smaller N is, the more pronounced this feature becomes. If $N \to 0$, Eq 4-43 gives

$$f(\tau) \to 0, \tau > 0 \qquad \text{(Eq 4-46)}$$

$$f(\tau) \to \infty, \tau = 0 \qquad \text{(Eq 4-47)}$$

$$\int_0^\infty f(\tau)\, d\tau = 1 \qquad \text{(Eq 4-48)}$$

Despite these properties, $f(\tau)$ does not approach $\delta(\tau)$ in the limit of large N, because $v_n \to \infty$ for $n > 1$, whereas $v_n = 0$ for $n > 1$ with the delta function. In the limit of the gamma distribution, all particles bypass completely, except for an infinitesimal few that stay in the system for such an extremely long time that the mean residence time retains its value of $\bar{t}$.

Fractional tank extension. The gamma-function extension has the advantage of mathematical simplicity, at least as far as the moment relations are concerned, but the parameter N lacks a physical interpretation, except when it is an integer. The fractional-tank extension to the tanks-in-series model has been proposed as an alternative that overcomes this problem. The fractional-tank extension envisions the system as consisting of $I + 1$ stirred tanks in series, I of which have identical volumes $V/(I + \beta)$ and one of which has the smaller volume $\beta V/(I + \beta)$, where $0 \le \beta < 1$ can be considered the relative size of the fractional tank. Note that I and β are constrained by

$$I + \beta = N \qquad \text{(Eq 4-49)}$$

Where:

I = an integer just less than or equal to N
β = a fractional number of stirred tanks

Thus, the fractional-tank extension depends only on the single parameter N. A system of 30 m^3 total volume fitted with this model, using $N = 1.2$, would be treated as two stirred tanks in series, one with a volume of 25 m^3 and one with a volume of 5 m^3. The washout function for $N > 1$ becomes a finite series of $I + 1$ terms:

$$\mathcal{W}(\tau) = \left(\frac{-\beta}{1-\beta}\right)^I \exp^{-N\tau/\beta} + \exp^{-N\tau} \sum_{i=0}^{I-1} \frac{N^i \tau^i}{i!}\left[1 - \left(\frac{-\beta}{1-\beta}\right)^{I-1}\right] \qquad \text{(Eq 4-50)}$$

This clearly reduces to Eq 4-40 when $\beta = 0$.

For $N < 1$, $I = 0$ and the fractional-tank extension takes the form of a single stirred tank with a fraction $1 - N$ of the fluid completely bypassing it. The washout function is

$$\mathcal{W}(\tau) = N \exp^{-N\tau} \qquad \text{(Eq 4-51)}$$

Use of Extended Models

If all the subsystems in a series were stirred vessels, only the basic tanks-in-series model would be needed. A major use of either the gamma-function or fractional-tank extension is to model systems that are physically compartmentalized into subsystems of agitated vessels. The extensions are used to account for relatively small departures from the ideal exponential distribution in each subsystem. The extended models also can be used to account for backmixing between stirred tanks in series. In this case they compete with more specialized backmixing models. In addition, the extensions can be used to model small deviations from ideality for a single stirred tank. In this case, the gamma function is probably preferred if $\sigma_\tau^2 > 1$, since it predicts partial bypassing, while the fractional-tank extension predicts complete bypassing for a fraction of the entering fluid. The best method of discriminating between competitive models remains the method of time domain fitting discussed later in this chapter under the heading "Data Analysis."

Crossflow, Stagnancy, and Side Capacity

Adding backflow to a model increases its ability to simulate mixing in the axial, or flow, direction. Another way of modifying simple models is to add regions that are more or less stagnant. This enhances the ability of the simple model to handle mixing in the radial direction (that is, perpendicular to the main flow). In reference to agitated-tank models, exchange of particles between an actively mixed region and a stagnant region, with no possibility of leaving directly from the stagnant region, is called crossflow. A simple version of this model, shown in Figure 4-7A, contains two adjustable parameters: q/Q and V_1/V_2 (q = recycle or subsystem flow rate; V_1, V_2 = volume of subsystems).

The density function for the model is

$$f(t) = \frac{(1 - a_1 t_2) \exp^{-a_1 t} - (1 - a_2 t_2) \exp^{-a_2 t}}{(a_2 - a_1)(1 + q/Q) t_1 t_2} \quad \text{(Eq 4-52)}$$

Where:

$t_1 = V_1/(Q + q)$
$t_2 = V_2/q$
a_1, a_2 = roots of the quadratic, given by

$$a_1, a_2 = \frac{1}{2}\left[\frac{t_1 + t_2}{t_1 t_2} \pm \sqrt{\frac{(t_1 + t_2)^2}{t_1^2 t_2^2} - \frac{4}{(1 + q/Q) t_1 t_2}}\right] \quad \text{(Eq 4-53)}$$

The side tank in Figure 4-7A causes partial stagnancy. A qualitative definition of stagnancy can be based on the residence time intensity function

$$\Lambda(t) = \frac{f(t)}{W(t)} \quad \text{(Eq 4-54)}$$

The only restriction on the intensity function $\Lambda(t)$ is that $\Lambda(t) \geq 0$ for $t \geq 0$. Given any $\Lambda(t)$, the corresponding density function can be found from

$$f(t) = \Lambda(t) \exp\left[-\int_0^t \Lambda(t') dt'\right] \quad \text{(Eq 4-55)}$$

The intensity function is a conditional probability density function: Given that a particle has age t, $\Lambda(t) dt$ is the probability that it will leave the system within the next time increment dt. The intensity function is an escape probability for particles in the system, and stagnancy exists whenever this escape probability is a decreasing function of t for any interval of t. Returning to the tanks-in-series model, $\Lambda(t)$ is constant for $N = 1$ and is monotone increasing for $N > 1$. Thus, no stagnancy exists in such systems. Figure 4-7B shows $\Lambda(t)$ for a crossflow system. Here, $\Lambda(t)$ is monotone decreasing so that the system does show stagnancy in a qualitative sense.

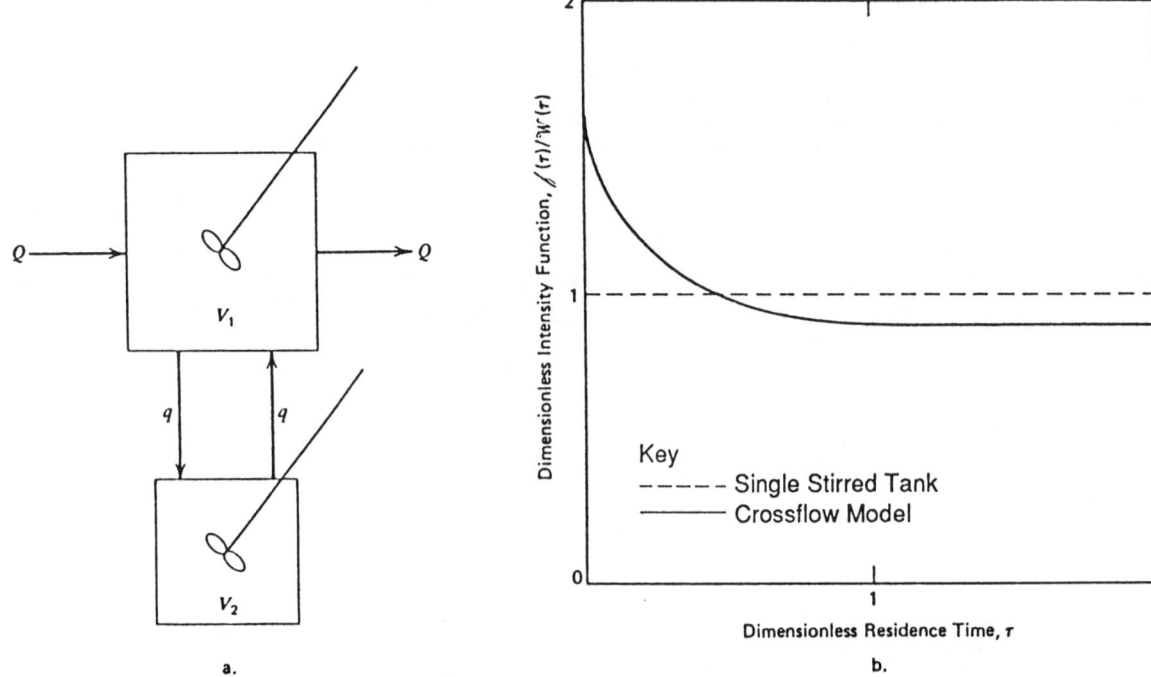

Source: Nauman, E.B. & Buffham, B.A. 1983. Mixing in Continuous Flow Systems. Copyright 1983, John Wiley and Sons, New York. Reprinted by permission of John Wiley and Sons, Inc.

Figure 4-7 Modeling stagnancy with crossflow: (A) stirred tanks with crossflow; (B) intensity function indicating stagnancy.

A still qualitative, but more physically based, definition of stagnancy regards a system as exhibiting stagnancy whenever it has a stagnant region like the lower stirred tank in the crossflow model (Figure 4-7A). A stagnant region allows circulation or diffusion of particles in the radial direction but has little, if any, net velocity in the axial, or streamwise, direction. Typically, particles rejoin the active region from the stagnant region at a position very close to where they left. More generally, the particles have experienced net, positive velocities much smaller than they would have experienced had they remained in the active zone.

Axial Dispersion

One of the best-known and most useful models in transport modeling is that for axial dispersion. The performance of many, more-or-less tubular and open-channel flow systems can be approximated by the one-dimensional, convective diffusion equation

$$\frac{\partial C}{\partial \theta} + \bar{u}\frac{\partial C}{\partial z} = D\frac{\partial^2 C}{\partial z^2} + \mathcal{R}(C) \qquad \text{(Eq 4-56)}$$

Where:

> θ = elapsed or clock time
> C = $C(\theta, z)$
> $\bar{u}$ = the average fluid velocity
> z = axial coordinate, or axial displacement
> D = a kind of effective diffusivity known as the axial dispersion coefficient

The coefficient D lumps the effects of a nonuniform velocity profile (shear dispersion), molecular diffusivity, and eddy diffusivity into a single parameter. The model provides a reasonably good fit for the transient and steady-state response of packed beds, turbulent flow in tubes of arbitrary cross section, and even laminar flow with molecular diffusion. Equation 4-56 is usually expressed with independent variables in dimensionless form as

$$\frac{\partial C}{\partial \tau} + \frac{\partial C}{\partial x'} = \frac{1}{\text{Pe}} \frac{\partial^2 C}{\partial x'^2} + \bar{t}\,\mathcal{R} \qquad \text{(Eq 4-57)}$$

Where:

> τ = dimensionless residence time, $\theta \bar{u}/L = \theta/\bar{t}$
> x' = dimensionelss axial coordinate z/L
> Pe = axial Peclet number* $\bar{u}L/D$

Where:

> L = length of system

It is this dimensionless group Pe, that determines the residence time distribution in a region of length L

$$\mathcal{W}(\tau) = \exp^{\text{Pe}/2} \sum_{i=1}^{\infty} \frac{8\omega_i \sin \omega_i \exp\left[-\left(\frac{\text{Pe}^2 + 4\omega_i^2}{4\text{Pe}}\right)\tau\right]}{\text{Pe}^2 + 4\text{Pe} + 4\omega_i^2} \qquad \text{(Eq 4-58)}$$

Where:

> ω_i = the positive roots of

$$\tan \omega_i = \frac{4\omega_i \,\text{Pe}}{4\omega_i^2 - \text{Pe}^2} \qquad \text{(Eq 4-59)}$$

This series converges rapidly for small Pe, but an asymptotic solution is necessary to avoid convergence problems for large Pe. For Pe > 16, the following result provides an excellent approximation:

$$\mathcal{W}(\tau) = 1 - \int_0^\tau \frac{\text{Pe}}{4\pi\,\theta^3} \exp\left[\frac{-\text{Pe}(1-\theta)^2}{4\theta}\right] d\theta \qquad \text{(Eq 4-60)}$$

---

*The inverse of Pe is called the axial dispersion number.

Figure 4-8 illustrates the behavior of $\mathcal{W}(\tau)$ for various values of Pe.

In the axial dispersion model there is no first appearance—the fastest particles move with infinite speed. Also, the tail of the distribution is exponential and moments of all positive orders exist. Many real systems have these characteristics, and it is often possible to correlate $Pe = \bar{u}L/D$ with dimensionless groups, such as the Reynolds number, Re, which allows scale-up with relative certainty. Standard texts, such as those by Levenspiel (1972), Himmelblau and Bischoff (1968), and Wen and Fan (1975), give correlations for packed beds and turbulent pipe flows. Wen and Fan also give extensive references to the experimental literature. Unfortunately, all existing correlations apply to tubes and other conduits and not to the open channels and tanks commonly used in water treatment plants.

The dimensionless variance corresponding to the axial dispersion model is

$$\sigma_\tau^2 = \frac{2}{Pe} - \frac{2}{Pe^2}(1 - \exp^{-Pe}) \qquad \text{(Eq 4-61)}$$

This result has the theoretical range of $0 < \sigma_\tau^2 < 1$ for $0 < Pe < \infty$. However, experience shows the axial dispersion model is best suited for situations approaching piston flow with $\sigma_\tau^2 < 0.5$.

In a steady-state reacting system, Eq 4-57 reduces to an ordinary differential equation

$$\frac{1}{Pe}\frac{d^2C}{dx'^2} - \frac{dC}{dx'} + \bar{t}\,\mathcal{R}(C) = 0 \qquad \text{(Eq 4-62)}$$

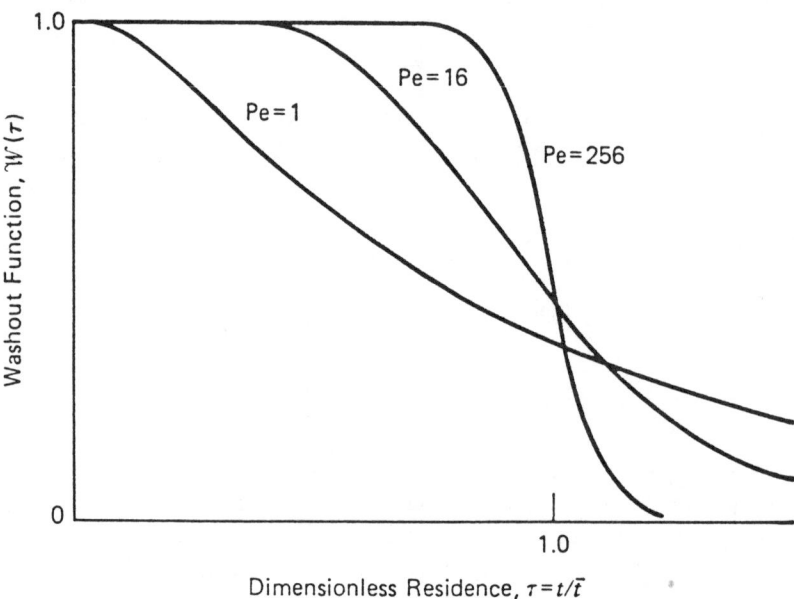

Source: *Nauman, E.B. & Buffham, B.A. 1983.* Mixing in Continuous Flow Systems. *Copyright 1983, John Wiley and Sons, New York. Reprinted by permission of John Wiley and Sons, Inc.*

Figure 4-8 Residence time distributions for the axial dispersion model.

For a first-order reaction with $\mathcal{R}(C) = -kC$ (k = reaction-rate constant), this equation is linear, has constant coefficients, and is homogeneous. The solution is subject to the so-called closed, or Danckwerts, boundary conditions. At the inlet

$$C_{in}(\tau) = C(\tau, 0-) = C(\tau, 0+) - \frac{1}{\text{Pe}} \frac{\partial C}{\partial x}\bigg|_{0+} \quad \text{(Eq 4-63)}$$

Where:

C_{in} = inlet concentration, steady state (average over entering flow)

and at the outlet there is the gradient boundary condition

$$\frac{\partial C}{\partial x}\bigg|_{L-} = 0 \text{ for all } \tau \quad \text{(Eq 4-64)}$$

The solution is

$$\frac{C_{out}}{C_{in}} = \bar{f}(k\bar{t}) = \frac{4a \exp[\text{Pe}(1-a)/2]}{(1+a)^2 - (1-a)^2 \exp(-\text{Pe}\, a)} \quad \text{(Eq 4-65)}$$

Where:

C_{out} = outlet concentration, steady state (average over exiting flow)

$$a = \sqrt{1 + \frac{4k\bar{t}}{\text{Pe}}} \quad \text{(Eq 4-66)}$$

This result gives the fraction unreacted C_{out}/C_{in} for a first-order reaction in a closed axial dispersion system. The solution contains the two dimensionless parameters Pe and $k\bar{t}$. The Peclet number Pe controls the level of mixing in the system. With Pe $\to 0$ (either small $\bar{u}$ or large D), diffusion becomes so important that the system acts as a perfect mixer. Thus,

$$\frac{C_{out}}{C_{in}} \to \frac{1}{1 + k\bar{t}} \text{ as } \text{Pe} \to 0 \quad \text{(Eq 4-67)}$$

As Pe $\to \infty$ (large $\bar{u}$ or small D), the system behaves as a piston-flow reactor, so that

$$\frac{C_{out}}{C_{in}} \to \exp(-k\bar{t}) \text{ as } \text{Pe} \to \infty \quad \text{(Eq 4-68)}$$

It is apparent from this that the parameter Pe controls the residence time distribution and that the distribution is independent of $k\bar{t}$.

The yield of a first-order reaction is the Laplace transform of the differential distribution function. In dimensionless form

$$\frac{C_{out}}{C_{in}} = \bar{f}(k\bar{t}) = \int_0^\infty \exp^{-k\bar{t}\tau} f(\tau)\, d\tau \quad \text{(Eq 4-69)}$$

Thus, Eq 4-65 gives the Laplace transform of the dimensionless density function when $k\bar{t}$ is interpreted as the transform parameter $\emptyset = k\bar{t}$. The time domain form $f(\tau)$ can, in principle, be found by Laplace transform inversion, but this is difficult because of the necessary algebraic manipulations. Fortunately, a time domain solution for $f(\tau)$ also can be found through mathematical transient-response techniques with inert tracers. The time domain washout function is given in Eq 4-58.

Motionless-Mixer Models

Motionless mixers can be defined as an order set of static, flow-diverting elements installed within the fluid stream. They redistribute fluid across the flow channel, promoting turbulence and mixing. In most water treatment applications for motionless mixers, the flow is in the turbulent regime, and there is little experimental data on the residence time distributions. Data are available for the Kenics static mixer (see chap. 11) in the laminar range, and this can be used to provide a conservative estimate for turbulent flows. (Here, conservative means the variance σ^2 calculated using this approximation should be larger than the variance for the real device.)

A mathematical approximation of the Kenics mixer can be made using the tanks-in-series model with N equal to the number of Kenics elements in the motionless mixer divided by four. Thus, a 20-element Kenics mixer should have a residence time distribution corresponding to five stirred tanks in series. The basic concepts behind this approximation are discussed by Nauman and Buffham (1983) and have since been shown valid for the flattened velocity profiles typical of pseudoplastic fluids in laminar flow. The tanks-in-series approximation is reasonably accurate for Kenics mixers containing 16 or more elements.

MEASUREMENT TECHNIQUES

This section describes techniques for measuring the residence time density function in simple, closed systems and also in multiport closed systems and open systems.

Inert Tracers

In a simple, closed system the residence time density function can be measured directly by injecting a unit impulse of conserved particles into the inlet of a simple, closed system. Strictly speaking, the measured distribution is that for the tracer particles themselves, but it is usually hoped that the measurement reflects the residence time distribution of some other particles in the fluid stream. A perfect tracer has exactly the same flow properties as the substance or particles it represents, and yet it is sufficiently different in some nonflow attribute that it can be detected by an analytical instrument. Ideally, the background concentration of the tracer should be essentially zero, and the detector should give a response that is linear with concentration. Lithium chloride has proven to be a suitable tracer for water treatment systems and has been used in large-scale treatment plant studies.

Once an inert tracer has been chosen, a suitable injection technique must be chosen. Impulse injection is conceptually simple but requires normalizing or scaling of the output response before it can be interpreted as a density function with $\mu_0 = 1$. The scaling takes the form

$$f(\theta) = \frac{C_{\text{out}}(\theta)}{\int_0^\infty C_{\text{out}}(\theta)\, d\theta} \quad \text{(Eq 4-70)}$$

Where:

$f(\theta)$ = the population of particles leaving the system at time θ

C_{out} must represent the amount by which the tracer concentration is in excess of any background concentration. The desirability of having negligible background concentrations is apparent.

In the positive-step-change experiment, C_{in} is changed instantaneously from some steady reference value (perhaps zero) to a higher value. The cumulative distribution function is defined directly by the outlet concentration divided by the ultimate value of the response, as shown in the following equation:

$$F(\theta) = \frac{C_{out}(\theta)}{C_{out}(\infty)} \qquad (Eq\ 4\text{-}71)$$

This has the fairly minor disadvantage of requiring a measurement of $C_{out}(\infty)$. As before, C_{out} represents the concentration above background.

The negative-step-change experiment is the inverse of the positive step change. Here the tracer is switched from a steady, high concentration to a lower concentration. The washout function is obtained directly when the outlet concentration is divided by the initial concentration:

$$W(\theta) = \frac{C_{out}(\theta)}{C_{out}(0^-)} \qquad (Eq\ 4\text{-}72)$$

When the background concentration is negligible, the negative step change provides the best possible method for defining the tail of the distribution and for accurately determining moments.

Positive and negative step changes can be used together to test for tracer density effects. Density effects are hard to detect with impulse experiments. They are less important in step-change experiments because the absolute tracer concentrations are lower. They are also easy to detect since the density gradients for positive and negative step changes will be in opposite directions.

Well-designed laboratory experiments will almost always use one of the ideal transient-response methods described here. With existing equipment, however, it may not be possible to apply a good impulse or step change to the inlet of the system.

In principle, $f(t)$ can still be determined using nonideal signals if both $C_{in}(\theta)$ and $C_{out}(\theta)$ are monitored. The theoretical justification of the statement is the convolution pair

$$C_{out}(\theta) = \int_0^\infty (\theta - \theta') \exp^{-k\theta'} f(\theta')\, d\theta' = \int_{-\infty}^0 C_{in}(\theta') \exp^{-k(\theta - \theta')} f(\theta - \theta')\, d\theta' \qquad (Eq\ 4\text{-}73)$$

which is written for a reactive tracer with first-order rate constant k. An inert tracer has $k = 0$. Actual deconvolution of the experimental data to recover $f(t)$ may be very difficult.

Multiport Closed Systems

A simple system has only a single inlet and a single outlet. A system that is not simple is called multiport and must have at least three connections to its environment. A multiport closed system becomes a simple closed system if the inlets and outlets are connected by inlet and outlet manifolds. When one is concerned with the residence times of all the material passing through the system, this ploy is a convenient way of using the simple, closed-system theory, especially when using mathematical models. When it comes to experimentation, the provision of manifolds is likely to be inconvenient, if indeed it is possible. Moreover, the presence of the manifolds causes difficulties because their necessarily finite volumes increase the mean residence time.

To be properly representative, tracer particles must replace flow particles in quantities proportional to the inlet flow rates. For a step-response experiment, this condition could be met by simultaneously replacing the inlet streams by tracer-containing streams flowing at the same rates as the streams they replace, and all having the same tracer concentration. If impulse injection of tracer is preferred, the number of particles injected at each inlet must be proportional to the flow rate at that inlet. The rates at which tracer emerges from the several outlets could be measured by monitoring the concentrations: The total rate at which tracer leaves the system would be calculated as the sum of the products of concentration and flow rate for the various streams. The distribution function for the multiport closed system could then be calculated in the same way as for a simple closed system.

Since the system responses are linear with respect to tracer concentrations and the flows are all held constant, the residence time behavior for the system may be synthesized from the responses obtained by manipulating the inputs individually. This allows one to consider the residence time distribution for a particular species when that species is present at different concentrations in the various streams entering the system. The overall residence time distribution for the system is a weighted sum of functions representing the way a given outlet responds to tracer flow at a given inlet.

Suppose that the flow rate into the ith inlet is $q_{i.}$ and that from the jth outlet is $q_{.j}$. By overall material balance, the throughflow is

$$Q = \sum_i q_{i.} = \sum_j q_{.j} \qquad \text{(Eq 4-74)}$$

Let q_{ij} be the flow rate of particles from the ith inlet to the jth outlet. Again, by material balance

$$q_{i.} = \sum_j q_{ij} \qquad \text{(Eq 4-75)}$$

and

$$q_{.j} = \sum_i q_{ij} \qquad \text{(Eq 4-76)}$$

The $q_{i.}$ and $q_{.j}$ can be measured directly, whereas the q_{ij} must be deduced, for example, from steady-state flow-rate and concentration measurements when the stream entering a particular inlet is the only one to contain tracer.

Particles take various times to flow from the ith inlet to the jth outlet. If $f_{ij}(t)\,dt$ is the fraction of the i-to-j flow that takes time $(t, t + dt)$, then $q_{ij} f_{ij}(t)\,dt$ is the rate of flow from inlet i to outlet j of particles with residence time $(t, t + dt)$.

The total flow through the system of particles with this residence time is found by summing over all the inlets and outlets and is also $Qf(t)dt$, so that

$$f(t) = Q^{-1} \sum_i \sum_j q_{ij} f_{ij}(t) \qquad \text{(Eq 4-77)}$$

As $f_{ij}(t)$ is a conditional density function, it can be found from the concentration measured at the jth outlet when an arbitrary quantity of tracer is injected at the ith inlet. If the injection is an impulse, then

$$f_{ij}(\theta) = \frac{C_j(\theta)}{\int_0^\infty C_j(\theta)\, d\theta} \qquad \text{(Eq 4-78)}$$

The analysis becomes simpler for the case of a single inlet with multiple outlets. A delta function or step change in concentration is made at the inlet. Then the tracer concentration is monitored at each of the outlets, either simultaneously using multiple detectors or sequentially using a single detector but with multiple experiments. The overall outlet concentration, if obtained by averaging with respect to flow rate, is

$$\overline{C}_{\text{out}}(\theta) = Q^{-1} \sum_j q_{\cdot j} \left[C_{\text{out}}(\theta) \right]_j \qquad \text{(Eq 4-79)}$$

The data analysis using $\overline{C}_{\text{out}}(\theta)$ is identical to that for a system with a single inlet and a single outlet.

Open Systems

Figure 4-9 illustrates the differences between open and closed systems. The transfer lines are small and have high velocities in closed systems, so that any backward flow of material in the transfer lines is negligible. Open systems are characterized by large transfer lines and fluid velocities similar to those in the system as a whole. In terms of the axial dispersion model, closed systems have $D_{\text{out}} = 0$ in the inlet and outlet transfer lines, while open systems have $D_{\text{in}} > 0$ and $D_{\text{out}} > 0$ in these lines. (D_{out} = value of D for $Z > L$; D_{in} = value of D for $Z < 0$.)

Material that once enters a closed systems stays in it until it finally exits, never to return. The residence time distribution in such systems is easily determined by inert tracer experiments.

In open systems, as in closed systems, particles have some initial entrance into the system and some final exit from it. In between, however, open systems allow temporary excursions outside the system boundaries, while closed systems do not. Residence stops outside the system boundaries; therefore, time spent outside does not contribute to the transient response of inert tracers because nonreactive tracers have no way of knowing whether or not they are inside the arbitrarily defined system boundaries.

The transient response is still given by a convolution pair

$$C_{\text{out}}(\theta) = \int_0^\infty C_{\text{in}}(\theta - \theta')\, g(\theta')\, d\theta' \qquad \text{(Eq 4-80)}$$

$$= \int_{-\infty}^\theta C_{\text{in}}(\theta')\, g(\theta - \theta')\, d\theta'$$

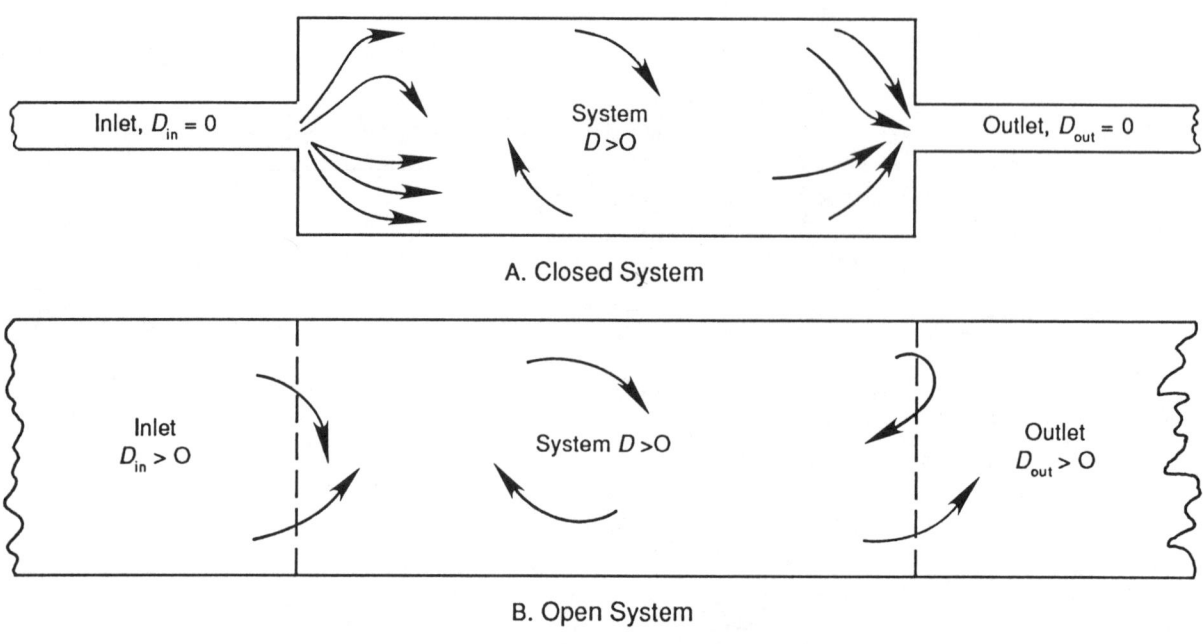

Adapted from Bisio, A.L. & Kabel, R.L., 1985. Scaleup of Chemical Processes: Conversion From Laboratory Scale Tests to Successful Commercial Size Design. Wiley-Interscience, New York.

Figure 4-9 D_{In}, D_{out} = axial dispersion coefficients: (A) closed system; (B) open system.

Where:

$g(\theta)$ = transient response to unit impulse

but $g(\theta)$ is no longer identical to $f(\theta)$. This situation has been discussed extensively elsewhere (Nauman and Buffham 1983). The true residence time distribution can be deduced mathematically from a model of the open system or deduced experimentally using reactive tracers. Such determinations are awkward at best, and it is fortunate that most water treatment systems are closed or nearly closed. In principle, truly open systems should be modeled using some form of the convective diffusion equation (Eq 4-56).

A simple test exists to determine whether a system is sufficiently open to cause significant error in measuring the residence time distribution. A transient experiment is made and analyzed as though the system were closed. The mean residence time is then calculated from the experimental data using Eq 4-14 or Eq 4-16. This value for $\bar{t}$ is then compared to that calculated from Eq 4-27 or Eq 4-28. If they agree, the system is approximately closed and the error in the measured residence time distribution should be negligible. Of course, this test also assumes that common errors, such as imperfect injection and detection techniques, have not been made. For the open system illustrated in Figure 4-9B, the mean calculated from Eq 4-14 or Eq 4-16 is

$$\mu_1 = \frac{L}{\bar{u}} + \frac{D_{in} + D_{out}}{u^2} = \bar{t} + \frac{D_{in} + D_{out}}{u^2} \qquad \text{(Eq 4-81)}$$

so that the two means agree when $D_{in} = D_{out} = 0$.

Measurement Techniques for Full-Scale Water Treatment Facilities

For the large tanks used in contemporary flocculation processes, the spatial distribution of both the incoming and outgoing flow is an important design characteristic. By using the correct flow-distribution device (for example, manifolds, baffles, or weirs), certain pathological flow behaviors, such as short-circuiting and density currents, can be minimized (Hudson 1981). The use of manifolds and weirs has an interesting drawback: The system becomes more complex with regard to residence time distribution measurement. In complex (multiport) systems, it is necessary to adopt the approaches used for multiport closed systems. If it is not possible to determine the conditional density function f_{ij} (Eq 4-78) and q_{ij}, then the following experimental constraints are required:

1. The flow through each inlet and outlet must be known or accurately approximated.

2. For multiple inlets, a method must be devised for injecting tracer into each of the flow channels such that the tracer concentration is uniform in each inlet.

3. For multiple outlets, tracer concentration must be measured simultaneously in each outlet, or in successive outlets using duplicate tracer-injection experiments.

4. For distributed inlets and outlets, such as serrated weirs, it may be sufficient to inject and measure tracer at a few representative points along the inlet and outlet structures.

Due to the difficulties inherent in setting up tracer experiments for multiple-inlet tanks, certain simplifications might be unavoidable. For example, if separate channels feed different sections of the weir, then the tank system might be redefined as the tank plus the inlet weir and distribution channels. One would then avoid the need to either measure the flow in each channel or develop a method of uniformly distributing tracer across the individual channel sections. An even more extreme redefinition of the flocculation tank involves injecting tracer prior to the rapid-mix unit. With all of these simplifications, one must be completely aware that the residence time distribution data are no longer strictly for the flocculation tank alone. However, when the volume represented by the inlet-flow-distribution system is small relative to the main tank, and when there are no inherent pathological mixing situations in the flow-distribution system, a small modification of the definition of the system may be a small price to pay to overcome a formidable experimental challenge. The above comments also hold for distributed-outlet systems.

Often, the settling tank and flocculation tank are actually part of one continuous tank. In this case, one has little choice but to consider the system as flocculation-plus-settling. In some of these systems it may be possible to measure tracer concentration between the flocculation and settling tanks, but it may be very difficult to properly inject tracer at the correct rate. However, if the tracer measurements between the tanks and at the exit from the settling tank are known, the methods previously described can yield the residence time distributions corresponding to the flocculation tank and the combined tanks. From these two distributions, the distribution corresponding to the settling tank alone can be found (in principle) by mathematical deconvolution. The recommended approach is to fit a simple model to the two measured distributions, take their Laplace transforms, perform the deconvolution in the Laplace domain, and invert to obtain the time domain distribution for the settling tank. Further details are given by Nauman and Buffham (1983).

DATA ANALYSIS

Suppose that $f(t)$ or $W(t)$ has been determined experimentally. The results will contain random errors in the form of scatter, they will be truncated at some finite value of t, and they may contain errors. The data can be used in a variety of ways.

- Moments of the residence time distribution can be determined directly from the data.
- The data can be fitted to a system model by identification of parameters other than the moments.
- The data can be used directly to predict reaction yields or to estimate the system response when subject to incoming concentration fluctuations.

The literature has stressed direct determination of the moments. Impulse response data always require determination of the zero moment to ensure that $\mu_0 = 1$, and it is useful to calculate $\bar{t}$ for any residence time measurements.

Determination of Moments

When moments are to be extracted from experimental or simulated residence time measurements, it is best to use the washout function $W(t)$. The zero moment condition is automatically satisfied: $W(0) = \mu_0 = 1$. Higher moments are found from Eq 4-15, which uses a t^{n-1} weighing factor in the integration to find μ_n. This minimizes the effect of the tail of the distribution compared to calculation from $f(t)$ data using Eq 4-13. Even so, the calculated moments will be heavily dependent on data taken at large values of t, and some means of extrapolating to $t = \infty$ is necessary to avoid sizable truncation errors. Fitting an exponential tail to the experimental data is usually appropriate, since diffusion ensures that the asymptotic form of the distribution will always be exponential (Nauman and Buffham 1983).

When the cumulative distribution is obtained experimentally, it is immediately converted to the washout function: $W(t) = 1 - F(t)$. When the differential distribution is measured, it is necessary to first scale the results according to Eq 4-70 so that $\mu_0 = 1$. An exponential extrapolation of the experimental concentration data is usually needed for this integration, and extrapolations are almost certainly needed for the higher moments, which use a weighting factor of t^n to calculate μ_n.

When central moments are desired, they are not computed directly from the data but are found using Eq 4-18 through Eq 4-22. They tend to be slightly more error-prone than the ordinary moments of the same order.

Parameter Estimation and Model Discrimination

Most mixing models express prior knowledge about the mixing mechanism. Incomplete knowledge is reflected by the presence of adjustable parameters in the model. The parameters are determined by comparing experimental data with model predictions.

The chemical engineering literature has stressed the use of moments to determine model parameters. For example, consider a composite system model that consists of a stirred tank in parallel with a piston-flow element, the two subsystems having identical values for $\bar{t}$. The dimensionless variance for the composite system is in the range $0 \leq \sigma_t^2 \leq 1$, and the numerical value of σ_t^2 corresponds to the fraction of the total system volume that is accounted for by the stirred tank. (Equivalently, it is the fraction of total flow rate fed to the stirred tank.) This fraction is V_s. Then the model can be fitted to experimental data by setting $V_s = \sigma_t^2 = \mu_2/(\bar{t})^2 - 1$, where $\bar{t}$ and μ_2 have been evaluated using the methods previously described. Assuming $\sigma_t^2 \leq 1$, this concludes the problem of parameter estimation. Now consider the question of model discrimination based on goodness of fit.

An infinite number of models have dimensionless variances in the range of 0 to 1. For example, stirred-tank and piston-flow elements in series also have $0 \leq \sigma_t^2 \leq 1$. An additional criterion is needed to distinguish between the competing models, such as the parallel and series configurations in the example under consideration. The method of moments is sometimes suitable for parameter estimation, given a fixed model. It is the normal approach used with the imperfect-impulse method, since this method usually gives nothing but the moments but is generally unsuited for model discrimination. The method of moments would use the skewness μ_3', for this purpose, but accurate third moments are seldom available. Fortunately, better methods for parameter estimation are available when the full $f(t)$ or $W(t)$ curve is known. The best, general-purpose method is time domain fitting.

Suppose a mathematical model of the system contains parameters $P_1, P_2, \ldots$ Then time domain fitting consists of choosing the parameters to minimize the sum of squares of differences, or sum-squared error (SSE), between the data and the model, as follows:

$$\text{SSE}(P_1, P_2, \ldots) = \sum_i [f_{\text{data}}(t_i) - f_{\text{model}}(t_i)]^2 \qquad (\text{Eq 4-82})$$

Discrimination between competing models can be based on the value of SSE after minimization. Equation 4-82 is written as a finite sum rather than an integral over $(0, \infty)$ because the experimental data are usually available at discrete values of t. Also, the form of Eq 4-82 emphasizes that the selection of $P_1, P_2, \ldots$ is a problem in (nonlinear) regression analysis.

Canned computer routines are available to calculate the parameters. These should be used with the caution appropriate in regression analysis (largely to avoid overfitting the data). There is the additional constraint that $f(t)$ must be positive. Also, it may be desired to pick the parameters subject to a fixed first moment, $\bar{t}$. These constraints are best satisfied by restricting the functional form of the model. Alternatively, one often can use a microcomputer to complete an efficient search for parameter values leading to the best model fit based on minimization of the sum-squared error (see example calculations later in this chapter).

Equation 4-82 has been written in terms of the differential distribution function $f(t)$. This is usually best for time domain fitting because impulse-response data are more sensitive to the internal structure of the system than step change data. However, $F(t)$ or $W(t)$ data also can be used, if desired. The constraints that $F(t)$ be nondecreasing or $W(t)$ be nonincreasing may be satisfied by restricting the form of the mathematical model.

Determining parameters for imperfect pulses. Numerical time domain fitting can be used to determine model parameters when responses to imperfect pulses have been measured. Suppose that the impulse response for the model is known analytically or numerically. For any set of parameters that describe the model, the output corresponding to the imperfect pulse can be calculated by numerical convolution. This is a stable operation. Some measure of the goodness of fit (for example, SSE) can be minimized by a suitable regression, optimization, or search routine. This method allows competing models to be compared (via the sum-squared error), whereas the moments method does not allow this comparison.

Direct Utilization of Data

Some of the most important uses of residence time data can be achieved without calculating moments or fitting models. The outlet concentration from a mixing system with or without first-order reaction can be found by numerical integration of Eq 4-73 when $f(t)$ is known or by integration of

$$C_{\text{out}}(\theta) = C_{\text{in}}(\theta) - \int_0^\infty [C'_{\text{in}}(\theta - t) + kC_{\text{in}}(\theta - t)] \exp^{-kt} W(t) dt \quad \text{(Eq 4-83)}$$

when $W(t)$ is known. Equation 4-83 was obtained from Eq 4-73 through integration by parts. Both equations are valid for the general case of reaction with unsteady inlet concentrations. They reduce to the prediction of steady-state reaction yields when $C_{\text{in}}(\theta) = C_{\text{in}} = $ constant. The result is $C_{\text{out}} = C_{\text{in}} \overline{f}(t)$:

$$\overline{f}(k) = \int_0^\infty \exp^{-kt} f(t) dt = 1 - k \int_0^\infty \exp^{-kt} W(t) dt \quad \text{(Eq 4-84)}$$

For yield prediction, k has some fixed value determined from laboratory kinetic data. When a model is to be fitted in the Laplace domain, various values of k are selected and a numerical integration of Eq 4-84 is done for each value of k to generate a set of experimental $\overline{f}(k)$ to be used in the fitting procedure.

To estimate a time-dependent output numerically, one picks a fixed value for $k \geq 0$ and a set of values for θ. Equation 4-73 or Eq 4-84 is then integrated for each θ. For the common case where $C_{\text{in}}(\theta)$ is periodic, for example, $C_{\text{in}}(\theta) = C_0 + A \sin \omega\theta, A \leq C_0$ the range on θ is one period.

It is worth mentioning that numerical differentiation of experimental $W(t)$ or $F(t)$ data is almost never needed. Integration by parts gives equivalent expressions that avoid $f(t)$ in the calculation of moments or steady-state reaction yields. For the unsteady case, Eq 4-83 is seen to involve a derivative of $C_{\text{in}}(\theta)$. This, of course, causes no problem when C_{in} is specified analytically. Should both $C_{\text{in}}(\theta)$ and $W(\theta)$ be known only numerically, one has a choice: Eq 4-73 can be used and $W(\theta)$ can be differentiated, or Eq 4-83 can be used and $C_{\text{in}}(\theta)$ differentiated.

Example Calculations

The following examples illustrate the processes used to achieve the best model fit.

Example 1. An impulse tracer experiment was performed on a prototype flocculation tank. The tank consisted of an underflow inlet baffle and two equal-sized flocculation compartments separated by a perforated baffle. The tank outlet also consisted of a perforated baffle. The flocculation impellers were of the horizontal-shaft type.

From the experimental data, the zero and first moment, as well as the second moment about the mean, can be determined. The first step is to determine the residence time density function using Eq 4-70. Note that the denominator is simply the area under the concentration-versus-time curve. The approximation to the denominator of Eq 4-70 using tracer concentration data measured at even time intervals Δt is

$$\text{``area''} = \sum_{I=0} [C(I) + C(I+1)]/2 * \Delta t \qquad \text{(Eq 4-85)}$$

Where:

I = number of tracer concentration measurements minus one
Δt = the time interval between tracer samples;
$\Delta t = t(I+1) - t(I)$

Once "area" has been calculated, the individual concentration values are normalized by "area," as indicated in Eq 4-70. This yields the discrete approximation to the residence time density function. The first moment is calculated using Eq 4-14. The discrete approximation of Eq 4-14 is

$$\bar{t} = \sum_{I=0} [f(I) + f(I+1)]/2 * [t(I) + t(I+1)]/2 * \Delta t \qquad \text{(Eq 4-86)}$$

The second moment about the mean is calculated using the discrete approximation of Eq 4-23

$$\sigma^2 = \sum_{I=0} [f(I) + f(I+1)]/2 * (\{[t(I+1)) + t(I)]/2\}^2 - \bar{t}) * \Delta t \qquad \text{(Eq 4-87)}$$

The dimensionless residence time density function can be determined by multiplying each value of $f(I)$ by $\bar{t}$. The normalized (dimensionless) variance is determined from Eq 4-31 by dividing σ^2 by $\bar{t}^2$. Finally, the dimensionless time τ corresponding to each value of the dimensionless residence time distribution is found from Eq 4-30. A computer printout showing the results of some of these calculations is provided in Table 4-1, and the dimensionless residence time distribution is plotted versus τ in Figure 4-10. Even though there may appear to be a discrete first appearance time at $\tau = 0.06$, the density function looks very much like that generated by the tanks-in-series model (Nauman and Buffham 1983)

$$f(\tau) = \frac{N^N \tau^{N-1} \exp^{-N\tau}}{(N-1)!} \qquad \text{(Eq 4-88)}$$

The dimensionless variance of this model is given by Eq 4-42. Trying to match dimensionless variances of data (0.290) and the tanks-in-series model, the $N = 3$ tanks-in-series model yields the closest "fit" to the data, $\sigma_\tau^2 = 1/3 = 0.333$ (Eq 4-42). Figure 4-10 plots the residence time density function of the $N = 3$ tank-in-series model, and it appears quite close to the experimental data. Using the raw data and the $N = 3$ version of the tanks-in-series model, a simple microcomputer program checks the sum-squared error (Eq 4-82): SSE = 0.0343. As a comparison, the sum-squared error for the single-tank model is 4.273.

Although intuition might allow the model search to stop here, the gamma-function model Eq 4-43 might be used in an attempt to better fit the experimental data.

Table 4-1 Calculations for Residence Time Density Function in Example 1

Time	Dimensionless Time	C(I)	f(I)	Dimensionless f(I)
.0000000E+00	.0000000E+00	.0000000E+00	.0000000E+00	.0000000E+00
.5000000E+00	.6289998E-01	.1000004E+00	.7917686E-03	.6293871E-02
.1000000E+01	.1258000E+00	.1600000E+01	.1266825E-01	.1007016E+00
.1500000E+01	.1886999E+00	.3000000E+01	.2375297E-01	.1888154E+00
.2000000E+01	.2515999E+00	.6400000E+01	.5067299E-01	.4028061E+00
.2500000E+01	.3144999E+00	.8299999E+01	.6571653E-01	.5223892E+00
.3000000E+01	.3773999E+00	.9700001E+01	.7680126E-01	.6105031E+00
.3500000E+01	.4402998E+00	.1070000E+02	.8471892E-01	.6734416E+00
.4000000E+01	.5031998E+00	.1280000E+02	.1013460E+00	.8056123E+00
.4500000E+01	.5660998E+00	.1320000E+02	.1045131E+00	.8307878E+00
.5000000E+01	.6289998E+00	.1330000E+02	.1053048E+00	.8370815E+00
.5500000E+01	.6918997E+00	.1320000E+02	.1045131E+00	.8307878E+00
.6000000E+01	.7547997E+00	.1310000E+02	.1037213E+00	.8244939E+00
.6500000E+01	.8176997E+00	.1220000E+02	.9659541E-01	.7678493E+00
.7000000E+01	.8805997E+00	.1180000E+02	.9342833E-01	.7426738E+00
.7500000E+01	.9434996E+00	.1070000E+02	.8471892E-01	.6734416E+00
.8000000E+01	.1006400E+01	.1010000E+02	.7996833E-01	.6356785E+00
.8500000E+01	.1069300E+01	.9100000E+01	.7205067E-01	.5727400E+00
.9000000E+01	.1132200E+01	.8799999E+01	.6967536E-01	.5538584E+00
.9500000E+01	.1195100E+01	.8400000E+01	.6650831E-01	.5286831E+00
.1000000E+02	.1258000E+01	.8100000E+01	.6413301E-01	.5098016E+00
.1050000E+02	.1320899E+01	.7400000E+01	.5859065E-01	.4657446E+00
.1100000E+02	.1383799E+01	.6600000E+01	.5225653E-01	.4153939E+00
.1150000E+02	.1446700E+01	.6200001E+01	.4908947E-01	.3902185E+00
.1200000E+02	.1509599E+01	.5900000E+01	.4671416E-01	.3713369E+00
.1250000E+02	.1572499E+01	.5400000E+01	.4275534E-01	.3398677E+00
.1300000E+02	.1635399E+01	.5000000E+01	.3958828E-01	.3146923E+00
.1350000E+02	.1698299E+01	.4200000E+01	.3325415E-01	.2643415E+00
.1400000E+02	.1761199E+01	.3400000E+01	.2692003E-01	.2139907E+00
.1450000E+02	.1824099E+01	.3100000E+01	.2454474E-01	.1951093E+00
.1500000E+02	.1886999E+01	.2800000E+01	.2216944E-01	.1762277E+00
.1550000E+02	.1949899E+01	.2600000E+01	.2058591E-01	.1636400E+00
.1600000E+02	.2012799E+01	.2500000E+01	.1979414E-01	.1573462E+00
.1650000E+02	.2075699E+01	.2200000E+01	.1741884E-01	.1384646E+00
.1700000E+02	.2138599E+01	.1900000E+01	.1504354E-01	.1195831E+00
.1750000E+02	.2201499E+01	.1600000E+01	.1266825E-01	.1007016E+00
.1800000E+02	.2264399E+01	.1400000E+01	.1108471E-01	.8811382E-01
.1850000E+02	.2327299E+01	.1200000E+01	.9501185E-02	.7552614E-01
.1900000E+02	.2390199E+01	.1000000E+01	.7917656E-02	.6293847E-01
.1950000E+02	.2453099E+01	.8999996E+00	.7125887E-02	.5664459E-01
.2000000E+02	.2515999E+01	.8000002E+00	.6334126E-02	.5035078E-01
.2050000E+02	.2578899E+01	.6999998E+00	.5542357E-02	.4405691E-01
.2100000E+02	.2641799E+01	.5000000E+00	.3958828E-02	.3146923E-01
.2150000E+02	.2704699E+01	.3999996E+00	.3167059E-02	.2517536E-01
.2200000E+02	.2767599E+01	.1999998E+00	.1583530E-02	.1258768E-01

Table continues next page.

Table 4-1 Calculations for Residence Time Density Function in Example 1 (continued)

.2250000E+02	.2830499E+01	.1000004E+00	.7917686E-03	.6293871E-02
.2300000E+02	.2893399E+01	.0000000E+00	.0000000E+00	.0000000E+00
.2350000E+02	.2956299E+01	.0000000E+00	.0000000E+00	.0000000E+00
.2400000E+02	.3019199E+01	.0000000E+00	.0000000E+00	.0000000E+00
.2450000E+02	.3082099E+01	.0000000E+00	.0000000E+00	.0000000E+00
.2500000E+02	.3144999E+01	.0000000E+00	.0000000E+00	.0000000E+00

NOTE: $\bar{t}$ = .7949129E+01; σ_τ^2 = .2905725E+00

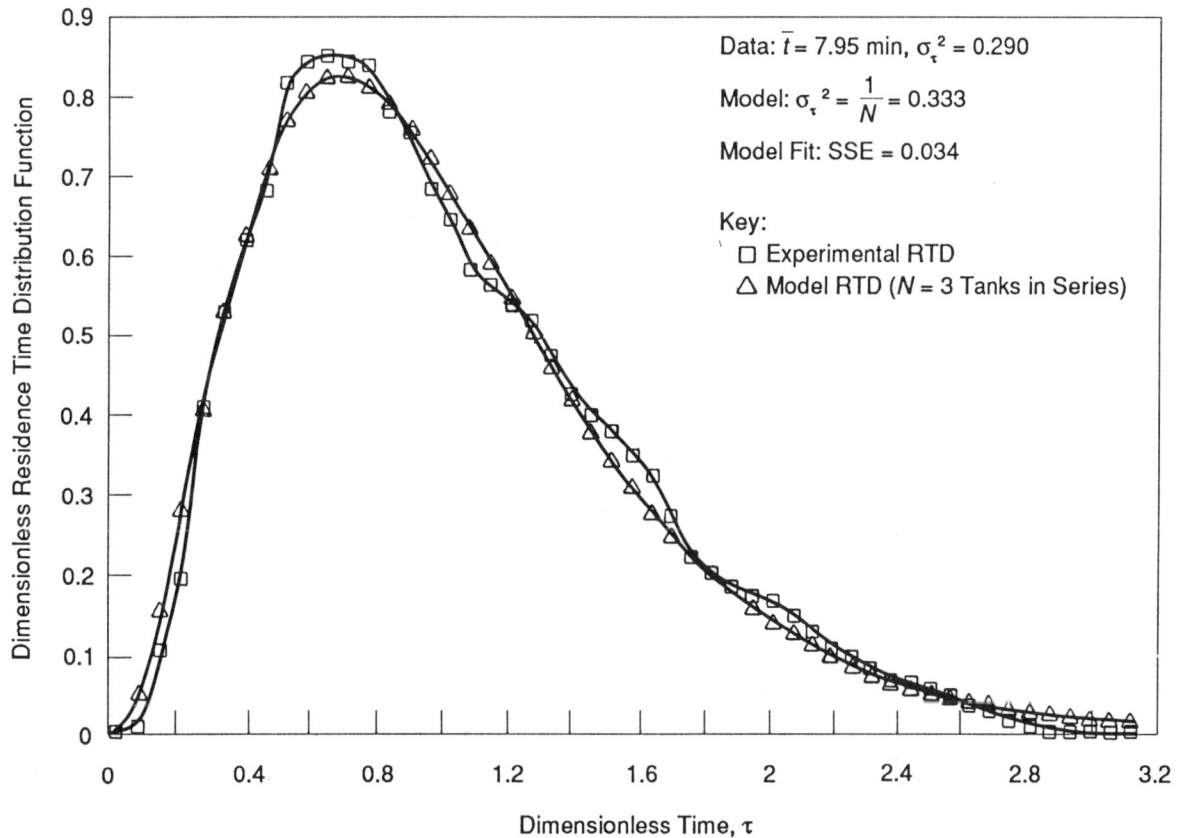

Figure 4-10 Dimensionless RTD and model fit for example 1.

A simple computer-search technique that tries various N-values (N is continuous in the gamma-function model) in the region of $N = 3$ yields the optimal gamma-function model for $N = 3.06$. However, this only reduces the sum-squared error to 0.0328. Therefore, either the $N = 3$ tanks-in-series model or the $N = 3.06$ gamma-function model seems quite appropriate for representing the experimental data.

One of the hazards of fitting models using only the second moment (dimensionless variance) is evident from one further test using the gamma-function model. If $N = 3.442$ (corresponding to the measured dimensionless variance of the data [Eq 4-42]) is used in the gamma model the result is a sum-squared error of 0.0664—more than twice the error for the optimal N-value determined using a simple search method.

Example 2. An impulse tracer experiment was performed on another prototype flocculation tank consisting of a submerged inlet baffle, two equal-sized flocculation compartments separated by a simple submerged baffle, and an outlet structure consisting of an under–over baffle. Using the same techniques as in example 1, the calculated mean residence time is found to be 5.22 min, and the experimental dimensionless variance is 0.678. Figure 4-11 shows that the dimensionless density function, calculated from the tracer data, indicates performance similar to mixed flow of a single stirred tank.

Fitting the nondimensionalized form of Eq 4-8 to the data, the single stirred tank gives an SSE of 1.082. The fractional tubularity model may decrease the SSE because it removes the contribution to the SSE by the time = 0 point. The density function for the fractional tubularity model is

$$f(\tau) = \frac{1}{\tau_s} \exp\left[\frac{1}{\tau_s}(\tau_p - \tau)\right] \qquad \text{(Eq 4-89)}$$

(Nauman and Buffham 1983). Using this model with τ_p in the range 0.001–0.09 substantially reduces the SSE to the approximate range 0.08–0.11. Figure 4-11* uses $\tau_p = 0.06$ (SSE = 0.108) as an appropriate representation of the data and shows the plot. The hazard of the method of moments is again apparent in this example when an attempt is made to fit the fractional tubularity model with the same dimensional variance (0.678) as the data. This approach gives a poor fit and an SSE of 0.808.

CHEMICAL REACTIONS AND MICROMIXING

The steady-state yield of the first-order, isothermal, homogeneous reaction is uniquely determined by the residence time distribution:

$$\frac{C_{\text{out}}}{C_{\text{in}}} = \int_0^\infty \exp^{-kt} f(t)\, dt = k \int_0^\infty \exp^{-kt} W(t)\, dt \qquad \text{(Eq 4-90)}$$

---

*Based only on the SSE-value, it is difficult to select the correct τ_p for the fractional tubularity model because there is no clear optimal value in the range $0 < \tau_p < 0.096$ (0.096 corresponding to the first nonzero data point). This is because discretized data are being used, and the SSE calculation does not take into account the apparent values of the density function for $\tau < 0.096$. Thus, $\tau_p = 0.06$ is used because it leads to a residence time distribution cutoff that seems to split the difference in the range $0 < \tau_p < 0.096$.

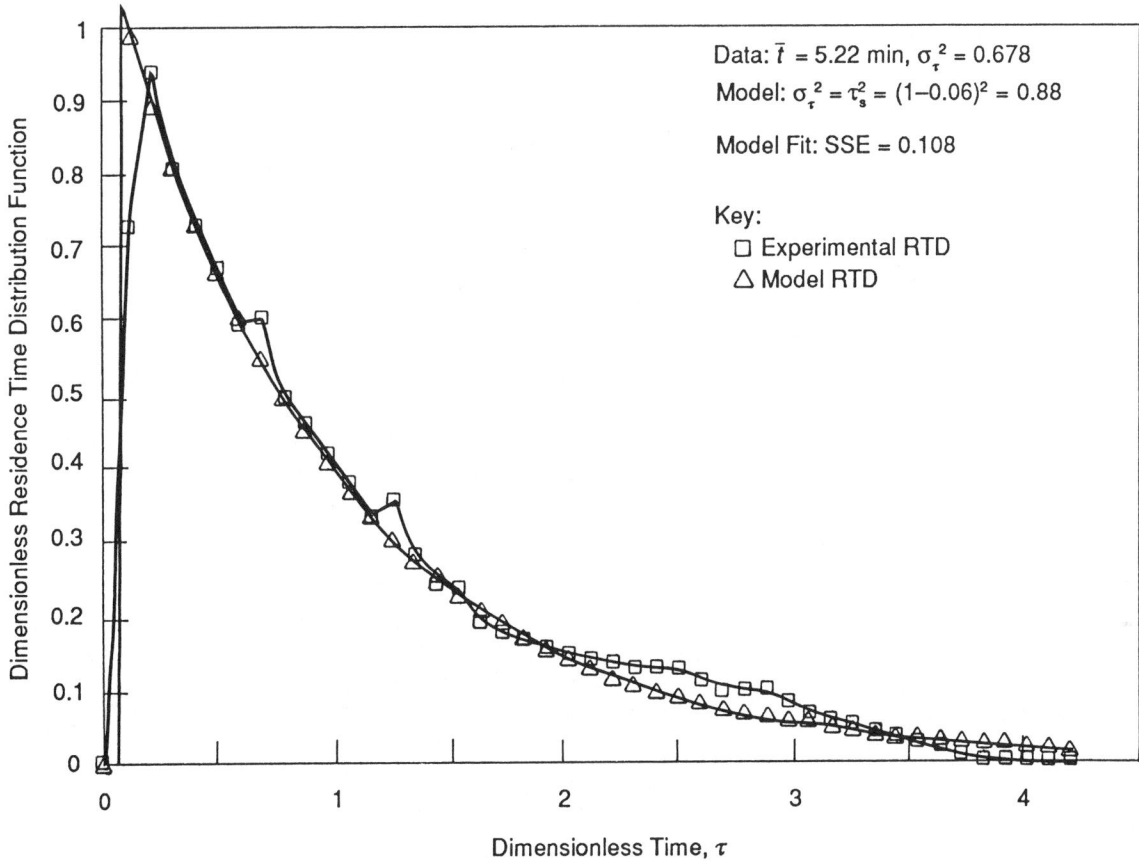

Figure 4-11 Dimensionless RTD and model fit for example 2.

For reactions of order other than first, the yield prediction is not unique, but knowledge of the residence time distribution can usually provide quite close bounds on the yield.

The Bounding Theorem

A reaction is called simple when the rate of reaction of a key component depends only on the concentration of that component

$$\text{rate} = \mathcal{R}(C) \qquad \text{(Eq 4-91)}$$

Where:

C = concentration of key component

Thus, unimolecular reactions are simple. Multimolecular reactions may be treated as simple if the reactants have similar flow and mixing patterns in the vessel so that stoichiometry is locally preserved. This requirement is usually satisfied in single-phase, premixed feed systems where the various reactants have similar diffusivities.

Given a simple, isothermal, homogeneous reaction, suppose also that the rate expression is either a concave upward or concave downward function of concentration:

$$\mathcal{R}''(C) = \frac{d^2R}{dC^2} < 0 \quad \text{concave downward} \qquad \text{(Eq 4-92)}$$

$$\mathcal{R}''(C) = \frac{d^2R}{dC^2} > 0 \quad \text{concave upward} \qquad \text{(Eq 4-93)}$$

Suppose one of these conditions is satisfied for all C in the range $0 \leq C \leq C_{in}$ (assume the component is consumed by the reaction). Then residence time theory may be used to place limits on the yield of the reaction. These limits correspond to the theoretical extremes of molecular-level mixing, usually called micromixing, that are possible with a given residence time distribution.

One extreme of micromixing is called complete segregation. The other is called maximum mixedness.

Complete Segregation

In complete segregation, fluid travels through the reactor in discrete, noninteracting packets. These packets constitute the particles of residence time theory. They are small compared to system dimensions but contain a large number of molecules. A residence time t is associated with each packet and with all conserved entities within the packet. The packets behave as small batch reactors. All have the same entering concentration C_{in}, but have differing exit concentrations $C_{batch}(t)$, depending on the particular value for t. The fraction of packets with residence times $(t, t + dt)$ is $f(t)dt$; the average exit composition is found by summing over all packets:

$$C_{out} = \int_0^\infty C_{batch}(t) f(t) \, dt \qquad \text{(Eq 4-94)}$$

Complete segregation represents an extreme form of mixing (or lack thereof) on a molecular scale. Molecules enter mixed and leave mixed, but there is no mixing at all within the reactor. Any residence time distribution is possible with complete segregation. Completely segregated reactors can be modeled as piston-flow reactors in parallel or as a single piston-flow reactor with side exits (Figure 4-12). These models are equivalent. The parallel-reactor model conforms to the definition of complete segregation, since each parallel element corresponds to the path followed by a packet of molecules. The single-reactor model has mixing in the radial direction and thus has mixing between packets in violation of the intuitive definition of complete segregation. However, this mixing within the reactor occurs only between molecules having the same age and thus the same composition. Such mixing has no effect on conversion, and the outlet concentration for either model is given by Eq 4-94.

Completely segregated reactors may be imagined to have mixing between packets with differing ages, but this mixing takes place only at the reactor exit, where the various packets merge to form the outlet stream. The mixing occurs when the age of a particle α_t equals t for all packets (but different packets will have different values of $\alpha_t = t$). Thus, whatever mixing there is between molecules with differing ages will occur as late as possible, namely, at the reactor outlet.

Maximum Mixedness

Maximum mixedness is the opposite extreme with regard to mixing between molecules with differing ages; with maximum mixedness, mixing occurs at the reactor inlet and thus is as early as possible.

An arbitrary residence time distribution can be modeled as a single piston-flow reactor with a number of side entrances. The situation, which is merely a reversal of the flow direction in Figure 4-12B, is illustrated in Figure 4-13. The details of molecular mixing for this model are far different than for the segregated model of

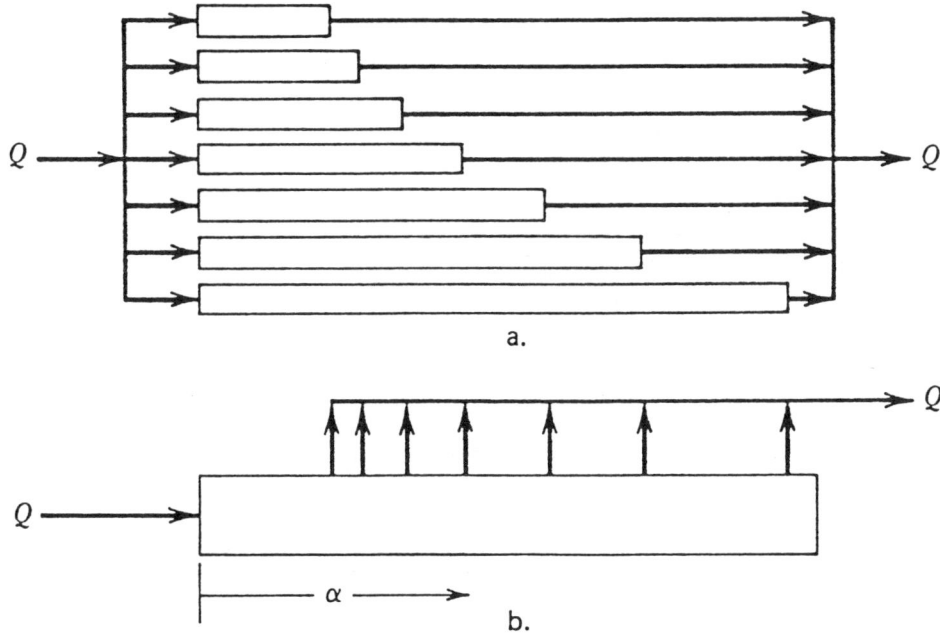

Source: Nauman, E.B. & Buffham, B.A. 1983. Mixing in Continuous Flow Systems. Copyright 1983, John Wiley and Sons, New York. Reprinted by permission of John Wiley and Sons, Inc.

Figure 4-12 Models for segregated reactors: (A) piston-flow elements in parallel; (B) single piston-flow element with side exits.

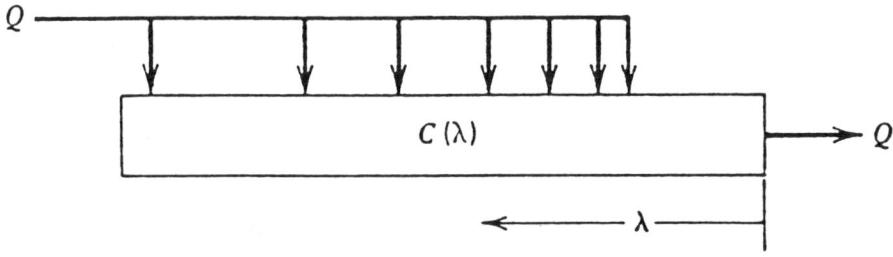

Source: Nauman, E.B. & Buffham, B.A. 1983. Mixing in Continuous Flow Systems. Copyright 1983, John Wiley and Sons, New York. Reprinted by permission of John Wiley and Sons, Inc.

Figure 4-13 Model for maximum-mixedness reactor: single piston-flow element with side entrance.

Figure 4-12B. In the maximum mixedness case, fresh feed enters the system with age $\alpha_t = 0$ and immediately mixes with material (already in the reactor) of many different ages. When mixing occurs between two groups of molecules with different ages, one of the groups will have age $\alpha_t = 0$. Thus, mixing between ages occurs as early as possible.

Replacing the conventional axial coordinate z in Figure 4-13 is a new variable λ, known as the residual life, which has units of time and is defined by

$$\alpha_t + \lambda = t \qquad \text{(Eq 4-95)}$$

A molecule enters a system with $\alpha_t = 0$ and $\lambda = t$ and leaves with $\alpha_t = t$ and $\lambda = 0$. For the maximum-mixedness reactor of Figure 4-13, molecules at a given axial position may have different ages, but all have the same residual life. In contrast, the completely segregated model uses α_t rather than λ as a surrogate for z. Molecules at a given axial position may have different residual lives, but all have the same age.

The outlet concentration from a maximum-mixedness reactor is found by solving the following differential equation:

$$\frac{dc}{d\lambda} + \frac{f(\lambda)}{W(\lambda)} [C_{\text{in}} - C(\lambda)] + \mathcal{R}(C) = 0 \qquad \text{(Eq 4-96)}$$

The boundary condition associated with Eq 4-96 is that C must be bounded at finite time for all λ. This usually takes the form

$$\lim_{\lambda \to \infty} \frac{dC}{d\lambda} = 0 \qquad \text{(Eq 4-97)}$$

but special forms are required if there is a maximum possible residence time $t_{\max}$. Once Eq 4-96 has been solved subject to the appropriate boundary condition, the conversion may be calculated from $C_{\text{out}} = C(\lambda = 0)$. Owing to the nonlinearity of typical rate expressions, analytical solutions are possible in only a few cases, and numerical solution is generally required.

Equation 4-94 and Eq 4-96 provide upper and lower limits for the yield of simple, isothermal, homogeneous reactions, provided one of the inequalities in Eq 4-93 is satisfied. If $\mathcal{R}''(C) < 0$, Eq 4-94 predicts the highest possible conversion for a given $f(t)$ and Eq 4-96 gives the lowest possible conversion. If $\mathcal{R}''(C) > 0$, the converse is true.

As an example of simple kinetics, consider the nth-order reaction

$$\mathcal{R}(C) = -kC^n \qquad \text{(Eq 4-98)}$$

then

$$\mathcal{R}''(C) = -n(n-1)kC^{n-2} \qquad \text{(Eq 4-99)}$$

which is negative for $n > 1$ and positive for $n < 1$. Thus, reactions of order higher than first have greater conversions in completely segregated systems than in maximum-mixedness systems. Normally, the bounds on the yield provided by Eq 4-94 and Eq 4-96 are quite close. For a second-order reaction, the maximum possible difference in conversion is only 7 percent. This maximum difference applies in the case of an exponential residence time distribution and is smaller for other $f(t)$.

Micromixing becomes unimportant as the residence time distribution approaches piston flow.

Importance of Micromixing

The performance of many industrial reactors is determined almost completely by batch kinetics and residence time distribution. Micromixing effects tend to be small except under rather specific conditions, including the following:

- The residence time distribution must show significant departures from piston flow.
- The reaction kinetics must be highly nonlinear and relatively fast (that is, conversions must be relatively complete; at low conversions all reactions will appear pseudo first order).
- The physical nature of the system must allow segregation as a practical possibility (most gas- and liquid-phase reactors operate near the limit of maximum mixedness).

A qualitative but conceptually useful representation of micromixing is shown by the mixing-space diagram of Figure 4-14. The x axis, labeled macromixing, measures the breadth of the residence time distribution. A possible scale of measurement for this axis is the dimensionless variance σ_τ^2. The y axis, labeled micromixing, is really a measure of how important micromixing effects can be. They are unimportant for piston-flow reactors and have maximum importance for stirred-tank reactors. Properly designed reactors usually lie within the normal region shown in Figure 4-14 and will be bounded in performance by the three special reactors at the apexes of the normal region: piston flow, perfect mixing, and segregated stirred

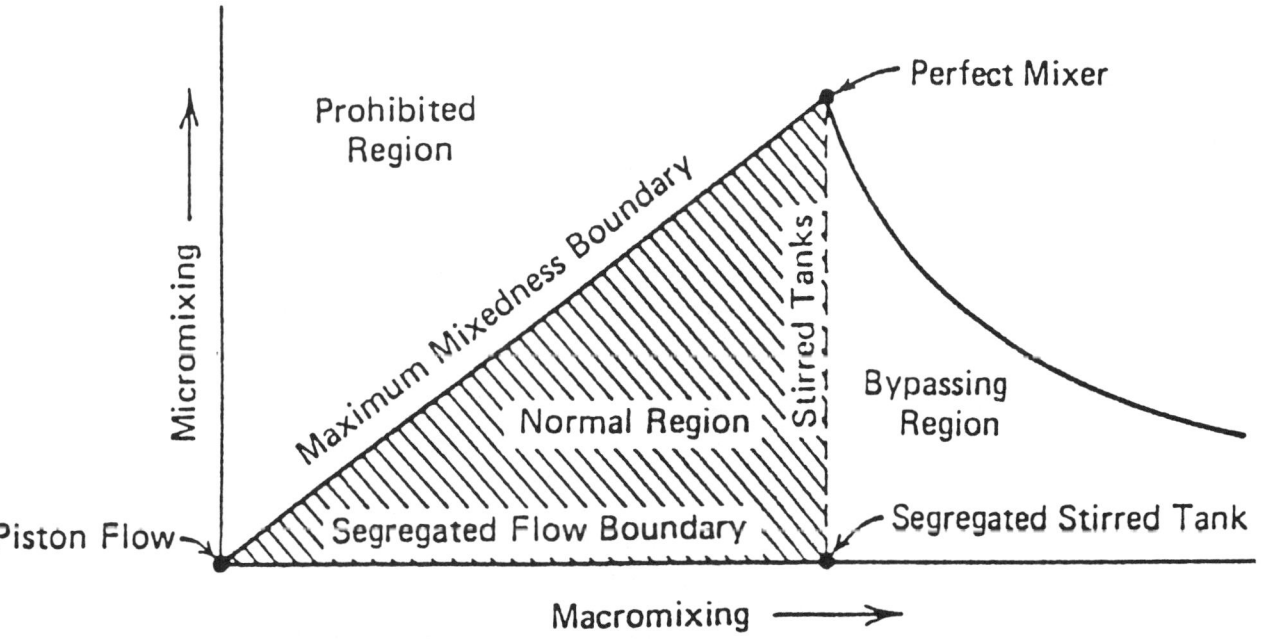

Source: Nauman, E.B. & Buffham, B.A. 1983. Mixing in Continuous Flow Systems. Copyright 1983, John Wiley and Sons, New York. Reprinted by permission of John Wiley and Sons, Inc.

Figure 4-14 Schematic representation of mixing space.

tank. Only very rarely will system performance within the normal region be better than it is at one of the apexes. The rare exceptions involve the optimization of intermediate-product yields in complex kinetic schemes. Interestingly, one of these rare exceptions may be the case of coagulant precipitation in the rapid-mix unit, discussed in chap. 5.

If a reactor is known to operate within the normal region, a first approximation for the conversion can be made from knowledge of the batch laboratory kinetics and the mean residence time. One merely calculates the conversion at the three apexes in Figure 4-14. The spread in conversions is usually less than 20 percent. An improved estimate can be made using knowledge of $f(t)$. This confines the real reactor to operate somewhere on a vertical line in Figure 4-14 (known level of macromixing, unknown level of micromixing). The uncertainty in conversion will be reduced to less than 7 percent for realistic kinetic schemes. Typically, the uncertainty will be much less than 5 percent since many reactions are pseudo first order, in which case the uncertainty is zero. For nonlinear kinetics a precise conversion estimate requires a quantitative description of micromixing in the real system. Methods for doing this are discussed in chap. 5. Although knowledge of micromixing will usually eliminate only a small uncertainty in conversion, it can eliminate a large uncertainty in reactor volume needed to obtain a given conversion.

DESIGNING FOR GOOD MIXING

This section discusses a general methodology for achieving desirable residence time distributions in pilot-scale equipment and for scaling-up the pilot results to full-scale units. This represents an admittedly narrow view of design and scale-up problems in liquid processing. The performance of some processes is only vaguely related to the residence time distribution or even to the mean residence time. Even when a strong causal relation exists, it is not always possible to control or scale-up residence time distributions. However, in most processes it is both possible and appropriate to treat design and scale-up in the conceptual framework of residence time theory. In those cases where the theory does not seem applicable, it should at least be examined. In particular, any scale-up that alters the mean residence time between pilot and production units should be handled very cautiously.

Achievement of Mixing Extremes

Optima in process performance are seldom found at intermediate states of macro- or micromixing. Water treatment equipment usually operates best when the residence time distributions are close to the delta distribution of piston flow. (See discussion in chap. 8.)

Process steps that operate best when the residence time distribution is exponential are comparatively rare. Rarer still are processors that operate best at an intermediate level of micromixing. System performance is almost always best at one of the three apexes that define the normal region of mixing space, as shown in Figure 4-14. This section briefly considers methods for achieving operation at or near one of these three points.

Since velocity gradients or velocity differences always exist in real flow systems, piston flow and the associated delta function residence time distribution can only be approximated. Almost-piston flow is a common design objective, and there are two generic approaches to achieving it:

- minimize the magnitude of the velocity differences between streamlines, or
- compensate for velocity gradients by flow redistribution between streamlines.

The first approach is carried to its logical extreme in paper machines and certain food, pharmaceutical, and polymer processing equipment where the fluid being processed is conveyed by a moving belt or web. It is approximated in highly turbulent flows.

The second approach, using flow redistribution to equalize residence times, is generally useful. The basic idea is to move particles across the flow field so that an individual particle experiences many different velocities during its stay in the system. This can be achieved by the radial velocity components that exist in highly turbulent flows, by special flow devices that promote radial flow (such as static mixers), or in very small systems, by radial diffusion. Radial motion displaces the particles from their original streamlines and thus causes them to change axial velocity. As the particle moves back and forth across the streamlines, it samples many different velocities and travels with an average velocity close to $\bar{u}$. As a practical matter, piston flow can be closely approached by using flow channels that have a large length-to-diameter (width) ratio. It can also be approached by staging the process, with intense mixing between the stages.

When desired, an exponential distribution of residence times is easily achieved in a recycle system. Except for certain pathological situations, the only requirement is that the ratio of recycle flow to net throughput be reasonably high, say $q/Q > 16$.* This means that a well-designed, agitated-tank reactor will have a residence time distribution that is virtually indistinguishable from the exponential distribution. It is quite another story, however, to achieve a close approximation to perfect mixing, that is, to achieve maximum mixedness, given that $f(t)$ is exponential. The theoretical requirement for a close approximation to perfect mixing is that the identity of individual fluid elements must be periodically destroyed. This can be accomplished only by molecular diffusion, which in turn requires that the scale of segregation be reduced to some adequately small value. Chapters 3 and 5 discuss the limits on micromixing that can be achieved through turbulence.

Scale-up Considerations

The previous section gave a broad approach to achieving optimal residence time distributions in pilot-scale equipment. Once a desired type of operation has been achieved in the pilot scale, the problem becomes one of scale-up. Often, but not always, this requires maintaining a constant residence time distribution, independent of scale. *Scaleup in the Chemical Process Industries* (Bisio and Kabel 1984) treats general aspects of scale-up. The present section draws largely on chap. 8 of that work.

Basic situations where residence time distributions can be preserved during scale-up are treated with a high degree of confidence. These situations occur when

- the pilot system is an open tube or open channel with a residence time distribution that approximates piston flow, or
- the pilot system is a stirred tank or recycle loop with an exponential distribution of residence times.

These two situations encompass many, perhaps most, scale-up cases. They correspond to what are normally the optimal residence time distributions, and these are commonly encountered in well-designed processing equipment. In most scale-up attempts, the pilot plant will have operated close to one of these macromixing

---

*For simple problems, such as the yield of a first-order reaction, $q/Q > 5$ gives an adequately close approach to the exponential distribution.

extremes, and the desire is that the production unit operate close to that limit. It is also the usual desire to hold $\bar{t}$ constant upon scale-up.

Define a scale-up factor S as the ratio of process flow rates

$$S = \frac{(Q_{in})_{fs}}{(Q_{in})_{pp}} \qquad \text{(Eq 4-100)}$$

Where:

the subscript *pp* denotes the pilot plant
the subscript *fs* denotes the full-scale plant

The constancy of $\bar{t}$ implies that the inventory in the system scales by the same factor

$$\frac{I_{fs}}{I_{pp}} = S \qquad \text{(Eq 4-101)}$$

Where:

I_{fs} = material inventory in full-scale unit
I_{pp} = material inventory in pilot-plant unit

Equation 4-100 holds by definition. Equation 4-101 may or may not hold, depending on the intent of the scale-up. Usually, however, it is desirable that Eq 4-101 be satisfied and indeed that

$$\frac{V_{fs}}{V_{pp}} = S \qquad \text{(Eq 4-102)}$$

Where:

V_{fs} = volume of full-scale plant
V_{pp} = volume of pilot plant

Equation 4-102 is equivalent to Eq 4-101 whenever the particle density is independent of scale.

Constancy of $\bar{t}$ during scale-up is natural for chemical reactors, since any alteration will change the reaction yield and selectivity. In those comparatively rare cases where $\bar{t}$ changes upon scale-up, it may be desirable to retain the dimensionless form of the residence time distribution. It is usually desirable to retain the dimensionless form when it approximates one of the limiting distributions, nearly exponential or nearly a delta function. It is rarely desirable to retain the exact form of an intermediate distribution. Instead, upon scale-up, one usually attempts to push an intermediate distribution toward one of the limiting forms.

Consider the scale-up of an open tube, open channel, or packed bed that has an almost-delta function residence time distribution. This situation is closely approximated by the axial dispersion model. Thus, the dimensionless residence time distribution is a function only of geometric ratios, such as L/d_t, and of the Reynolds number. For turbulent flows, a scale-up with geometric similarity and constant $\bar{t}$ will be slightly conservative with respect to narrowness of the residence time distribution. This form of scale-up has $\bar{u}$, L, and d_t increasing as $S^{1/3}$, which means that Re will increase as $S^{2/3}$. The radial Peclet number ud_t/D increases weakly with

Re for well-developed turbulent flow. Since the aspect ratio L/d_t is constant, this means that the axial Peclet number uL/D will increase slightly and that the residence time distribution will narrow. Thus, the system will more closely approach piston flow as the scale is increased.

The disadvantage of scaling-up with geometric similarity and constant $\bar{t}$ is that the pressure drop will increase from the pilot plant to the production unit. The magnitude of the increase is easily estimated by using the Blasius correlation of (Fanning) friction factor with Reynolds number

$$\text{Fa} = \frac{1}{2} \frac{\Delta P}{L} \frac{d_t}{\rho \bar{u}^2} = 0.079 \text{Re}^{-1/4} \qquad \text{(Eq 4-103)}$$

Where:

ΔP = pressure drop across system
P = density of fluid

The case of constant L/d_t and $\bar{t}$ gives ΔP increasing as $S^{1/2}$. This increase may be reasonable in liquid-phase systems, particularly if the pilot-scale experiment uses pressure drops that will be reasonable when scaled-up.

An exponential distribution normally means that there are high levels of fluid recirculation within the system. This recirculation may be explicit, as in the recycle loop shown in Figure 4-15A, or it may be implicit, as in the continuous-flow agitated tank illustrated in Figure 4-15B. Whichever the form, the exponential distribution may be retained upon scale-up by maintaining high recirculation rates.

To scale-up the residence time distribution of a recycle system, geometric similarity and constant values for $\bar{t}$ and q/Q are maintained. This approach is not completely conservative, since the Reynolds number in the flow element will increase by a factor of $S^{2/3}$, which will mean a slightly higher value for uL/D. Thus,

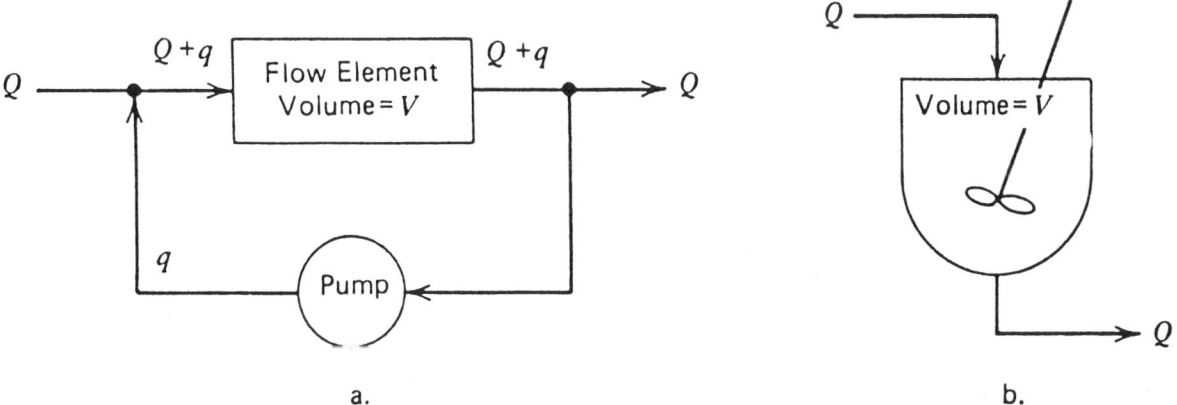

Source: Nauman, E.B. & Buffham, B.A. 1983. Mixing in Continuous Flow Systems. Copyright 1983, John Wiley and Sons, New York. Reprinted by permission of John Wiley and Sons, Inc.

Figure 4-15 Common means for achieving an exponential distribution: (A) external recirculation: the recycle loop reactor; (B) internal recirculation: the continuous-flow stirred-tank reactor.

the per-pass residence time distribution will be somewhat closer to piston flow, and the composite distribution for the recycle system will be somewhat farther from ideal mixing. However, an exponential distribution for the system is obtained in the limit of high q/Q even when the per-pass $f(t)$ is a perfect delta function. Thus, if q/Q remains high, say $q/Q > 16$, the recycle system will remain well stirred upon scale-up. In this scale-up, Q and q will both increase as S^1, while the pressure drop across the flow element will increase approximately as $S^{1/2}$. Total power input to the system will thus increase as $S^{3/2}$, while power per unit volume will increase as $S^{1/2}$.

The most important industrial example of a nearly exponential distribution is the continuous-flow agitated tank illustrated in Figure 4-15B. This system behaves much as a recycle loop, since the impeller generates an internal flow that has magnitude q, while the tank acts as a distributed-flow element. Consider scale-up with geometric similarity and constant $\bar{t}$. For Newtonian fluids in the turbulent regime, pumping capacity q scales as $N_I D_I^3$. This means that maintaining a constant q/Q requires a constant rotational velocity for the impeller. Total power input scales as $N_I^3 D_I^5$ or, with geometric similarity and constant N_I, as $S^{5/3}$. Power per unit volume thus increases as $S^{2/3}$, compared to $S^{1/2}$ for the recycle model. Within the accuracy of the analysis, these scale factors may be considered the same.

A common approach to scale-up is to maintain constant power per unit volume. Constant power per unit volume is reasonable from the viewpoint of micromixing, since it maintains the Kolmogoroff scale of turbulence (see chap. 3 and 5). It is also reasonable from the viewpoint of macromixing if q/Q remains sufficiently high in the production unit, say $q/Q > 16$. Most pilot-plant units operate at considerably higher values of q/Q so that some reduction is possible upon scale-up.

It is rarely desirable or feasible to scale-up a residence time distribution that is intermediate between those of piston-flow or stirred-tank reactors. Exceptions to this generalization are stirred-tank reactors in series, mixer/settlers, agitated vessels with multiple impellers, and similar cases of physical compartmentalization. Such composite systems can be scaled-up in terms of the individual components.

CONCLUSIONS

For simple, slow reactions, conversion of reactants is fairly tightly controlled within certain narrow bounds—even for the mixing extremes of complete segregation and

Table 4-2 Summary of Simple Models for Mixing

Model	Model Parameter	Dimensionless First-Appearance Time	Dimensionless Variance
Plug flow	None	1	0
Stirred tank	None	0	1
Fractional tubularity	τ_p	τ_p	$(1 - \tau_p)^2$
Tanks-in-series (including both fractional-tank and gamma extensions)	N	0	$1/N$
Axial dispersion	Pe	0	$2/\text{Pe} - 2/\text{Pe}^2 (1 - \exp^{-\text{Pe}})$

maximum mixedness. If flocculation is considered a simple, slow "reaction," then it could be concluded that the challenge of reactor design and scale-up in flocculation is not very great. However, flocculation performance is not insensitive to design, which, among other things, implies that the flocculation "reaction" is not a simple one. Stagnancy, flow distribution, internal recycle, and the dimensionless variance of the residence times distribution all can affect the flocculation "reaction." Flocculation is in reality a complex and consecutive reaction, with both reversible and irreversible aspects (see chap. 8). The reaction rate is dependent upon a suite of variables, several of which are related to mixing, fluid flow, and turbulence. Therefore, modeling and measurement of the residence time distribution should be an important aspect of design, analysis, and scale-up. There is a scarcity of good residence time data in the water treatment field, particularly data from experiments in which inlet and outlet conditions are correctly or rigorously treated.

By way of summary, Table 4-2 provides a brief description of the simple models most commonly used for fitting experimental data.

References

BISIO, A.L. & KABEL, R.L., eds. 1984. *Scaleup in the Chemical Process Industries*. Wiley-Interscience, New York.

———. 1985. *Scaleup of Chemical Processes: Conversion From Laboratory Scale Tests to Successful Commercial Size Design*. Wiley-Interscience, New York.

HIMMELBLAU, D.W. & BISCHOFF, K.B. 1968. *Process Analysis and Simulation*. John Wiley and Sons, New York.

HUDSON, H.E. JR. 1981. *Water Clarification Processes: Practical Design and Evaluation*. Van Nostrand Reinhold, New York.

LEVENSPIEL, O. 1972. *Chemical Reaction Engineering*, 2nd ed. John Wiley and Sons, New York.

NAUMAN, E.B. & BUFFHAM, B.A. 1983. *Mixing in Continuous Flow Systems*. John Wiley and Sons, New York.

WEBER, W. 1972. *Physicochemical Processes for Water Quality Control*. Wiley-Interscience, New York.

WEN, C.Y. & FAN, L.T. 1975. *Models for Flow Systems and Chemical Reactors*. Marcel Dekker, Inc., New York.

5

Micromixing Models and Application to Aluminum Neutralization Precipitation Reactions

René David, Ph.D., Ingenieur, Directeur de Recherche,
 Laboratoire des Sciences du Génie Chimique,
 CNRS-ENSIC—Institut National Polytechnique de Lorraine,
 Nancy Cedex, France

Mark M. Clark, Ph.D., Assistant Professor, Department of Civil
 Engineering, University of Illinois at Urbana–Champaign,
 Urbana, Ill. USA

WHAT IS MICROMIXING?

To understand micromixing, one must first consider a mixing device containing a single-phase flowing fluid that is fed by two or more feed streams. This device can be operated continuously or batchwise (see Figure 5-1). The local concentrations of the compounds to be mixed vary from point to point within the apparatus, and at a given point the concentration of every component fluctuates with time, provided that the flow is turbulent or periodically laminar. The second law of thermodynamics implies that in the absence of chemical reaction, all concentration fluctuations disappear at a finite rate. The two goals for such a mixing device are (Figure 5-1)

- to reduce the average concentration gradients in a given volume of aging fluid by convection. This is called macromixing. Perfect macromixing implies an exponential residence time density function (see chap. 4).
- to reduce the concentration fluctuations of the aging fluid by turbulence or stretching, followed by diffusion. This second goal is called micromixing.

The following chapter is restricted to liquid flows. Their main characteristics compared with gas flows are

- incompressibility;
- dilute reactants in a dense solvent, leading to appreciable reaction rates with minor thermal effects; and
- uncoupling of velocity and concentration microgradients.

Two types of mixing devices are generally used: turbulent pipes, packed beds, or static mixers, where the mean age of the fluid (time elapsed since it has entered the device) is increasing along the flow axis; and stirred tanks, where uniform

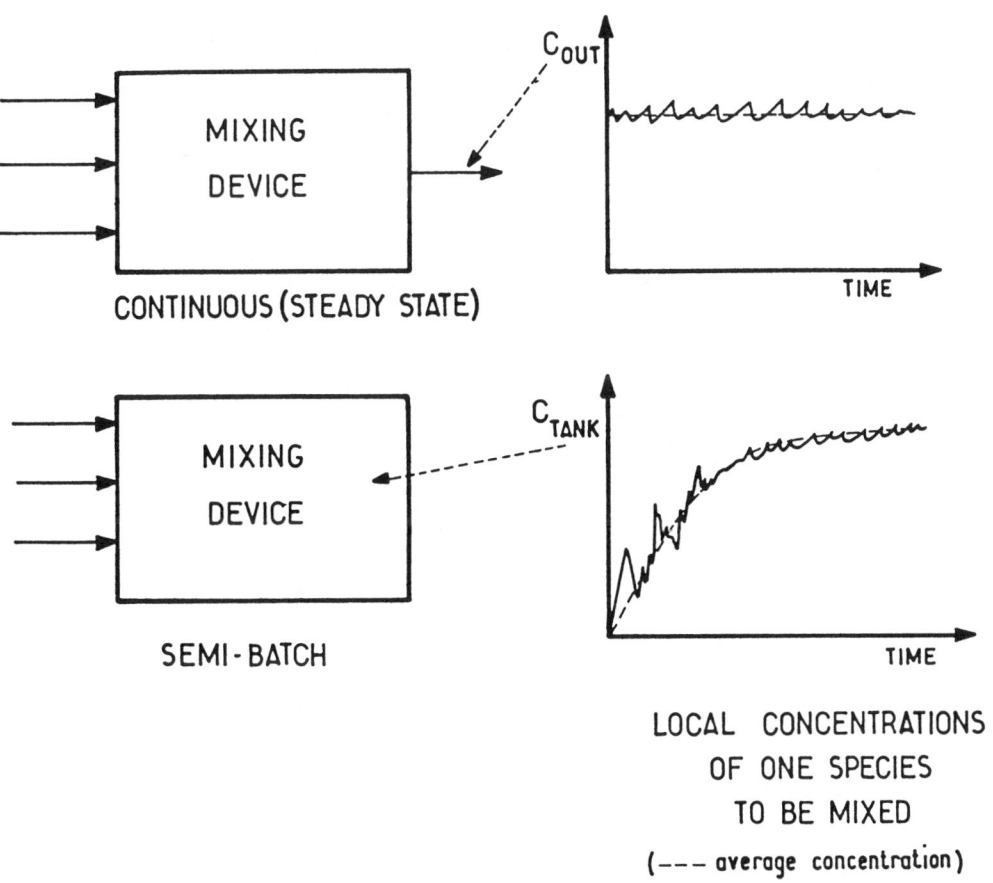

Figure 5-1 Principle of mixing operation.

macromixing and thus equal mean ages are wanted throughout the tank. In-line mixers with high external recycle flow rates also fall in this category. Intermediate mixing devices with partial recycle of the flow can be found. They are described by macromixing models; see Wen and Fan (1975) and chap. 4 of this book.

The aim of this chapter is to review and discuss micromixing models, and to show how these models can be applied to problems of mixing in coagulation processes.

THE MECHANISM OF MIXING

Mechanisms of Energy Dissipation in Vessels

The mixing of conservative or reactive species proceeds via four consecutive steps with respect to the age of the fluid (see Table 5-1). These steps are

- macroscopic distribution of inlet flow in the mixing device by the average velocity field;
- shearing and cutting by turbulence (sometimes called turbulent diffusion);
- laminar stretching of smaller fluid eddies and formation of vortices; and
- molecular diffusion when the eddies become small enough.

These steps are linked to the dissipation of the mechanical energy and characteristic length scales associated with mixing. For instance, in stirred tanks, the mechanisms of power and segregation dissipation are illustrated by Figure 5-2 (see Placek et al. 1986).

When two solutions containing different amounts of an inert tracer are mixed, concentration gradients arise. By means of locally sensitive probes, the tracer concentration $C_i(t) = \overline{C}_i + c_i(t)$ can be measured. The average concentration $\overline{C}_i$ indicates the quality of macromixing (it should be uniform in a well-macromixed tank), whereas the fluctuations $c_i(t)$ indicate the micromixing behavior. As is done classically with velocity fluctuations, these fluctuations can be recorded and expressed using amplitude probability distributions, autocorrelation functions, or Fourier transformation (spectral analysis). It is also possible to deduce the standard deviation $\sqrt{\overline{c_i^2}}$ and characteristic length scales (macroscale, dissipation scale) from the fluctuations. Until now, the following two different physical methods have been used for recording such microfluctuations of concentration in liquids:

- microconductimetry (see Gibson and Schwartz 1963; Truong and Methot 1974; Brodberger, Valentin, and Stork 1983; Mahouast et al. 1986); and
- fluorescence (see Patterson et al. 1982; Gaskey et al. 1988).

Extreme States of Segregation

Microfluid. Microfluid is a physical state where the different species of the feed streams that are macromixed (uniform average concentration in space) are also mixed on the molecular level. Thus, the mass balances are written only between average concentrations $\overline{C}$. For instance, for a well-macromixed stirred tank

$$Q_{in} \overline{C}_{j,o} + v_j rV = Q_{out} \overline{C}_{j,\,out} + V \frac{d\overline{C}_{j,\,out}}{dt} \quad \text{(Eq 5-1)}$$

Table 5-1 Mixing Stages and Corresponding Length Scales and Physical Models

Stage	Process	Characteristic Scale of Process	Model
1	Recirculation flow	Macroscale	Internal recirculation (deviation from perfect macromixing)
2	Reduction of turbulent scales	Velocity dissipation scale (Taylor microscale)	Erosion; turbulent diffusion
3	Stretching and swirling of small eddies	Kolmogoroff microscale; concentration dissipation scale (Corrsin microscale)	Laminar stretching of eddies; reincorporation of fluid by microvortices
4	Diffusion in small eddies	Batchelor microscale	Reaction–diffusion in an eddy; IEM exchange

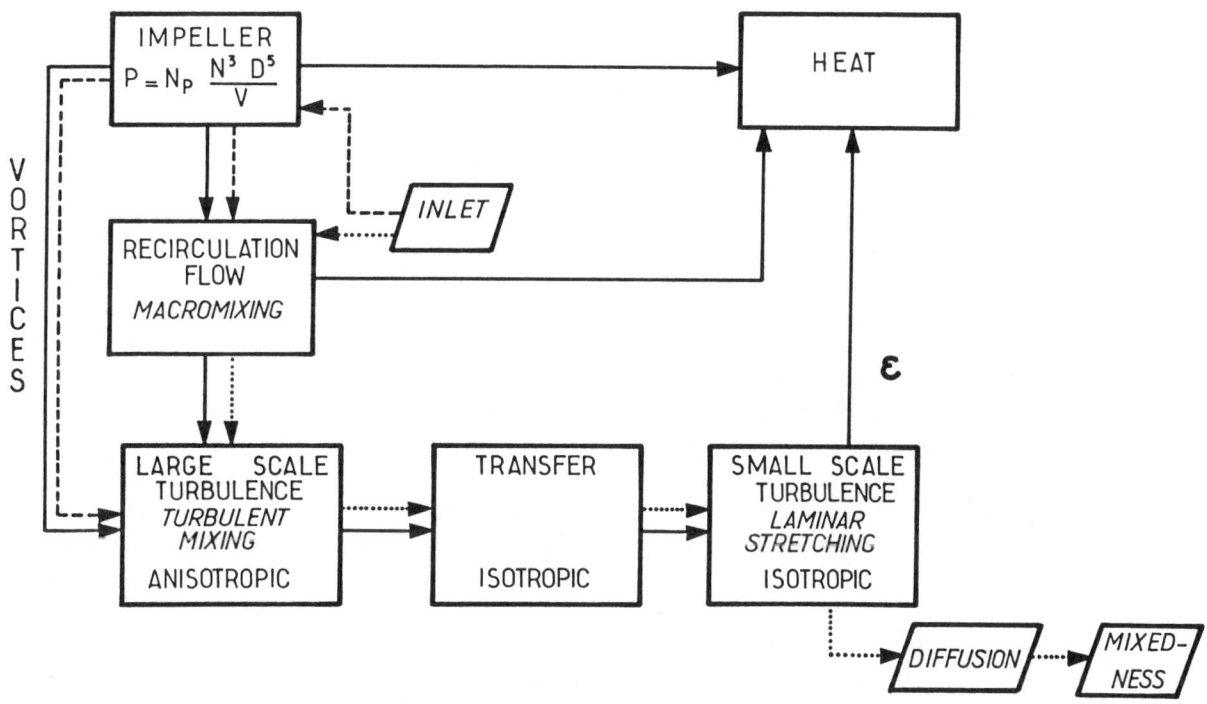

Source: Placek, J.; Tavlarides, L.L.; Smith, G.W.; & Fort, I. 1986. Turbulent Flows in Stirred Tanks, Part II: A Two-Scale Model of Turbulence. AIChE Jour., 32:11:1771–1786. Reproduced by permission of the American Institute of Chemical Engineers. Copyright 1986, AIChE.

Figure 5-2 Model of dissipation of mechanical energy and segregation in a stirred vessel.

If species j undergoes any reaction, the reaction rate r, expressed as a function of only the average concentrations, is included in the balance. For a well-macromixed tank at steady state

$$Q_{\text{in}} \overline{C}_{j,o} + \nu_j rV = Q_{\text{out}} \overline{C}_{j,\text{out}} \qquad \text{(Eq 5-2)}$$

All concentration fluctuations in this case are equal to zero.

Complete segregation. Complete segregation, also called macrofluid, is the state in which reactants initially contained in an eddy cannot escape the eddy. Any reaction takes place within "batch" eddies. In an eddy of age α_t at time t

$$\frac{dC_j}{dt} = \nu_j rV \quad \text{with } C_j(\alpha_t = 0) = C_{j,o}(t - \alpha_t) \qquad \text{(Eq 5-3)}$$

The outlet concentration is averaged through the residence time distribution f of the mixing device

$$C_{j,\text{out}}(t) = \int_0^\infty C_j(\alpha_t, t) f(\alpha_t) d\alpha_t \qquad \text{(Eq 5-4)}$$

If the reactants are not premixed, such a model leads to zero conversion of the reactants. Microfluid and macrofluid are often called the extreme states of segregation or micromixing.

Micromixing and Reaction

Interaction of micromixing and chemical reactions. As the reduction of all concentration microgradients occurs at a finite rate, a rapid reaction step can generate additional local microgradients by consuming the reactants and producing reaction products. In turn, these microgradients change the yield or the selectivity of the reactions by influencing the reaction rate.

Monomolecular reactions—generally first order with respect to the reactant—are not dependent on their chemical environment and are, therefore, not sensitive to micromixing (see chap. 4). On the contrary, the cases most sensitive to micromixing effects are those where chemical feeding maintains the macroscopic gradients of reactants and where there is a multiple-reaction system involving at least one rapid step whose extent influences the product selectivity.

Basically, temperature undergoes the same macro- and micromixing phenomena as concentration. For the sake of simplicity, it can be assumed in the rest of this chapter that all feeds are the same temperature. Although temperature fluctuations can be generated by reactions, the large values of Lewis numbers (Le $\gg$ 1) indicate that these fluctuations are smoothed out more rapidly than concentration fluctuations.

A simplified sketch of the interactions between chemical reaction, turbulence, and average residence time is given in Figure 5-3. A simple reaction A (white) + B (black) $\rightarrow C$ (crosses) occurs at different first Damköhler numbers (ratio of average residence time to reaction time) and second Damköhler numbers (ratio of micromixing time and average residence time). The case without reaction (blending) corresponds to the column on the left. Increasing the reaction rate leads to a higher yield of C. Segregation (i.e., micromixing times are large compared to residence times) causes a decrease of C. Good sensitivity of reaction yield to micromixing is achieved when the first Damköhler number is high enough and the second Damköhler number has an intermediate value.

A difficulty arises when one tries to define the relevant micromixing time. If only blending is desired, the micromixing time should be equated to the largest time constant out of the four steps in Table 5-1. However, if reaction takes place between species A and B, these species must contact on the molecular level, and the relevant

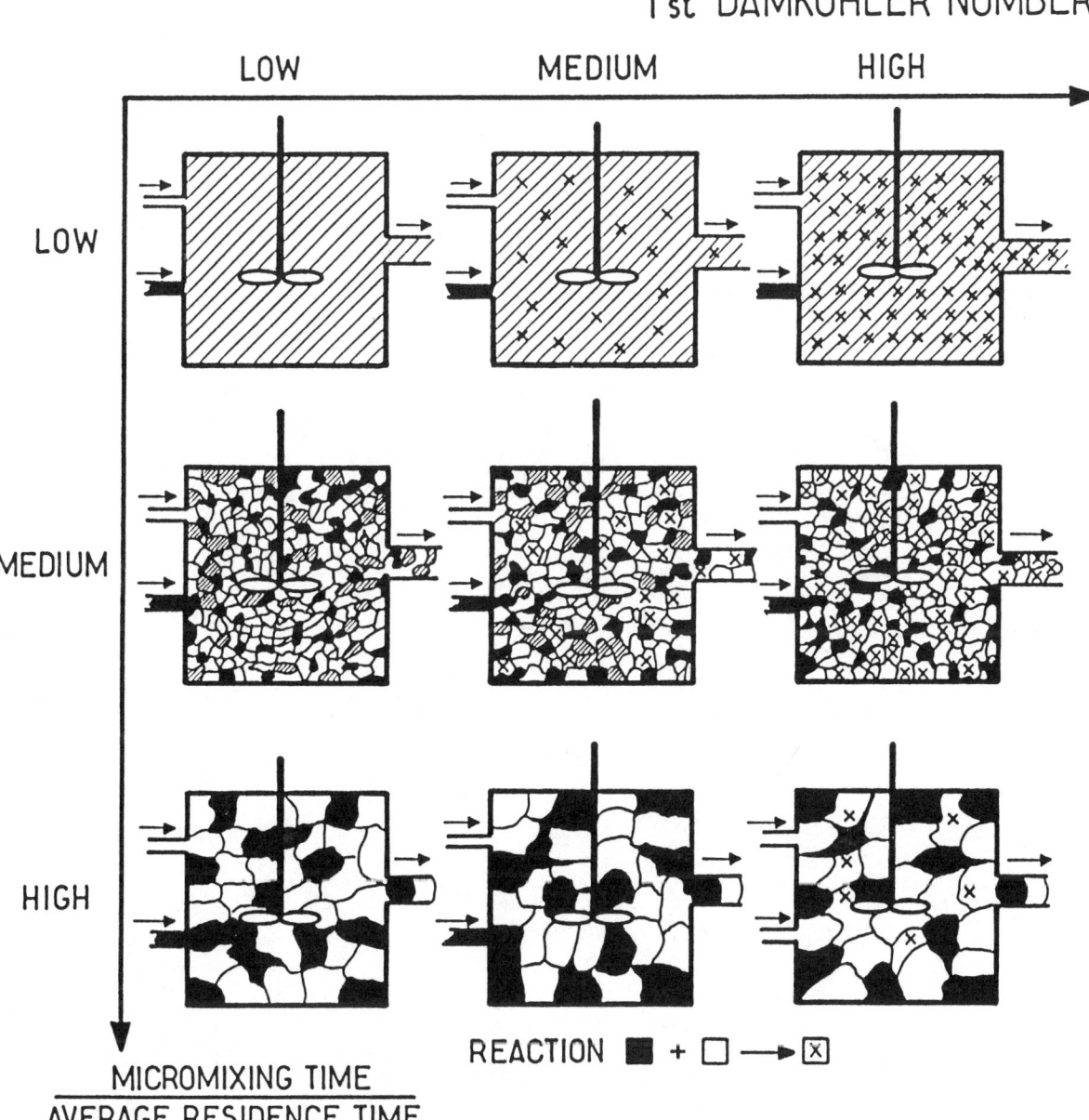

Source: David, R. Lecture notes in Engineering, 40:387, Springer-Verlag, New York (1989).

Figure 5-3 Interaction between micromixing, flow, and reaction in a stirred tank.

micromixing times are those corresponding to the last steps, stretching and molecular diffusion. This point is discussed further in the section on micromixing models.

Single-step reactions. The simplest reactions that are sensitive to micromixing are second-order reactions between two reactants (see Figure 5-4A). Instantaneous acid–base reactions have often been used with a color pH indicator to show a qualitative picture of the reaction zone (see Bourne and Tovstiga 1988; Mao and Toor 1970; Karagozian et al. 1987). Pohorecki and Baldyga (1983a) and Mao and Toor (1970) have based models on such experimental observations. Second-order reactions with finite reaction rates have been used as micromixing indicators by Treleaven and Tobgy (1973), Aubry and Villermaux (1975), Zoulalian and Villermaux (1974), and Klein et al. (1980). Figure 5-5 shows some results of Klein et al. (1980), who mixed sodium hydroxide and nitromethane in a stirred tank at different average residence times and different stirring speeds. Two effects are clearly shown here. At low values of τ, the mixing is essentially devoted to reducing the gradients of reactants entering the tank. At higher values of τ, the mixing changes the physical microstructure of the fluid and destroys segregation between the smallest eddies. A similar study was made by Plasari et al. (1978) for a zero-order reaction, which is also sensitive to micromixing (Figure 5-4B). However, the study of mixing in stirred tanks using single-step reactions that do not go to completion has three main disadvantages:

- Instantaneous conversion has to be deduced from the experimental concentration of reactants or products, and this requires locally specific sensors with short response times.
- Effects of segregation on conversion are often small, and the kinetics have to be known with accuracy.
- Maximum sensitivity is achieved when the reaction time is of the same order of magnitude as the average residence time (or space time) and/or the micromixing time (Figure 5-4); consequently, the number of available reactions is limited and unusually short space times may be required in the liquid phase.

Therefore, the following conditions are required of a chemical system designed to be sensitive to micromixing:

- Two or more reactants are involved so that contact between them is controlled by mixing.
- The reactants are fed separately into the tank to create initially large and well-defined concentration gradients.
- At least one reaction must be much faster than the mixing process itself. This practically implies that reaction times for the fastest step are less than 0.1 s in liquids.
- An indicator of mixing history must be kept in the system in the form of one or several stable products.

The last criterion eliminates single reactions that have to be stopped at the reactor outlet and can be used only with given space times. Conversely, fast multi-step reactions leading to a distribution of products frozen by the consumption of one of the major reactants are well-suited for micromixing studies. The space time is then generally much longer than the shortest reaction time and has no direct

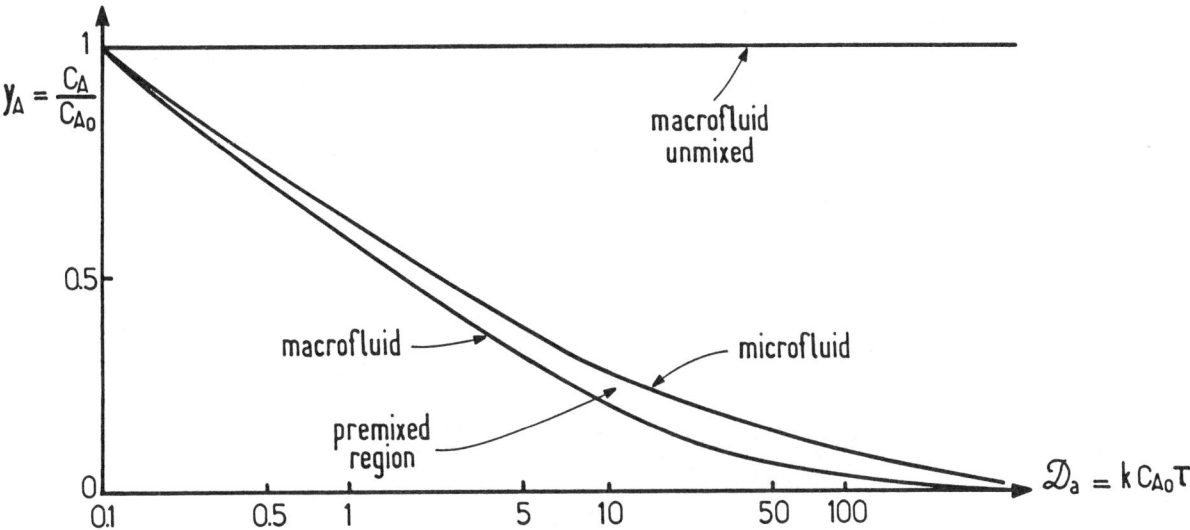

NOTE: Micromixing extremes: $A + B \rightarrow$ products; $r = k_2 C_A C_B$; initial concentrations after mixing $C_{A0} = C_{B0}$.

Figure 5-4A Residual fraction of reactant versus first Damköhler number for a second-order reaction.

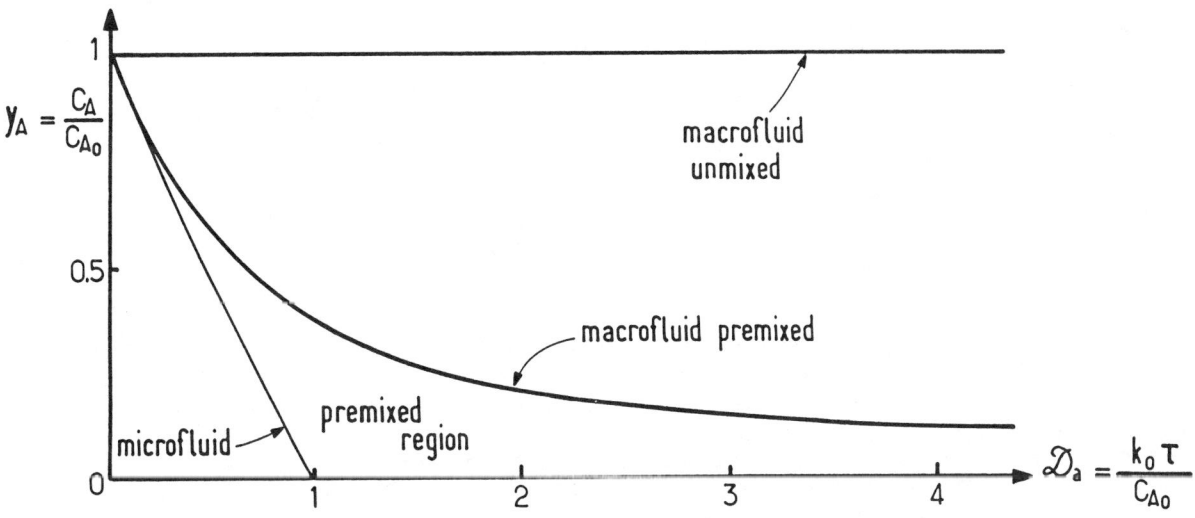

NOTE: Micromixing extremes: $A + B \rightarrow$ products; $r = k_0$; initial concentration after mixing C_{A0}.

Figure 5-4B Residual fraction of reactant versus first Damköhler number for a zero-order reaction.

influence on the selectivity, so the method may be used in both continuous and semibatch reactors.

Multistep reactions. Only the consecutive–competing reaction example will be treated here. A second classical example, polymerization and polycondensation reactions, is not discussed here. Details about micromixing and polymerization can be found in Nauman (1974), Villermaux and Blavier (1984), and Villermaux (1986).

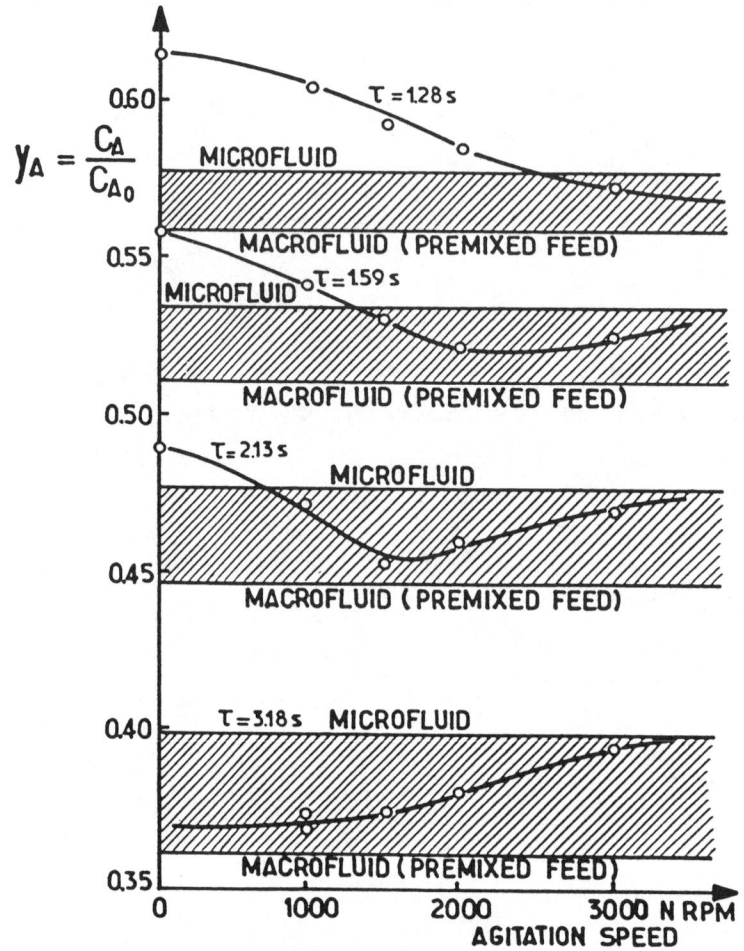

NOTE: Experimental remaining fraction of nitromethane versus stirring speed at different space times. Experimental conditions: alkaline hydrolysis of nitromethane; 5% polyethylene glycol, $C_{A0} = 33.1$ mol/m^3, $C_{B0} = 49.7$ mol/m^3; $k_1 = 0.016$ m^3 mol^{-1}s^{-1}; $\nu = 2.32 \times 10^{-6}$ m^{+2} s^{-1}; $\tau = 19.2$ °C. Microfluid means conversion obtained when the reactants are allowed to react in a state of microfluid (whatever the kind of feed, unmixed or premixed); macrofluid means conversion obtained when the reactants are initially premixed and then allowed to react in a state of total segregation. The shaded area represents the conversion obtainable with premixed reactants and a variable state of segregation (premixed region). The region between $y_A = 1$ and the microfluid limit represents the conversions obtainable with unmixed feed and a variable state of segregation (unmixed region).

Source: Klein, J.P. et al. 1980. Interpretation of Experimental Liquid Phase Micromixing Phenomena in a Continuous Stirred Reactor With Short Residence Times. Ind. Engrg. Chem. Fund. 19:373–379. Copyright 1980, American Chemical Society. Reprinted with permission.

Figure 5-5 Experimental residual fraction of nitromethane versus stirring speed at different space times.

Consecutive–competing reactions have been extensively used by Paul and Treybal (1971), Zoulalian and Villermaux (1974), Bourne et al. (1981), Barthole et al. (1982b), and Bourne and Tovstiga (1988). The general scheme is

$$A + B \xrightarrow{t_{R1}} R \qquad \text{(Eq 5-5)}$$

$$R + n_o B \xrightarrow{t_{R2}} S \qquad \text{(Eq 5-6)}$$

The most interesting case is for $t_{R2} \gg t_{R1}$ when a limited amount of B (in stoichiometric defect) is added to A, so that B is totally consumed at the end of both reactions. If the mixing of B occurs instantaneously, no secondary product S is formed. If the fluid is partly segregated, some R stays in contact with B and is converted into S. Both reactions are then stopped by total consumption of B.

The amount of S formed is an index of segregation. Bourne et al. (1981) have defined this index as

$$X_s = \frac{(n_o + 1)C_s}{C_{B0}} \qquad \text{(Eq 5-7)}$$

Villermaux and David (1983) have chosen

$$\beta' = \frac{X_S - X_{SM}}{1 - X_{SM}} \qquad \text{(Eq 5-8)}$$

where X_{SM} is the yield observed in a well-micromixed reactor (microfluid). Two reactions have been used with success: azo-coupling of 1-naphthol (A) with diazotized sulfanilic acid (B) (Bourne et al. 1981) and precipitation of barium sulfate (S) from barium complexed by ethylenediaminetetraacetic acid (EDTA) in an alkaline medium (reactant A) under the influence of an acid (reactant B) (Barthole et al. 1982b). For the latter reaction, X_S is written

$$X_S = \frac{(2n_o + 1)C_S}{n_o C_{B0}} \qquad \text{(Eq 5-9)}$$

Figure 5-6 shows typical experimental results for X_S versus stirring speed obtained by Bourne and Dell'Ava (1987) in a stirred tank of 2.5 dm^3 with the azo-coupling reaction system.

A further interesting study of micromixing with the same reactions has been made in T-mixers by Tosun (1987). Similarly, Bourne and Tovstiga (1988) have studied the mixing of a laminar central flow with a major turbulent stream in a tubular reactor.

MICROMIXING MODELS

Since 1958, many authors have tried to model experimental micromixing results. These models have become increasingly more sophisticated and realistic.

Empirical Single-Parameter Models

Empirical single-parameter models rely on the following assumptions:
- perfect macromixing;

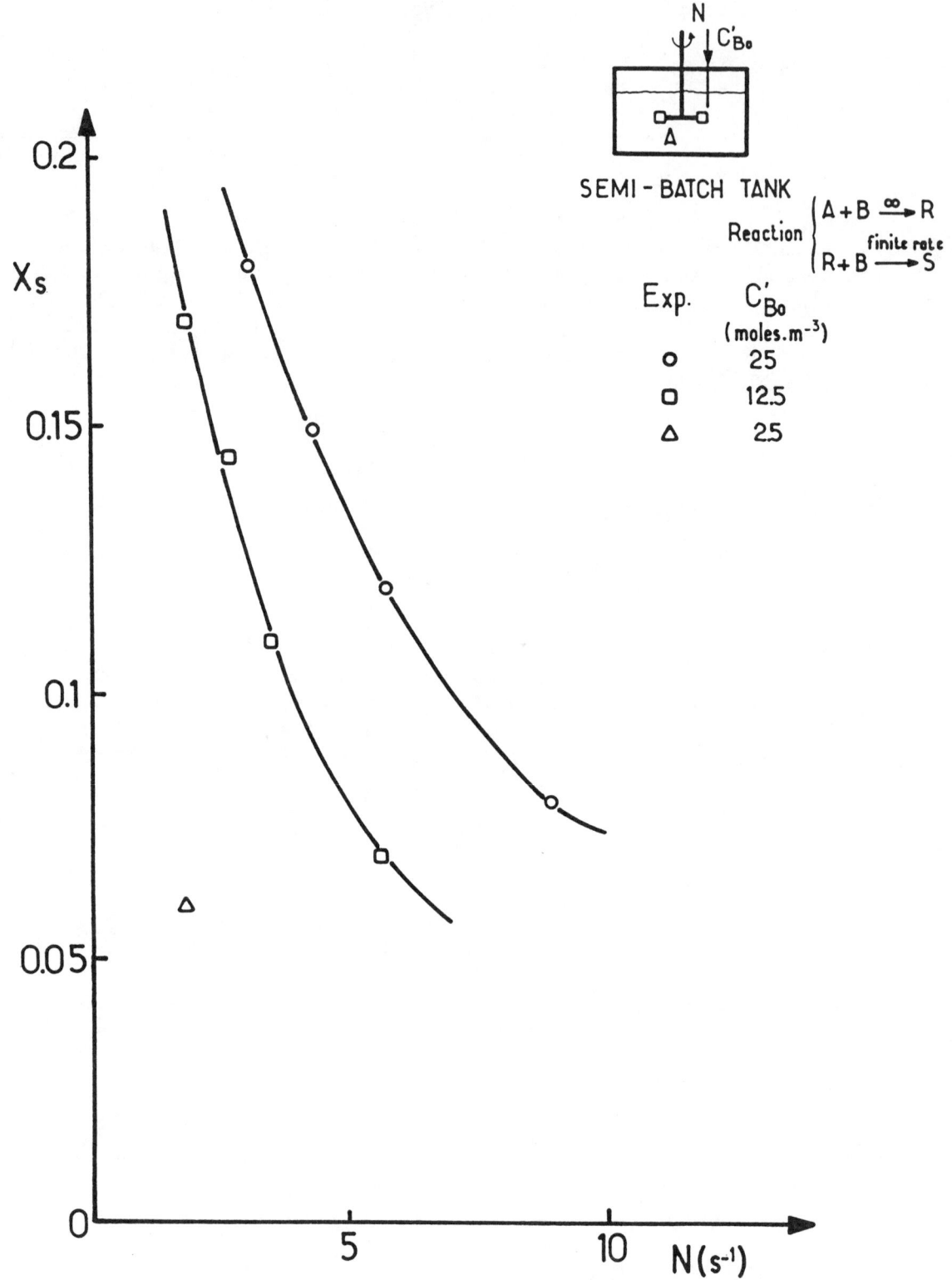

Figure 5-6 Experimental segregation index versus stirring speed for a consecutive-competing reaction in a semibatch stirred tank of 2.5 dm^3 volume.

- uniform concentration fluctuations throughout the tank;
- division of the fluid into small fluid particles or aggregates with uniform internal concentration, with mass transfer occurring between the particles; and
- fluid particles or aggregates identified by their age (Lagrangian point of view).

The single-parameter models assume a single micromixing process starting from either of two opposite sets of initial conditions: one feed of premixed reactants, or two or more feeds of unmixed reactants.

The concept of mixing earliness. The behavior of fluid aggregates in a closed system by an internal age α_t (time elapsed since entering the tank) and a life expectancy or residual lifetime λ (time remaining until leaving the tank) must first be defined. The sum $\alpha_t + \lambda = t_s$ is the residence time within the tank (see chap. 4). Obviously, all entering aggregates have the same α_t ($\alpha_t = 0$) and all leaving aggregates have the same λ ($\lambda = 0$). The aggregates have different residence times t_s, and generally do not simultaneously have the same α_t and the same λ. They start in an entering environment (EE), which is a state of minimum mixedness (particles having different residence time never mix), and switch to a leaving environment (LE), which is a state of maximum mixedness (all the aggregates have the same probability to leave and thus the same λ). The EE and LE should not be confused with macro- and microfluid, which denote a physical structure. In Figure 5-7, these phenomena are modeled as a "bundle of parallel tubes," in which the reactor volume

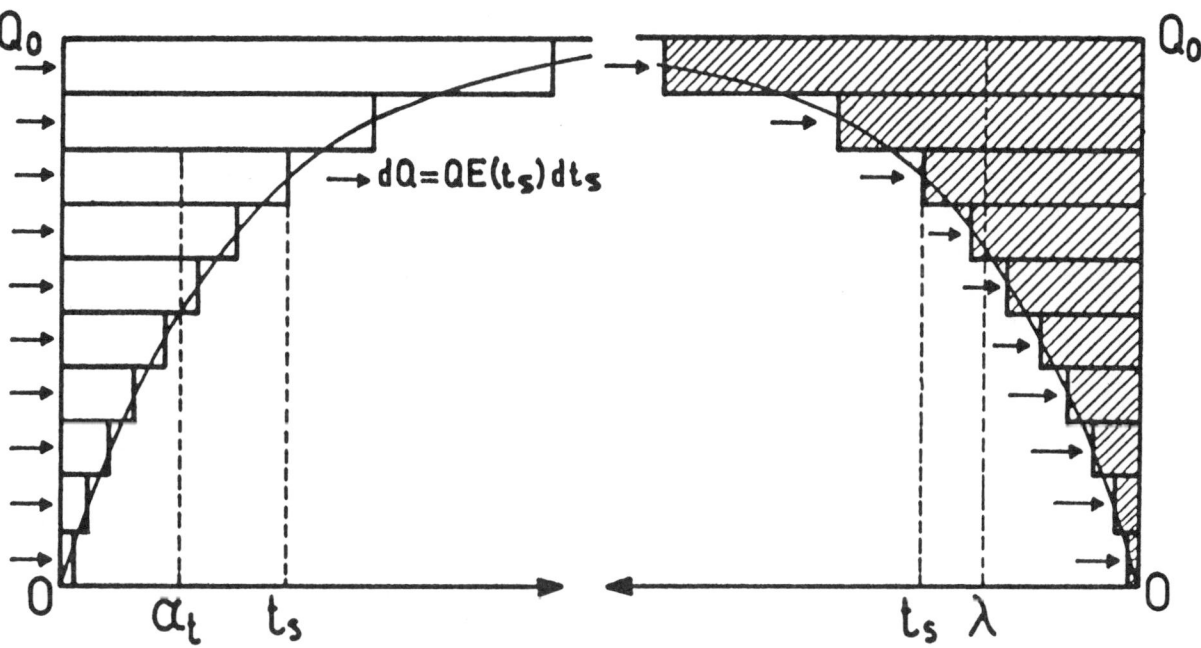

Source: From Encyclopedia of Fluid Mechanics, *Vol. 2, by Nicholas P. Cheremisinoff. Copyright © 1986 by Gulf Publishing Company, Houston, Texas. Used with permission. All rights reserved.*

Figure 5-7 Representation of micromixing models by a bundle of parallel tubes. Left: minimum mixedness state. Right: maximum mixedness state; mixing is achieved on a vertical line.

is reorganized in a bundle of small tubes of increasing length equivalent to t_s. The fluid flows with a rate

$$dQ = Q_o f(t_s) dt_s \qquad \text{(Eq 5-10)}$$

through each tube, and the loci of the extremity of the tubes are

$$F(t_s) = \int_0^{t_s} f(t_s) dt_s$$

In the EE on the left, the tubes are piled in such a way that aggregates of the same α_t are on the same vertical line. On the right (LE), aggregates with the same λ are on the same vertical line. The micromixing models differ in the way they transfer the aggregates from the EE to the LE. Spencer et al. (1980) have defined a general "segregation function" $s(t_s, \lambda)$ or $s(\alpha_t, \lambda)$, which is actually the fraction of fluid of a given λ and t_s (or α_t), which is in the EE ($1 - s$ is in the LE). Other models that have been proposed are summarized by Villermaux (1983, 1986). For instance, the first micromixing models were imagined by Zwietering (1959) and Danckwerts (1958) and describe the extreme mixing earliness situations shown in Figures 5-8A and 5-8B. Danckwerts (1958) and Zwietering (1959) treated the case where only minimum mixedness exists in the EE (no LE), which corresponds to Figure 5-8A. A microfluid is then achieved only in the individual tubes. In Figure 5-8A, a local mass balance of the volume dV gives

$$QC + \nu_j r(C) dV = (Q - dQ)(C + dC) + C dQ \qquad \text{(Eq 5-11)}$$

which yields

$$Q \frac{dC}{dV} = \nu_j r \qquad \text{(Eq 5-12)}$$

and, because $dt_s = d\alpha_t = dV/Q$

$$\frac{dC}{d\alpha_t} = \nu_j r \qquad \text{(Eq 5-13)}$$

The outlet concentration is

$$C_{\text{out}} = \frac{1}{Q_o} \int_0^{Q_o} C(\alpha_t) dQ = \int_0^\infty C(\alpha_t) f(\alpha_t) d\alpha_t \qquad \text{(Eq 5-14)}$$

The above equations are the same as for the macrofluid (complete segregation) defined in the discussion of complete segregation, in the section "Extreme States of Segregation."

Zwietering (1959) also treated the case of maximum mixedness in the leaving environment only (Figure 5-8B). All fluid aggregates in tubes on the same vertical line are in a microfluid state. In the volume dV, the life expectancy is $\lambda = t_s = \int_{V_R}^V \frac{dV}{Q}$. The entering flow rate through the sidewall is

$$dQ = \lambda Q_o f(t_s) dt_s \qquad \text{(Eq 5-15)}$$

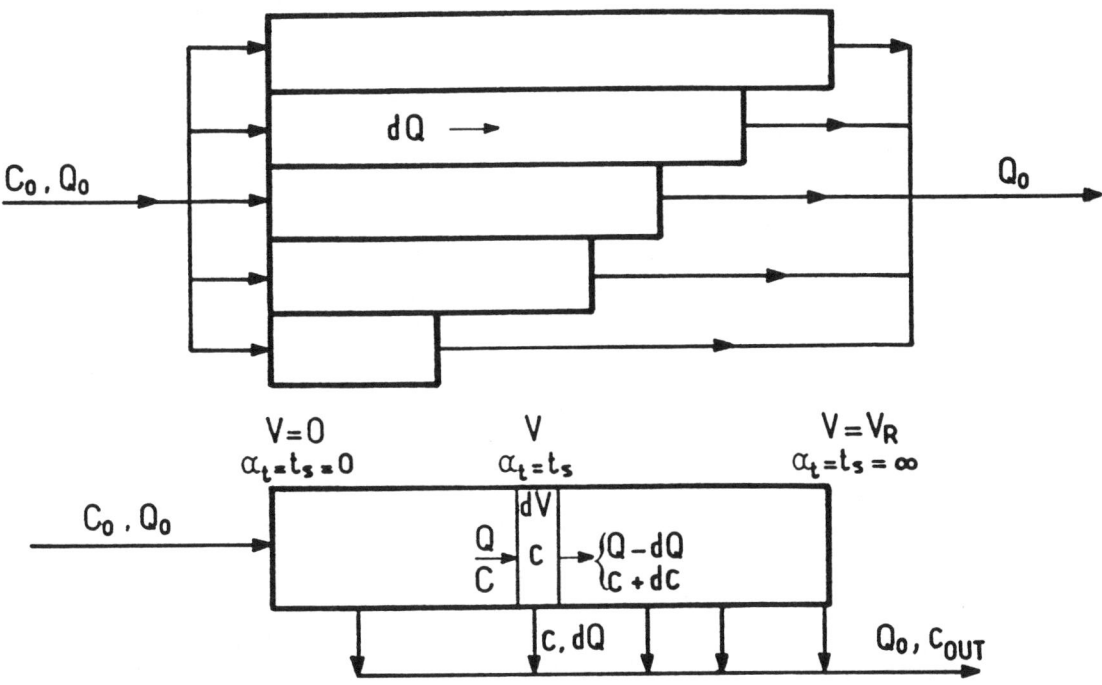

Figure 5-8A Flow model of Danckwerts and Zwietering for minimum mixedness.

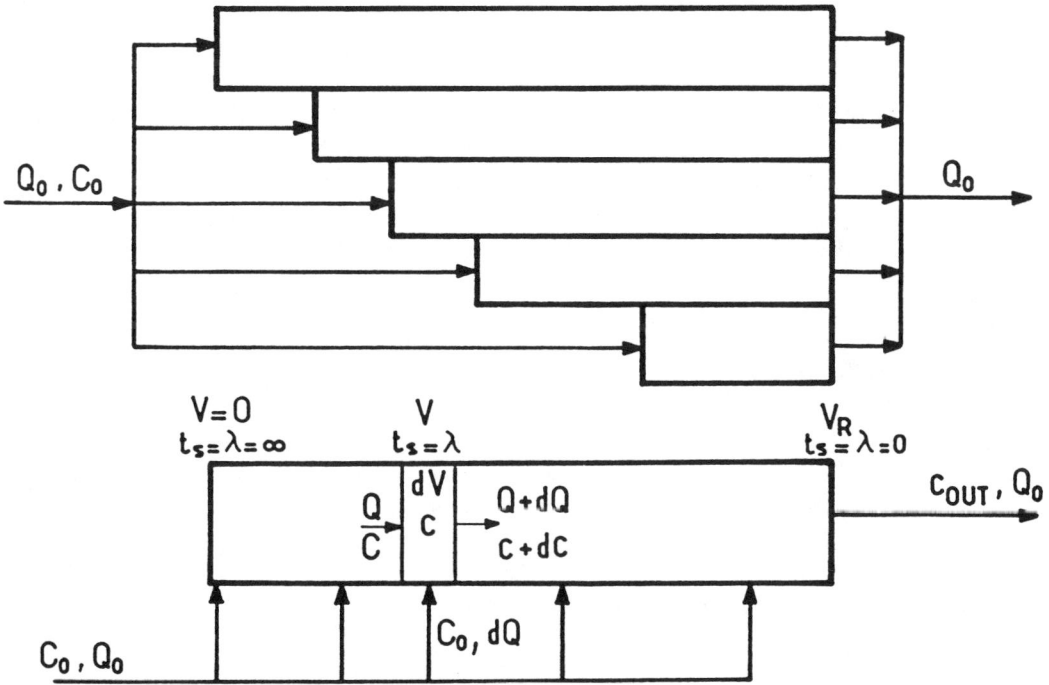

Figure 5-8B Flow model of Zwietering for maximum mixedness.

and the overall flow rate Q through dV is written

$$Q = \int_\infty^{t_s} dQ = \int_{t_s}^\infty Q_o f(t_s) dt_s = Q_o [1 - F(t_s)] \quad \text{(Eq 5-16)}$$

where $F(t_s) = \int_0^{t_s} f(t_s) dt_s$

The mass balance over dV is

$$QC + C_o dQ + \nu_j r(C) dV = (Q + dQ)(C + dC) \quad \text{(Eq 5-17)}$$

which results in

$$(C_o - C)\frac{dQ}{dV} + \nu_j r = Q\frac{dC}{dV} \quad \text{(Eq 5-18)}$$

and because $dt_s = -dV/Q$

$$\frac{dC}{dt_s} = -r\nu_j - (C_o - C)\frac{f(t_s)}{1 - F(t_s)} \quad \text{(Eq 5-19)}$$

This differential equation should be integrated with $\frac{dC}{dt_s} = 0$ at high values of the ratio t_s/τ (in practice, about 10) where τ is the space time ($= V/Q$) of the reactor. C_{out} is the solution of the above equation at $t_s = 0$.

Interaction by exchange with the mean (IEM) model. As the rigorous treatment of coalescence–dispersion (CD) models leads to complicated equations (see next section), Villermaux and Devillon (1972) and Costa and Trevissoi (1972), starting from the idea of Harada et al. (1962), suggested mass transfer between aggregates and a fictitious average concentration in the tank, which accounts for multiple contact with other aggregates. Villermaux and Devillon (1972) and Costa and Trevissoi (1972) have proposed simultaneously the basic developments of this model. The mass exchange between aggregates is characterized by a time constant t_m, which is generally taken as uniform throughout the tank; $t_m = 0$ is the case of the microfluid, whereas $t_m \to \infty$ is the case of the macrofluid (complete segregation). The basic equation for the variation of the concentration C_j of species j in an aggregate of age α_t is written

$$\frac{dC_j}{d\alpha_t} = \frac{\overline{C_j} - C_j}{t_m} + R_j \quad \text{(Eq 5-20)}$$

R_j is the rate of production of j by reactions and C_j is defined by the condition that the net sum of all exchange fluxes over the tank is zero. In a continuous stirred-tank reactor (CSTR), for example, since the probability of encountering aggregates of age α_t is $e^{-\alpha_t/\tau}/\tau$, $\overline{C_j}$ is written

$$\overline{C_j} = \frac{1}{\tau}\int_0^\infty C_j e^{-\alpha_t/\tau} d\alpha_t \quad \text{(Eq 5-21)}$$

Results of calculations using the IEM (interaction by exchange with the mean) model are presented in Figures 5-9 and 5-10 (Villermaux 1986) for second-order and consecutive–competing reactions in a CSTR.

Coalescence–dispersion model. Proposed initially by Curl (1963), the CD model was adapted by Spielman and Levenspiel (1965). It assumes aggregates of equal size that coalesce with neighboring aggregates and produce upon redispersion two aggregates of the same size as the original ones but with averaged concentrations. The frequency ω of the CD process is the model parameter. Chemical reaction occurs batchwise in the aggregates during the time between CD events. Obviously, ω = 0 corresponds to complete segregation, and ω → ∞ corresponds to microfluid. The distribution of concentration obtained in the tank can be discrete or continuous (Curl 1963). In open reactors, inlet and outlet feed streams are simulated by adding and removing aggregates from the coalescing population. These models are often solved by the Monte Carlo method, which consists of picking out aggregates at random from a large number of aggregates and making them coalesce and redisperse; this process is repeated ω times per simulated time unit.

The first attempt to relate the coalescence frequency to turbulence parameters was made by Canon et al. (1977). An approximate equivalence could be found between CD and IEM models when taking David and Villermaux (1975)

$$\omega = 4/t_m \qquad \text{(Eq 5-22)}$$

For instance, Eq 5-22 seems appropriate for the consecutive–competing reactions in Figure 5-10.

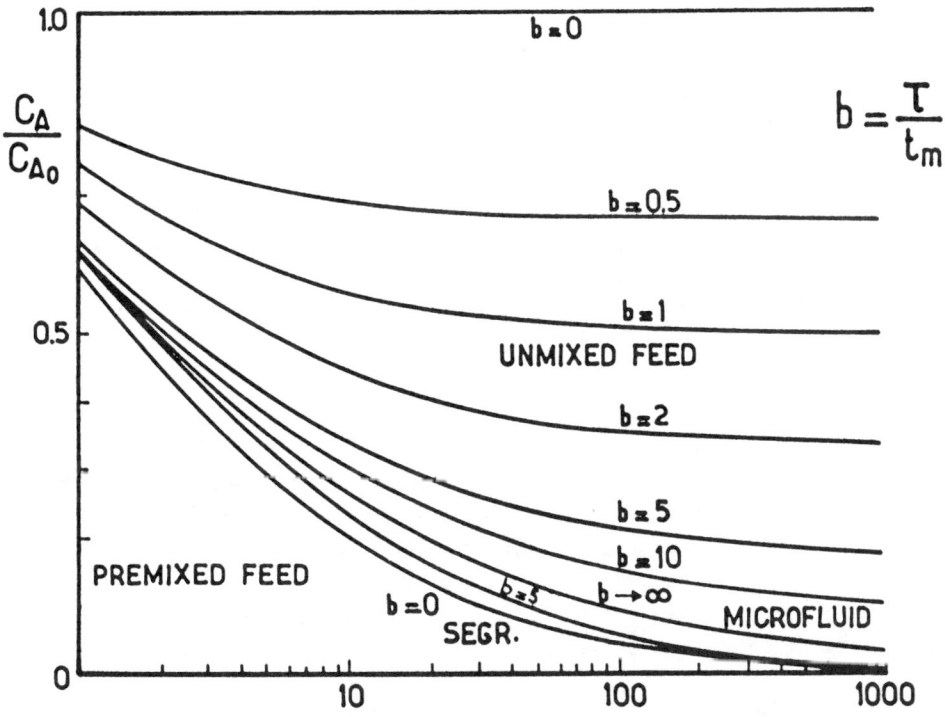

Source: From Encyclopedia of Fluid Mechanics, *Vol. 2, by Nicholas P. Cheremisinoff. Copyright © 1986 by Gulf Publishing Company, Houston, Texas. Used with permission. All rights reserved.*

Figure 5-9 Prediction of residual fraction of reactant versus first Damköhler number by the IEM model: second-order reaction $A + B \rightarrow$ products. Influence of the micromixing parameter $b = \tau/t_m$ on residual fraction.

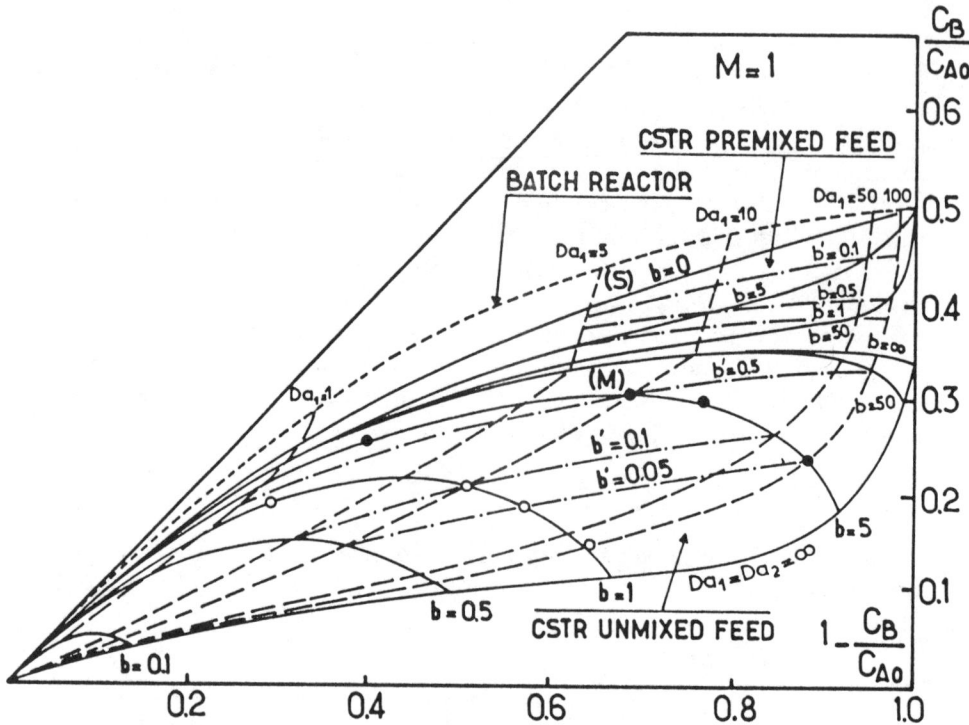

Source: From Encyclopedia of Fluid Mechanics, *Vol. 2, by Nicholas P. Cheremisinoff. Copyright © 1986 by Gulf Publishing Company, Houston, Texas. Used with permission. All rights reserved.*

Figure 5-10 Prediction of the yield of species R versus conversion of reactant B by the IEM model: consecutive-competitive reactions: $A + B \xrightarrow{k_1} R$, $R + B \xrightarrow{k_2} S$.; $k_2/k_1 = 0.5$; $C_{B0}/C_{A0} = 1$; calculations by the CD model showing excellent agreement with the IEM model. $Da_1 = k_1 C'_{A0} \tau$; $Da_2 = k_2 C_{A0} \tau$.

All these phenomenological models (CD, IEM, mixing earliness) are approximately equivalent in predicting conversions or yields within the macro- and microfluid limits. The major criticism of the models is their lack of physical relevance to the flow structure. In addition, the single parameter leads one to ignore the spatial distribution of the micromixing intensity in the tank. Therefore, this parameter must be fitted to the experimental data. Finally, the models are unsuitable for reactor scale-up because of the empirical character of the parameters. Therefore, a second direction of modeling was to start from the physical aspect of mixing. In mixing with reaction, it is generally possible to select the most important mixing stage (from those discussed under "Mechanisms of Energy Dissipation in Vessels") for the process and to represent it by a single-stage model.

Models Using the Concept of Segregation in Physical Space

The four stages in the physical mixing process have already been defined. Starting from these stages, a comprehensive model should provide a quantitative description of the partial segregation and a physical interpretation of the model parameters. Typical models take into account one or more stages, according to the importance of a given mixing stage for the conversion or yield of chemically reacting species. The micromixing process is then characterized by a single time constant t_m depending on the mechanism involved. As the chemical process is also characterized by a reaction time (e.g., $t_R = 1/kC_0^{n-1}$ for nth-order reactions), the parameter controlling the process is the ratio t_R/t_m, which allows the description of all the intermediate states between microfluid ($t_R/t_m \to \infty$) and macrofluid ($t_R/t_m = 0$). The authors cited below have tried to estimate the micromixing time from turbulence theory.

Erosion (second stage). This model is sketched in Figure 5-11A for a continuous stirred tank with two feed streams. It assumes that the volume V_a of an aggregate generated by the inlet flow is reduced during the aging of the aggregate. Plasari et al. (1978) chose

$$V_a = V_{a_o}\left(1 - \frac{\alpha_t}{t_e}\right)^3 \qquad \text{(Eq 5-23)}$$

whereas Pohorecki and Baldyga (1983a) assumed

$$V_a = V_{a_o} e^{-\alpha_t/t_e} \qquad \text{(Eq 5-24)}$$

The time constant t_e was correlated with kinetic or mechanical energy dissipated in the tank. In the case of the erosion law of Pohorecki and Baldyga, the segregated volumes of Figure 5-11 are given by

$$V_1 = \frac{V}{2\left(1 + \dfrac{\tau}{t_e}\right)} \qquad \text{(Eq 5-25)}$$

and flow rates by

$$Q_1 = \frac{Q}{2}\frac{1}{\left(1 + \dfrac{\tau}{t_e}\right)} \qquad \text{(Eq 5-26)}$$

The overall concentration $\overline{\overline{C_A}}$ of reactant A is expressed by

$$\overline{\overline{C_A}} = \frac{Q_1 C_{AO}}{Q} + \overline{C_A}\frac{Q_2}{Q} \qquad \text{(Eq 5-27)}$$

$\overline{C_A}$ is the solution of a classical CSTR mass balance with microfluid flow structure.

Stretching (third stage). In the theory of Ottino (1980), the striation thickness δ of lamellae of fluid is determined by the stretching process. δ is a function of a stretching time, which is defined as

$$t_\delta \equiv -\delta\left(\frac{d\delta}{dt}\right)^{-1} \qquad \text{(Eq 5-28)}$$

These authors further complemented their model by including reduction of the concentration gradients in the slab by diffusion when δ is small enough (fourth stage). This aspect has been developed by Lee et al. (1986).

Molecular diffusion (fourth stage). Here the main process is a classical diffusion process, which can be described by modeling reaction–diffusion in aggregates of size L. This was done initially by Nauman (1975) and Bourne et al. (1981), starting from the classical mass balance with partial differential equations.

The IEM model can also be used to model diffusion. It has been shown to be almost equivalent to the reaction–diffusion model (Villermaux and David 1983). The

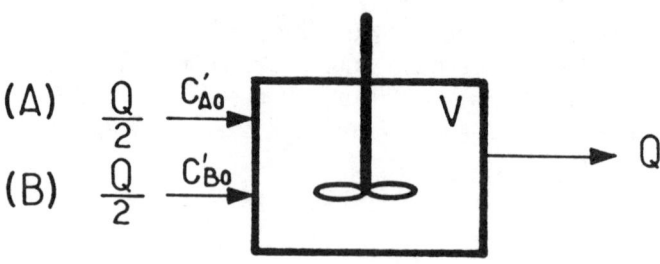

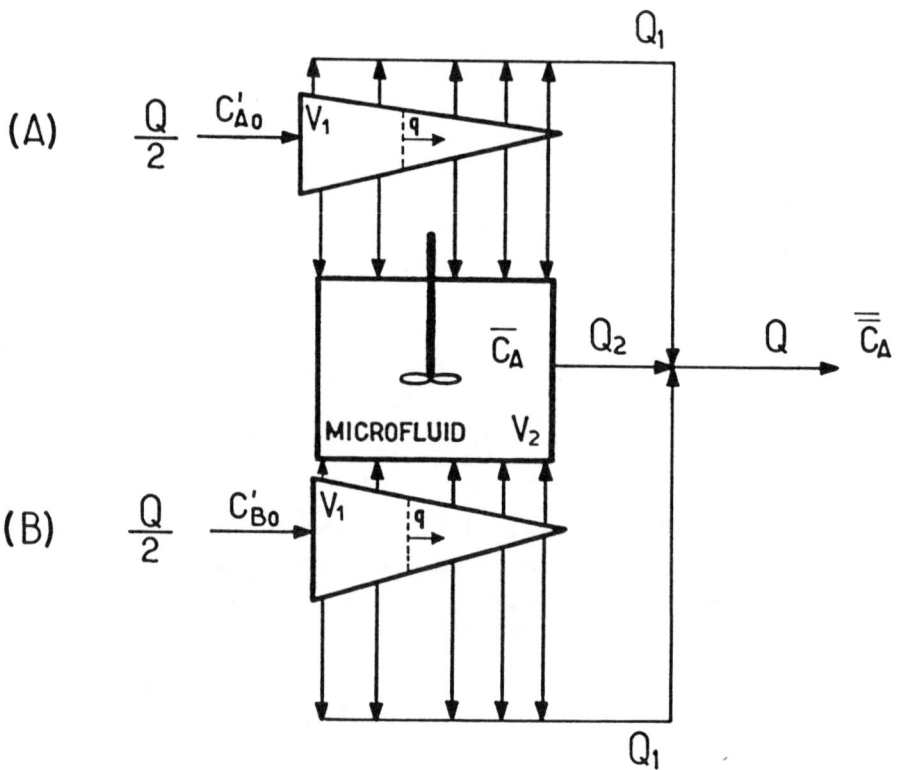

Figure 5-11 Erosion model applied to a CSTR with two feed streams ($2 V_1 + V_2 = V$; $2 Q_1 + Q_2 = Q$).

equivalence between time constants of both models is $t_m = t_D = \mu L_A^2/D_m$ (D_m = diffusivity, μ = shape factor).

An interesting property is revealed by the simulation reported in Figure 5-12. When one assumes that a real tank with perfect macromixing can be represented by a macrofluid (volume fraction β') and a microfluid (volume fraction $[1 - \beta']$ of the reactor), it turns out that the ratio $(1 - \beta')/\beta'$ is close to t_R/t_D for several different reacting conditions (Figure 5-12). On the average:

$$\frac{1 - \beta'}{\beta'} \cong 2\left(\frac{t_R}{t_D}\right)^{0.8} \quad \text{(Eq 5-29)}$$

Nevertheless, numerous experimental results such as those of Zoulalian and Villermaux (1974), Klein et al. (1980), Bourne et al. (1981), Mehta and Tarbell (1983), Chang et al. (1986), and David and Villermaux (1987) have clearly shown

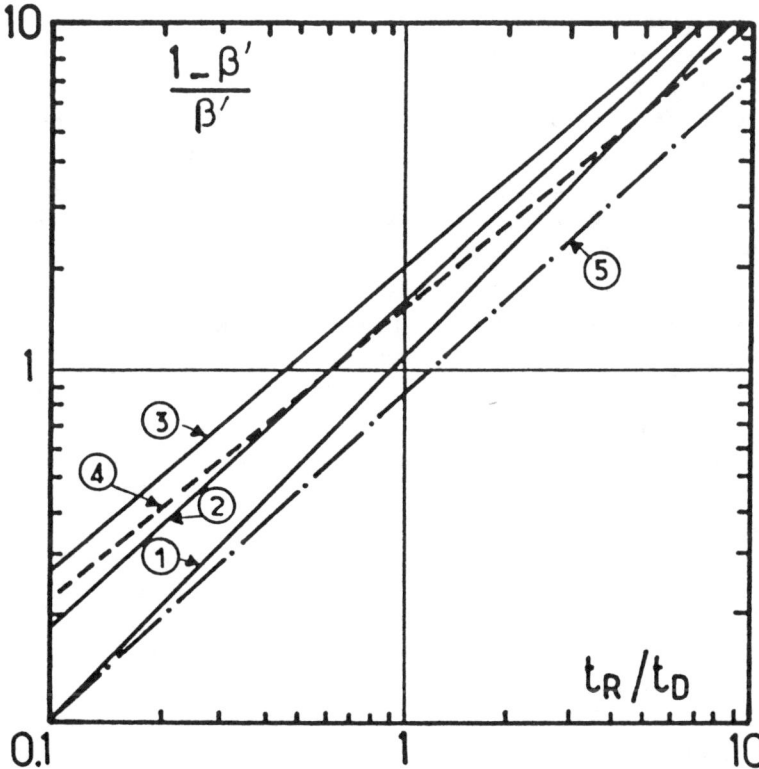

Source: Villermaux, J. & David, R. 1983. Recent Advances in the Understanding of Micromixing Phenomena in Stirred Reactors. Chem. Engrg. Comm., 21:105–122. Reprinted with permission.

Figure 5-12 Micromixedness ratio $(1 - \beta')/\beta'$ versus ratio of reaction time to micromixing time curves 1 to 4: simulations with IEM model. Second-order reaction with premixed feed; $C_{A0} = C_{B0}$.
$A + B \rightarrow$ products: (1) Da = 2; (2) Da = 5; (3) Da = 10.
Consecutive-competitive reactions $A + B \xrightarrow{k_1} R$; $R + B \xrightarrow{k_2} S$; $k_2/k_1 = 0.5$ (4). Reaction and diffusion in a slab (5).

that models relying on a single mixing stage were not sufficient to explain the dependence of the results on parameters like stirring speed, flow rate or space time, reactor size, and reactor operation mode. Therefore, it is now believed that models accounting for different mixing stages are necessary. These models correspond to different stages of the mixing process (last column of Table 5-1). The two main directions of improvement of the single-stage models will now be considered.

Multistage Semiempirical Models

Starting from the mixing earliness concept and the models of Zwietering (1959) and Danckwerts (1958), multiple-environment models have been proposed. These models are single-parameter models, but the overall model integrates several mixing stages. They are summarized in Figure 5-13.

Kattan and Adler (1972) and Treleaven and Tobgy (1973) first generalized the mixing earliness concept to reactors with two reactant inlets. Kattan and Adler (1972) proposed the model (Figure 5-13A) where both feed streams, each with its own residence time distribution (RTD), are in the flow pattern of maximum mixedness. The mixing is achieved by coalescence–redispersion between both populations of the same life expectancy.

The two-environment (2E) model of Ng and Rippin (1965), which belongs to the family of models described by Spencer et al. (1980), has been developed by Ritchie and Tobgy (1979) into the three-environment (3E) model. Two EEs, one for each reactant feed stream, supply a single LE (see Figure 5-13B). The EEs are in a minimum mixedness flow pattern, whereas the LE is in maximum mixedness with internal microfluid flow pattern. The RTDs f_1 and f_2 of the two feed streams can be different. The unique parameter of the model is $R_s = 1/t_m$, which represents the time constant of transfer between the EEs and the LE. R_s was related to the theory of the isotropic mixer of Corrsin (1964) by setting t_m equal to $(L^2/\varepsilon)^{1/3}$. Obviously, the EEs account for the first two stages of mixing, dispersion and size reduction of eddies, and the LE represents the ultimate diffusion (here supposed instantaneous) and reaction stages.

Mehta and Tarbell (1983) refined and extended the 3E model by using two LEs, one for each feed stream. These LEs are exchanging mass with the same time constant as between each EE and LE. This model is called the four-environment (4E) model. More recently, these authors have compared critically the predictions of the 4E model with single-stage models such as IEM, CD, slab diffusion, and with the 3E model for different reactions and reactors (Chang et al. 1986; Mehta and Tarbell 1987). They showed that the 4E model gave the best fit of previous experimental results.

Zwietering (1984) published a new model, which described the dilution of entering fluid streams with older surrounding fluid. Each volume V is diluted at a rate

$$\frac{dV_a}{d\alpha_t} = pV_a \qquad \text{(Eq 5-30)}$$

where p is the model parameter, which is allowed to vary between $p = 0$ (minimum mixedness) and $p \to \infty$ (maximum mixedness). Older volumes dilute younger volumes, and so on. A fraction of each volume is picked up to build an outlet stream corresponding to the RTD of the reactor.

All of these models give a better representation of experimental results than the empirical single-parameter models and models of segregation in physical space, but certain criticisms of their empirical character still apply.

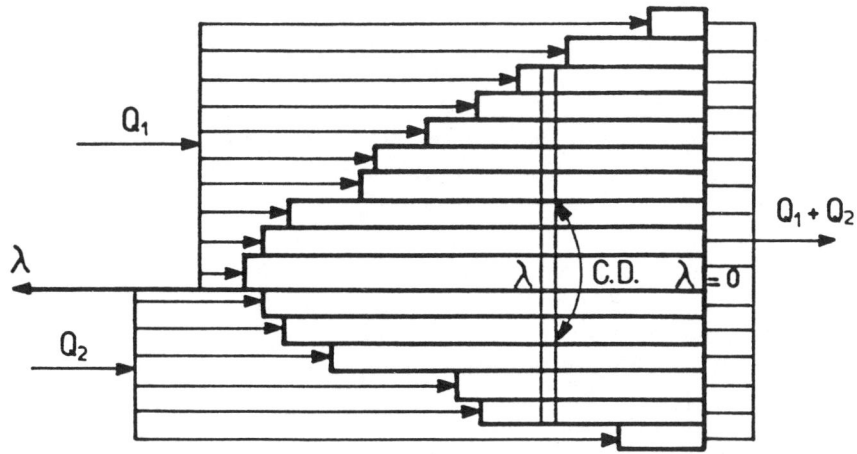

Figure 5-13A Model of Kattan and Adler.

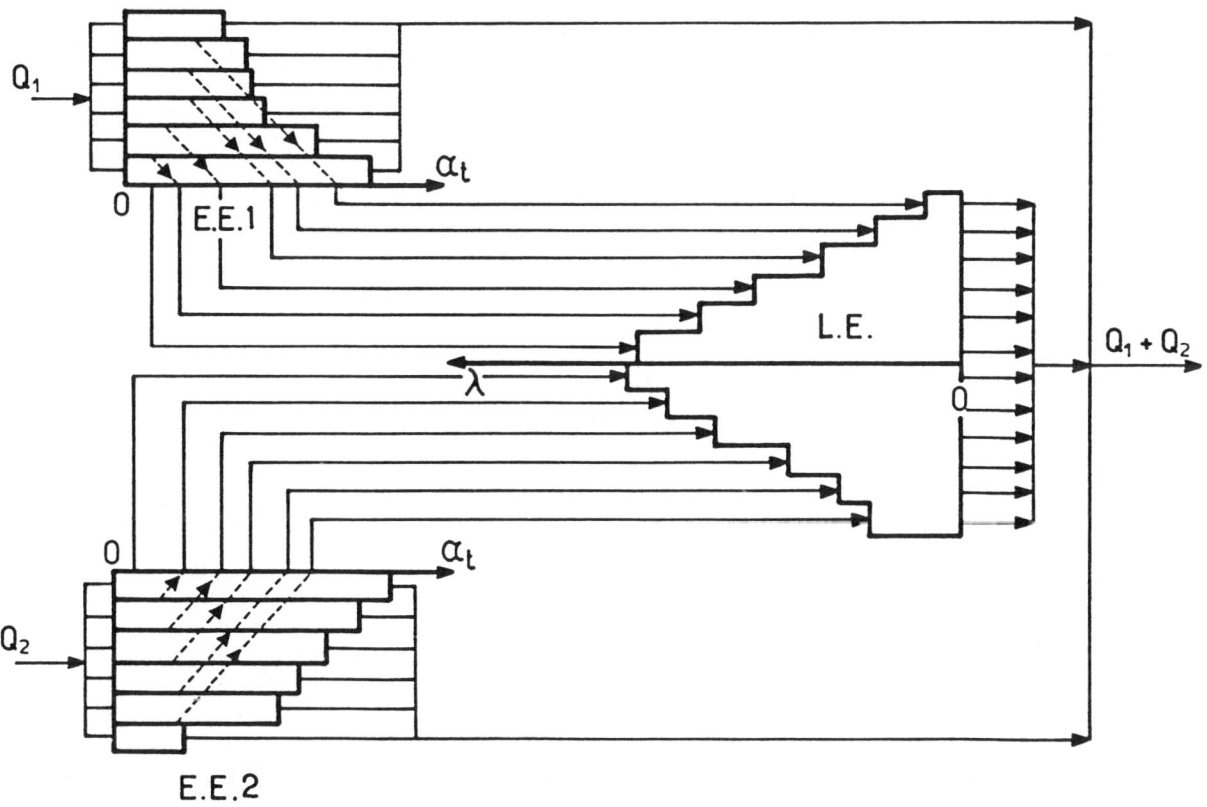

Source: Ritchie, B. & Tobgy, A.M. 1979. A Three Environment Micromixing Model for Chemical Reactors With Arbitrary Separate Feedstreams. Chem. Engrg. Jour., 17:173–182.

Figure 5-13B Model of Ritchie and Tobgy.

Physical Multistage Models

Models including different mixing stages with physical meaning have been developed by Klein et al. (1980), Pohorecki and Baldyga (1983a), Bourne and co-workers (Bourne et al. 1981; Bourne 1984; Baldyga and Bourne 1984; David and Villermaux 1987). Recently, David and Villermaux (1987) and Baldyga and Bourne (1984) took into account the internal inhomogeneity of turbulent mixing in the stirred tank. This inhomogeneity can be expressed in terms of deviations from ideal macromixing and/or deviations from perfect micromixing. The in-line mixer has been studied by Bourne and Tovstiga (1988) and Baldyga and Rohani (1987). Mao and Toor (1970, 1971) and Li and Toor (1986) have tried to solve the micromixing problem by closure models, starting from the local mass balances of reactants and products. Using various assumptions, they related the reaction terms to the concentration fluctuations of reactants obtained from inert tracer mixing studies. This type of model is described in chap. 3.

Model of David and Villermaux. The model of David and Villermaux (1987) involves modeling the four stages of mixing. It is concerned with a stirred tank showing a small deviation from perfect macromixing combined with inhomogeneous micromixing. The internal circulation process assumed by these authors is explained in Figure 5-14.

Mixing stage 1. An injected volume V_{Bo} is added in m fractions of volume $V_1 = V_{Bo}/m$, m being equal to 1 in the case of a pulse injection. Each fraction is initially associated with a volume V_2 of surrounding fluid to form the initial reacting cloud. The cloud is convected along the bulk recirculation streamlines according to the trajectories determined by average velocity measurements. The cloud passes across two zones, one far from the stirrer (zone 1: low mixing intensity) and the other close to the stirrer (zone 2: high mixing intensity). The time for closing the recirculation loop is t_c. An injection point is thus defined by the time required by the cloud to reach the next zone. In more complicated models, the tank is divided in a network of cells (Middleton et al. 1986; Knysh and Mann 1984), each cell having different mixing parameters and exchange flows with neighboring cells.

Mixing stages 2 and 3. The reaction cloud grows by two mechanisms. The velocity fluctuations cause the invasion of the whole tank by the cloud. The parameter is the turbulent diffusion D_T

$$D_T \sim u'D \qquad \text{(Eq 5-31)}$$

A second growth mechanism was suggested by Baldyga and Bourne (1984). Due to vorticity, small eddies of fluid form striated structures that incorporate fresh fluid of equal volume at regular time intervals t_{inc}

$$t_{\text{inc}} \sim \left(\frac{\nu}{\varepsilon}\right)^{1/2} \qquad \text{(Eq 5-32)}$$

D_T and t_{inc} change according to which zone the cloud is passing through (David et al. 1985). Actually, both mechanisms occur and the reacting cloud volume is determined by that mechanism which at a given time leads to the smallest growth of the cloud volume.

Mixing stage 4. The reacting cloud consists of a collection of added volumes V_i and consequently has a microstructure. The microscopic mass exchange is represented by an IEM model. For species j in the volume V_i, the mass balance is written

$$\frac{dC_{ji}}{dt_i} = \frac{\overline{C}_j - C_{ji}}{t_{mk}} + \sum_l \nu_{lj} r_{li} \qquad \text{(Eq 5-33)}$$

Where:

- t_i = the age of fluid element i
- l = the reaction index
- t_{mk} = the micromixing time in zone k (k = 1,2), often taken as the time constant of an eddy whose dimension is equal to the Kolmogoroff microscale (see Baldyga and Bourne 1984)

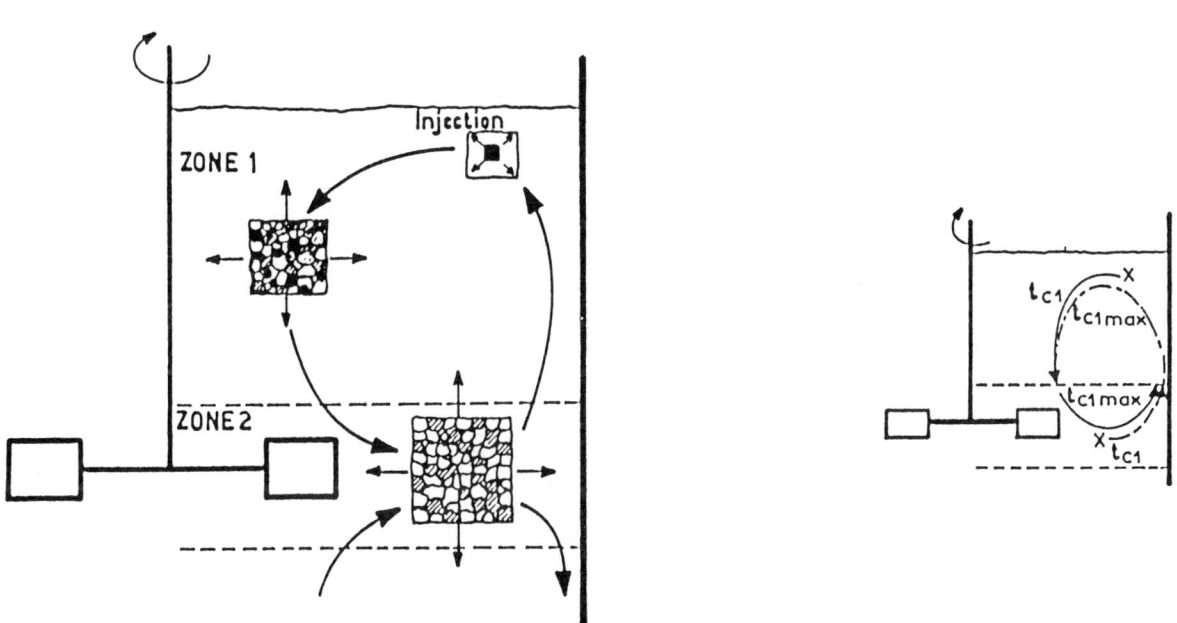

Source: David, R. & Villermaux, J. 1987. Interpretation of Micromixing Effects on Fast Consecutive–Competing Reactions in Semi-Batch Stirred Tanks by a Simple Interaction Model. Chem. Engrg. Comm., 54:333–352.

Figure 5-14 Four-stage mixing model: principle.

Initial conditions are given by the composition of the volume V_i and the concentration of species initially present in the tank. Generally, the fourth mixing stage can be considered rapid if the fluid is of low viscosity (Bourne and Dell'Ava 1987).

The average concentration in the cloud is calculated as

$$\overline{C}_j = \frac{\sum_i C_{ji} V_i}{\sum_i V_i} \quad \text{(Eq 5-34)}$$

Overall concentrations in the tank are calculated from $\overline{C}_j$ and initial concentrations in the rest of the tank. The calculations are stopped when the key reactant is totally consumed, or the final equilibrium between reactants and products is approximately reached over the entire tank. The different characteristic scales in Table 5-1 vary with viscosity, diffusivity, stirrer speed, and tank size (see David and Villermaux 1987 and chap. 8 of this book).

Results for consecutive–competing reactions. The test reaction developed by Barthole et al. (1982a) is based on the precipitation of $BaSO_4$ (S) from a basic EDTA–barium complex ($[BaY^{2-}]n_o$, OH^-) = A by addition of H_3O^+ ions (B) in the presence of sulfate ions (U):

$$A + B \overset{\infty}{\to} R + 2H_2O \quad \text{(Eq 5-35)}$$

$$n_o U + R + 2n_o B \to n_o S + n_o YH_2^{2-} + 2H_2O \quad \text{(Eq 5-36)}$$

$$2A + YH_2^{2-} \overset{\infty}{\to} 2R + Y^{4-} + 2H_2O \quad \text{(Eq 5-37)}$$

$$\text{where } R = [BaY^{2-}]n_o \quad \text{(Eq 5-38)}$$

The segregation index X_S is defined as indicated under "Multistep Reactions"; n_o is the initial ratio $[Ba^{2+}]/[OH^-]$. The experimental reactor had a volume of 140 dm^3 and was stirred by a Rushton turbine with tank-diameter/turbine-diameter ratio equal to 2. It was also possible to use a Mixel TT* axial stirrer (diameter 0.36 m) instead of the Rushton turbine. A volume of 0.1 dm^3 of $1.2N$ hydrochloric acid was added either dropwise or as an injected pulse at five different places in the tank (Figure 5-15). After completion of all three reactions, which requires between 0.2 t_c and 0.5 t_c, the concentration of S could be measured by simple light-absorption spectrophotometry, and X_S could be calculated.

In Figure 5-15, X_S is plotted against N for dropwise addition. The best mixing is achieved when injection takes place near the stirrer (point 3) at a given stirring speed. Figure 5-16 shows X_S versus the power delivered by the stirrer to the liquid for both modes of addition of HCl. Dropwise addition leads to

$$X_S \sim P^{-\frac{1}{2}} \sim N^{-1.5} \sim \text{(dissipated energy)}^{-\frac{1}{2}} \quad \text{(Eq 5-39)}$$

---

*Mixel Co., Lyon, France.

Pulse addition leads to the relation

$$X_S \sim P^{-1/3} \sim N^{-1} \sim \text{(velocities in the tank)}^{-1} \qquad \text{(Eq 5-40)}$$

The four free parameters (D_t, t_m, t_{inc}, V_2/V_1) of the model were fitted using the experimental data.

Simulations have shown that with dropwise addition the size of the reaction cloud was determined by vorticity (stage 3), whereas with pulse addition the size depended on turbulent diffusion (stage 2). This could explain the dependencies on P reported above. Moreover, the good quality of mixing achieved by the axial stirrer under pulse addition can be related to the well-known efficiency of this type of stirrer for transforming the stirrer power into large average velocities instead of turbulence, as the radial stirrer does. Thus, the use of an axial stirrer improves the efficiency of mixing stages 1 and 2, which tend to control selectivity for pulse addition. The model also explains well all the X_S versus N data of Figure 5-16. Further experiments were made in a 0.1-dm^3 tank (David et al. 1987) and scale-up rules were determined (David and Villermaux 1987).

It should be noted that the zones where the reactions take place vary with impeller speed N and size of the tank. This illustrates the difficulty in determining and scaling-up the relevant stages and scales responsible for the overall mixing performance of a tank. For example, with liquids of high viscosity the fourth stage will become dominant. These observations support the use of multistage models.

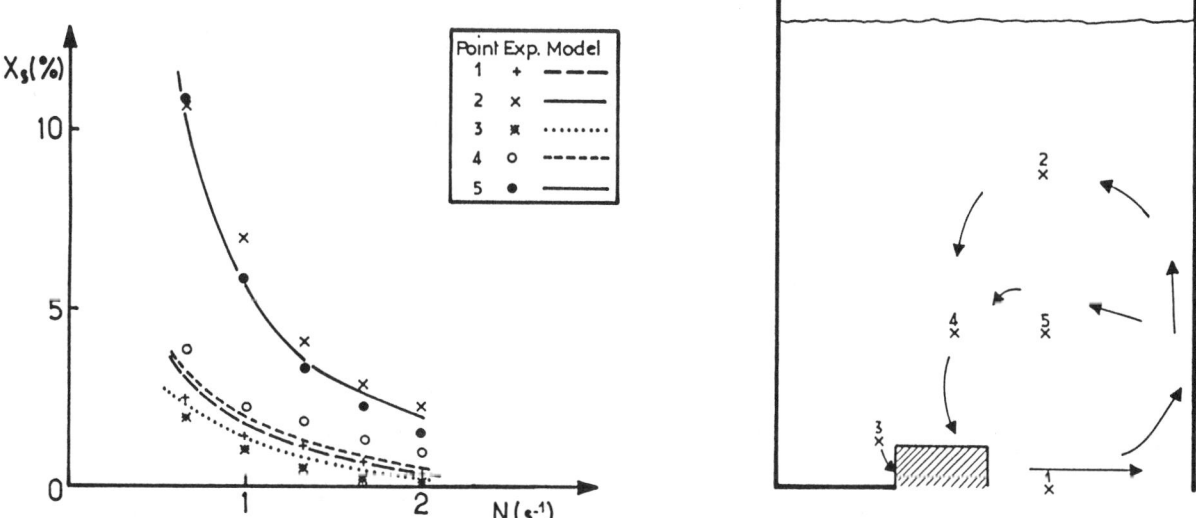

Source: Reprinted by permission of VCH Publishers, Inc., 220 East 23rd St., New York, N.Y. 10010 from: "Chemie Ingenieur Technik," Vol. 59: 254–255 (1987).

Figure 5-15 Four-stage mixing model: segregation index versus rotation speed N in a stirred tank for five different points of addition of reactant HCl.

196 MIXING

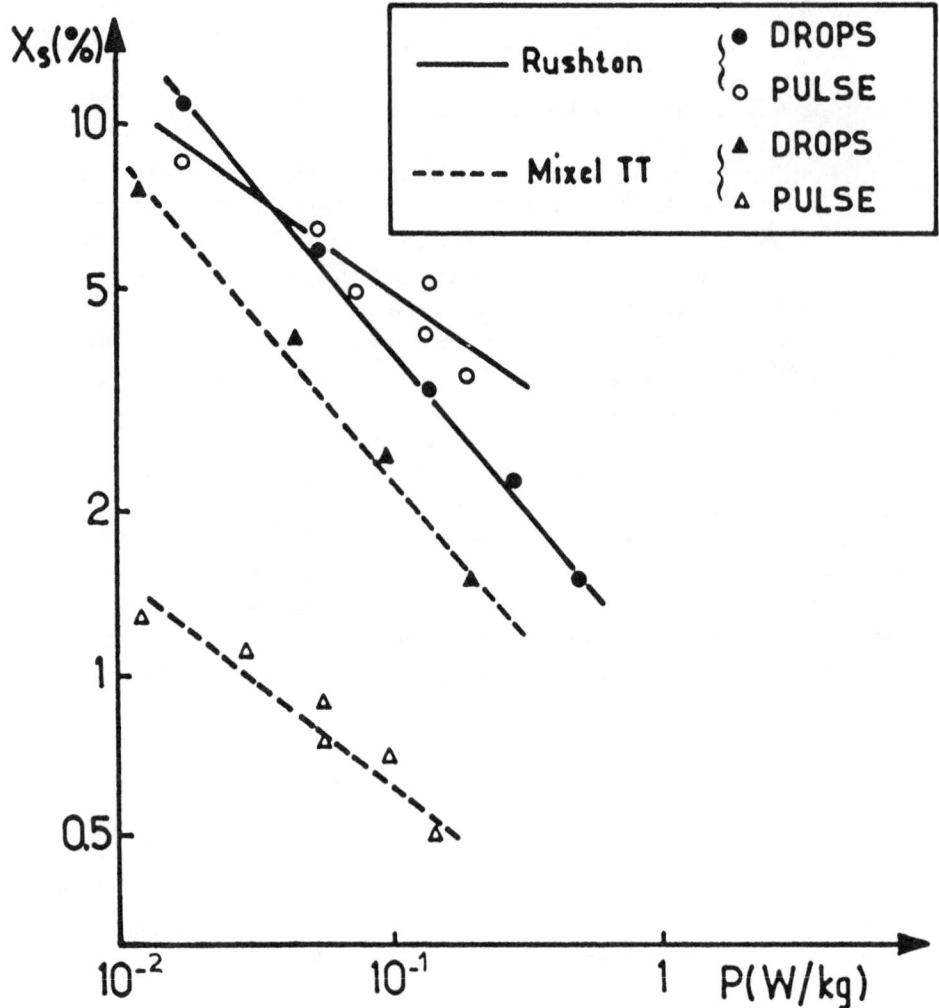

Source: Reprinted by permission of VCH Publishers, Inc., 220 East 23rd St., New York, N.Y. 10010 from: "Chemie Ingenieur Technik," Vol. 59: 254–255 (1987).

Figure 5-16 Four-stage mixing model: segregation index X_S versus power delivered to the unit mass of fluid for addition at point 5 (defined in Figure 5-15). Comparison of the performance of two types of stirrers.

PRECIPITATION OF SPARINGLY SOLUBLE SALTS AND MICROMIXING

A precipitation process can be represented by the following reaction scheme:

$$A + B \underset{\leftarrow}{\overset{K_e}{\longrightarrow}} R \xrightarrow{\text{crystallization}} C \qquad \text{(Eq 5-41)}$$
$$\text{chemical} \qquad\qquad\qquad \text{solid crystal}$$
$$\text{reaction}$$

The solubility of the intermediate R (less soluble than A and B) is C_{R^*}. The concentrations of A, B, and R are always linked by an equilibrium relationship

$$K_e C_A C_B = C_R \quad \text{(Eq 5-42)}$$

In some cases, especially for inorganic salts, R may not even exist. Then one writes the equilibrium in terms of the solubility product P_S

$$P_S = C_{A^*} C_{B^*} \quad \text{(Eq 5-43)}$$

Even in this case, it is possible to imagine an intermediate R whose solubility C_{R^*} can be defined as

$$C_{R^*} = K_e C_{A^*} C_{B^*} = K_e P_S \quad \text{(Eq 5-44)}$$

The driving force of crystallization is supersaturation

$$S_u^i = \frac{C_R}{C_{R^*}} - 1 = \frac{C_A C_B}{P_S} - 1 \quad \text{(Eq 5-45)}$$

Reduced concentrations are expressed as $x = C/P_S^{1/2}$.

Crystallization proceeds mainly via two processes: nucleation and growth. Nucleation means creation of small crystals called nuclei. Nucleation rate is given by

$$r_N = \underbrace{k_N S_u^i}_{\substack{\text{primary} \\ \text{nucleation}}} + \underbrace{k_N' S_u^{i'} C_c^k}_{\substack{\text{secondary} \\ \text{nucleation}}} = k_N S_u^i \left(1 + k_N' S_u^{i'-i} \frac{C_c^k}{P_S^{k/2}} \right) \quad \text{(Eq 5-46)}$$

with $1 \leq k \leq 2$. Primary nucleation dominates when starting from a solution without crystals. In suspensions, secondary nucleation provoked by the impinging of crystals on walls, stirrer, or other crystals dominates.

Crystal growth refers to the growth of nuclei. A distinction must be made between "chemical" and "diffusional" growth. The chemical growth rate G_r is defined from the surface supersaturation S_{uS}

$$G_r = k_C Su_S^j P_S^{j/2} \text{ or } G_r = k_C' C_R^{*j} Su_S^j = k_c Su_S^j \quad \text{(Eq 5-47)}$$

Generally, G_r is independent of the crystal size L. But, to reach the surface, the reacting species (R or $A + B$) have to diffuse through the external layer surrounding the crystal. The balance between the chemical growth and the mass-transfer fluxes leads to the equation (Garside 1985)

$$h Su_S^j + Su_S - Su = 0 \quad \text{(Eq 5-48)}$$

where $h = \dfrac{k_c' C_R^{*j-1}}{k_{DR}}$ when R diffuses

and $x_A - x_{AS} = x_B - x_{BS} = h(x_A x_B - 1)^j$

where $h = \dfrac{k_c P_S^j}{k_{DAB} P_S^{1/2}}$ when both A and B diffuse.

From the nucleation and growth equations, it is possible to define time and length scales

$$t* = (k_c^3 k_N)^{1/4} \; ; \; L* = (k_c/k_N)^{1/4}$$

The chemical yield can be replaced by the crystal size distribution (CSD) $\psi(L_s, t)$, which is the solution of the crystal population balance over a volume V for crystal size L_s

$$\frac{\partial(\psi V)}{\partial t} + Q_{out}\psi_{out} + \frac{\partial(VG\psi)}{\partial L_s} = Q_{in}\psi_{in} + Vr_N\delta(L_s - L_N) \qquad \text{(Eq 5-49)}$$

where Q_{in} and Q_{out} are, respectively, the inlet and outlet fluxes of crystals of size L_s, and δ is the Dirac function.

The moments of the CSD are often used, where $\mu_{0,\psi}$ = number of particles per unit volume or mass of suspension, and $\overline{L_s} = \mu_{1,\psi}/\mu_{0,\psi}$ = average size (in number) of crystals. The standard deviation of ψ is defined as

$$VAR = \left[\frac{\mu_{2,\psi}\mu_{0,\psi}}{\mu_{1,\psi}^2} - 1\right]^{1/2} = \text{standard deviation of } \psi \qquad \text{(Eq 5-50)}$$

The concentration of C can be deduced from $\mu_{3,\psi}$,

$$C_c = b\mu_{3,\psi} \text{ with } b = \Phi_V\rho_C/M_C \qquad \text{(Eq 5-51)}$$

Garside and Tavare (1985) demonstrated that the CSD (Figure 5-17) is very sensitive to the micromixing state. The reasons are the rapidity of primary nucleation and its sensitivity to supersaturation, which in turn is determined by the contacting of the reacting species.

Simulations With Precipitation of Sparingly Soluble Salts

The first attempt to use micromixing models to explain CSD results during precipitation of sparingly soluble salts was by Pohorecki and Baldyga (1983b). The same model of the four stages of mixing was applied by Villermaux and David (1988) to the precipitation of a sparingly soluble salt by adding the ion B dropwise in a solution of ion A, making a so-called "single-jet precipitation" (Figure 5-18C). Precipitation was assumed to occur without intermediate R. They took reasonable values of the reaction parameters and of the mixing parameters $\theta_{inc} = t_{inc}/t*$ and $\theta_m = t_m/t*$; V_2/V_1 was set to its optimal value of 10 determined with consecutive–competing reactions. For the sake of simplicity, no distinction was made between the two zones. Typical results are found in Figures 5-18A and 5-18B. Micromixing seems to be very important (Figure 5-18B). Efficient mixing leads to more primary nucleation and, therefore, more crystals of smaller size, and to a wider distribution of crystal sizes (Figure 5-17). External transfer limitation (h increasing, see Eq 5-48) results in smaller crystals. Secondary nucleation (high K_N') has the same effect.

ALUMINUM PRECIPITATION

Aluminum salts, such as aluminum sulfate, aluminum chloride, and prepolymerized aluminum preparations (O'Melia 1985), have been used extensively in the municipal and industrial water treatment industry, as discussed in some of the earlier

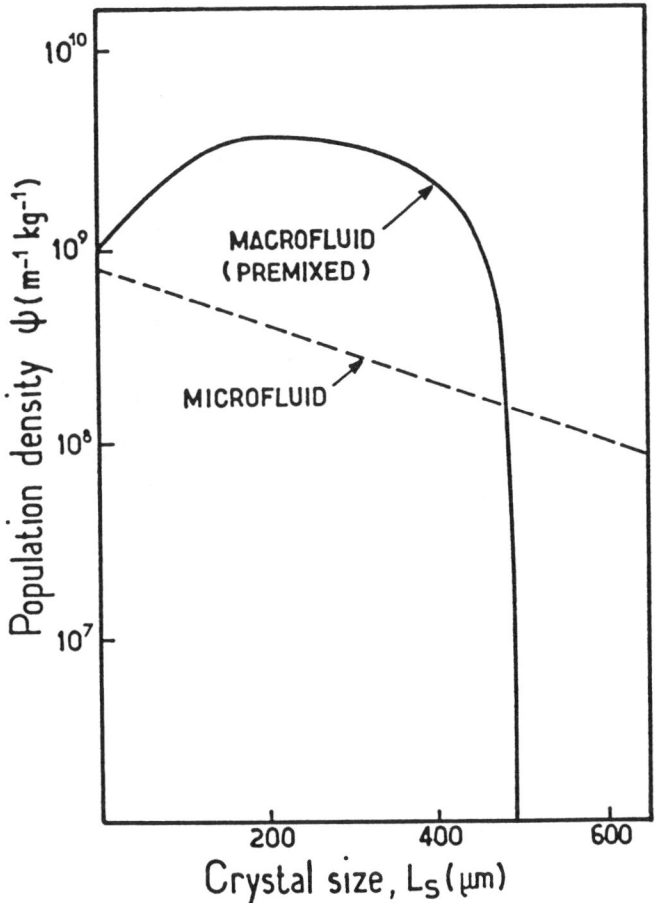

Source: Garside, J. & Tavare, N.S. 1985. Mixing, Reaction and Precipitation: Limits of Micromixing in an MSMPR Crystallizer. Chem. Engrg. Sci., *40:1485–1493.*

Figure 5-17 Comparison of crystal size distribution for microfluid and premixed macrofluid after precipitation.

chapters. The conventional understanding of the efficacy of these coagulants has been through consideration of the solubility of the various supposed destabilizing species in extremely simplified systems (Weber 1972). Given the actual complexity of the natural waters to be treated, it is interesting that this approach has been so useful in describing the approximate pH and aluminum concentration required for successful water treatment (e.g., Amirtharajah and Mills 1982; Dentel 1987). Nevertheless, several rather vexing problems remain in understanding how these coagulants work in water treatment applications. One important problem is the effect of mixing conditions on the speciation of aluminum.

In typical water treatment processes, the coagulant is added to the main flow in an intense, short-duration premixing step called "initial" or "rapid" mixing. It has been suggested that intense premixing is required due to the speed of the hydrolysis and precipitation reactions (Vrale and Jorden 1971; Weber 1972). Clearly, fast reactions can become transport limited. Figure 5-19 compares typical mixing, molecular diffusion, and metal ion hydrolysis characteristic times in a hypothetical 100-L Rushton reactor over a range of operational conditions described by the characteristic velocity gradient

$$\overline{G} = (\overline{\varepsilon}/\nu)^{1/2} \qquad \text{(Eq 5-52)}$$

Here $\overline{\varepsilon}$ is the unit mass energy dissipation rate, and ν is the kinematic viscosity. For a wide range of $\overline{G}$, which spans typical values used in water treatment practice, reaction times tend to be shorter than mixing or diffusive time scales. Hence, for aluminum hydrolysis reactions in typical mixers, one would expect transport limitation; small-scale mixing ("micromixing") is required for the reaction to proceed. Further, in fast, complex, consecutive hydrolysis reactions, it is expected that the details of micromixing will influence the final distribution of hydrolysis and precipitation products; e.g., the relative distribution of polymeric aluminum and aluminum hydroxide. As shown earlier in this chapter (and chap. 4), this is in contrast to slower elementary reactions, in which even with the two extremes of micromixing, product distribution is fairly tightly controlled. Unfortunately, no techniques are currently available to environmental engineers for dealing with the coupled problems of fast reaction and mixing.

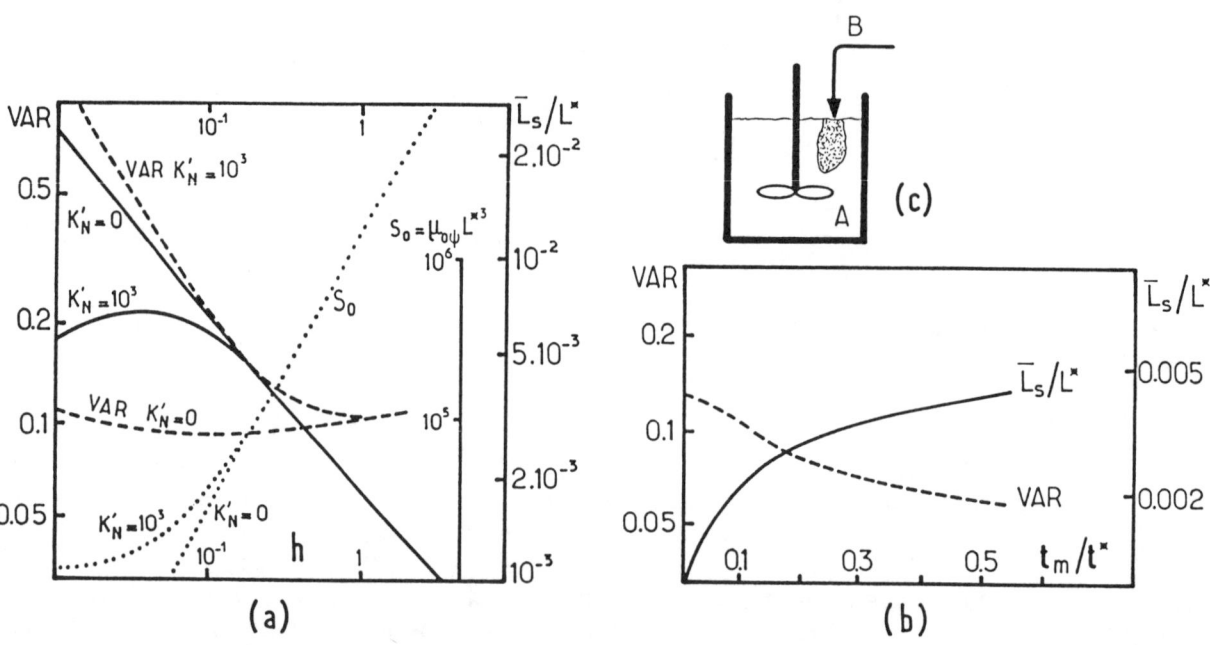

Source: Villermaux, J. & David, R. 1988. Effet du Micromelange sur la Precipitation. Jour. Chim. Phys., 85:273–279.

Figure 5-18 Mixing and precipitation $A + B \rightarrow C$ (solid). Data of the simulation: $j = 2$; $i = 6$; $k = 2$; $i' = 2$; $b = 10^4$; $C_{A0} = C_{B0} = 100\ P_S^{1/2}$; $V_{A0} = V_{B0} = 1\ \text{dm}^3$; $t_{inc}/t^* = 0.1$; $t_m/t^* = 0.1$ (a); $h = 1$ (b).

MICROMIXING MODELS 201

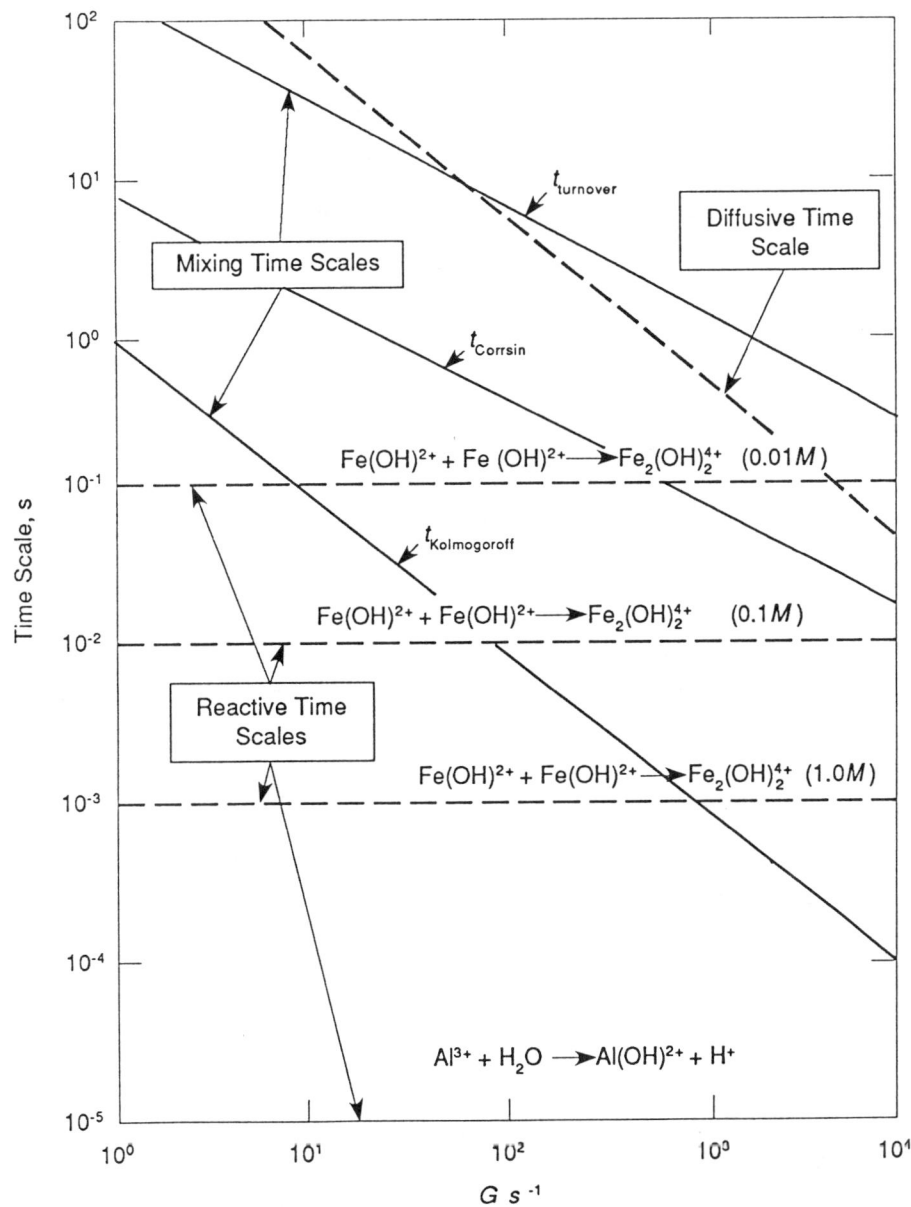

NOTE: Mixing time scales: (1) average turnover time of vessel contents due to impeller pumping, $t_{turnover}$ (Uhl and Gray 1966); (2) local mixing time in isotropic turbulence, due to S. Corrsin, $t_{Corrsin}$ (Villermaux and David 1983); and (3) local Kolmogoroff turnover time, $t_{Kolmogoroff}$ (Hinze 1975). Diffusive time scale: time for Al^{3+} ion to diffuse over distance corresponding to one-half the Kolmogoroff microscale (Bourne et al. 1981). Reactive time scale: (1) based on rate of ion polymerization at various monomer concentrations (Wendt 1962); and (2) hydrolysis rate of aluminum ion (Holmes, Cole, and Eyring 1968).

Source: Mark M. Clark, "Micromixing Effects in Fast Aluminum Neutralization Precipitation," research proposal submitted to National Science Foundation, Washington, D.C. Jan. 15, 1988.

Figure 5-19 Comparison of mixing, diffusive, and reactive time scales in 100-dm^3 Rushton backmix reactor.

In summary, from simple equilibrium chemistry and years of operating experience, an approximate coagulant dosage can be estimated for a given raw water type. Yet there is little organized understanding of the effects of mixing on the precipitation of a common coagulant such as aluminum.

Chemistry of Aluminum

It is generally accepted that at low pH, the trivalent aluminum ion, Al^{3+}, exists as an aquaion, coordinating six water molecules. Because of this hydration, aluminum acts as a polyprotic acid (Stumm and Morgan 1970). The progressive hydrolysis of aluminum that occurs as the pH increases has been studied previously. Frink and Peech (1963) used pH and conductivity measurements in dilute aluminum solutions at low pH and concluded that the simple hydrolytic mechanism

$$Al^{3+} + OH^- \underset{k_2}{\overset{k_1}{\rightleftarrows}} Al(OH)^{2+} \qquad (Eq\ 5\text{-}53)$$

was most consistent with their data. Holmes, Cole, and Eyring (1968) used electric-field jump techniques to determine hydrolysis rates, and found that the initial hydrolysis reaction (Eq 5-53) was quite fast, the forward reaction speed being on the order of 10^{-5} s. The existence of a second monomer, $Al(OH)^{2+}$, has long been suggested, although recent work indicates that it may be a minor component at typical pH and pAl (Bottero et al. 1980). No work suggests that aqueous aluminum hydroxide, $Al(OH)_3(aq)$, exists in measurable amounts, while, at pH > 9, it is agreed that the aluminate ion, $Al(OH)_4^-$, determines the solubility of aluminum (e.g., Van Stratton, Holtkamp, and DeBruyn 1984). There is considerable agreement about the equilibrium constants for these monomeric forms of aluminum, and these are provided in several works, such as Baes and Mesmer (1976).

There is somewhat less certainty about what form of solid-phase aluminum hydroxide, $Al(OH)_3(s)$, the monomers should be considered in equilibrium with. Initially, upon precipitation aluminum hydroxide is an amorphous material, which, depending on physical and chemical effects and aging, can be converted into a number of solids including bayerite and gibbsite (Baes and Mesmer 1976). Dempsey (1987) has pointed out that 500-fold variations in the concentration of aluminum result from assumptions about the solid phase. Parthasarathy and Buffle (1985) attribute a 300-fold variation in equilibrium aluminum concentration due to the size (hence, solubility) of aluminum "microsolids."

Polymeric forms of aluminum are also thought to exist over a wide range of pH and pAl. Matijevic, Janauer, and Kerker (1964) argued for the occurrence of $Al_8(OH)_{20}^{4+}$ based on analysis of coagulation data, and Hayden and Rubin (1974) found that "Al_8" was consistent with their aluminum titration data. Bottero et al. (1980) reported both potentiometric data and nuclear magnetic resonance (NMR) measurements supporting the occurrence of the $Al_2(OH)_2^{4+}$ and $Al_{13}O_4(OH)_{24}^{7+}$ polymers. Parthasarathy and Buffle (1985) recently concluded from their extensive literature survey that the "Al_{13}" polymer is the smallest and most stable of the undissociable polymers consistently isolated. In contrast to these investigators, rather than argue for a specific polymeric structure (e.g., Al_{13}), Stol, Van Helden, and DeBruyn (1976) interpreted data for the light scattering of base-neutralized aluminum solutions to indicate a continual increase in degree of polymerization with amount of added base.

Data for the titration of aluminum with base are often plotted as pH versus the ligand number or formation ratio r'

$$r' = \frac{[\text{moles base added}]}{[\text{total moles Al}]} \quad \text{(Eq 5-54)}$$

Figure 5-20 is such a titration for $8.5 \times 10^{-3} M$ aluminum chloride titrated with $0.51 M$ NaOH (Clark, David, and Wiesner 1987). In region I, the initial steep incline of the titration curve for $r' < 0.3$ is due to the titration of a small amount of HCl formed during initial aluminum hydrolysis, and the completely reversible formation of aluminum monomers such as those discussed above. On the plateau, region II, light-scattering studies (Stol, Van Helden, and DeBruyn 1976) and NMR analysis (Bottero et al. 1980) indicate the formation of polymeric aluminum species. Parthasarathy and Buffle (1985) used ultrafiltration techniques and NMR to study dilute aluminum solutions at $r' = 2.5$. For $[Al] = 0.12 M$, their work suggested that 80 percent of the aluminum occurred as Al_{13} polymer of average size between 10 and 20 Å. They also detected microsolids ranging in size from 10 to 2000 Å. Beyond the plateau region in Figure 5-20, aluminum hydroxide precipitates as the formation ratio approaches 3 (value for aluminum hydroxide). In this region, as less of the hydroxide complexes with aluminum and the pK of water is approached, the titration curve heads sharply upwards (region III). Clark, David, and Weisner (1987)

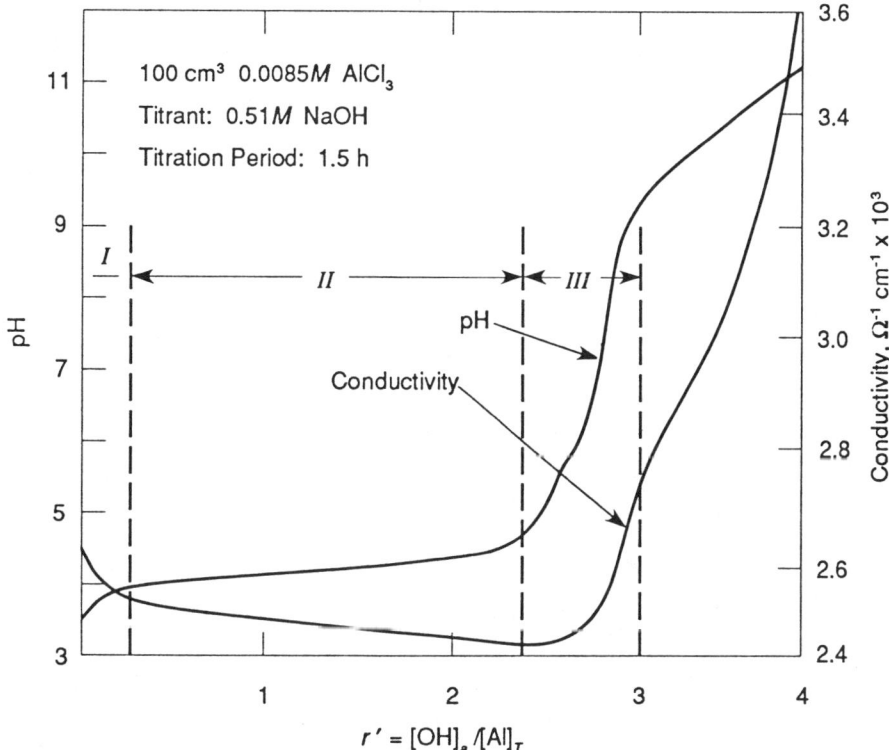

Source: Clark, M.M., David, R. & Wiesner, M.R. 1987. Effect of Micromixing on Product Selectivity in Rapid Mix. Proc. AWWA Annual Conf., Kansas City, Mo.

Figure 5-20 Slow automatic titration of aluminum chloride with sodium hydroxide under N_2 atmosphere.

have used conductimetry to monitor the speciation of aluminum during slow addition of base. In Figure 5-20, it is shown that the conductimetric titration curve roughly mirrors the pH titration curve below $r' = 2$, while breaking steeply upward at a point ($r' = 2.7$) slightly beyond the break in the pH curve. By subtracting the conductivity due to Na+, H+, Cl−, and OH− from the total conductivity, the apparent conductivity of aluminum species remains. Figure 5-21 shows that during neutralization, the conductivity of aluminum species continually decreases up to $r' = 2.7$. For most simple ionic species, conductivity or mobility is directly and predictably related to charge and hydrodynamic radius. Therefore, it can be imagined that on the plateau ($0.3 < r' < 2.7$) as the small and relatively highly charged (hence, highly conductive) monomeric aluminum species further hydrolyze, polymerize, and aggregate (becoming larger and less highly charged; see Bottero et al. 1987), the total conductivity due to aluminum species steadily declines. The apparent steady increase in conductivity of aluminum compounds for $r' > 2.7$ is probably due to the formation of aluminate, $Al(OH)_4^-$.

Several writers have commented on the effect of base addition rate and mixing intensity on the titration of aluminum. Vermeulen et al. (1975) compared dropwise and slow, continuous "homogeneous" base addition methods and found that, with

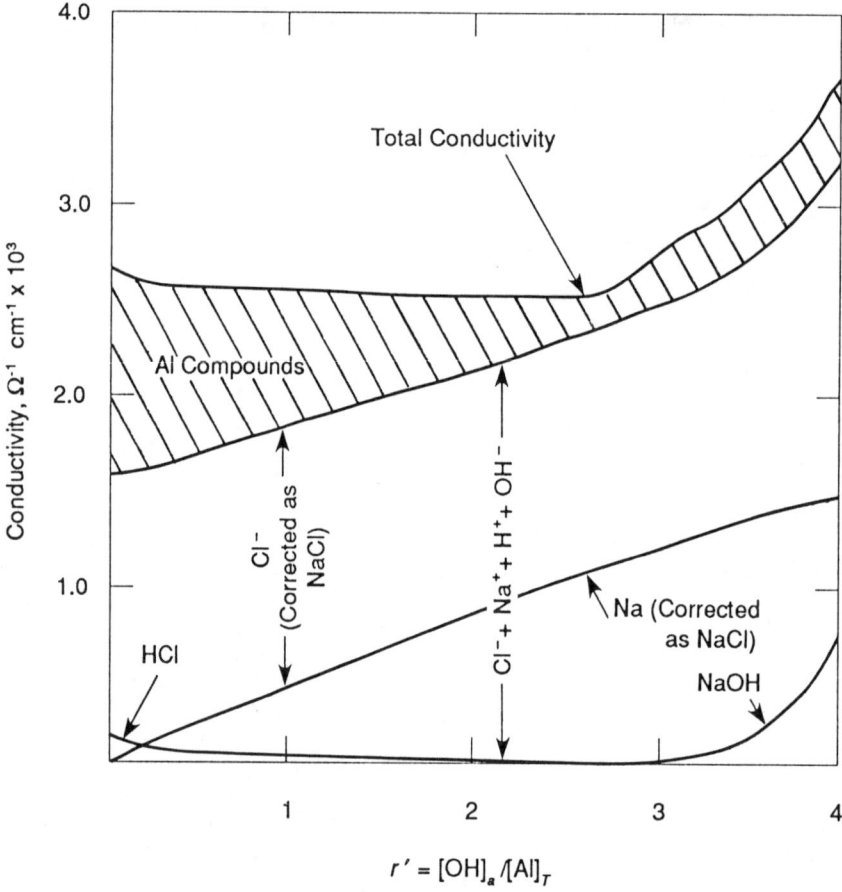

Source: Clark, M.M., David, R. & Wiesner, M.R. 1987. Effect of Micromixing on Product Selectivity in Rapid Mix. Proc. AWWA Annual Conf., Kansas City, Mo.

Figure 5-21 Contribution to conductivity by various ions, $[Al]_T = 8.5 \times 10^{-3}$M.

dropwise addition, the typical small plateau at $r' = 2.7$ in Figure 5-21 did not appear. Parthasarathy and Buffle (1985) claimed that a larger proportion of slowly reacting aluminum species ("microsolids") were formed with "inefficient" mixing. On the other hand, Letterman and Asolekar (1987) claim that a continuous "pulse" titrant injection method using a simple magnetic stirring apparatus yielded titration results equivalent to the classical slow homogeneous titration results of Vermeulen et al. (1975). Schneider (1984), commenting on the effects of mixing on iron hydrolysis (a system similar to aluminum), said that owing to the fundamental properties of the mononuclear iron species there resulted an "unavoidable variety of intermediate products due to overlapping time scales of mixing and substitution as well as protonation reactions." This assessment is probably accurate for the situation in aluminum base titration as well.

Aluminum as a Coagulant

As opposed to the continuous laboratory aluminum titrations reported above, in the typical water treatment plant, the scale of aluminum "titration" is immense; reactant injection is both sped up and inverted, with a relatively small concentrated coagulant flow being quickly blended with a large baseline flow of varying natural or adjusted buffering capacity. If the solubility of aluminum hydroxide is exceeded, it is believed that the polymeric species reported above form as either intermediates in the precipitation of aluminum hydroxide, or act themselves as the contaminant destabilizer, being readily adsorbed at interfaces (Weber 1972). For colloid destabilization due to charge neutralization by aluminum polymers, it might be expected that the required polymer dose would be stoichiometrically related to the contaminant surface area. This and the related phenomenon of restabilization or "overdosing" have been observed for both model colloidal systems (Hahn and Stumm 1968) as well as for a wide variety of natural waters containing both dissolved and particulate contaminants (e.g., Amirtharajah and Mills 1982). If the solubility product of aluminum hydroxide is exceeded to a large enough degree, a voluminous precipitate will form that can destabilize suspensions through adsorption or "enmeshment" in the precipitate (Weber 1972).

Understanding of the efficacy of aluminum as a coagulant has been based primarily on the principles of equilibrium chemistry in simplified or ideal systems. Yet the multitude of complexities existing in applications in real waters has not been resolved. These problems include the interaction of the coagulant hydrolysis products with other ligands and surfaces (Amirtharajah and Trusler 1986), as well as the effect of mixing on mass transport and contaminant destabilization. Several problems disturb the simple approach of homogeneous equilibrium chemistry. For example, it has never been clear what type of mixing is required during rapid mix. Moffet (1968) found using jar-test results that 27 percent more aluminum coagulant was required to achieve zero zeta potential when the coagulant was injected near the surface than at the agitator level, suggesting that the more intense turbulence near the mixing paddle was beneficial. On the other hand, in a recent study of the coagulation of a soil humic acid with a prepolymerized aluminum coagulant, Charles et al. (1987) found that the worst mixing studied (surface injection at low mixing intensity) resulted in the best complexation of the humic acid as well as the lowest residual turbidity following settling. These two examples from the literature seem to correspond to stories reported from the field in which a plant operator, through some *ad hoc* modification of mixing, finds a beneficial effect on either coagulant demand, final water clarity, or sludge characteristics.

Modeling of Mixing and Aluminum Precipitation

For unpremixed reactants, it is useful to describe both the macro- and micromixing phenomena. One can imagine that when a volume V_1 of reactant is injected into a reactor as a pulse or drop, an expanding reacting cloud is formed that is convected throughout the vessel by the gross flow patterns. The bulk flow patterns for the Rushton mixer have been well documented and can be approximated for modeling purposes (Villermaux and David 1983). Two descriptions of cloud growth have been previously utilized. Based on the work of Nagata (1975), it can be assumed that due to turbulent dispersion, the cloud volume increases as

$$dV_c/dt = D_T V_c^{1/3} \qquad \text{(Eq 5-55)}$$

where D_T is the turbulent diffusivity, which was assumed by Nagata to be proportional to the root mean square (rms) velocity fluctuation. Within the cloud, Baldyga and Bourne (1984) suggested that due to small-scale stretching and vorticity, striated structures are formed that fold over themselves and incorporate equal volumes of fresh fluid. The volume of these structures increases over some characteristic time interval

$$V_s = V_1 \times 2^{t/t_{\text{inc}}} \qquad \text{(Eq 5-56)}$$

where t_{inc} is the local turnover time, Eq 5-32. David and Villermaux (1975) have assumed that these two mechanisms work in series: the initial volume of the added reactant, which is strongly segregated on the large scale, is transported and dispersed by the large-scale motion, followed by smaller-scale stretching and vorticity, which produce smaller-scale segregation.

Clark, David, and Wiesner (1987) used the above approach in their modeling of the neutralization/precipitation of aluminum with NaOH. The 120-mL batch Rushton vessel was divided into three reaction zones: an impeller zone, a zone directly outside the impeller zone, and a zone comprising the rest of the tank. Bulk circulation through the vessel was described by characteristic circulation times for the three zones.

Clark, David, and Wiesner (1987) represented the microscopic exchange between eddies in the reacting cloud, using the IEM model described above (Eq 5-33 and 5-34) and published values for micromixing times, turbulent diffusivity, and bulk circulation times. Their kinetic reaction scheme includes Eq 5-53 and the following:

$$Al(OH)^{2+} + OH^- \rightleftarrows Al(OH)_2^+ \qquad \text{(Eq 5-57)}$$

$$Al(OH)_2^+ + OH^- \rightleftarrows Al(OH)_3(aq) \rightarrow Al(OH)_3(s) \qquad \text{(Eq 5-58)}$$

$$0.5n\ Al(OH)_2^+ + 0.5n\ Al(OH)_3(aq) \rightarrow Al_n(OH)_{2.5n}^{0.5n+} \qquad \text{(Eq 5-59)}$$

where published values of the equilibrium constants were used, and it was assumed that

$$r_p = k_p [Al(OH)_2^+] [Al(OH)_3(aq)] \qquad \text{(Eq 5-60)}$$

and r_s is given by analogy to the expression for the nucleation and precipitation of barium sulfate (Barthole et al. 1982a)

$$r_s = k_s ([Al]_T)^{2/3} [Al^{3+}][OH^-]^3 \qquad \text{(Eq 5-61)}$$

In the preliminary simulations of Clark, David, and Wiesner (1987), k_p and k_s were selected to give nominally correct results at known reactant conversions.

Some representative simulations of aluminum speciation and solution pH in a 1-L Rushton mixing vessel are summarized from Clark, David, and Wiesner (1987). In Figures 5-22A, B, and C, slow dropwise and instantaneous pulse neutralization (NaOH) of a dilute aluminum chloride solution is compared for two mixing speeds (102 and 17 rps) and three injection points. In Figure 5-22A, it is seen that for pulse injection at the highest mixing speed, the injection point has little impact on the final pH values. It was suggested that with very good micromixing—approaching instantaneous molecular mixing—the injection point should have little impact on selectivity and final pH. With good micromixing, essentially homogeneous nucleation of the solid phase (Al[OH]$_3$[s]) may occur with a subsequent small amount of aluminum polymer. Conversely, if mixing speed is lowered, the model predicts sensitivity to both location of injection point and the type of injection. In Figure 5-22B (17 s^{-1}) for pulse injection, the steep section of the pH curve moves to the left as the injection point is moved from position 1 to 2, and to 3. To explain these results, one must recall both the macro- and micromixing effects built into the model, and the integrated effect of segregation along the circulation path. Although the local mixing intensity increases moving from injection points 1 to 2, and to 3 (hence, predicting a decreasing trend in *local* segregation), base injected at position 3 actually ends its reactive trip in the less well-mixed zone near the top of the vessel. Hence, the integrated segregation effect actually increases moving from injection point 1 to 2, and to 3, and more polymer is formed as the injection point is changed in this manner. Because polymers are less completely neutralized than Al(OH)$_3$, the titration curve begins to slope upward at a lower pH. Bottero et al. (1987) believe that in actual solutions, aggregated polymers may be temporarily shielded from hydrolysis, which would also lead to an earlier rise of the titration curve.

Figures 5-22B and C show that a dropwise addition causes an earlier rise in the pH curve in comparison to pulse injection. It is believed that this is also due to the greater segregation effects for dropwise addition.

In Figure 5-22C, the effect of mixing speed on conversion for reactant injected at point 1 by the dropwise method is shown. The increasing segregation due to decreasing mixing speed causes the titration curve to move to the left, indicating more polymer formation.

Clark, David, and Wiesner (1987) were able to find qualitative consistency between the above model predictions and experimental results for the fast base neutralization of $1.8 \times 10^{-3} M$ aluminum solutions. In these experiments, the entire quantity of base was quickly injected (0.2 to 0.5 s) in order to arrive at a given neutralization ratio r'. A titration curve similar to those above resulted, except that each point on the curve represents a separate experiment at a different r' value. Figure 5-23 from Clark, David and Wiesner shows that for this type of neutralization at high mixing speed (51 s^{-1}), the final pH values comprised a curve very similar in shape to the plots in Figure 5-22—with good mixing (fast injection, higher mixing speed), the upward swing in the "titration" curve could be delayed to r' values very close to 3. With poorer mixing (slower injection rate, slower mixing speed), there was both a somewhat earlier upswing in the pH curve, and a good deal

208 MIXING

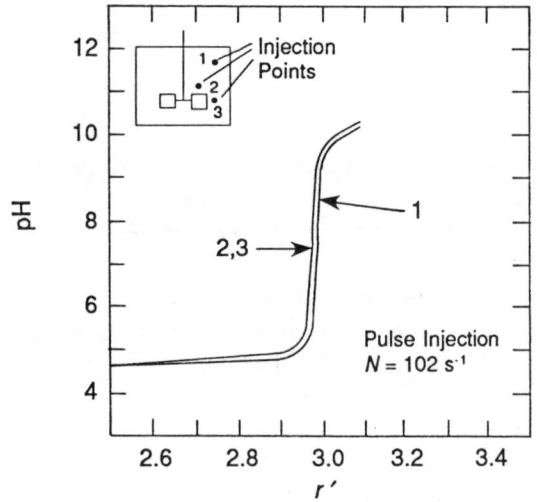

Figure 5-22A Simulation of final pH for three injection points, $N = 102$ s^{-1}, $[Al]_\tau = 1.8 \times 10^{-3} M$.

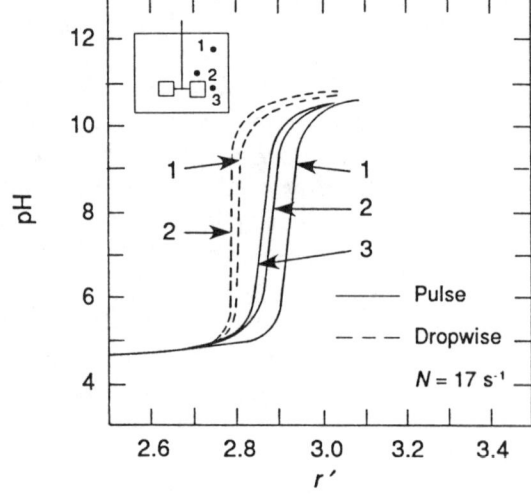

Figure 5-22B Simulation of final pH for three injection points, dropwise and pulse injection $[Al]_\tau = 1.8 \times 10^{-3} M$.

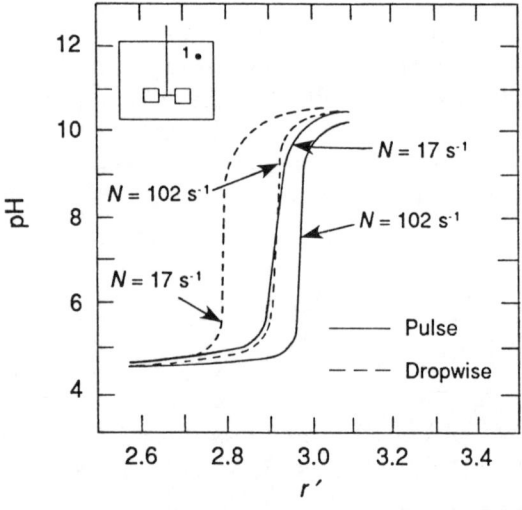

Figure 5-22C Simulation of final pH for dropwise and pulse injection at two mixing speeds $[Al]_\tau = 1.8 \times 10^{-3} M$.

Source: Clark, M.M., David, R. & Wiesner, M.R. 1987. Effect of Micromixing on Product Selectivity in Rapid Mix. Proc. AWWA Annual Conf., Kansas City, Mo.

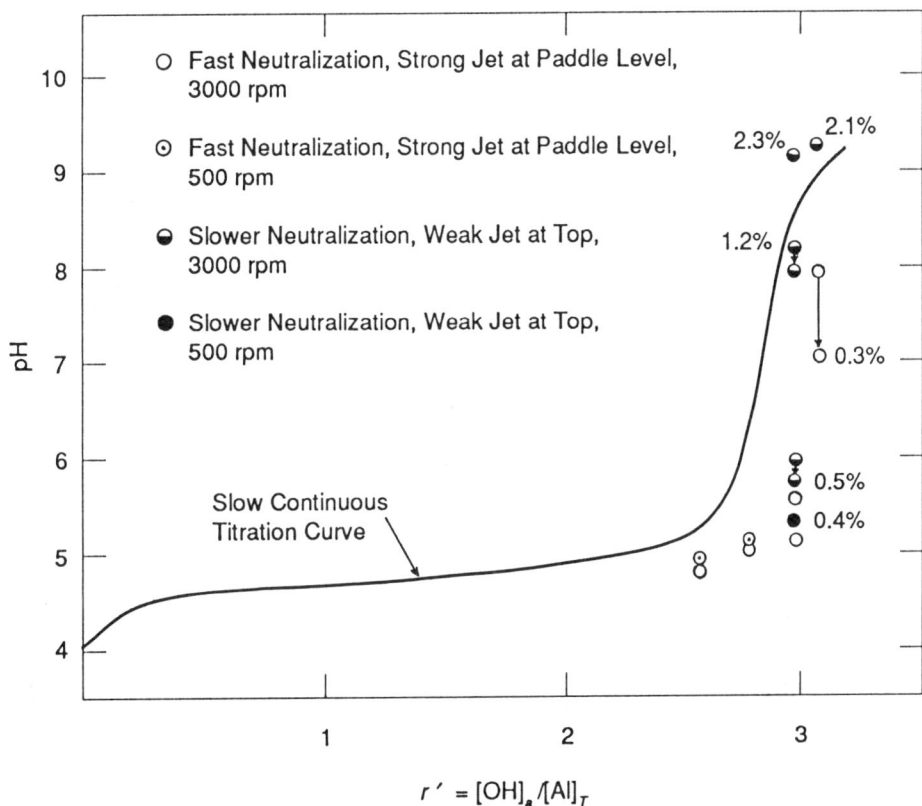

Source: Clark, M.M., David, R. & Wiesner, M.R. 1987. Effect of Micromixing on Product Selectivity in Rapid Mix. Proc. AWWA Annual Conf., Kansas City, Mo.

Figure 5-23 Effect of mixing on final pH in quickly neutralized $AlCl_3$ solutions, $[Al]_T = 1.8 \times 10^{-3} M$. Percentages are percent absorbance at $\lambda = 650$ nm.

more variation in final pH values; i.e., two neutralization experiments at a certain r' value in the range $2.8 < r' < 3.0$ could result in widely different pH values. Apparently, at lower mixing speeds in this neutralization region, stochastic variations in the timing of the pulse relative to impeller blade position lead to fairly large variations in final product distribution and pH. Therefore, these experiments indicate that micromixing is very important during fast aluminum neutralization in the neutralization range, $2.8 < r' < 3.0$.

CONCLUSIONS

Although many micromixing models have been proposed, it is believed that those based on physical descriptions of mixing are to be preferred. Also, models incorporating several different mixing stages may be particularly useful when one is not sure which mixing stage is controlling reactant conversion. Nevertheless, many problems still wait to be solved, including micromixing in non-Newtonian fluids, multiphase mixing, and scale-up (and scale-down).

This chapter has also reviewed the first work aimed at using a multistage micromixing model to predict the precipitation of aluminum, a common coagulant in the water treatment field. Although initial results are encouraging, much work remains to be done in verifying reaction stoichiometry and kinetics, and expanding the model to include the formation of second-phase materials in more realistic and complex solutions.

References

AMIRTHARAJAH, A. & MILLS, K.M. 1982. Rapid-Mix Design for Mechanisms of Alum Coagulation. *Jour. AWWA*, 74:4:210.

AMIRTHARAJAH, A. & TRUSLER, S.L. 1986. Destabilization of Particles by Turbulent Rapid Mixing. *Jour. Envir. Engrg., Div. ASCE*, 112:6:1085.

AUBRY, C. & VILLERMAUX, J. 1975. Representation du Melange Imparfait de Deux Courants de Reactifs dans un Reacteur Agite Continu. *Chem. Engrg. Sci.*, 30:457.

BAES, C.F. & MESMER, R.E. 1976. *The Hydrolysis of Cations*. Wiley-Interscience, New York.

BALDYGA, J. & BOURNE, J.R. 1984. A Fluid Mechanical Approach to Turbulence Mixing and Chemical Reaction. *Chem. Engrg. Comm.*, 28:231.

———. 1988. Calculation of Micromixing in Inhomogeneous Stirred Tank Reactors. *Chem. Engrg. Res. Des.*, 66:33.

BALDYGA, J. & ROHANI, S. 1987. Micromixing Described in Terms of Inertial–Convective Disintegration of Large Eddies and Viscous–Convective Interactions Among Small Eddies. *Chem. Engrg. Sci.*, 42:2597.

BARTHOLE, J.P. ET AL. 1982a. Cinetique Macroscopique de al Precipitation de Sulfate de Barium en Presence d'EDTA: Une Reaction Chimique-Test pour la Caracterisation de al Qualite du Melange dans les Reacteurs Industriels. *Jour. Chim. Phys.*, 79:10:719.

BARTHOLE, J.P. ET AL. 1982b. A New Chemical Method for the Study of Local Micromixing Conditions in Industrial Stirred Tanks. *ACS Symp. Ser.*, 196:545.

BOTTERO, J.Y. ET AL. 1980. Studies of Hydrolyzed Aluminum Chloride: I, Nature of Aluminum Species and Composition of Aqueous Solutions. *Jour. Phys. Chem.*, 84:2933.

———. 1987. Mechanism of Formation of Aluminum Trihydroxide From Keggin Al_{13} Polymers. *Jour. Colloid Inter. Sci.*

BOURNE, J.R. 1984. Micromixing Revisited. *Inst. Chem. Engrs. Symp. Ser.*, 87:797.

BOURNE, J.R. ET AL. 1981. Mixing and Fast Chemical Reaction. *Chem. Engrg. Sci.*, 36:10:1643.

BOURNE, J.R. & DELL'AVA, P. 1987. Micro- and Macro-Mixing in Stirred Tank Reactors of Different Sizes. *Chem. Engrg. Res. Des.*, 65:180.

BOURNE, J.R.; HILBER, C.; TOVSTIGA, G. 1985. Kinetics of the Azo-Coupling Reactions Between 1-Naphthol and Diazotised Sulfanilic Acid. *Chem. Engrg. Comm.*, 37:293.

BOURNE, J.R. & TOVSTIGA, G. 1988. Micromixing and Fast Chemical Reactions in a Turbulent Tubular Reactor. *Chem. Engrg. Res. Des.*, 66:26.

BRODBERGER, F.; VALENTIN, G.; & STORCK, A. 1983. Utilisation d'Une Microsonde Conductiometrique Pour l'Etude des Phenomenes de Melange au Sein de Cuves Agités. *Entropie*, 112:20.

CANON, R.M. ET AL. 1977. Turbulence Level Significance of the Coalescence–Dispersion Rate Parameter. *Chem. Engrg. Sci.*, 32:1349.

CHANG, L.J. ET AL. 1986. An Evaluation of Models of Mixing and Chemical Reaction With a Turbulence Analogy. *Chem. Engrg. Comm.*, 42:139.

CHARLES, P. ET AL. 1987. Influence des Conditions de Melange Initial. AGHTM Conference, Nice, France.

CLARK, M.M.; DAVID, R.; & WIESNER, M.R. 1987. Effect of Micromixing on Product Selectivity in Rapid Mix. Proc. AWWA Annual Conf., Kansas City, Mo.

CORRSIN, S. 1964. The Isotropic Turbulent Mixer: Part II, Arbitrary Schmidt Number. *AIChE Jour.*, 10:870.

COSTA, P. & TREVISSOI, C. 1972. Reactions With Nonlinear Kinetics in Partially Segregated Fluids. *Chem. Engrg. Sci.*, 27:2041.

CURL, R.L. 1963. Dispersed Phase Mixing: Theory and Effects in Simple Reactors. *AIChE Jour.*, 9:175.

DANCKWERTS, P.V. 1958. The Effect of Uncomplete Mixing on Homogeneous Reactions. *Chem. Engrg. Sci.*, 8:93.

DAVID, R. Lecture notes in *Engineering*, 40:387, Springer-Verlag, New York (1989).

DAVID, R. ET AL. 1985. Interpretation of Micromixing Experiments Involving Consecutive–Competitive Reactions in a Stirred Tank by a Simple Interaction Model. Proc. Fifth European Conf. on Mixing. Wurzburg, Germany.

———. 1987. Untersuchung der Lokalen Mikrovermischungsverhaltnisse in Ruhrbehaltern Durch eine Chemisched Methode. *Chem. Ing. Tech.*, 59:254.

DAVID, R. & VILLERMAUX, J. 1975. Micromixing Effects on Complex Reactions in a CSTR. *Chem. Engrg. Sci.*, 30:1309.

———. 1987. Interpretation of Micromixing Effects on Fast Consecutive–Competing Reactions in Semi-Batch Stirred Tanks by a Simple Interaction Model. *Chem. Engrg. Comm.*, 54:333.

DEMPSEY, B.A. 1987. Chemistry of Coagulants. AWWA Sem. on Influence of Coagulation on the Selection, Operation, and Performance of Water Treatment Processes. AWWA Annual Conf., Kansas City, Mo.

DENTEL, S. 1987. Influence of Coagulation on the Selection, Operation, and Performance of Water Treatment Facilities. Proc. AWWA Coagulation Committee, AWWA Annual Conf., Kansas City, Mo.

FRINK, C.R. & PEECH, M. 1963. Hydrolysis of the Aluminum Ion in Dilute Aqueous Solution. *Inorg. Chem.*, 2:3:473.

GARSIDE, J. 1985. Industrial Crystallization From Solution. *Chem. Engrg. Sci.*, 40:3.

GARSIDE, J. & TAVARE, N.S. 1985. Mixing, Reaction and Precipitation: Limits of Micromixing in an MSMPR Crystallizer. *Chem. Engrg. Sci.*, 40:1485.

GASKEY, S. ET AL. 1988. Investigation of Concentration Fluctuations in a Continuous Stirred Tank by Space Resolved Fluorescence Spectroscopy. Proc. Sixth European Conf. on Mixing. Pavia, Italy.

GIBSON, C.H. & SCHWARTZ, W.H. 1963. Detection of Conductivity Fluctuation in a Turbulent Flow Field. *Jour. Fluid Mechanics*, 16:357.

HAHN, H.H. & STUMM, W. 1968. Kinetics of Coagulation With Hydrolyzed Al (III). *Jour. Colloid Inter. Sci.*, 28:1:134.

HARADA, M. ET AL. 1962. Micromixing in a Continuous Flow Reactor. *The Memoirs of the Faculty of Engineering, Kyoto Univ.*, 24:431.

HAYDEN, P.L. & RUBIN, A.J. 1974. Systematic Investigation of the Hydrolysis and Precipitation of Aluminum (III). In *Aqueous Environmental Chemistry of Metals*, ed. by A.J. Rubin. Ann Arbor Science, Ann Arbor, Mich.

HOLMES, L.P.; COLE, D.L.; & EYRING, E.M. 1968. Kinetics of Aluminum Hydrolysis in Dilute Solutions. *Jour. Phys. Chem.*, 72:1:301.

KARAGOZIAN, A.R. ET AL. 1987. Experimental Studies in Vortex Pair Motion Coincident With a Liquid Reaction. Proc. French–American Workshop on Turbulent Reactive Flows, NSF–CNRS, Rouen, France.

KATTAN, A. & ADLER, R.J. 1972. A Conceptual Framework for Mixing in Continuous Chemical Reactors. *Chem. Engrg. Sci.*, 27:1013.

KLEIN, J.P. ET AL. 1980. Interpretation of Experimental Liquid Phase Micromixing Phenomena in a Continuous Stirred Reactor With Short Residence Times. *Ind. Engrg. Chem. Fund.*, 19:373.

KNYSH, P. & MANN, R. 1984. Utility of Networks of Interconnected Backmixed Zones to Represent Mixing in a Closed Stirred Vessel. Symp. Ser., Inst. Chem. Engrs., *Fluid Mixing II*, 89:127.

LEE, C.S. ET AL. 1986. An Evaluation of the Lamellar Stretch Description of Mixing With Diffusion and Chemical Reaction. *AIChE Jour.*, 32:1043.

LETTERMAN, R.D. & ASOLEKAR, S. 1987. A Comprehensive Equilibrium Model for Aluminum Hydrolysis and Precipitation, unpublished manuscript, Dept. of Civil Engineering, Syracuse University, Syracuse, N.Y.

LI, K.T. & TOOR, H.L. 1986. Turbulent Reactive Mixing With a Series-Parallel Reaction: Effect of Mixing on Yield. *AIChE Jour.*, 32:1312.

MAHOUAST, M. ET AL. 1986. Caracterisation des Champs Hydrodynamique et de Concentration dans une Cuve Agitee Standard Alimentee en Continu. *Entropie*, 133:7.

MAO, K.W. & TOOR, H.L. 1970. A Diffusion Model for Reactions With Turbulent Mixing. *AIChE Jour.*, 16:49.

———. 1971. Second-Order Chemical Reactions With Turbulent Mixing. *Ind. Engrg. Chem. Fund.*, 10:192.

MATIJEVIC, E.; JANAUER, G.E.; & KERKER, M. 1964. *Jour. Colloid Sci.*, 19:333.

MEHTA, R.V. & TARBELL, J.M. 1983. An Experimental Study of the Effect of Turbulent Mixing on the Selectivity of Competing Reactions. *AIChE Jour.*, 33:1089.

———. 1987. Four Environment Model of Mixing and Chemical Reaction. *AIChE Jour.*, 29:320.

MIDDLETON, J.C. ET AL. 1986. Computations of Flow Fields and Complex Reaction Yield in Turbulent Stirred Reactors, and Comparison With Experimental Data. *Chem. Engrg. Res. Des.*, 64:18.

MOFFET, J.W. 1968. The Chemistry of High Rate Water Treatment. *Jour. AWWA*, 60:11:1255.

NAGATA, S. 1975. *Mixing.* Halsted Press, Tokyo.

NAUMAN, E.B. 1974. Mixing in Polymer Reactors. *Jour. Macromol. Sci. Revs. Macromol. Chem.*, C10:75.

———. 1975. The Droplet Diffusion Model for Micromixing. *Chem. Engrg. Sci.*, 30:1135.

NG, D.Y.C. & RIPPIN, D.W.T. 1965. The Effect of Incomplete Mixing on Conversion in Homogeneous Reactions. Proc. 3rd European Symp. on Chemical Reactions in Engineering, Amsterdam.

O'MELIA, C. 1985. Polymeric Inorganic Flocculants. Engineering Foundation Conf. on Flocculation, Sedimentation, and Consolidation, Sea Island, Ga.

OTTINO, J.M. 1980. Lamellar Mixing Models for Structured Chemical Reactions and Their Relationship to Statistical Models: Macro- and Micromixing and the Problem of Averages. *Chem. Engrg. Sci.*, 35:1377.

PARTHASARATHY, N. & BUFFLE, J. 1985. Study of Polymeric Aluminum (III) Hydroxide Solutions for Application in Waste Water Treatment. Properties of the Polymer and Optimal Conditions of Preparation. *Water Res.*, 19:1:25.

PATTERSON, G.K. ET AL. 1982. Measurements of Mixing Effects on Local Reaction-Conversion in Stirred Tank. Proc. 4th European Conf. on Mixing. Leeuwenhorst, the Netherlands.

PAUL, E.L. & TREYBAL, R.E. 1971. Mixing and Product Distribution for a Liquid-Phase, Second-Order, Competitive-Consecutive Reaction. *AIChE Jour.*, 17:718.

PLACEK, J. ET AL. 1986. Turbulent Flows in Stirred Tanks, Part II: A Two-Scale Model of Turbulence. *AIChE Jour.*, 32:1771.

PLASARI, E. ET AL. 1978. Micromixing Phenomena in Continuous Stirred Reactors Using a Michaelis–Menten Reaction in the Liquid Phase. *ACS Symp. Ser.*, 65:126.

POHORECKI, R. & BALDYGA, J. 1983a. New Model of Micromixing in Chemical Reactors. I, General Development and Application to a Tubular Reactor. II, Application to a Stirred Tank Reactor. *Ind. Engrg. Chem. Fund.*, 22:392.

———. 1983b. The Use of a New Model of Micromixing for Determination of Crystal Size in Precipitation. *Chem. Engrg. Sci.*, 38:79.

RITCHIE, B. & TOBGY, A.M. 1979. A Three Environment Micromixing Model for Chemical Reactors With Arbitrary Separate Feed Streams. *Chem. Engrg. Jour.*, 17:173.

SCHNEIDER, W. 1984. Hydrolysis of Iron (III): Chaotic Olation versus Nucleation. *Comments Inorg. Chem.*, 3:205.

SPENCER, J.L. ET AL. 1980. Identification of Micromixing Mechanisms in Flow Reactors: Transient Input of Reactive Tracers. *Ind. Engrg. Chem. Fund.*, 19:135.

SPIELMAN, L.A. & LEVENSPIEL, O. 1965. A Monte Carlo Treatment for Reacting and Coalescing Dispersed Phase Systems. *Chem. Engrg. Sci.*, 20:247.

STOL, R.J.; VAN HELDEN, A.K.; DEBRUYN, P.L. 1976. Hydrolysis–Precipitation Studies of Aluminum (III) Solutions: II, A Kinetic Study and Model. *Jour. Colloid Inter. Sci.*, 57:1:115.

STUMM, W. & MORGAN, J. 1970. *Aquatic Chemistry.* Wiley-Interscience, New York.

TOSUN, G. 1987. A Study of Micromixing in Tee-Mixers. *Ind. Engrg. Chem. Res.*, 26:1184.

TRELEAVEN, C.R. & TOBGY, A.H. 1973. Residence Times, Micromixing and Conversion in an Unpremixed Feed Reactor. II, Chemical Reaction Measurements. *Chem. Engrg. Sci.*, 28:413.

TRUONG, K.T. & METHOT, J.C. 1974. Etude Experimentale du Micromelange a l'Aide d'un Traceur Non-reactif dans un Reacteur Agité. *Can. Jour. Chem. Engrg.*, 52:767.

VAN STRATTEN, H.A.; HOLTKAMP, T.W.; & DEBRUYN, P.L. 1984. Precipitation From Supersaturated Aluminate Solutions: I, Nucleation and Growth of Solid Phases at Room Temperature. *Jour. Colloid Inter. Sci.*, 98:2:342.

VERMEULEN, A.C., ET AL. 1975. Hydrolysis–Precipitation Studies of Aluminum (III) Solutions: I, Titration. *Jour. Colloid Inter. Sci.*, 51:449.

VILLERMAUX, J. 1983. Mixing in Chemical Reactors. *ACS Symp. Ser.*, 226:135.

———. 1986. Micromixing Phenomena in Stirred Reactors. *Encyclopedia of Fluid Mechanics*. Gulf Publishing, West Orange, N.J.

VILLERMAUX, J. & BLAVIER, L. 1984. Free Radical Polymerization Engineering. *Chem. Engrg. Sci.*, 39:87.

VILLERMAUX, J. & DAVID, R. 1983. Recent Advances in the Understanding of Micromixing Phenomena in Stirred Reactors. *Chem. Engrg. Comm.*, 21:105.

———. 1988. Effet du Micromelange sur la Precipitation. *Jour. Chim. Phys.*, 85:273.

VILLERMAUX, J. & DEVILLON, J.C. 1972. Representation de la Coalescence et de la Redispersion. Les Domaines de Segregation dans un Fluide par un Modele d'Interaction Phenomenologique, Proc. 2nd Intl. Symp. Chemical Reactions Engineering, Amsterdam.

VRALE, L. & and JORDEN, M. 1971. Rapid Mixing in Water Treatment. *Jour. AWWA*, 63:52.

WEBER, W.J. 1972. *Physicochemical Processes for Water Quality Control*. Wiley-Interscience, New York.

WEN, C.Y. & FAN, L.T. 1975. *Models for Flow Systems and Chemical Reactors*. Marcel Dekker, Inc., New York.

ZOULALIAN, A. & VILLERMAUX, J. 1974. Influence of Chemical Parameters on Micromixing in a Continuous Stirred Tank Reactor. *Adv. Chem. Ser., Chem. React. Engrg.* II, 133:348.

ZWIETERING, T.N. 1959. The Degree of Mixing in Continuous Flow Systems. *Chem. Engrg. Sci.*, 11:1.

———. 1984. A Backmixing Model Describing Micromixing in Single-Phase Continuous Flow Systems. *Chem. Engrg. Sci.*, 39:1765.

Application

6

Particle Destabilization and Flocculation Reactions in Turbulent Pipe Flow

Rudolf Klute, Dr. Ing., Institut für Siedlungswasserwirtschaft, University of Karlsruhe, Karlsruhe, Germany

Appiah Amirtharajah, Ph.D., P.E., Professor, School of Civil Engineering, Georgia Institute of Technology, Atlanta, Ga. USA

INTRODUCTION

The aggregation of particles plays a decisive role in water and wastewater treatment as a preliminary step in phase separation processes. Aggregation consists of three steps. First, by addition of chemicals—usually hydrolyzing ions, such as Al^{3+} and Fe^{3+}, or synthetic organic polymers—the surface charge characteristics of the particles are changed and the particles are destabilized. Immediately after particle destabilization microflocs form. In the final step, macroscopic flocs develop, which can be separated by sedimentation, filtration, or flotation.

In addition to the chemical parameters, flow conditions also control the reaction rate of the coagulation and flocculation process and floc characteristics. From an operational point of view, to achieve optimum overall process efficiency the three steps of particle aggregation require distinct conditions with respect to turbulence intensity and mixing time. Since the reactions involved in the destabilization process for charge neutralization are fast and essentially irreversible, a high intensity of turbulence is crucial during the very first seconds after coagulant addition. On the other hand, the formation of macroscopic flocs requires mixing conditions characterized by a moderate and uniform intensity of turbulence.

In practice, particle destabilization and floc formation are accomplished most commonly in reaction chambers with mechanical mixers. These so-called backmix reactors are surprisingly inefficient in achieving particle destabilization by charge neutralization. A complete and homogeneous dispersion of the coagulants in the raw water usually cannot be achieved within 1 s or less because of short-circuiting of flow and mass rotations of water. Because of the disadvantages of backmix reactors, there is currently a trend toward the use of in-line blenders, which have proven to be more efficient rapid-mix devices.

This chapter is concerned with particle destabilization reactions and aggregation processes in pipelines. It contains an overview of rapid-mix devices for pipes, defines performance and design criteria for rapid mixing, and presents a review of literature on practical experience with pipe mixers for particle destabilization and aggregation in turbulent pipe flow.

STRUCTURE OF TURBULENT PIPE FLOW

The flow conditions most commonly encountered in pipes and conduits of water and sewage works are turbulent. Turbulent flow is characterized by a distribution of different eddy sizes. In turbulent pipe flow, the largest eddies are of the order of the pipe diameter. They carry most of the kinetic energy of the main flow. Inertial forces in the system give rise to a transfer of energy from the large eddies to smaller ones in the form of a continuous cascade. The eddy-size-based Reynolds number is large for the whole spectrum of eddies except for the smallest, which have a Reynolds number of about one. In this smallest size range of the spectrum, viscous forces become important and the kinetic energy is dissipated as heat (Figure 6-1). According to theoretical considerations first presented by Kolmogoroff, eddies in this so-called universal equilibrium range are considered to depend only on the rate of energy dissipation or energy input ε and on the kinematic viscosity ν. By dimensional analysis, the size of these eddies is of the order of:

$$\eta = (\nu^3/\varepsilon)^{1/4} \qquad (Eq\ 6\text{-}1)$$

However, the smallest eddy size is large in terms of molecular dimensions. Thus, a rapid and homogeneous dispersion of the coagulant by high-intensity turbulence is a necessary requirement for complete utilization of destabilizing chemicals.

Dispersion in pipelines is accomplished by the inherent mixing available from pipeline turbulence, by jets injected into the pipe, or by a combination of both effects. The rate of energy dissipation per unit mass ε in pipelines can be calculated from the measurement of head loss H_T. The head loss in a steady, fully developed turbulent pipe flow is given by the Darcy equation

$$H_T = C_D \frac{L}{d_t} \frac{v^2}{2g}$$ (Eq 6-2)

Where:

C_D = pipe friction factor
L = length
d_t = pipe diameter
v = velocity
g = acceleration due to gravity

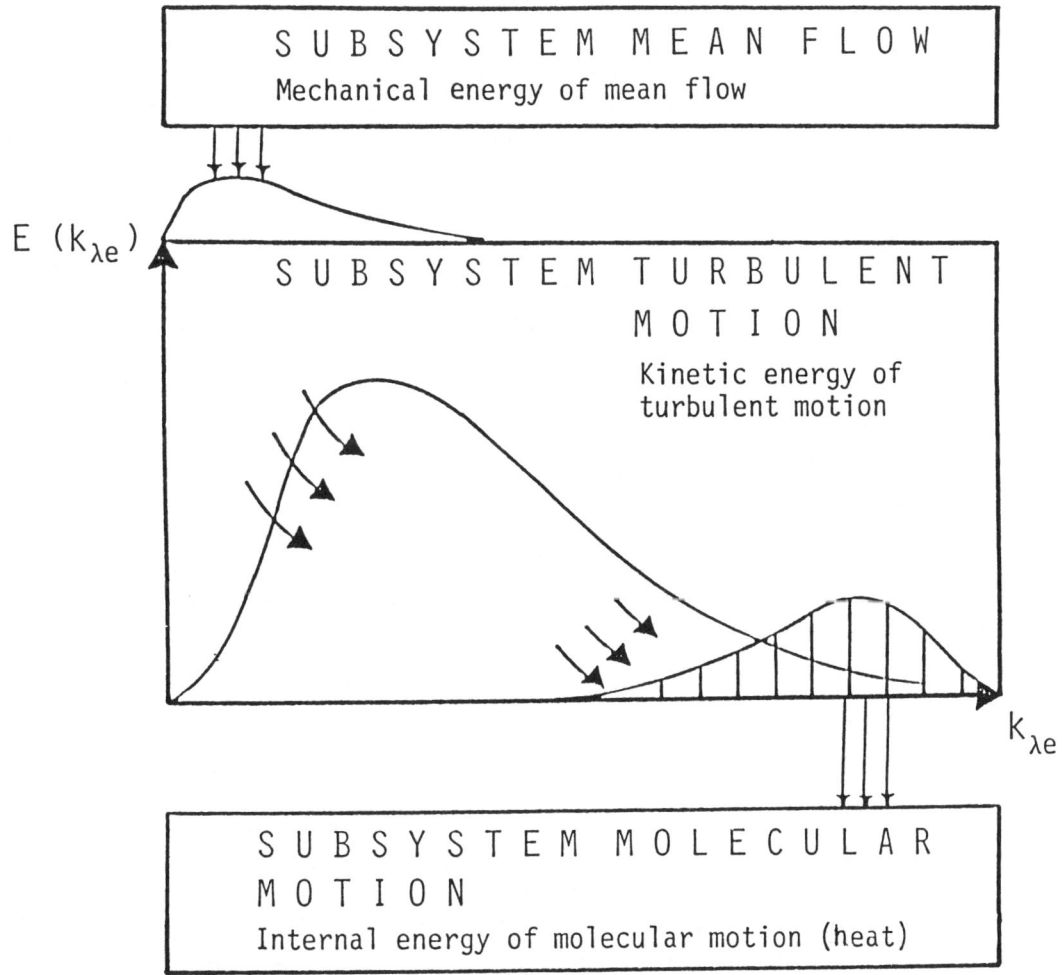

Figure 6-1 Schematic representation of energy transfer in turbulent flow.

The friction factor C_D depends on wall roughness and Reynolds number Re, where:

$$\text{Re} = \frac{vd_t}{\nu} \qquad \text{(Eq 6-3)}$$

with ν = kinematic viscosity.

For smooth pipes, experimental results from many investigations have shown that the following empirical relation holds for the Reynolds number in the range $4 \times 10^3 \leq \text{Re} \leq 2 \times 10^6$:

$$1/\sqrt{C_D} = 2.0 \log \text{Re}\sqrt{C_D} - 0.8 \qquad \text{(Eq 6-4)}$$

An empirical relation according to Blasius holds up to Re = 10^5:

$$C_D = 0.316 \, \text{Re}^{-1/4} \qquad \text{(Eq 6-5)}$$

However, investigations by Laufer (1954) have shown that the energy available from pressure drop is partly converted to heat by direct dissipation and partly converted into turbulent energy. Figure 6-2 illustrates the distribution of energy given up by the main flow according to Laufer's data. As stated by Rotta (1972), the analytical calculation of the two energy fractions is difficult. Moll (1985) presented a calculation as a function of Reynolds number showing a decrease of the directly dissipated energy from 49 percent to 9 percent of the total energy input if the Reynolds number increases from Re = 10^3 to Re = 10^8.

TYPES OF PIPE MIXERS

A very simple method for mixing a coagulant–flocculant chemical (C) with the raw water (W) in a pipe is injection perpendicular or under a certain angle to the pipe axis. Examples of such mixing devices, which are commonly used in water technology, are shown in Figures 6-3 and 6-4. Another method is the parallel injection of chemicals illustrated in Figure 6-5. In large conduits, pipes, or open channels a multiple injection device as shown in Figure 6-6 may be used to obtain more rapid distribution of the chemicals over the pipe or channel cross section.

The mixing effect of this type of device mainly depends on the large-scale turbulence of the main flow, and it improves with increasing flow rate. However, its mixing performance is only moderate, and concentration gradients can be observed even at distances of 40 to 50 pipe diameters downstream of the injection point. Since the reactions controlling the destabilization of particles are extremely fast and essentially irreversible, the nonuniform distribution of chemicals will result in poor destabilization of a fraction of colloids and an overdosing and possible restabilization of another fraction of the particles. The mixing can be improved by enhancing the turbulence intensity and promoting smaller-scale turbulence.

Dividing up the main stream into two or more substreams and injecting the chemicals at the junction point is considered an efficient method to promote mixing. Three types of these mixing devices are shown in Figures 6-7, 6-8, and 6-9.

To increase turbulence intensity, baffles are often installed in pipes and open channels at an angle perpendicular to the axis, as shown in Figure 6-10. The installation of static mixers in pipes is practiced extensively in chemical engineering. An example of this type of mixing device, the Komax static mixer, is shown in

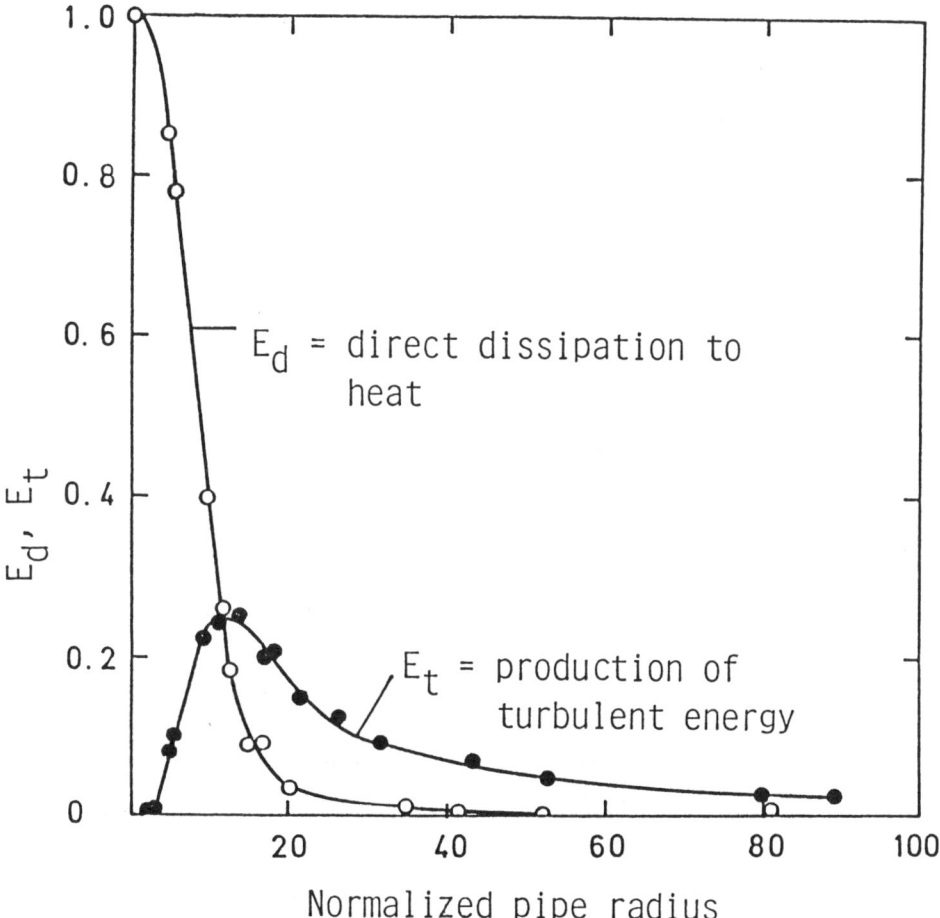

Source: Laufer, J. 1954. The Structure of Turbulence in Fully Developed Pipe Flow. Rept. NACA TR-1174, US National Advisory Committee for Aeronautics, Washington, D.C.

Figure 6-2 Conversion of kinetic energy of the mean flow in pipes into heat E_d and turbulent energy E_t.

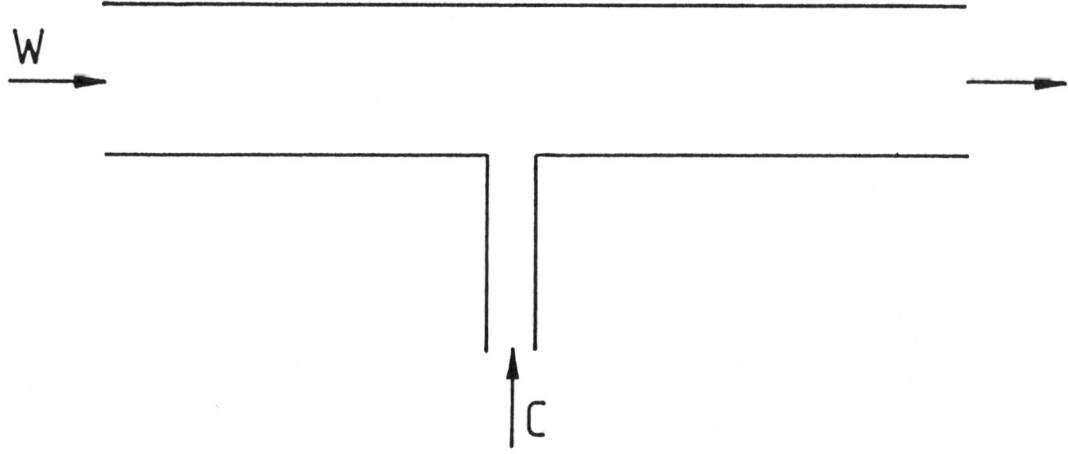

Figure 6-3 Side-injection mixing device.

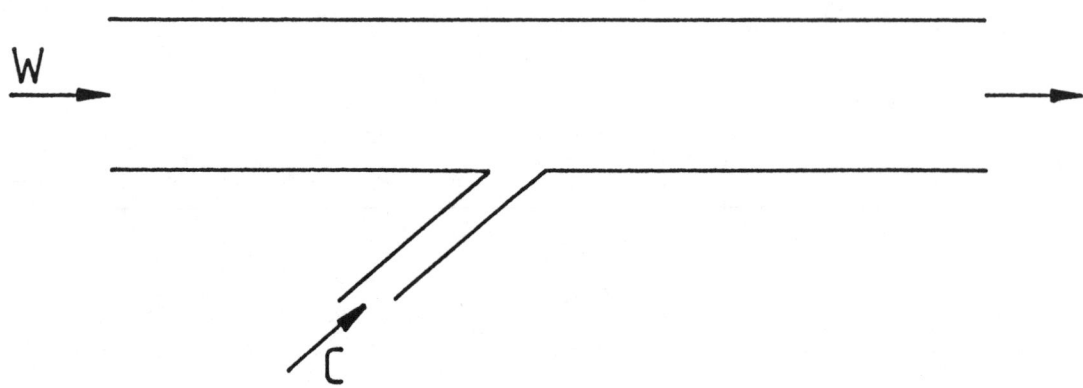

Figure 6-4 Side-injection mixing device.

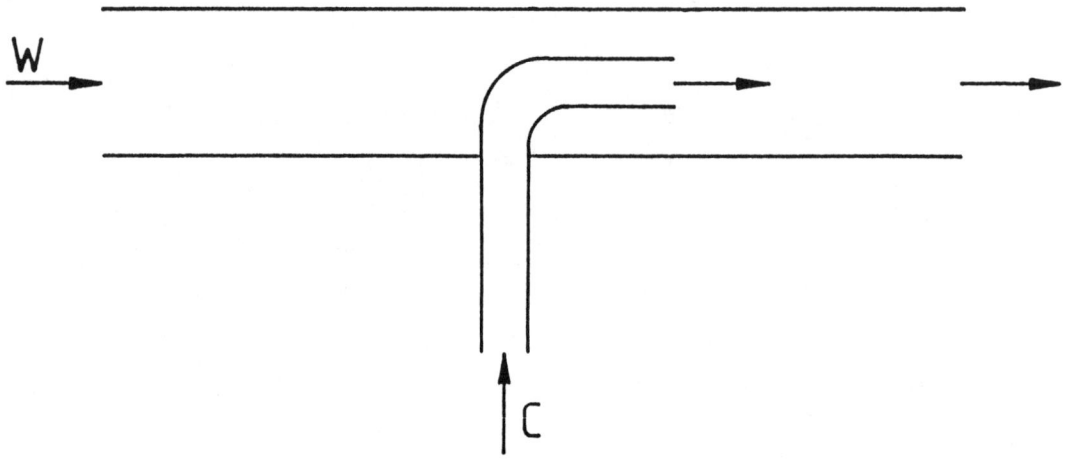

Figure 6-5 Pipe-core injection mixer.

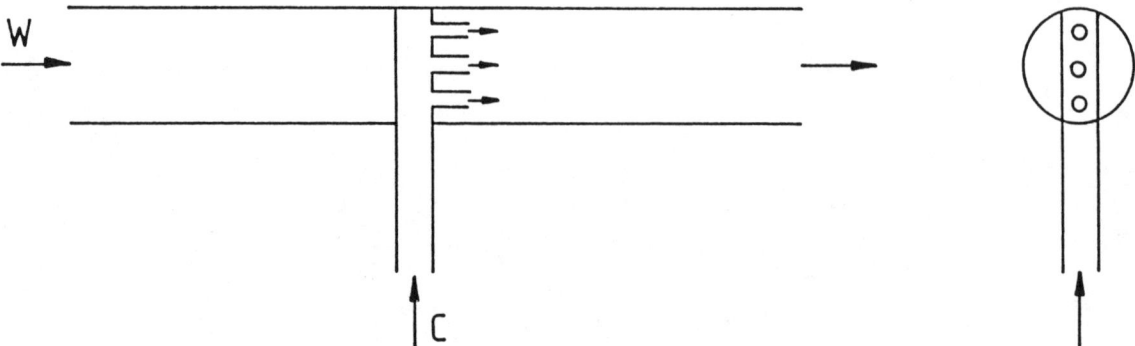

Figure 6-6 Multiple-feed injection mixer.

Figure 6-11. Mixing is achieved by dividing and recombining the fluid downstream of the injection point. However, the application of such mixers in coagulation and flocculation processes has not been reported in the literature. Stenquist and Kaufman (1972) used a biplane square-mesh grid of bars placed in a pipe as a mixing device. More details about static mixers and their application in fluid mixing are presented in chap. 2.

A constriction of the pipe and injection of chemicals into the backwash zone from an annular ring has been described by Vrale and Jorden (1971). This mixing device is illustrated in Figure 6-12. In a similar mixing system used by Klute (1985) the chemicals were injected at the pipe center, and the point of injection was located at a flow time of 0.1 s upstream of the section of pipe expansion. An elastic rubber ring fixed to the injector improved radial distribution of chemicals. Some details of this mixing device are shown in Figure 6-13. Injection of chemicals by a ring of nozzles installed at the pipe expansion was proposed by Dittmann et al. (1988). This mixing device was constructed for a large water treatment plant and is illustrated in Figure 6-14.

A mixing unit that was designed for several large water treatment plants is shown in Figure 6-15. Kawamura (1976) stated that this device operates with a velocity gradient $\overline{G}$ = 750 to 1000 s^{-1} and a mixing time t = 1 s. Ventresque and Bablon (1988) reported on the application of a coagulant injection system, by which the chemicals are mixed into the water through a multiport valve, installed in the center of a pipe as shown in Figure 6-16. In this mixing device the coagulant is prediluted 0.1 s prior to injection into the raw water.

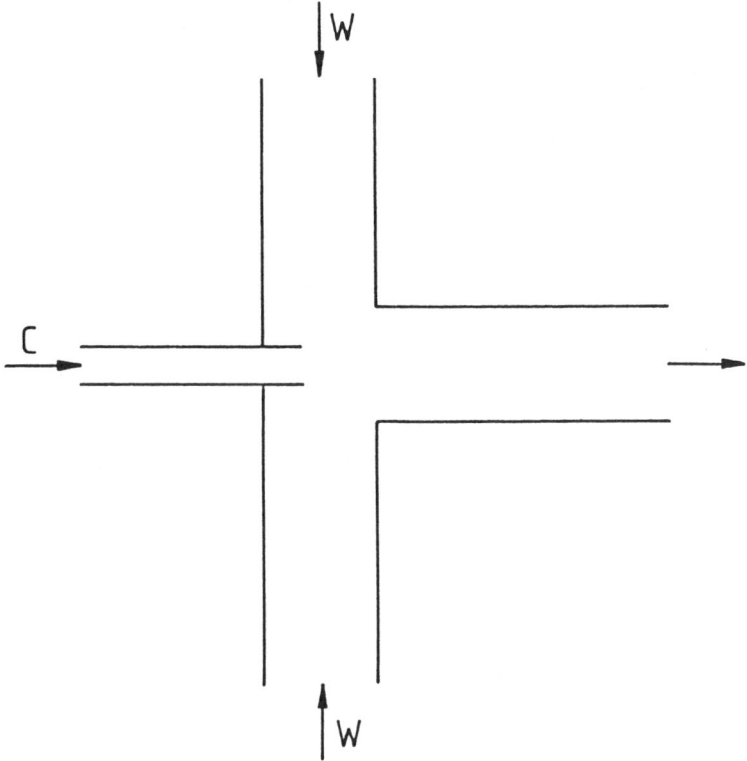

Figure 6-7 Opposed-flow mixer.

Whereas mixing in the devices described so far is based predominantly on the turbulence generated by the main flow, mixing by the high-velocity injection system is mainly accomplished by the reflection of the jet at the opposite pipe wall and the subsequent dispersion of the chemical over the pipe cross section. This system was described by Grohmann, Haesselbarth, and Langer (1977).

A turbine mixing device, which is installed into a pipe or a channel, was described by Stendahl (1985). The chemicals are injected just ahead of the turbine and are dispersed by the motor-driven cone, as shown in Figure 6-17.

Additional information about mixing operations in pipelines, and types and design of pipe mixers can be found in the work of Brodkey (1966), Olson and Stout (1967), Brodkey (1975), Simpson (1975), and more recently Gray (1986). These authors provide in their review articles a complete summary of theoretical and practical aspects of mixing operations in pipes, especially from a chemical engineering point of view. However, many of the aspects dealt with may be successfully applied in design and operation of coagulation and flocculation processes in pipes.

PIPE MIXING PERFORMANCE CRITERIA

If the coagulant is injected into a pipe for particle destabilization and aggregation, a characteristic concentration pattern is formed downstream of the injection point. This pattern depends on the mixing device, the flow rates of the main stream and the coagulant solution, and the diffusion characteristics of the coagulant and its

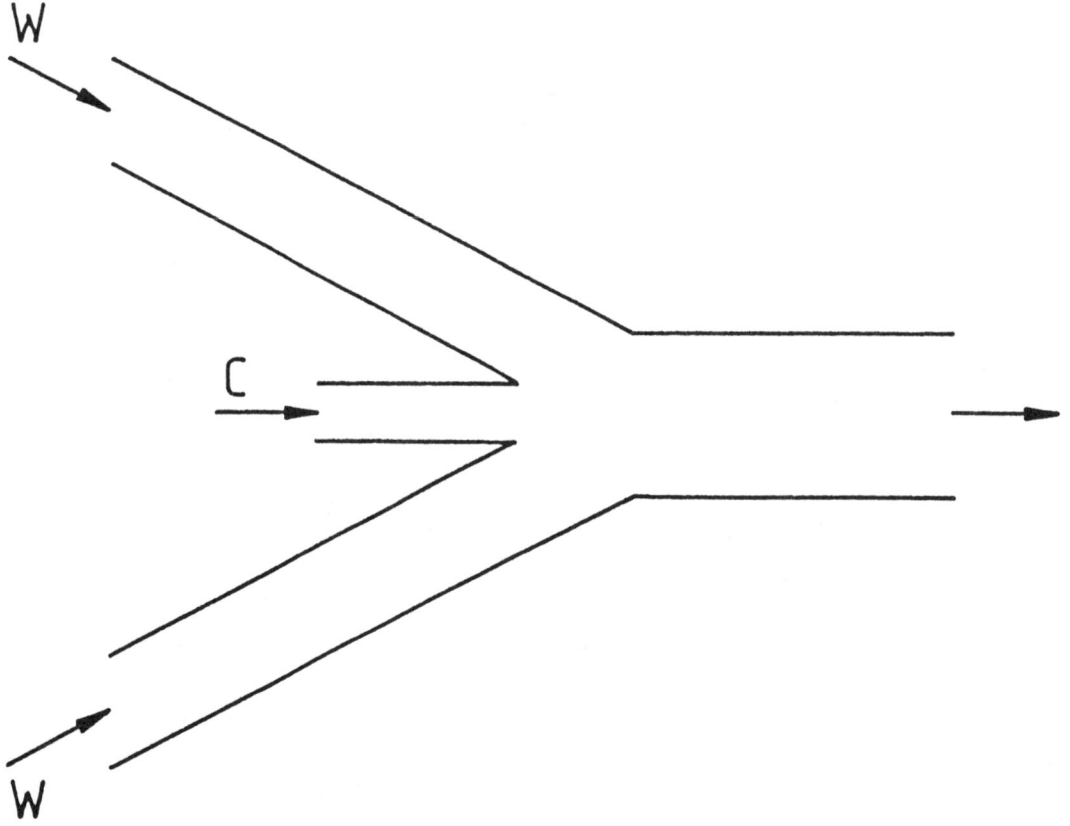

Figure 6-8 *Y*-flow mixer.

PARTICLE DESTABILIZATION 225

hydrolysis species. Close to the injection point there is a large concentration gradient of the coagulant over the pipe cross section. Farther downstream, the concentration gradient gradually decreases as the fluid moves along the pipe, and finally the concentration gradient disappears. The mixing performance of pipe flow mixing devices can be expressed in two different ways.

1. In chemical engineering the number of pipe diameters required to achieve a defined concentration gradient is often used as a criterion for mixing. Instead of

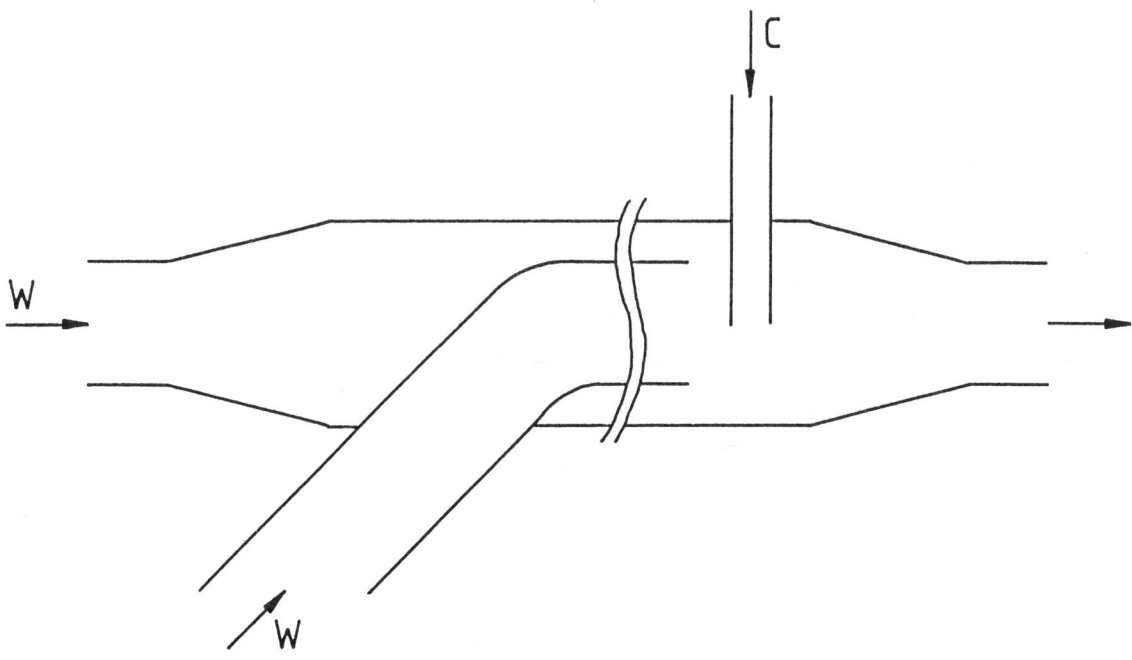

Figure 6-9 Parallel-flow mixer.

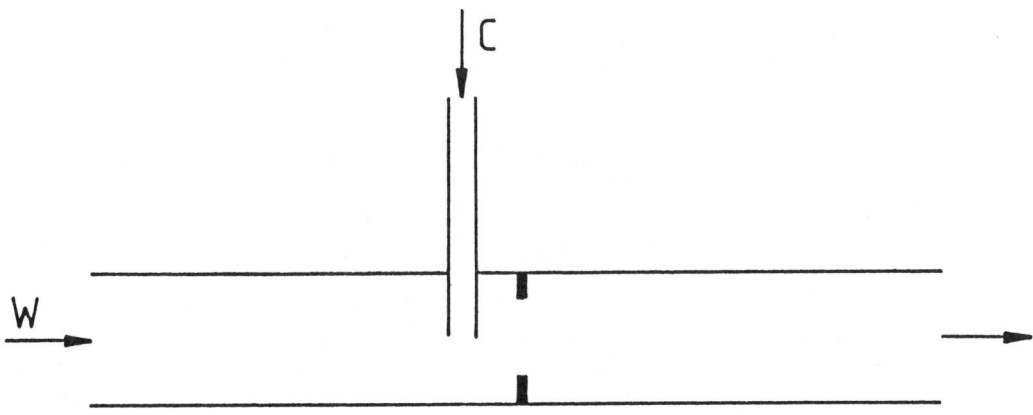

Figure 6-10 Baffle mixing device.

226 MIXING

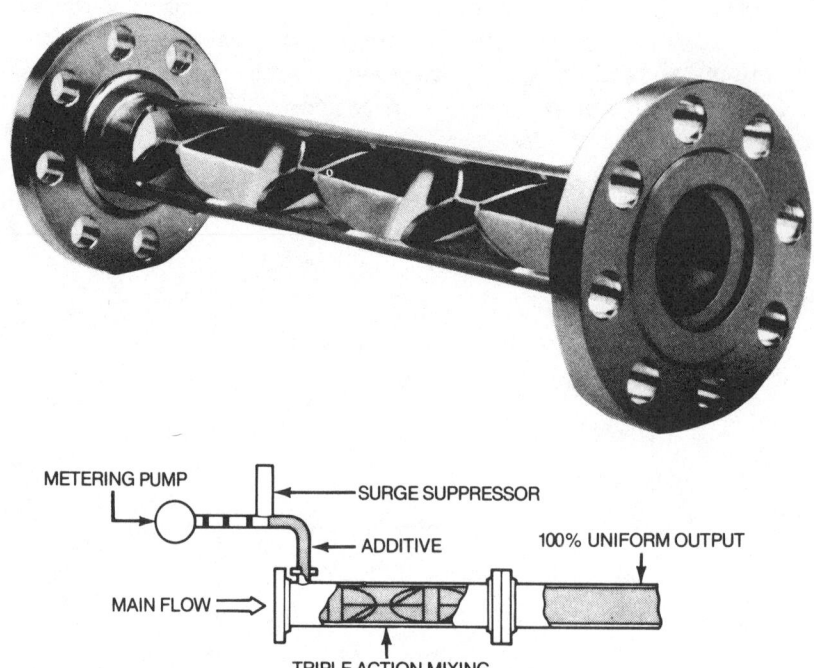

Courtesy Komax Systems, Inc., Wilmington, Calif.

Figure 6-11 Komax® static mixer.

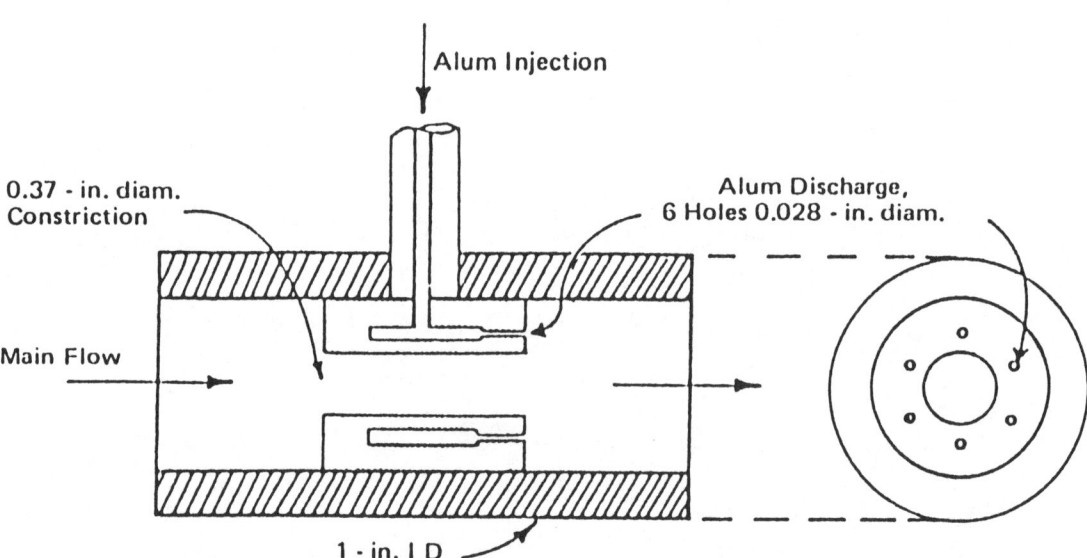

Source: Vrale, L. & Jorden, R.M. 1971. Rapid Mixing in Water Treatment. Jour. AWWA, 63:52.

Figure 6-12 Pipe-constriction mixing device.

the concentration gradient, the standard deviation of the concentration gradient σ_c may be used as a measure of mixing performance. σ_c is expressed as

$$\sigma_c = \sqrt{\sum_{i=1}^{n} (c_i - \overline{C})^2/(n-1)} \qquad \text{(Eq 6-6)}$$

Where

c_i = concentrations at points i
$\overline{C}$ = $\Sigma c_i/n$, average concentration of the chemical used for destabilization

The variation coefficient $C_v = \sigma_c/\overline{C}$ can be used as a measure of uniformity. Instead of the coagulant, other conservative substances—acids, bases, salts, and dyes, for example—can be injected as a tracer to evaluate the performance of the mixing device. For these tracers, pH, conductivity, and light absorption measurements are used for characterization and analysis of the effectiveness of mixing.

2. The second method for evaluating the mixing efficiency is to measure directly the degree of particle destabilization or aggregation at a defined distance downstream of the injection point or after a certain reaction time. Alternately, the

Source: Klute, R. 1985. Rapid Mixing in Coagulation–Flocculation Processes: Design Criteria. In Chemical Water and Wastewater Treatment, *ed. by A. Grohmann, H.H. Hahn, and R. Klute. Gustav Fischer Verlag, Stuttgart, Germany.*

Figure 6-13 Pipe-expansion mixing device.

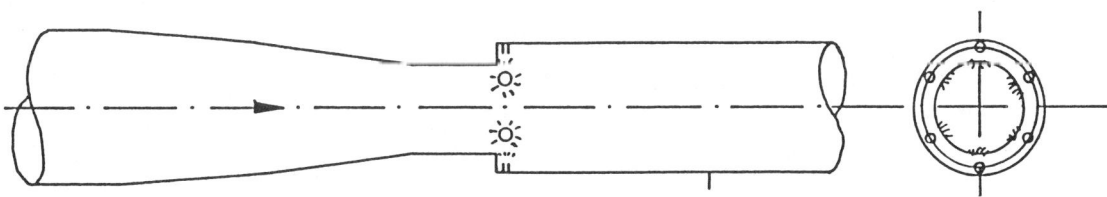

Source: Dittmann, W. et al. 1988. Mixing Systems for Pipe Reactors in Water Treatment. Vom Wasser, *70:129.*

Figure 6-14 Pipe-expansion mixing device.

228 MIXING

mixing performance can be evaluated on the overall removal efficiency after phase separation. In contrast to the methods described under section 1, destabilization and aggregation measurements are usually carried out discontinuously. However, these methods allow a more precise determination of the effect of different mixing devices on the destabilization performance and the particle aggregation efficiency.

The degree of particle destabilization accomplished by the different mixing devices can be determined by measuring the zeta potential or the electrophoretic

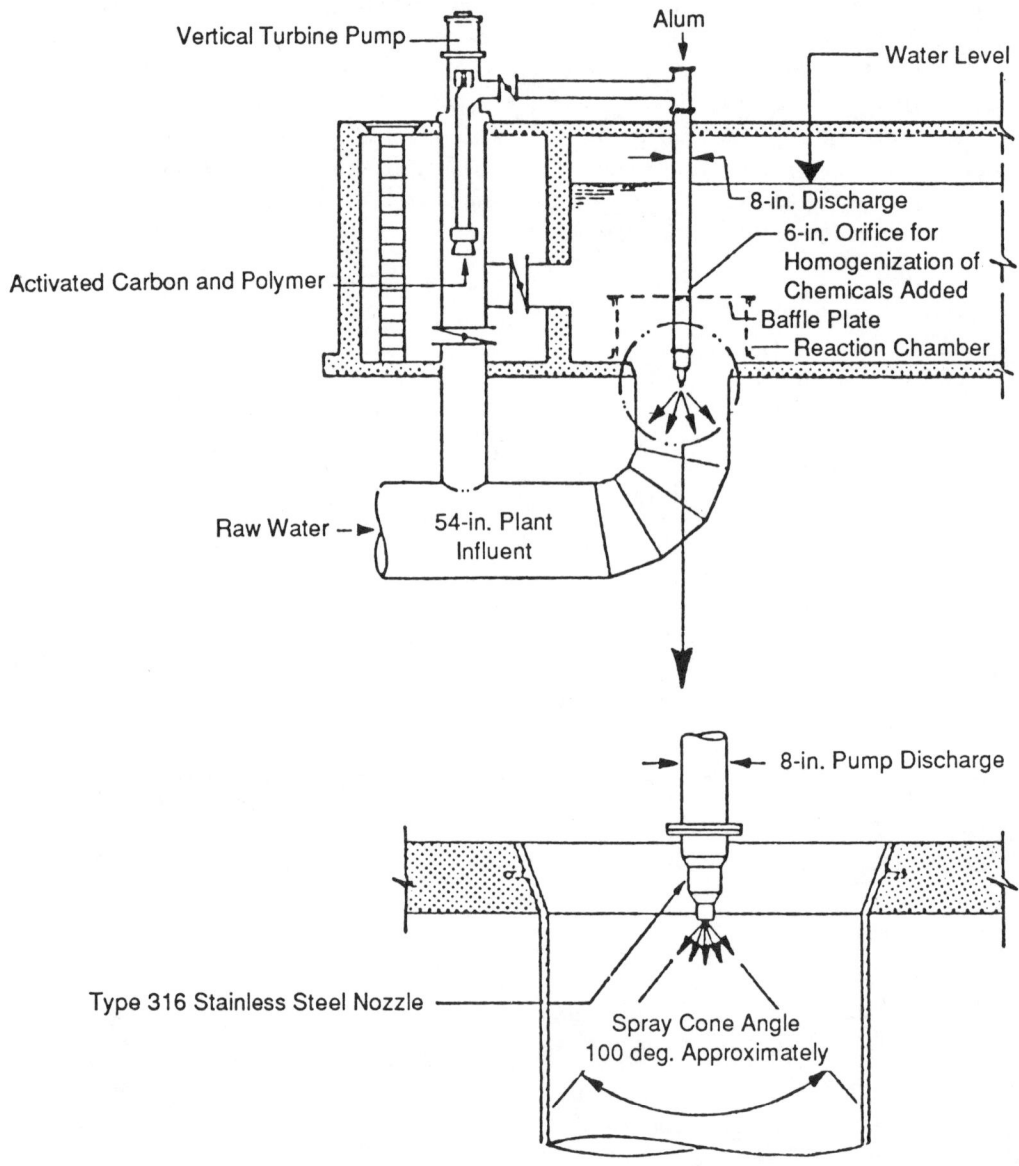

Source: Kawamura, S. 1976. Considerations on Improving Flocculation. Jour. AWWA, 68:328.

Figure 6-15 Pipe-flow flash mixing facility.

mobility of particles in samples withdrawn from the pipeline downstream of the injection point. Samples from different points along the length of pipe and over its cross section present a precise picture of particle destabilization as a function of mixing device, turbulent flow conditions, and reaction time. In addition, the mean value, the distribution, and the standard deviation of the electrophoretic mobility can be useful mixing criteria.

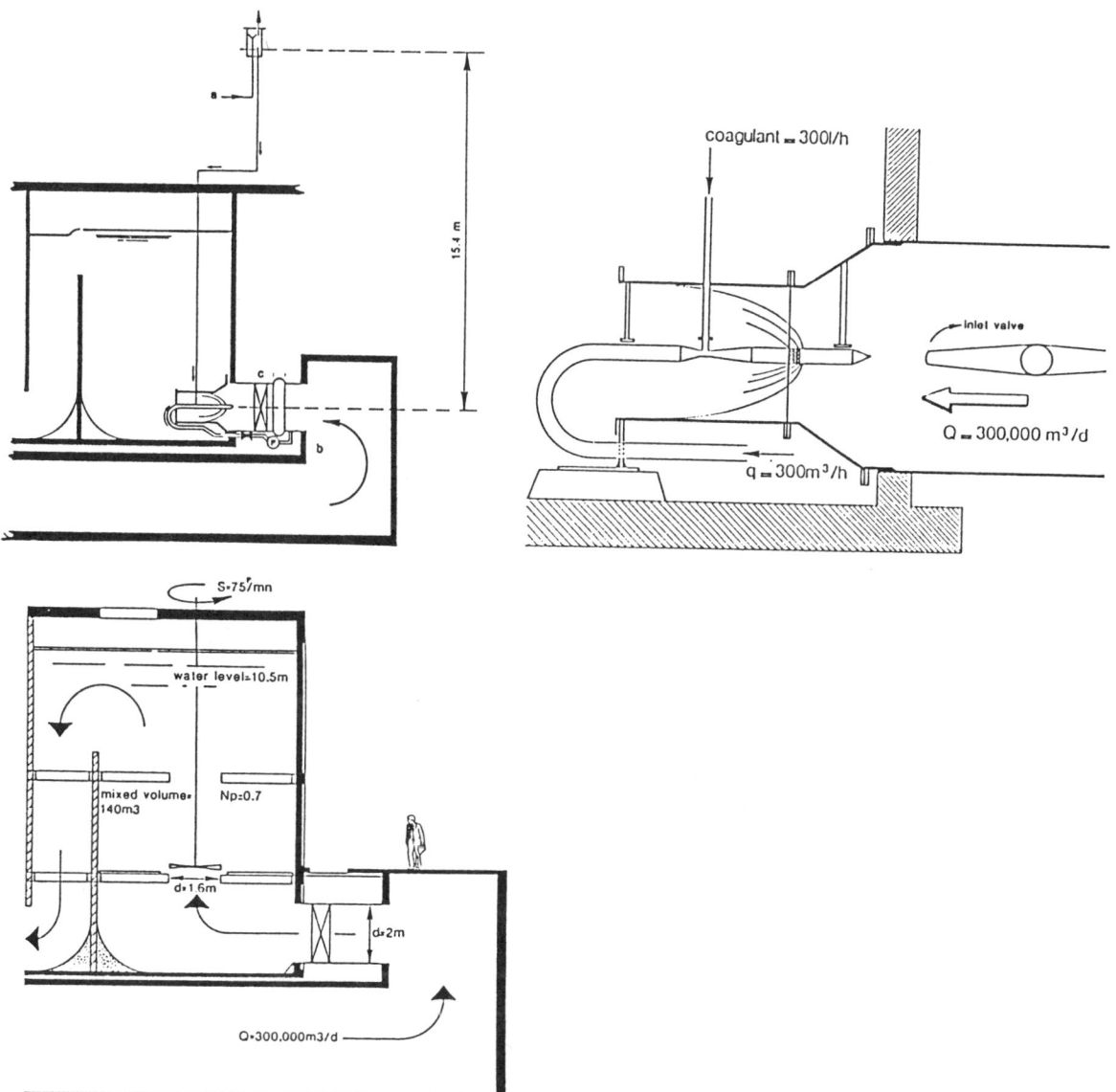

Source: Ventresque, C. & Bablon, G. 1988. *New Coagulant Injection Process. In* Pretreatment in Chemical Water and Wastewater Treatment, *ed. by H.H. Hahn and R. Klute. Springer-Verlag, Heidelberg, Germany.*

Figure 6-16 Coagulant injection system for large pipes.

230 MIXING

The particle aggregation efficiency as a function of different mixing devices or turbulent pipe flow conditions can be detected by turbidity measurements or particle counting techniques. These measurements may be performed either in situ or from samples taken from the pipe reactor. In addition, the effect of mixing on the phase separation characteristics of flocs can be detected by including sedimentation, flotation, or filtration in the analysis procedure.

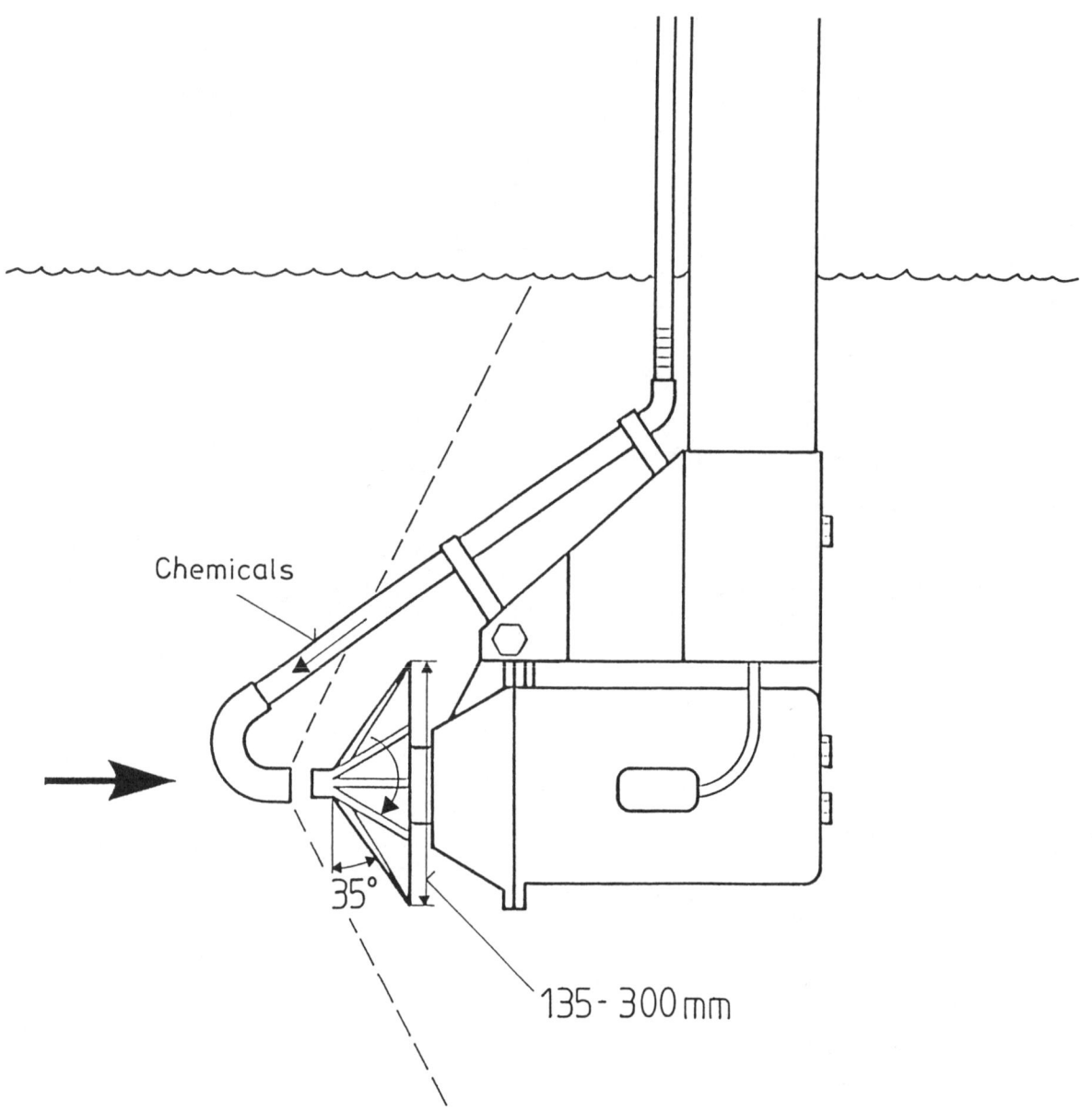

Source: Adapted from an original from Boliden/Kemira Turbo Mixer®.

Figure 6-17 Turbo mixer®.

DESIGN CRITERIA OF PIPE MIXERS

Jet Injection Mixing in Pipe Flow

The flow field produced by a turbulent jet discharged perpendicularly into a uniform cross flow has been described by several authors. Pratte and Baines (1967) and Chen, Lin, and Kennedy (1976) distinguished three principal subregions on the basis of experiments with jets of oil aerosols using air in wind tunnels: the near-field or potential zone, the curvilinear zone or zone of maximum deflection, and the far-field or vortex zone (Figure 6-18).

In the near-field zone, which is close to the point of discharge, the jet is deflected only minimally by the cross flow. However, the interaction between jet and cross flow leads to the formation of a quasistagnant zone in front of the jet and a wake behind it, similar to the flow pattern caused by a rigid cylinder, resulting in increased pressure ahead of the jet and decreased pressure in its wake. The flow around the jet periphery and into the wake creates a pair of counterrotating vortices in the wake region. According to Chen, Lin, and Kennedy (1976) the near-field zone can be considered jet dominated, in the sense that the effects of the cross flow on the jet are not significant; the jet trajectory continues to be nearly perpendicular to the plane of its origin and the velocity distribution is not greatly different from that of a simple jet.

In the second flow zone downstream of the jet, the interaction between jet and cross flow causes an intensive mass and momentum transfer from the cross flow into the jet. The mixing effect is augmented by the wake vortices, which expand and are the dominant flow feature at the end of the curvilinear zone. Within this zone, the mean velocity of the jet is rapidly decreased and the jet trajectory is deflected into the direction of the cross flow.

In the far-field zone the jet and the counterrotating vortices are moving downstream at a velocity almost equal to the cross flow velocity. The vortices are

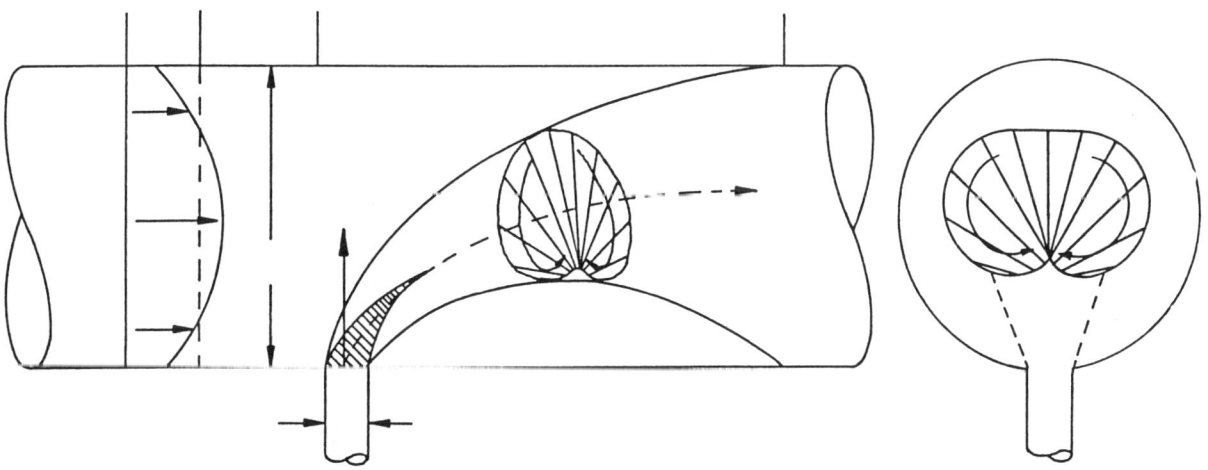

Source: Reprinted with permission of American Society of Civil Engineers, 345 East 47th Street, New York N.Y. 10017, from Pratte, B.D. & Baines, W.D. 1967. Profiles of the Round Turbulent Jet in a Cross Flow. Jour. Hydraulics Div. ASCE, 92:HY6:53.

Figure 6-18 Formation of flow zones by a turbulent jet entering a cross flow.

still increasing in size, whereas their circulation velocity decreases. In this final subregion the behavior of the jet is dominated by the cross flow.

Ger and Holley (1976) compared the mixing effect of three different single-point injections in pipe flow—a centerline source, a wall source, and a jet perpendicular to the pipewall—by determining the variations of the standard deviation σ_c with $C_D^{1/2} \cdot L$. The authors stated that the source located at the pipe wall gave a longer mixing distance than the other injection schemes (Figure 6-19). Mixing distance could be reduced by using a centerline source. The shortest mixing distance of the systems studied was obtained by using a jet rather than a simple source at the pipe wall or center. The mixing distance for a jet perpendicular to the wall was minimized if the ratio M_r of momentum fluxes of the jet and the pipe flow was optimum; i.e., $M_r = 0.0156$.

Furthermore, Ger and Holley (1976) found that the relation

$$C_D^{1/2} \cdot L = A \log (I/\sigma_c) \qquad \text{(Eq 6-7)}$$

Where:

L = dimensionless longitudinal distance
C_D = friction factor

can be used in predicting the mixing distance for each of the aforementioned injection schemes, with the parameters A and I (slope and intercept on linear form of

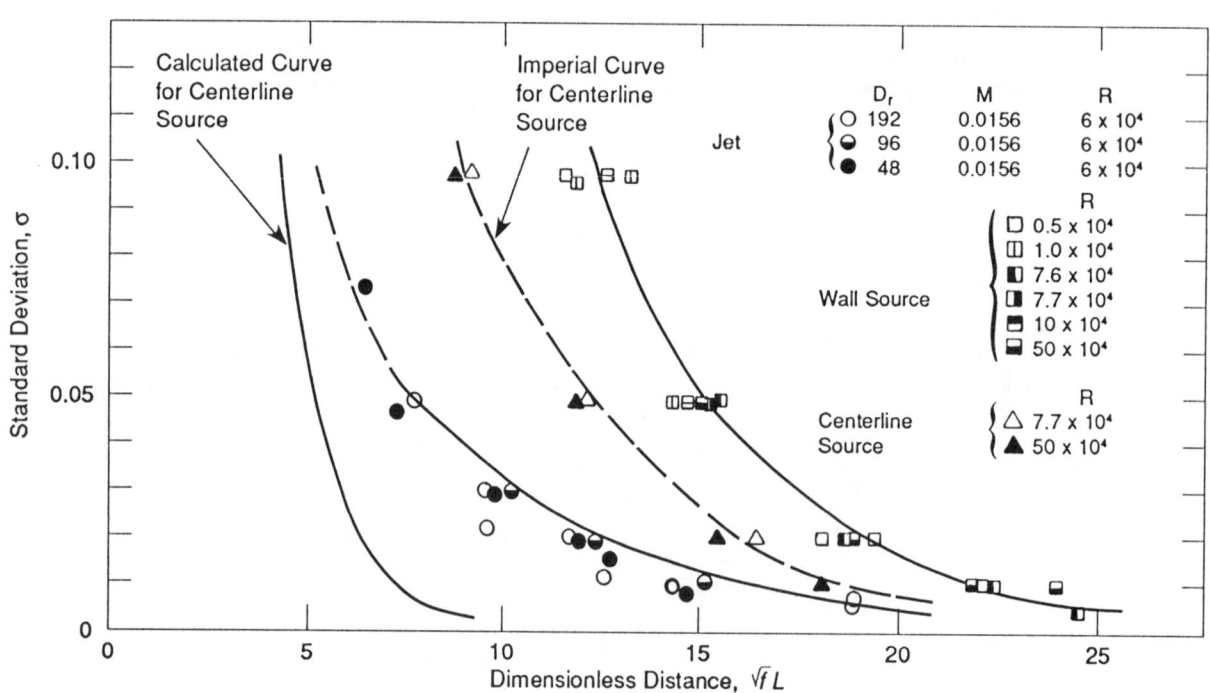

Source: Reprinted with permission of American Society of Civil Engineers, 345 East 47th Street, New York, N.Y. 10017, from Ger, A.M. & Holley, E.R. 1976. Comparison of Single-Point Injections in Pipe Flow. Jour. Hydraulics Div. ASCE, 102:HY6:731.

Figure 6-19 Comparison of different single-point injections.

Eq 6-7) determined either theoretically or empirically. The agreement between Eq 6-7 and the experimental findings was good for a simple edge source and a jet source located at the pipe wall if the densiometric Froude number $Fr_d \geq 50$ for the jet. However, the use of Eq 6-7 for prediction of L for centerline injections was not recommended because of significant differences between the theoretical curve and many experimental results.

Fitzgerald and Holley (1981) investigated several jet injection systems in pipe flow to determine their mixing characteristics and optimum operating conditions. In addition, they studied the effects of secondary currents on mixing. The coefficient of variation of the concentration distribution C_v defined as

$$C_v = \left[\frac{1}{A_p} \int_{A_p} \left(\frac{C}{\overline{C}} - 1\right)^2 dA_p\right]^{1/2} \quad \text{(Eq 6-8)}$$

was used as a measure of the degree of mixing at a given cross section

Where:

C = tracer concentration at a point in the cross section
$\overline{C}$ = cross-sectional average concentration
A_p = area of the pipe

For uniform pipe flow with a given number and orientation of jets, the ratio M_r of the jet momentum flux to the pipe-flow momentum flux was the parameter indicating the mixing induced by the jets. Increasing the angle of injection from 90° (cross flow) to 150° (almost counterflow) decreased the optimum dimensionless mixing distance Z_m, which was defined as flow distance required to obtain a specific C_v, but at the cost of a large increase in the momentum ratios and the power requirement.

Experiments with two diametrically opposed 90° jets in a uniform flow showed that optimum mixing was obtained with the momentum ratio for each jet, being on the order of one-half of the optimum momentum ratio for the single jet injection. Thus, even though the use of two jets in uniform flow did not produce a significant decrease in the mixing distance compared to a single jet, the two jets would require less power than a single jet. The authors concluded that the most rapid possible mixing with jet injection in a pipe flow was achieved by using two or more 90° jets equally spaced around the pipe periphery, with each jet having M_r values of 0.006 to 0.010, in which M_r = momentum flux ratio.

$$M_r = Q_r \cdot V_r = (V_r D_r)^2 \quad \text{(Eq 6-9)}$$

Where:

Q_r = volume flux ratio Q_j/Q_t
V_r = velocity ratio v_j/v_t
D_r = diameter ratio d_j/d_t

The subscripts j and t refer to the jet and the pipe, respectively.

Chao and Stone (1979) stated that the jet-type mixer provides a number of advantages over mechanical mixers and wake-type arrangements for rapid mixing. The liquid jets form an axisymmetric jet screen and generate their own wake, entrainment, and spreading, which provide efficient rapid mixing with the receiving stream without creating significant head loss. It is suggested by the authors that the chemical be injected perpendicular to the receiving flow stream, using a multi-orifice jet. Furthermore, it is assumed that a jet introduced at the center of the pipe, and terminating in the wall region instead of being introduced at the wall, would receive more mixing and would be transported back into the center of the pipe more effectively, because for a turbulent pipe flow the turbulent velocity intensities in the longitudinal, radial, and circumferential directions generally increase from the center of the pipe toward the wall (Daily and Harleman 1966). On the basis of empirical equations, derived by Pratte and Baines (1967) for jet profiles in a cross flow, Chao and Stone (1979) presented an example for the design of an in-line jet injection system, placed in the center of the pipe. However, no experimental data to verify this approach have been reported by the authors.

Guven and Benefield (1983) reviewed the available information on the initial mixing characteristics of turbulent jets in crossing pipe flow and discussed criteria for the optimum design of in-line jet injection blenders. From a study of the literature on jet behavior the authors concluded that centrally located multiple jets placed symmetrically around the pipe's axis and discharging radially outward produce better initial mixing than single or multiple side jets placed around the circumference, because the jets issue radially outward and do not merge at the center of the pipe. The faster far-field mixing due to axial symmetry of the central injection scheme was considered to be a second advantage over a single side jet, which is characterized by an asymmetric injection scheme. These advantages may outweigh such disadvantages as additional head loss and vibration problems. Guven and Benefield recommend that both the velocity ratio $V_r = v_j/v_t$ and the diameter ratio $D_r = d_j/d_t$ should be considered as design variables in the design of in-line jet injection blenders. They presented an analysis to indicate how the design proposal by Chao and Stone (1979) for the centrally injected jets can be improved so that not only D_r but also V_r can be considered as variables. As a result, it was concluded that optimum mixing conditions for small diameter ratios are characterized by a constant value for the momentum ratio M_r, which, by definition, is equal to $V_r^2 \cdot D_r^2$. For large diameter ratios, the optimum momentum ratio is not a constant and depends on the diameter ratio.

Kinetic Aspects of Particle Destabilization in Turbulent Pipe Flow

The destabilization of colloidal particles with hydrolyzing aluminum or iron salts requires an analysis of the time scale of the reactions that precede particle aggregation. The formation of hydroxocomplex species and their adsorption on the particle surface take place extremely quickly, within microseconds for monomeric species and within 1 s if polymeric species are involved (Hahn and Stumm 1968). The formation of the aluminum or iron hydroxide for entrapment of colloidal particles is completed within the range of 1 to 7 s (Letterman, Quon, and Gemmell 1973).

As stated by Amirtharajah and Mills (1982), the difference between the rates of these reactions and the two modes of coagulation in terms of rapid mixing is not delineated in the literature. According to these authors it is imperative that the coagulants be dispersed in the raw water stream as rapidly as possible (in less than 0.1 s) so that the hydrolysis products that develop in 0.01 to 1 s will be adsorbed and hence cause destabilization of the colloid. Extremely short dispersion times and high intensities of mixing are not as crucial as in adsorption–destabilization, if coagulation is caused by entrapment of the colloids in the metal hydroxide precipitate.

Recent experiments on particle destabilization in turbulent pipe flow by Klute and Dierschke (1990) have confirmed the recommendations made by Amirtharajah (1978). For this research it was necessary to have a reactor that allows destabilization measurements within defined reaction times of about 1 s under nearly homogeneous conditions throughout the liquid. In stirred reactors, the detention time of different fluid elements is significantly different, as is well-known from chemical reactor theory (Levenspiel 1972). In addition, stirred reactors produce non-homogeneous and anisotropic turbulence. A fully developed pipe flow, at sufficiently high flow rates, most closely accomplishes the requirements for stationary and homogeneous turbulence in the core region. This type of reactor makes it possible to study the kinetics of high-rate reactions.

The overall design of the reactor system is shown schematically in Figure 6-20. The water was taken from a reservoir and, after passing through two filters, was split into two identical streams into which the model suspension and the alum solution were dosed. The distance between the mixing module and the dosing point for the alum solution could be varied by the installation of pipelines of varying length. With this experimental setup it was possible to allow the formation of different aluminum hydroxocomplex species. Depending on the reaction time Δt, predominantly monomeric species—e.g., of the type $Al(OH)^{2+}$ or $Al(OH)_2^+$—or polymerized species—e.g., of the type $Al_8(OH)_{20}^{4+}$, $Al_{13}(OH)_{34}^{5+}$, or $Al(OH)_3(s)$—are formed, before the particle destabilization process is initiated. Thus, differences in the degree of destabilization may be attributed to the type of hydroxocomplex species that are predominant in solution.

In Figure 6-21 the electrophoretic mobility (EM) of the silica particles is plotted against the hydrolysis time Δt for three different aluminum concentrations. These results were obtained under identical conditions: concentration of the aluminum stock solution, particle concentration, pH, flow rate, and Reynolds number. Obviously, Δt has a pronounced influence on particle destabilization. An increase of Δt causes lower electrophoretic mobilities of the colloidal particles and hence a reduced destabilization efficacy. The most substantial effect is observed for hydrolysis times in the range $0 \leq \Delta t \leq 1$ s, whereas for $\Delta t \geq 6$ s almost no influence of Δt on EM can be detected.

The experimental results confirm the predictions of Amirtharajah and Mills (1982) on critical dispersion times for adsorption–destabilization, which were based on earlier attempts of Hahn and Stumm (1968) to establish the rate of the reactions between the hydrolysis products of aluminum and a colloidal suspension. It is obvious that the coagulants must be dispersed in the raw water stream in less than 0.1 s to achieve the most efficient particle destabilization. Longer dispersion times will result in the formation of hydroxocomplex species of gradually decreasing charge-neutralization capacity. It is interesting to note that hydrolysis times $\Delta t > 6$ s cause no further change of the electrophoretic mobility. Thus, earlier assumptions (Amirtharajah and Mills 1982) that extremely short dispersion times and high intensities of mixing are not crucial for sweep coagulation as compared with adsorption–destabilization are confirmed.

Some interesting conclusions about the significance of mixing for the particle destabilization process can be drawn from the data presented in Figure 6-22. In this figure, the aluminum doses required for three different degrees of destabilization are plotted against the concentration of the aluminum stock solution. The different degrees of destabilization are characterized by an EM of the particles of -0.5, ± 0, and $+0.5$ $\mu m \cdot cm/V \cdot s$, which correspond to incomplete destabilization, optimum destabilization, and restabilization. To achieve the same degree of destabilization or restabilization when the stock solution is more concentrated, higher aluminum doses are required. This is apparently due to reduced formation of those hydrolysis species, which have the highest destabilization capacity; i.e., polymerized species of high positive charge density. The deficiency of these species has to be compensated by higher aluminum doses. For example, the necessary aluminum dose to achieve

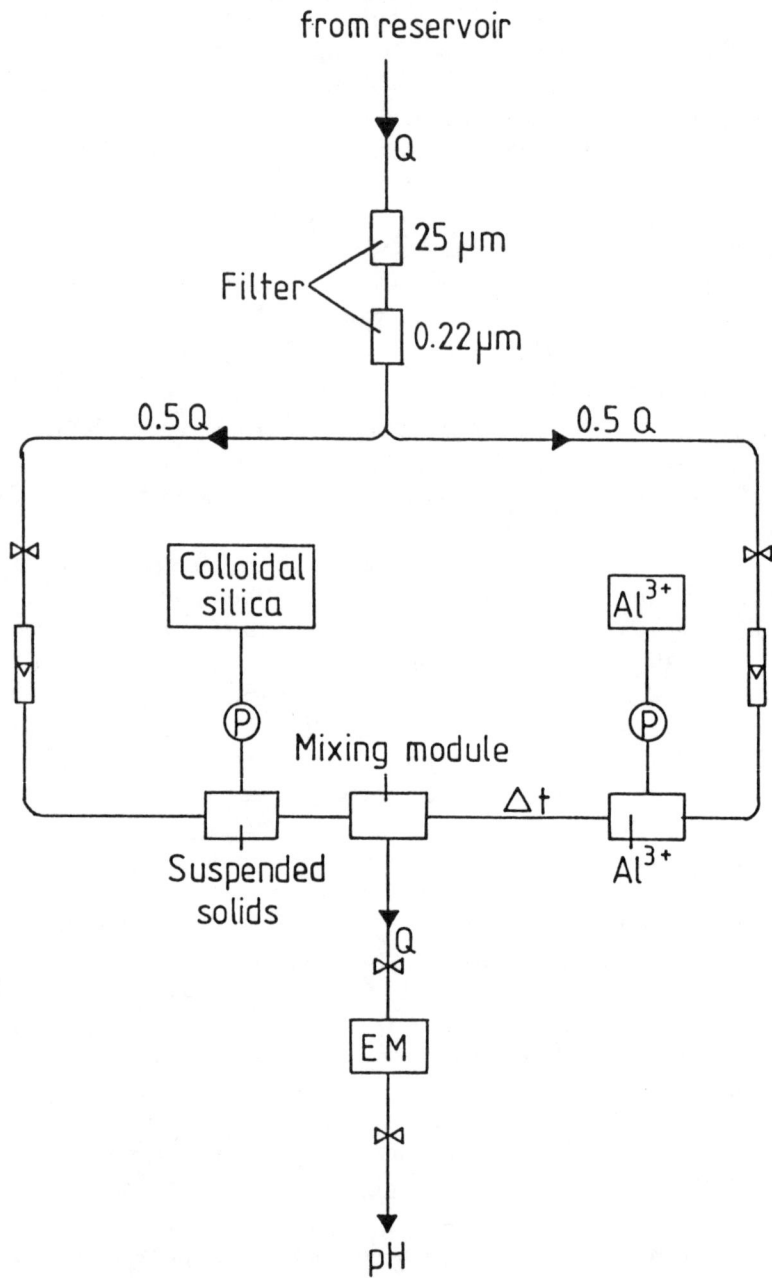

Source: Klute, R. & Dierschke, M. 1990. Kinetics of Destabilization With Al(III) in Turbulent Pipe Flow. Z. Wasser-Abwasser-Forschung, (in press).

Figure 6-20 Effect of hydrolysis time Δt on particle destabilization: experimental setup.

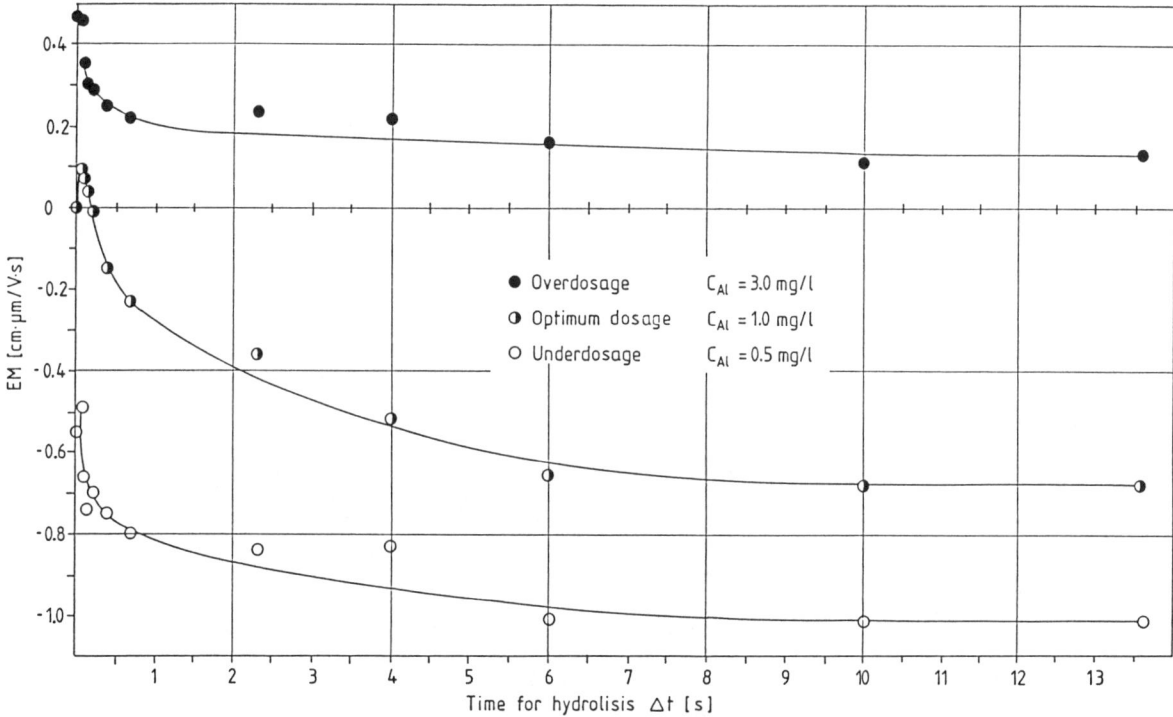

Source: Klute, R. & Dierschke, M. 1990. *Kinetics of Destabilization With Al(III) in Turbulent Pipe Flow.* Z. Wasser-Abwasser-Forschung, (in press).

Figure 6-21 Destabilization efficiency as function of hydrolysis time Δt.

complete destabilization (EM = ± 0 μm · cm/V · s) increases by a factor of approximately four if the concentration of the stock solution is raised from 0.5 to 15 g Al^{3+}/L.

It is interesting to note that the curve for EM = +0.5 μm · cm/V · s shows a steeper increase than those for EM = ± 0 and EM = –0.5. However, the total amount of Al^{3+} required for destabilization or restabilization increases by a factor of approximately four for all curves, if the concentration of the aluminum stock solution is raised from 0.5 to 15 g/L. Since the experiments were performed under identical flow conditions, it can be concluded that there is a direct dependence between the formation of aluminum hydrolysis species and their adsorption onto the particle surface and the turbulence intensity.

Doll (1986) studied the kinetics of adsorption of polyelectrolytes on silica particles in turbulent pipe flow. The adsorption or charge neutralization was monitored by measurements of the electrophoretic mobility of the particles as a function of reaction time. The experiments were performed under different flow rates and with three different cationic polyelectrolytes, characterized by significant differences in molecular weight.

In Figure 6-23 the measured reaction rates $R_D = 1/t_{R_0}$ are plotted against the $\overline{G}$ values for each of the three polyelectrolytes. In the figure the value t_{R_0} is the time necessary to achieve complete neutralization of the surface charge and can be

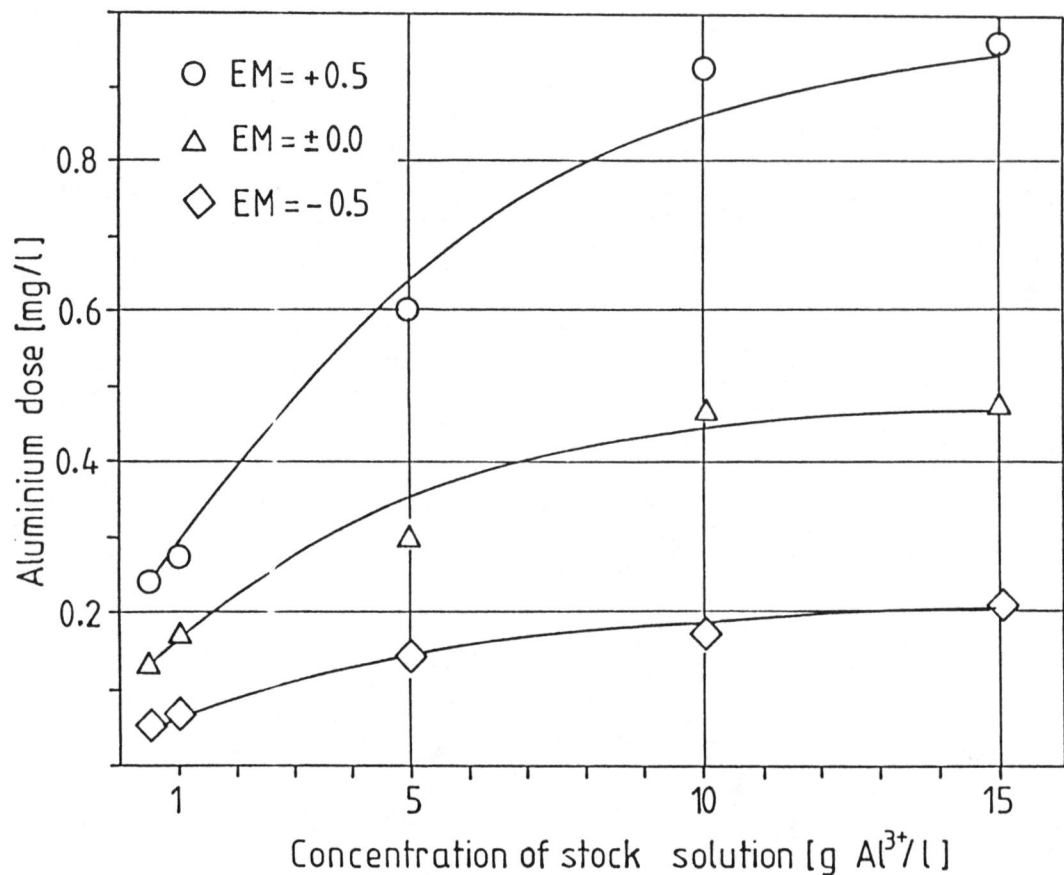

Source: Klute, R. & Dierschke, M. 1990. Kinetics of Destabilization With Al(III) in Turbulent Pipe Flow. Z. Wasser-Abwasser-Forschung, (in press).

Figure 6-22 Influence of stock solution concentration on destabilization efficiency.

considered as a characteristic reaction time for the specific destabilization process. $\overline{G} = (\varepsilon/\nu)^{1/2}$ is the turbulent root mean square velocity gradient.

As indicated by the results shown in Figure 6-23, a higher R_D value (shorter time to achieve complete neutralization of the negative surface charge of the particles) is caused by a higher flow rate under otherwise identical conditions. The shape of the curves suggests a nearly linear dependence of the reaction rate on the turbulent velocity gradient over the range of $\overline{G}$ values studied:

$$R_D = C_d \overline{G} \qquad \text{(Eq 6-10)}$$

However, the slopes of these lines, which describe the neutralization reaction, become flatter with lower molecular weights; i.e., longer reaction times are necessary for lower-molecular-weight polymers under otherwise identical conditions. This finding indicates that the sizes of the polyelectrolytes (represented through the molecular weight and the actual hydrodynamic coil parameter), the charge density, and the concentration (number basis) all influence the course of the reaction.

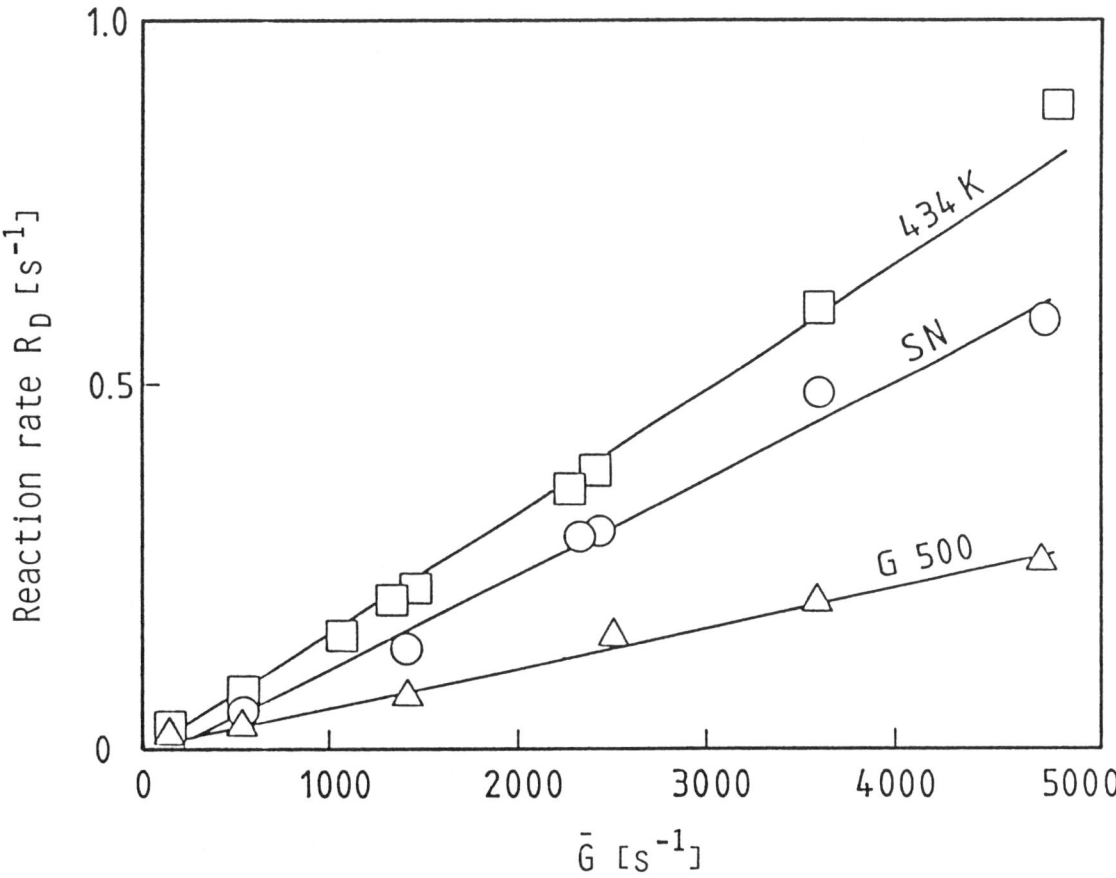

Source: Doll, B. 1986. *Die Kompensation der Oberflachenladung kolloidaler Silika-Suspensionen durch die Adsorption kationischer Polymere in turbulent durchstromten Rohrreaktoren.* Doctoral dissertation, Univ. Karlsruhe, Germany.

Figure 6-23 Reaction rate for particle destabilization with polyelectrolytes as function of $\overline{G}$ value.

The role of turbulence in the destabilization reaction is both to mix the polyelectrolyte throughout the suspension and to bring the polyelectrolyte molecules and the particles together. Turbulent motion consists of different scales of motion; the turbulent energy is dissipated primarily within eddies that have the size of the smallest scale of turbulence, known as the Kolmogoroff microscale. Since the experimental results showed a linear dependence of the reaction rate R_D on $\overline{G}$, this Kolmogoroff microscale, or the dissipation length η, was chosen as the scale of influence for the characteristic reaction rate R_D in Figure 6-24. This length scale decreases with increasing turbulence, since it is calculated according to Eq 6-1.

The values of the dissipation length were normalized with the average particle diameter on the abscissa of Figure 6-24. This normalization permitted a dimensionless representation; a limiting value for the ratio is one, since, at lower eddy sizes, the particle would not fit within an eddy. It is obvious in the figure that the reaction rate decreased dramatically as the dissipation rate decreased and approached the particle size. Extrapolation to the value of $\eta/d_p = 1$ suggests that the reaction rate could have been 15 to 20 times higher for various polymers. An extension of this interpretation suggests that a maximum rate of reaction—i.e., the highest efficiency

240 MIXING

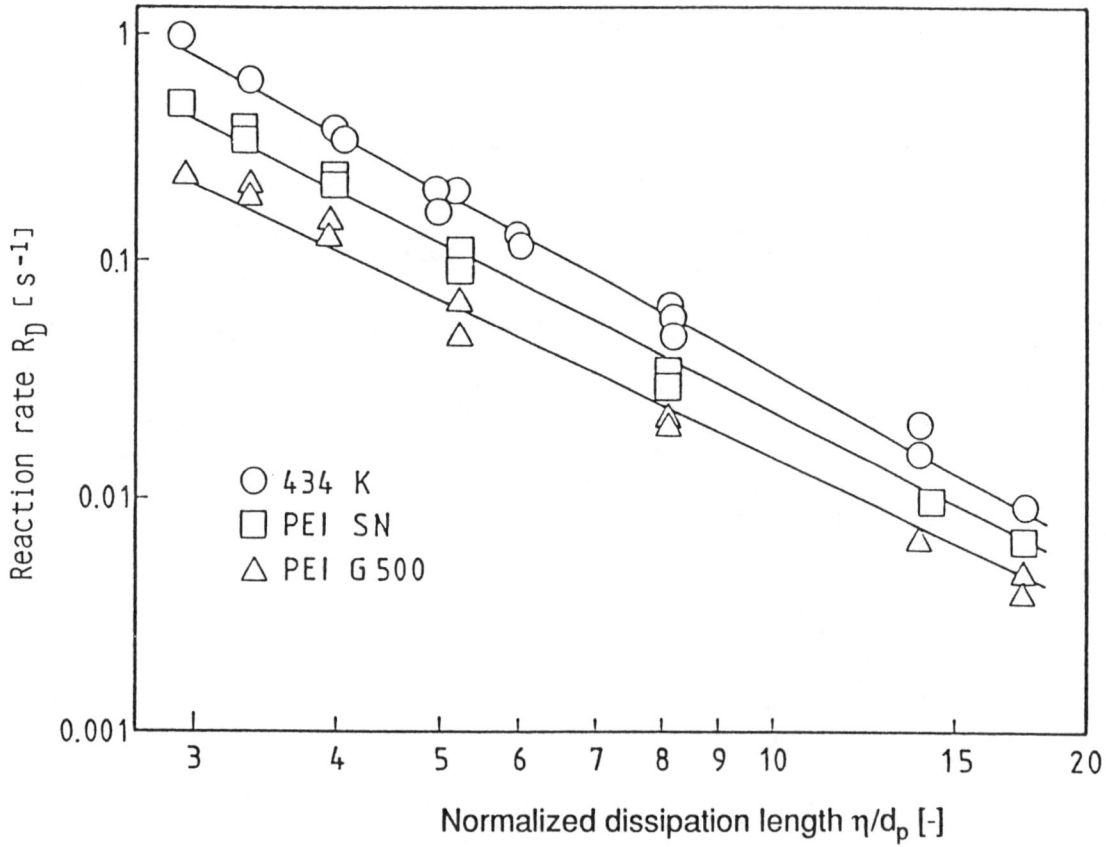

Source: Doll, B. 1986. Die Kompensation der Oberflachenladung kolloidaler Silika-Suspensionen durch die Adsorption kationischer Polymere in turbulent durchstromten Rohrreaktoren. Doctoral dissertation, Univ. Karlsruhe, Germany.

Figure 6-24 Reaction rate for particle destabilization with polyelectrolytes as function of normalized dissipation length η/d_p.

in particle destabilization with polymers—would occur as the dissipation length η, which is inversely related to the degree of turbulence, approaches the size of the diameter of the particles. This leads to the conclusion that a high-performance rapid-mix device has to satisfy the following conditions:

- The turbulent flow conditions during the destabilization reaction must result in a dissipation length that is the same size as the particle diameter.
- Local overdose conditions must be avoided since they lead to a possible restabilization of a certain proportion of the particles and hence a reduced destabilization efficiency.

These results are contradictory to the research of Amirtharajah and Trusler (1986), who indicated that a minimum destabilization rate occurred with inorganic coagulants when the ratio of microscale to particle diameter (η/d_p) was in the range of 2.0 to 1.33 in the zone of maximum turbulence around the mixing blades of a backmix reactor (see chap. 1). The differences in the results may be due to the nonisotropic nature of the turbulence in the backmix reactor as compared to turbulent pipe flow, or it may be due to different destabilization mechanisms dominating with inorganic coagulants as contrasted with organic polymeric coagulants. Additional well-designed experimental research is needed to reconcile these differences.

Kinetic Aspects of Particle Aggregation in Turbulent Pipe Flow

There are only a few investigations reported in the literature on particle aggregation processes in turbulent pipe flow. Tambo and Ogasawara (1970) were the first to use a pipe flow reactor for coagulation and flocculation studies. Delichatsios and Probstein (1975a) carried out coagulation rate experiments in fully developed turbulent pipe flow in order to test a theoretical coagulation model developed for destabilized particles in an isotropic turbulent flow. In the coagulation experiments an approximately monodisperse latex dispersion was used in which the particle sizes were less than the Kolmogoroff microscale. In the experiments the flow rate, volume fraction of the dispersed phase, and the Ca^{2+} concentration used for particle destabilization were varied. The coagulation efficiency for partially destabilized systems appeared to be independent of the particle transport mode—i.e., Brownian motion or turbulent flow. On the basis of these results Delichatsios and Probstein (1975b) presented scaling laws that delineated the regimes in which coagulation, breakup, and sedimentation were predominant and proposed a new design concept for a coagulation and clarification system using turbulent pipe flow. To reduce the length of pipe required for the system, the design provided for a fractional recirculation of the coagulated dispersion.

Klute (1977) performed experiments on the flocculation of silica particles with a cationic polyelectrolyte in turbulent pipe flow, using a pipeline 200 ft (60 m) long and 0.03 ft (10 mm) in diameter for particle destabilization and aggregation. He found that a variation of the pipe flow Reynolds number from Re = 8000 to Re = 16,000 resulted in an increased reaction rate for particle aggregation (Figure 6-25). The aggregation process during the initial reaction phase could be represented by a first-order rate law with rate constants in the order of $k = 10^{-2}$ s^{-1}. These rate constants were two orders of magnitude higher than those for flocculation processes in jar test reactors.

Pfeifer (1986) extended the flocculation experiments into the region of higher Reynolds number with $Re_{max} = 43,000$. The increase in reaction rates with increasing flow rates could be confirmed for the range of lower Reynolds numbers. However, it is obvious from the data presented in Figure 6-26 that in the region of highly turbulent flow the rate constants for particle aggregation decrease again. This effect may be attributed to a reduced collision efficiency of the primary particles and/or disruption of microflocs if the turbulence intensity in the pipe reactor exceeds a certain critical value.

Grohmann (1981, 1985) reported on flocculation experiments in pipes of 0.03 to 2 ft (8 to 600 mm) diameter. In these experiments colloidal dispersions (humic acid, silica particles, activated carbon) were destabilized by hydrolyzing salts and an anionic polyelectrolyte. Under steady-state conditions a decreased floc size was observed with an increasing flow velocity. The addition of an anionic flocculation aid resulted in the formation of larger flocs. Typical values obtained were flocs of 0.08 to 0.12 in. (2 to 3 mm) diameter with a settling velocity of 30 ft/h (10 m/h). According to these results, a $\overline{G}$ value of 70 s^{-1} and an overall detention time of 160 s were identified as proper design parameters for the systems studied.

242 MIXING

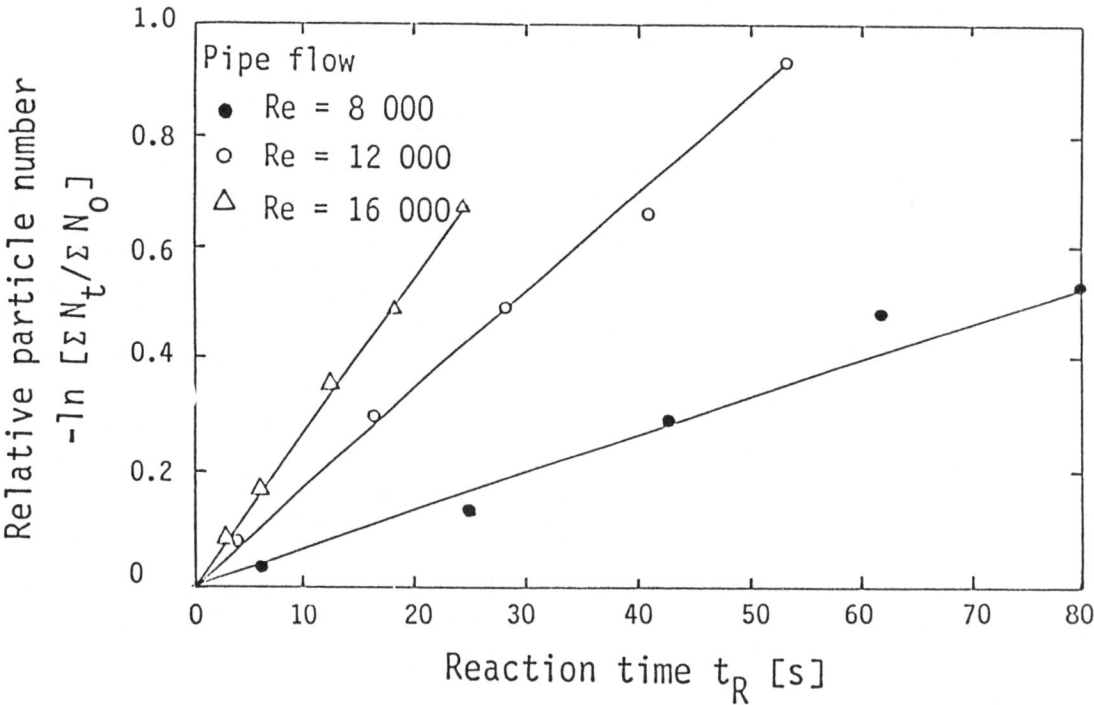

Source: Klute, R. 1977. Adsorption von Polymeren an Silicaoberflachen bei unterschiedlichen Stromungsbedingungen. Doctoral dissertation., Univ. Karlsruhe, Germany.

Figure 6-25 Particle aggregation in turbulent pipe flow as function of reaction time.

In conclusion, design parameters for coagulation and flocculation processes in turbulent pipe flow are summarized in Table 6-1 on the basis of data reported in the literature. Until now, there has only been limited experience in pipe coagulation and flocculation, especially in plant-scale application, since it is difficult to measure particle destabilization and microfloc formation within the short reaction times involved in these processes. In addition, the exact characterization of turbulence intensity in large-diameter pipes complicates the analysis. However, the magnitude of the design parameters presented in Table 6-1 are valid within an order of magnitude.

There seems to be sufficient experimental basis for the design data proposed in Table 6-1 for particle destabilization. However, it may be difficult to transfer these data based on laboratory-scale experiments to real plants. One of the limitations in using these results for plant design is that the high $\overline{G}$ values required can only be obtained by an adequate pressure drop. On the other hand, complete mixing between chemicals and raw water within short time intervals, which has proven to be crucial for optimum destabilization, will be extremely difficult to realize in practice (*see* Dittmann et al. 1989). Therefore, for practical purposes it seems reasonable to try to achieve complete mixing within the smallest time interval that is technically feasible.

With regard to microfloc formation and the generation of macroscopic flocs, the proposed design data for $\overline{G}$ values and reaction times denote only an approximate range of most favorable values. These values would depend on the physicochemical

PARTICLE DESTABILIZATION 243

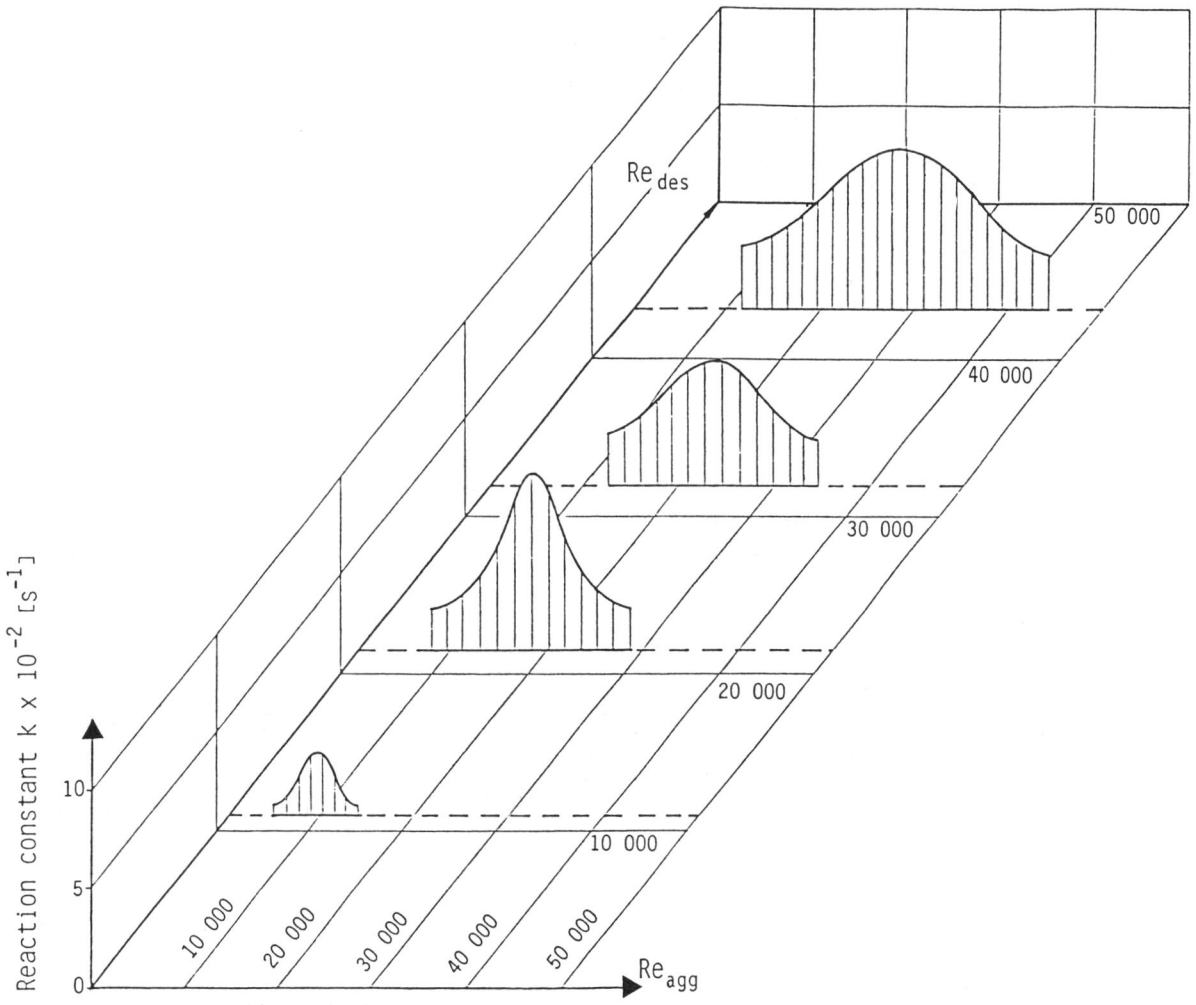

Source: Pfeifer, R. 1986. Particle Aggregation in Turbulent Pipe Flow. Unpublished data.

Figure 6-26 Reaction constants for particle aggregation in turbulent pipe flow.

Table 6-1 Design Parameters for Coagulation and Flocculation Processes in Pipes

Reaction	$\overline{G}$-Value, s^{-1}	Reaction Time, s
Particle destabilization with		
—hydrolyzing salts	>1000*	<0.1*
—polyelectrolytes	>1000†	~1‡
Particle aggregation to		
—microflocs	200–500	15–60§
		95**
—macroflocs	50–100	120–240

*Klute and Dierschke (1990)
†Doll (1986)
‡Klute (1977)
§Klute (1985)
**Grohmann (1981)

boundary conditions imposed by the raw water quality as well as the floc separation process. It is recommended that the optimum design data be determined on the basis of pilot-plant experiments with the specific raw water (see chap. 10).

PRACTICAL EXPERIENCE WITH PIPE MIXERS

Pipe Expansion

A review of the literature on particle destabilization and aggregation in turbulent pipe flow shows that until now only a very limited number of plant-scale studies have compared the efficiency of different designs of rapid-mix units. Klute (1985) reported on an investigation in the small water works of Neckartailfingem in Germany, where water from the river Neckar and from a eutrophic lake is treated in order to supplement the scarce groundwater. After an initial purification step consisting of filtration using a microstrainer with a 20-µm sieve, the water was introduced through a pipe with a diameter of 3 in. (80 mm) into a pressure filter. The filter had a diameter of 8 ft (2.5 m) and was operated under a pressure of 3.5 bar and a filter velocity of 13 ft/h (4 m/h). Polyaluminum chloride was used as coagulant and dosed into the pipe leading to the filter. Destabilization and the formation of microflocs occurred in the pipe during the given detention time of approximately 60 s. The growth of the macroscopic floc began only at the head of the filter (detention time approximately 4 min). The effect of different coagulant injection configurations was evaluated on the basis of electrophoretic mobility measurements and suspended solids removal. The electrophoretic mobility of the destabilized particles and microflocs was determined in samples withdrawn from the pipe about 9 s after coagulant feed. Samples taken at regular time intervals from the raw water and the filter effluent were also analyzed for turbidity and total number of algae by microscopic counting.

To clarify the role of turbulence by pipe expansion in determining particle destabilization and aggregation behavior, two different mixing methods were compared:

- injection of the coagulant into the wall region of a straight pipe with a diameter of 3 in. (80 mm), and
- injection of the coagulant in the pipe center 0.1 s upstream of a pipe expansion from 2 in. to 3 in. (50 mm to 80 mm).

In Figure 6-27, data for turbidity reduction and algae removal by coagulation and direct filtration are presented. The results, obtained under identical conditions except for the type of mixing unit, indicate that higher removal for particulate material occurs with the pipe-expansion mixer, which can be attributed to the more uniform and rapid mixing of the coagulant into the main flow. This conclusion is also supported by the electrophoretic mobility data presented in Figure 6-27, indicating a direct correlation between removal efficiency and remaining negative surface charge on the particles.

As already noted, a critical aspect in the design of rapid-mix devices in pipes is the turbulence intensity at the point of injection of the coagulants and the lag between the dosing point and zones of high turbulence intensity in the plug-flow reactor. To estimate this influence under conditions of practical operation, the coagulant was injected into the pipeline 5.6 ft (1700 mm) and 1 ft (300 mm) upstream of the pipe expansion. This corresponds under the chosen steady-state flow conditions (flow rate $Q = 20$ m^3/h) to a lag of 0.7 s and 0.1 s respectively until coagulant solution and surrounding water arrive at the zone of high turbulence.

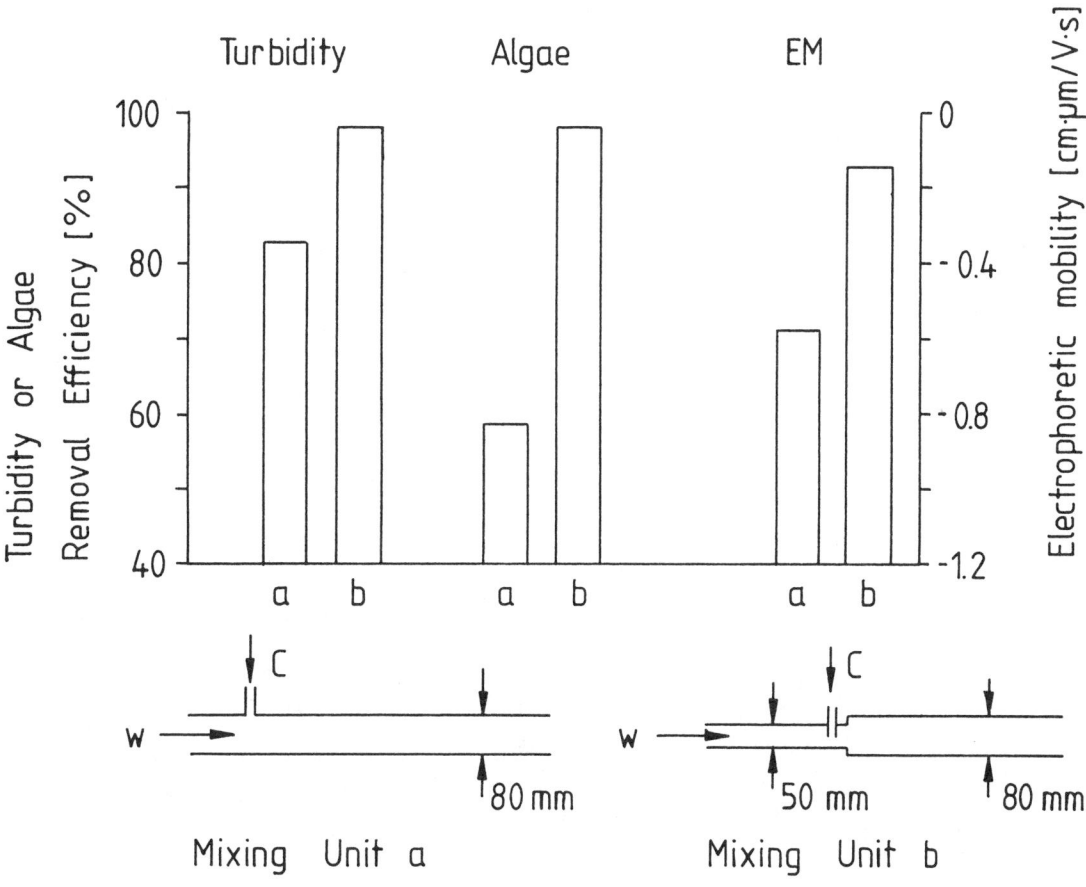

Source: Klute, R. 1985. Rapid Mixing in Coagulation–Flocculation Processes: Design Criteria. In Chemical Water and Wastewater Treatment, *ed. by A. Grohmann, H.H. Hahn, and R. Klute. Gustav Fischer Verlag, Stuttgart, Germany.*

Figure 6-27 Comparison of direct filtration removal efficiency for different dosing modes in turbulent pipe flow.

In Figure 6-28, turbidity and algae number in the filter effluent are plotted as functions of coagulant dose. In addition, the corresponding electrophoretic mobility values of the flocs are shown. The data clearly indicate the importance of instantaneous mixing after the injection of the coagulant solution.

If the complete dispersion of the coagulant in the raw water stream is accomplished with a lag of 0.7 s instead of 0.1 s after injection, a lower removal efficiency with respect to turbidity and algae number can be observed. It is obvious from the electrophoretic mobility data in Figure 6-28 that this is due to an incomplete destabilization of the particles. By application of a higher coagulant dose, the deficiency in destabilization can be compensated to a certain degree. However, this will increase the chemical costs significantly.

Comparison of Agitator and Ejector System

Ventresque and Bablon (1988) reported on practical experience with a coagulant injection system, installed in a water works at Neuilly-sur-Marne in France. This water works is designed for the treatment of 210 mgd (800,000 m³/day) from the

246 MIXING

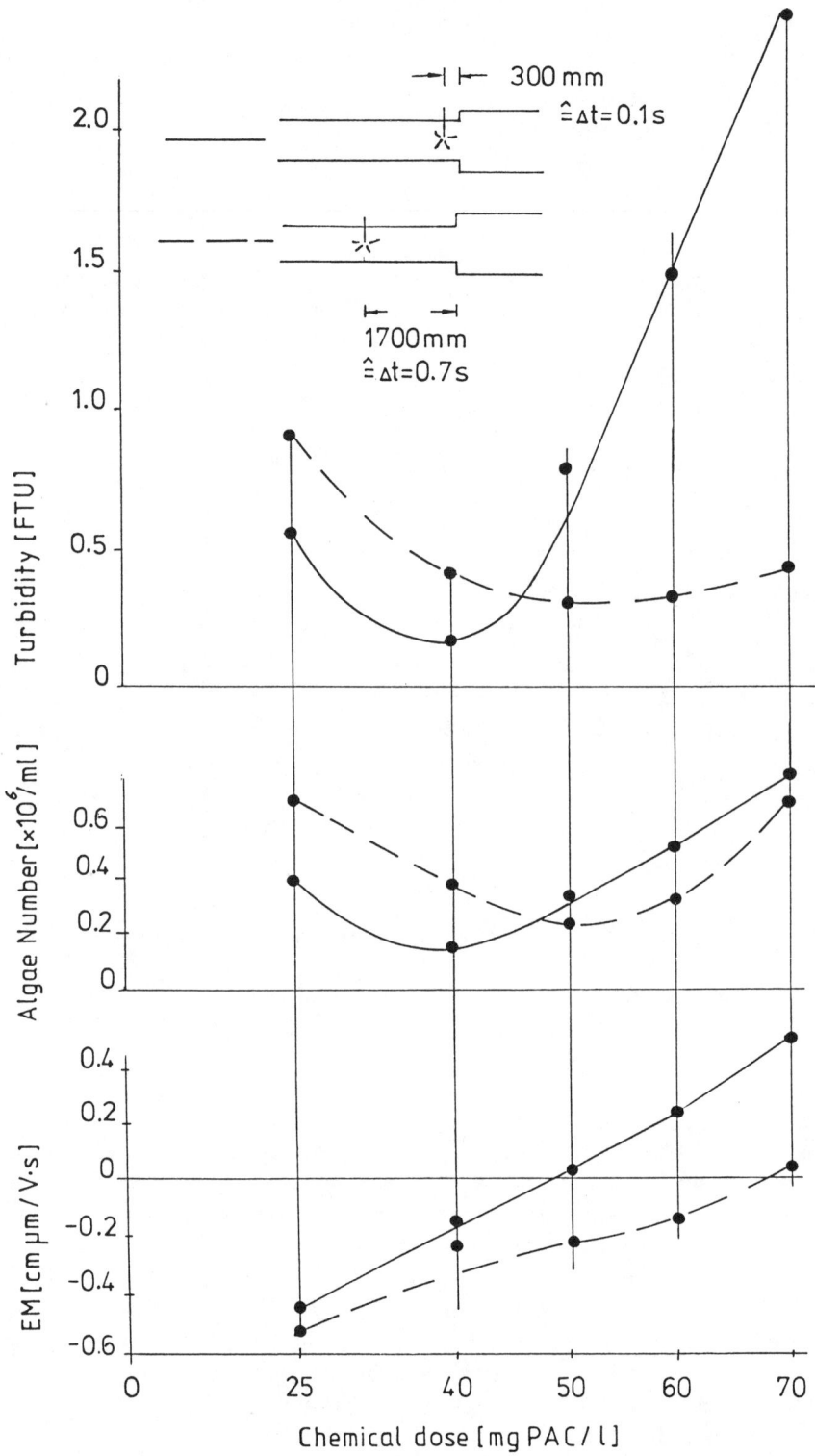

Source: Klute, R. 1985. Rapid Mixing in Coagulation-Flocculation Processes: Design Criteria. In Chemical Water and Wastewater Treatment, ed. by A. Grohmann, H.H. Hahn, and R. Klute. Gustav Fischer Verlag, Stuttgart, Germany.

Figure 6-28 Particle destabilization and direct filtration removal efficiency as function of chemical dose.

river Marne and supplies nearly 2 million inhabitants of the eastern suburbs of Paris.

In the past, the coagulant for the pretreatment process was fed into an agitated tank with a volume of 37,000 gal (140 m^3) and a detention time of about 30 s (see Figure 6-16). The average velocity gradient $\overline{G}$ generated was 400 s^{-1}, calculated on the basis of the power consumption of the agitator. This coagulant mixing device turned out to be somewhat ineffectual, especially during periods of algae growth. To improve destabilization, a coagulant injection system (Figure 6-16) was installed, based on the system recommended by Chao and Stone (1979). The coagulant was diluted prior to dosing with injection water to ensure the applicability of the system in large-diameter pipes. It is reported that mixing of the coagulant with the injection water was achieved in less than 0.1 s to avoid hydrolysis of the coagulant before it came into contact with the water to be treated.

Since the plant could be divided into several parallel trains, it was possible to run experiments with the agitator and ejector mixing device simultaneously. This was done by connecting two pilot sand filters with an area of 22 ft^2 (2 m^2) and a filter velocity of 20 ft/h (6 m/h) to the agitator and ejector treatment trains. Samples withdrawn from the filter at different depths showed lower aluminum concentrations and residual turbidity values for the ejector system (see Figure 6-29). A comparison of the zeta potential suggests that the improved quality of the treated water was due to a higher destabilization efficiency.

Comparative Studies With Different Mixing Units

Investigations into the effect of different pipe mixing units on the distribution of coagulants over the pipe cross section were reported by Dittmann et al. (1988). The full-scale experiments (flow V_R = 53 ft^3/s [1.5 m^3/s]) were performed at the phosphorus elimination plant at Berlin-Tegel, which was constructed for the removal of phosphorus and particulate material from the Lake Tegel inflow. Lake Tegel is a recreation area and a drinking water reservoir of importance for the population of

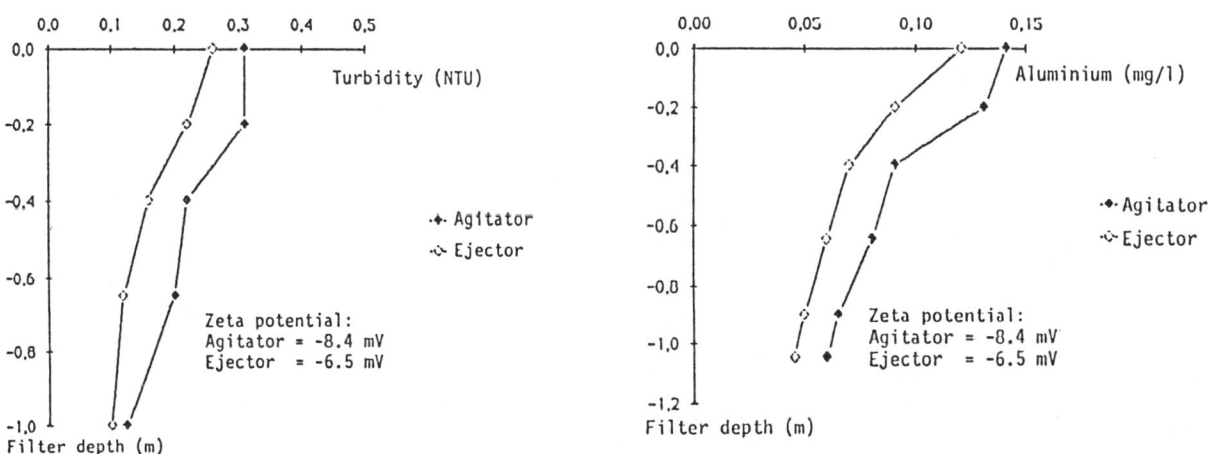

Source: Ventresque, C. & Bablon, G. 1988. *New Coagulant Injection Process.* In Pretreatment in Chemical Water and Wastewater Treatment, ed. by H.H. Hahn and R. Klute. Springer-Verlag, Heidelberg, Germany.

Figure 6-29 Turbidity and total aluminum profiles in filter in terms of depth: comparison between two mixing devices.

West Berlin. The flow sheet of the treatment plant with respect to coagulation and precipitation includes the following stages:

1. injection and mixing of coagulants with the raw water and formation of microflocs in two pipes (diameter of 3 ft [1 m], retention time of approximately 30 s);

2. formation of macroflocs in pipes (retention time approximately 100 s, $Re_{max} = 1 \times 10^6$);

3. sedimentation in Sulzer Rotopur high-rate sedimentation basins (retention time 12 to 30 min); and

4. filtration by a multilayer pressure filter.

Aluminum or iron salts are used as coagulants. The additions of cationic and anionic polyelectrolytes are carried out after the microflocculation and the sedimentation steps respectively.

The purpose of the investigation was to compare, under normal operating conditions, the mixing efficiency of five different mixing units installed into pipelines of 3.3 ft (1 m) diameter. These five mixing units were selected on the basis of experimental results reported in the literature. The five units are described in the following, and schematic figures of the units are shown in Figure 6-30.

High-velocity jet injection. The dispersion of chemicals within the water to be treated was achieved by the velocity difference between the main stream and the jet of chemicals, which was introduced by small nozzles. For nozzle diameter $d_n > 1$ mm, jet velocities v_j equal to 10 to 30 times the pipe flow velocity v_p, and a pipeline diameter d_p of 100 mm, the jet is reflected by the opposite wall of the pipeline, thus leading to the best mixing effect, as reported by Grohmann, Hasselbarth, and Langer (1977). With smaller nozzle diameters the necessary jet velocity increases exponentially.

Turbo mixer. The in-line turbo mixer was installed in the center of the pipe. The coagulant was injected in front of a cone disk, which rotates at high speed and brings about a rapid dispersion of the coagulant over the cross section of the pipe. Stendahl (1985) performed pilot studies comparing the overall process efficiency of different mixing devices. He reported that the turbo mixer had proven to be superior to the reference propeller mixer with respect to turbidity, color, suspended solids, and residual aluminum concentration.

Orifice plate. The installation of an orifice plate within the pipe was proposed by Bratby (1980). The function of this plate was the generation of an elevated velocity gradient at and downstream of the point of coagulant addition. According to Faison, Davis, and Achenbach (1967), the mixing process takes place in a space $L \approx 4$ to 5 d_t (d_t = pipe diameter). Complete mixing is only achieved for a orifice-to-pipe ratio of $d_o/d_t < 0.5$ (d_o = orifice diameter). However, the orifice diameter is limited by the resulting head loss. The coagulant was injected at the pipe center 1.6 ft (50 cm) upstream of the plate, which had an orifice diameter of 3.3 ft (70 cm) and was installed within a pipe of 3.3 ft (1 m) diameter.

Pipe expansion with center injection. The mixing characteristics downstream of a sudden pipe expansion are similar to those of an orifice within a pipe. As already discussed, the highest effluent quality by direct filtration was achievable, if the coagulant injection point was installed in the pipe center immediately before the pipe expansion (Klute 1985).

The coagulant source was placed in the pipe center 0.7 ft (20 cm) upstream of the pipe expansion from 2.3 to 3.3 ft (700 to 1000 mm). With the given flow of $Q = 53$ to 106 ft^3/s (1.5 to 3 m^3/s), the conditions for rapid dispersion of the coagulant (≤ 0.1 s) were met by this arrangement.

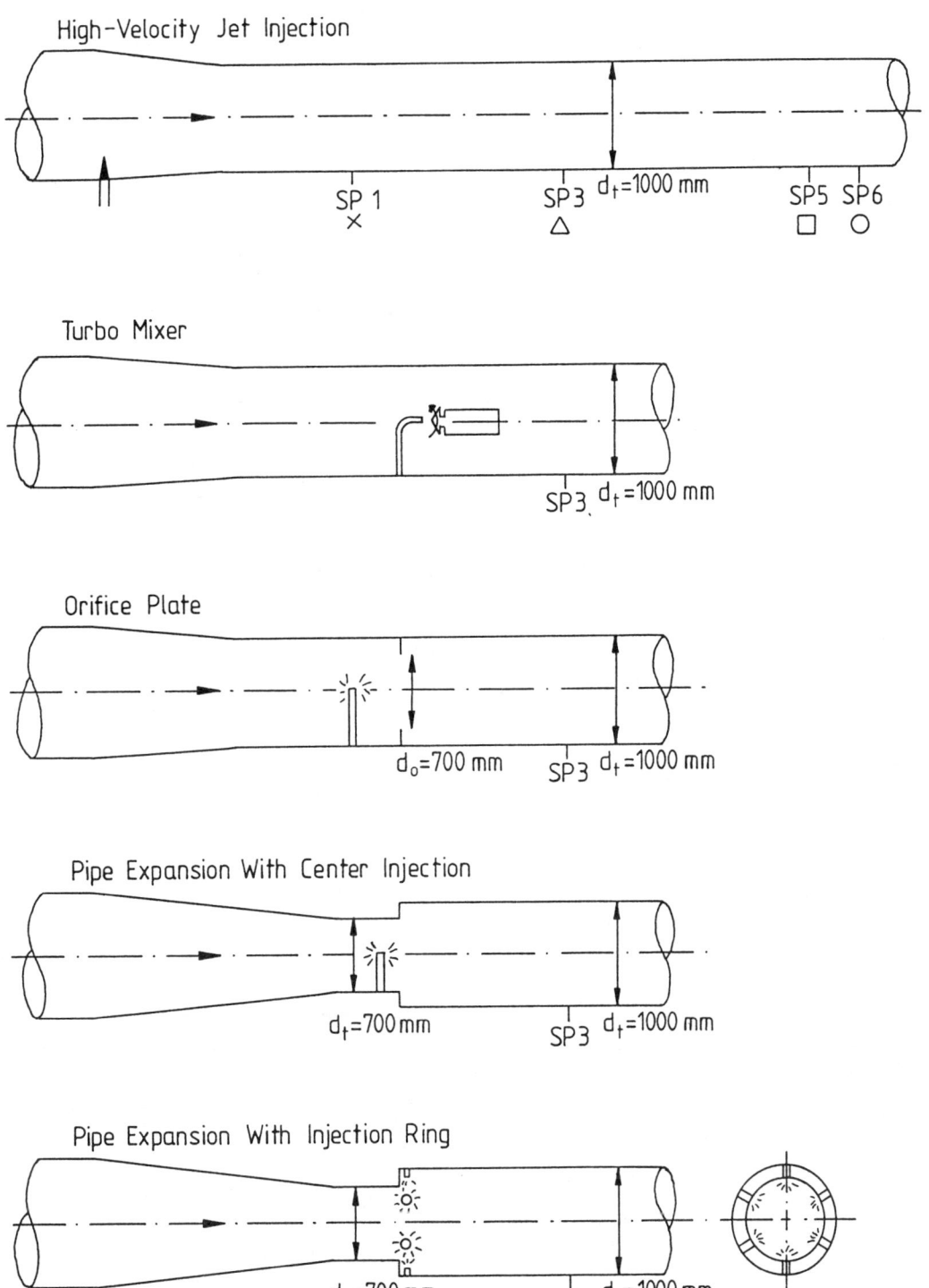

Source: Dittmann, W. et al. 1989. Vergleichende Untersuchungen uber den Einfluss von Mischprozessen auf die Wirkung von Flockungsmitteln bei der Wasseraufbereitung. Rept. BMFT 02-WT 438, Federal Ministry of Research, Bonn, Germany.

Figure 6-30 Mixing devices for comparative plant-scale studies.

Pipe expansion with injection ring. In large-diameter pipes, a rapid and complete dispersion of coagulant is difficult to achieve. Vrale and Jorden (1971) reported that an increased coagulation efficiency was obtained by injecting the coagulant via a six-hole ring placed within the expansion zone of a pipe. Therefore, in the full-scale comparative study a ring with six injection nozzles was installed at the 2.3/3.3-ft (700/1000-mm) pipe expansion. Contrary to the mixing device of Vrale and Jorden (1971) with parallel injection, the coagulant was injected perpendicular to the main flow direction to improve dispersion.

A detailed picture of the distribution of a coagulant solution injected into a pipeline can be obtained by pH measurements downstream of the injection point. The addition of acidic Al or Fe solution leads to a decrease of the pH as a function of the alkalinity of the raw water and coagulant concentration. Based on pH measurements and raw water data, the concentration profile over the pipe cross section can be determined.

In Figure 6-31 the concentrations profiles for two of the mixing devices evaluated are presented. The pipe-expansion mixing unit distributes the chemical evenly over the pipe cross section at sampling points SP5 and SP6, about 13 ft (4 m) downstream of the injection point. In contrast to the pipe expansion mixing ring, the high-velocity jet injection device at these sampling points still shows a completely uneven distribution. This is mainly due to the fact that at the given diameter ratio for nozzle and pipe line of $D_r = 0.024$ the jet does not reach the opposite pipe wall but is turned around into the main flow direction. Thus, for example, with a velocity ratio for jet and main flow of $V_r = 5.6$, the jet penetrates only 1 ft (30 cm) into the stream, as can be concluded by the position of the concentration maxima in Figure 6-31.

The mixing efficiency of different mixing units can be compared by determining the variation coefficient C_v, which is derived from Eq 6-6:

$$C_v = \sigma_c/\overline{C} = \sqrt{\sum_{i=1}^{n} (c_i/\overline{C} - 1)^2 /(n - 1)} \qquad \text{(Eq 6-11)}$$

Where:

σ_c = standard deviation of concentrations
$\overline{C}$ = medium concentration

The calculation of C_v is based on the measurement of the Fe(III) concentration c_i in samples taken at points SP5 and SP6 over the pipe cross section. To compare the five mixing units, the variation coefficient C_v is plotted as function of the Euler number Eu in Figure 6-32. The dimensionless Euler number is the power necessary to achieve a given mixing quality divided by the kinetic energy of the flow:

$$\text{Eu} = \overline{P}/\rho \overline{u}^2 Q \qquad \text{(Eq 6-12)}$$

Where:

$\overline{P}$ = power dissipated in the pipeline
ρ = fluid density
$\overline{u}$ = mean velocity in axial direction
Q = volumetric flow rate

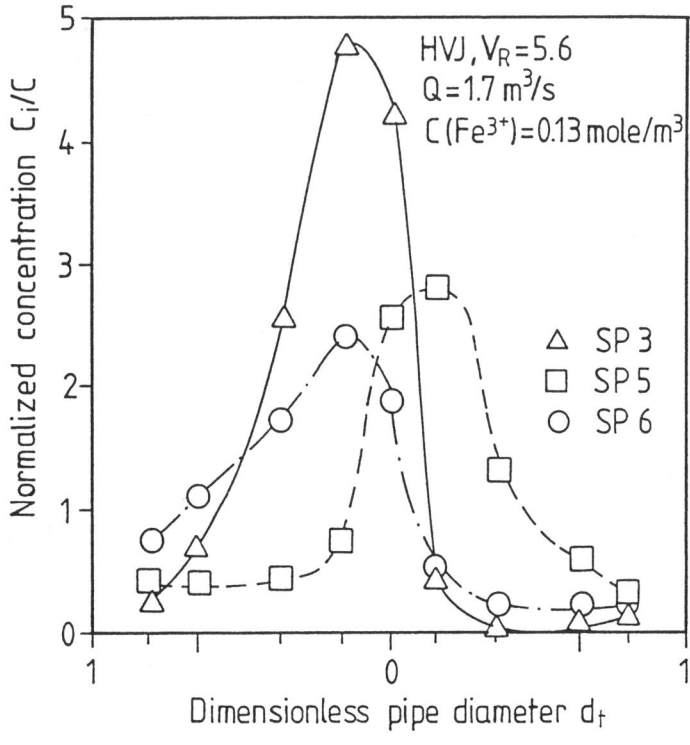

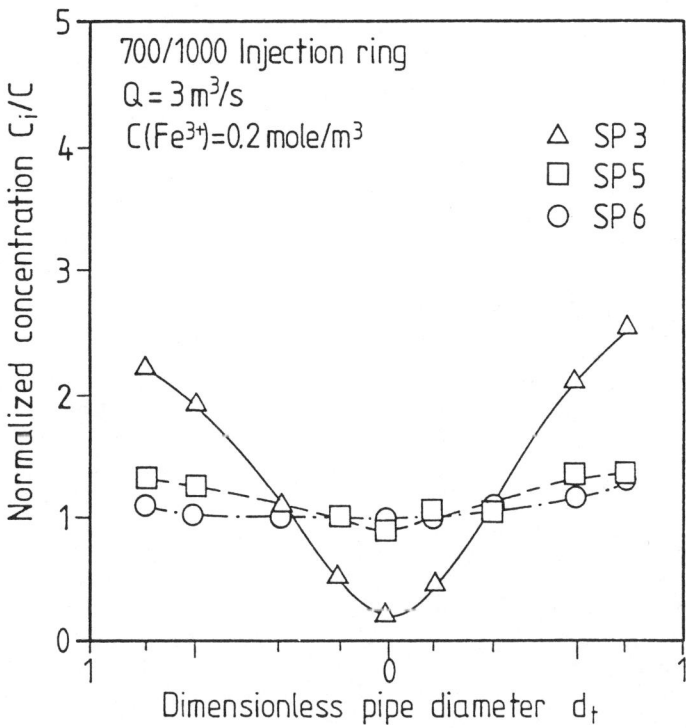

Source: Dittmann, W. et al. 1989. Vergleichende Untersuchungen uber den Einfluss von Mischprozessen auf die Wirkung von Flockungsmitteln bei der Wasseraufbereitung. Rept. BMFT 02-WT 438, Federal Ministry of Research, Bonn, Germany.

Figure 6-31 Concentration profiles over the pipe cross section.

252 MIXING

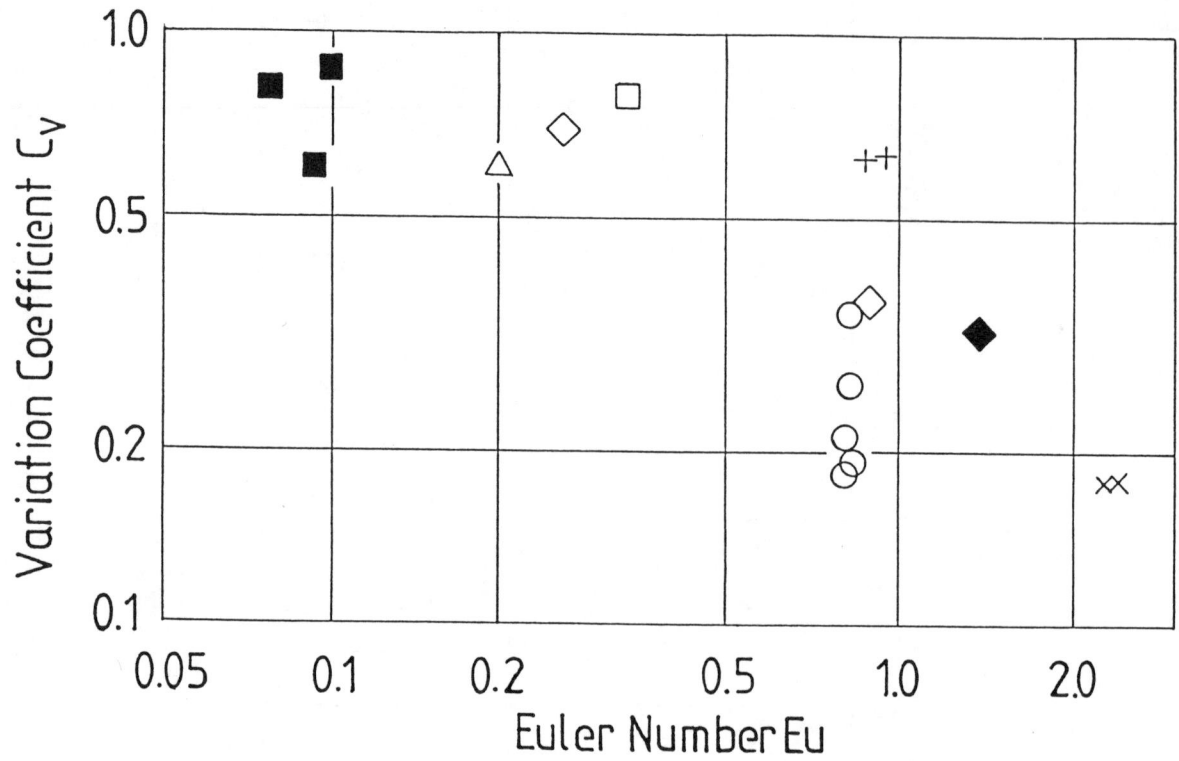

System	Q [m³/s]
■ HVJ, $V_r = 5.6$–6.8	1.5/1.7/2.5
□ HVJ, $V_r \neq 5.6$–6.8	1.5
◆ Turbo mixer 1200 UPM	1.5
◇ Turbo mixer 900 UPM	1.5/2.5
× Orifice plate 700	1.5/2.5
+ Pipe expansion 700/1000 (centre injection)	1.5/2.5
○ Pipe expansion 700/1000 (injection ring)	1.5/1.95/2.5/3.0
△ Injection ring	1.7

Source: Dittmann, W. et al. 1989. Vergleichende Untersuchungen uber den Einfluss von Mischprozessen auf die Wirkung von Flockungsmitteln bei der Wasseraufbereitung. Rept. BMFT 02-WT 438, Federal Ministry of Research, Bonn, Germany.

Figure 6-32 Variation coefficient at sampling points SP5 and SP6 with the Euler number.

For static mixers the power dissipation is given by

$$\overline{P} = \Delta p\, Q \qquad \text{(Eq 6-13)}$$

Where:

Δp = pressure drop

For jet mixers the pressure drops for all jets including the main flow have to be considered:

$$\overline{P} = \sum_{i=1}^{n} \Delta p_i Q_i \qquad \text{(Eq 6-14)}$$

Thus, the following equation results for the Euler number:

$$\text{Eu} = \Delta p / \rho \overline{u}^2 \qquad \text{(Eq 6-15)}$$

Mixing units with high mixing efficiency are characterized by a small variation coefficient achieved at a low Euler number.

The data presented in Figure 6-32 indicate that the mixing efficiency for the static mixing units (pipeline expansion, baffles) is independent of the energy input but influenced by geometry and nozzle arrangement. The variation coefficient for a specific mixing system is constant at a defined distance from the injection point and therefore independent of the flow rate Q. A comparison between a single nozzle at the pipe center and a six-nozzle ring shows that the variation coefficient at an identical Euler number of 0.8 is lower for multiple-nozzle injection. The same variation coefficient can be attained by pipe constriction at an orifice plate, but in that case a significantly higher Euler number of 2.2 and thus a higher pressure loss are required.

The efficient dispersion of chemicals by the high-velocity jet-injection device requires the reflection of the jet at the opposite pipe wall. Since the jet orifice to pipeline diameter is $D_r = 0.024$ in the existing device, a velocity ratio $V_r \geq 17$ will be necessary to fulfill this condition (Grohmann, Haesselbarth, and Langer 1977). Because the maximum attainable $V_r = 11.4$, the jet does not reach the opposite pipe wall. The system operates only as a transverse jet injection system and hence with the lowest mixing efficiency of the five mixing units under evaluation.

A homogeneous dispersion of the chemicals within the raw water stream in less than 0.1 s is considered essential to achieve complete destabilization of colloidal particles with hydrolyzing salts. The homogeneous dispersion is characterized by a variation coefficient $C_v = 0$ and cannot be realized in the five mixing systems in $t \leq 0.1$ s even with the highest flow rate of 3 m^3/s. This leads to the conclusion that only a certain part of the particulate material in the raw water is destabilized by the positively charged hydroxocomplex species, whereas sweep floc is involved as a destabilization mechanism in the remaining portion.

RESEARCH NEEDS

In general, there is only a limited data basis for the design of mixing devices for particle destabilization and aggregation processes in pipes and channels. Further data are needed from comparative studies with different mixing devices using EM measurements and particle counting techniques for the determination of process kinetics as well as for the overall process efficiency after phase separation. In particular, the conceptual distinction of the particle aggregation process into destabilization, formation of microflocs, and growth of macroscopic flocs should be considered in terms of specific mixing requirements.

The effects of coagulant solution concentration, of simple and multiple-point dosing, and of flocculation aids should be included in these studies. Of great interest is the effect of mixing on destabilization reactions with polyaluminum or polyiron salts.

Additional research is needed to determine the mixing intensity in terms of the Kolmogoroff microscale and its relationship to the size of particles to be destabilized. This research should be theoretical and experimental and include inorganic and organic polymers and inorganic salts as coagulants.

For the design of direct filtration unit operations with particle destabilization and aggregation in pipes as a preceding step, additional data on minimum reaction time and turbulence requirements are needed. The effect of baffles and other pipe installations resulting in an increased turbulence intensity should be researched in this context. As far as possible the studies should be performed at full-scale plants or at least in geometrically similar systems so that scale-up methods are applicable.

References

AMIRTHARAJAH, A. 1978. Design of Rapid-Mix Units. In *Water Treatment Plant Design for the Practicing Engineer*, ed. by R.L. Sanks. Ann Arbor Science, Ann Arbor, Mich.

AMIRTHARAJAH, A. & MILLS, K.M. 1982. Rapid-Mix Design for Mechanisms of Alum Coagulation. *Jour. AWWA*, 74:4:210.

AMIRTHARAJAH, A. & TRUSLER, S.L. 1986. Destabilization of Particles by Turbulent Rapid Mixing. *Jour. Envir. Engrg.*, 112:6:1085.

BRATBY, J. 1980. *Coagulation and Flocculation*. Uplands Press Ltd., Croydon, England.

BRODKEY, R.S. 1966. Fluid Motion and Mixing. In *Mixing: Theory and Practice*, Vol. 1, ed. by V.W. Uhl and J.B. Gray. Academic Press, New York.

———. 1975. Mixing in Turbulent Fields. In *Turbulence in Mixing Operations*, ed. by R.S. Brodkey. Academic Press, New York.

CHAO, J.L. & STONE, B.G. 1979. Initial Mixing by Jet Injection Blending. *Jour. AWWA*, 71:570.

CHEN, D.T.L.; LIN, J.T.; & KENNEDY, J.F. 1976. Entrainment and Drag Forces of Deflected Jets. *Jour. Hydraulics Div. ASCE*, 102:HY5:615.

DAILY, J.W. & HARLEMAN, D.R.F. 1966. *Fluid Dynamics*. Addison-Wesley, Reading, Pa.

DELICHATSIOS, M.A. & PROBSTEIN, R.F. 1975a. Coagulation in Turbulent Flow: Theory and Experiment. *Jour. Colloid Inter. Sci.*, 51:394.

———. 1975b. Scaling Laws for Coagulation and Sedimentation. *Jour. Wtr. Poll. Cont. Fed.*, 47:941.

DITTMANN, W. ET AL. 1988. Mixing Systems for Pipe Reactors in Water Treatment. *Vom Wasser*, 70:129.

———. 1989. Vergleichende Untersuchungen uber den Einfluss von Mischprozessen auf die Wirkung von Flockungsmitteln bei der Wasseraufbereitung. Rept. BMFT 02-WT 438, Federal Ministry of Research, Bonn, Germany.

DOLL, B. 1986. Die Kompensation der Oberflachenladung kolloidaler Silika-Suspensionen durch die Adsorption kationischer Polymere in turbulent durchstromten Rohrreaktoren. Doctoral dissertation, Univ. Karlsruhe, Germany.

FAISON, T.K.; DAVIS, J.C.; & ACHENBACH, P.R. 1967. Performance of Square-Edge Orifices and Orifice-Target Combination as Air Mixers. *Building Science Ser.*, 12. National Bureau of Standards, Washington, D.C.

FITZGERALD, S.D. & HOLLEY, E.R. 1981. Jet Injections for Optimum Mixing in Pipe Flow. *Jour. Hydraulics Div. ASCE*, 107:HY10:1179.

GER, A.M. & HOLLEY, E.R. 1976. Comparison of Single-Point Injections in Pipe Flow. *Jour. Hydraulics Div. ASCE*, 102:HY6:731.

GRAY, J.B. 1986. Turbulent Radial Mixing in Pipes. In *Mixing: Theory and Practice*, Vol. 3, ed. by V.W. Uhl and J.B. Gray. Academic Press, London.

GROHMANN, A. 1981. Uber die Anwendung der Flockenbildung in Rohren zur Wasserreinhaltung und Phosphatelimination. *Z. Wasser-Abwasser-Forschung*, 14:194.

———. 1985. Flocculation in Pipes: Design and Operation. In *Chemical Water and Wastewater Treatment*, ed. by A. Grohmann, H.H. Hahn, and R. Klute. Gustav Fischer Verlag, Stuttgart, Germany.

GROHMANN, A.; HAESSELBARTH, U.; & LANGER, W. 1977. Hochgeschwindigkeitsinjection bei der Wasseraufbereitung. *Vom Wasser*, 49:267.

GUVEN, O. & BENEFIELD, L. 1983. The Design of In-line Jet Injection Blenders. *Jour. AWWA*, 75:357.

HAHN, H.H. & STUMM, W. 1968. Kinetics of Coagulation With Hydrolyzed Al(III): The Rate Determining Step. *Jour. Colloid Inter. Sci.*, 28:1:134.

KAWAMURA, S. 1976. Considerations on Improving Flocculation. *Jour. AWWA*, 68:328.

KLUTE, R. 1977. Adsorption von Polymeren an Silicaoberflachen bei unterschiedlichen Stromungsbedingungen. Doctoral dissertation, Univ. Karlsruhe, Germany.

———. 1985. Rapid Mixing in Coagulation–Flocculation Processes: Design Criteria. In *Chemical Water and Wastewater Treatment*, ed. by A. Grohmann, H.H. Hahn, and R. Klute. Gustav Fischer Verlag, Stuttgart, Germany.

KLUTE, R. & DIERSCHKE, M. 1990. Kinetics of Destabilization With Al(III) in Turbulent Pipe Flow. *Z. Wasser-Abwasser-Forschung* (in press).

LAUFER, J. 1954. The Structure of Turbulence in Fully Developed Pipe Flow. Rept. 1174. National Advisory Committee for Aeronautics, Washington, D.C.

LETTERMAN, R.D.; QUON, J.E.; & GEMMELL, R.S. 1973. Influence of Rapid-Mix Parameters on Flocculation. *Jour. AWWA*, 65:716.

LEVENSPIEL, O. 1972. *Chemical Reaction Engineering*, 2nd ed. John Wiley and Sons, New York.

MOLL, H.G. 1985. Fluid Mechanical Principles of Flocculation in Pipes. In *Chemical Water and Wastewater Treatment*, ed. by A. Grohmann, H.H. Hahn, and R. Klute. Gustav Fischer Verlag, Stuttgart, Germany.

OLSON, J.H. & STOUT, L.E. 1967. *Mixing: Theory and Practice*, Vol. II, ed. by V.W. Uhl and J.B. Gray. Academic Press, New York.

PFEIFER, R. 1986. Particle Aggregation in Turbulent Pipe Flow. Unpublished data. ISWW, Univ. Karlsruhe, Germany.

PRATTE, B.D. & BAINES, W.D. 1967. Profiles of the Round Turbulent Jet in a Cross Flow. *Jour. Hydraulics Div. ASCE*, 92:HY6:53.

ROTTA, J.C. 1972. *Turbulente Stromungen*. Teubner, Stuttgart, Germany.

SIMPSON, L.L. 1975. Industrial Turbulent Mixing. In *Turbulence in Mixing Operations*, ed. by R.S. Brodkey. Academic Press, New York.

STENDAHL, K. 1985. Colloid Destabilization in Practice. In *Chemical Water and Wastewater Treatment*, ed. by A. Grohmann, H.H. Hahn, and R. Klute. Gustav Fischer Verlag, Stuttgart, Germany.

STENQUIST, R.J. & KAUFMAN, W.J. 1972. Initial Mixing in Coagulation Processes. SERL Rept. 72-2, Univ. California, Berkeley, Calif.

TAMBO, N. & OGASAWARA, M. 1970. Pipe Flocculator Studies. *Jour. Jpn. Wtr. Wks. Assn.*, 426:29 (in Japanese).

VENTRESQUE, C. & BABLON, G. 1988. New Coagulant Injection Process. In *Pretreatment in Chemical Water and Wastewater Treatment*, ed. by H.H. Hahn and R. Klute. Springer-Verlag Berlin Heidelberg, New York.

VRALE, L. & JORDEN, R.M. 1971. Rapid Mixing in Water Treatment. *Jour. AWWA*, 63:52.

7

Mixing, Breakup, and Floc Characteristics

Norihito Tambo, Dr. Eng., Professor, Department of Sanitary and Environmental Engineering, Hokkaido University, Sapporo, Japan

Roger J. François, D.Sc., Chemical Engineer, Technical Center Steel Cord—N.V. Bekaert, Waregem, Belgium

INTRODUCTION

A hydroxide floc is generally considered as an inert, more or less uniform object. However, in reality a floc has a very complicated structure. Hence, sophisticated measures are necessary to characterize a floc. Characteristics include porosity and density, strength and regrowth ability, structure, and homogeneity. These characteristics are determined by interacting phenomena observed during the life of a floc. Rather than reaching a final condition with fixed characteristics, a floc changes its nature continuously.

Knowledge of how floc characteristics can be manipulated by interacting phenomena yields the possibility of predicting floc characteristics. In this chapter, several characteristic parameters that describe floc nature such as particle size and shape, floc density and porosity, floc structure, floc strength and breakup, and aging with time will be described.

PARTICLE SIZE AND DIAMETER

The size of a spherical homogeneous particle can be defined simply by its diameter. However, almost all particles in a normal water treatment operation are irregular and heterogeneous. For those particles, two aspects of size must be evaluated—the size of the individual particles, and the size distribution of the particles. In this section, evaluation of individual particle sizes will be discussed.

For irregular particles such as flocs, size usually depends on the method of measurement and evaluation. Size is often denoted as a characteristic diameter or an equivalent diameter.

If it is possible to measure three dimensions of a particle such as length, breadth, and thickness, a characteristic diameter of the particle can be represented through a defined averaging calculation of the dimensions. However, in most cases floc particles are measured by a microscopic or microphotographic method that can only give two-dimensional patterns. Hence, use of a two-dimensional characteristic or equivalent diameter, called projected diameter, is common practice.

Among the projected diameters, projected area diameter, perimeter diameter, Feret diameter, and Martin diameter are useful for evaluating coagulated floc size. Those diameters are illustrated in Figure 7-1.

The projected area diameter is the diameter of a circle having the same area as the projected area of the particle. Measurements through photometric methods usually give this equivalent diameter, which is often called the Heywood diameter.

The perimeter diameter is useful in evaluating the shape factor of an irregular particle. However, since flocs are fractal agglomerates, as will be discussed in the following section, the length of the perimeter is a variable with precision of the measurement. The outer envelope of the projected floc figure can be used as the perimeter. For reasons of convenience, a linear dimension that is parallel to a fixed direction is often used to evaluate randomly oriented particles.

Feret and Martin diameters are often used to evaluate floc sizes when many flocs are measured. Feret diameter is defined as the perpendicular distance between two parallel tangents of the same direction to the projected outline of the particle. Martin diameter is the distance in a direction between the projected outline of the particle defined at the position where the projected area is divided equally into two

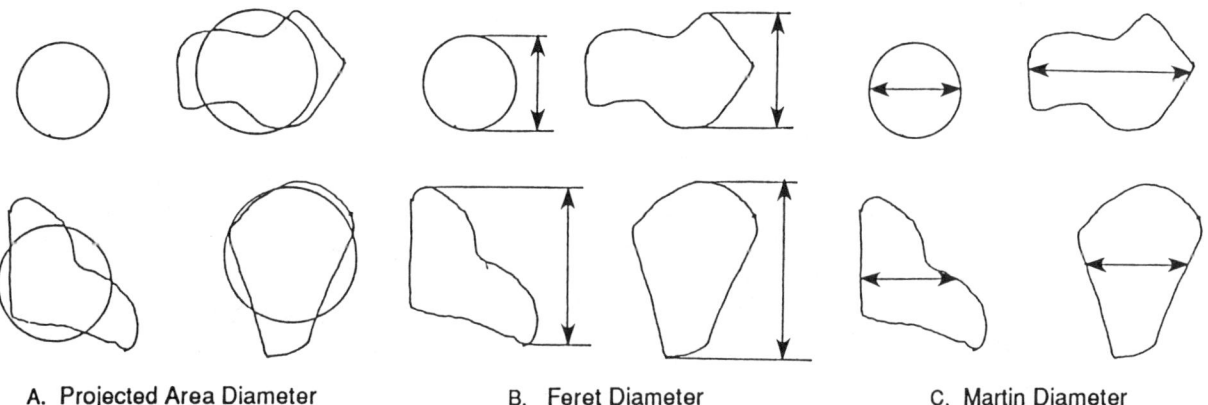

A. Projected Area Diameter   B. Feret Diameter   C. Martin Diameter

Figure 7-1 Example of the projected diameter.

parts. The relationship Feret diameter > projected area diameter > Martin diameter is observed in most cases.

Diameter of a sphere having the same density and free subsiding velocity as the measured particle in the same fluid is defined as the equivalent settling velocity diameter. In the laminar settling regime, the diameter is called the Stokes diameter. In the case of floc diameter evaluation, it is often very difficult to calculate Stokes diameter of a particle, because its apparent density changes with the diameter (see "Floc Density and Porosity," below).

PARTICLE SHAPE

Particle size is usually defined by a diameter. With the use of a characteristic or equivalent diameter, some quantitative indices are required to describe the diversity of the particle shape from a sphere.

One of the most popular shape factors is the Wadell sphericity (ψs) (Wadell 1932, 1933). The sphericity is defined by the following equation:

$$\psi s = \frac{\text{surface area of a sphere having the same volume as the particle}}{\text{surface area of the particle}}$$

In two-dimensional measurement the circularity (ψc) is defined as a shape factor:

$$\psi c = \frac{\text{projected area diameter}}{\text{projected outline length of the particle}}$$

The value of ψc is often used as an approximate value of ψs for a particle.

The concept of fractal nature proposed in 1975 by Mandelbrot (1982) is useful to describe ruggedness of flocs. If measurement of perimeters of a floc is performed with an accuracy defined by the smallest length of the measurement (i.e., resolution) r, a fractal relationship between the measured perimeter L and the projected area Ap is written as the following equation with fractal dimension Df:

$$Ap^{1/2} \propto L^{1/Df} \qquad \text{(Eq 7-1)}$$

If the measured particle is a sphere of nonfractal nature, the fractal dimension Df is unity. However, ordinary floc particles have the fractal dimension of about 1.2 to 1.8. An example of a recent measurement to show fractal nature is shown in Figure 7-2. Aratani et al. (1988) revealed that the fractal dimension of floc increases with the increase of the $\overline{G}$ value.

FLOC DENSITY AND POROSITY

Several investigators determined experimentally the floc density as a function of floc size. Tambo and Watanabe (1967, 1979a) and Lagvankar and Gemmel (1968) worked with aluminum hydroxide and iron hydroxide flocs.

Using the results reported by Vold (1963), Lagvankar and Gemmel (1968) estimated a size–density relationship for iron hydroxide flocs:

$$\rho e = c'' Ap^{-0.338} \propto d_f^{-0.676} \qquad \text{(Eq 7-2)}$$

Where:

ρe = effective floc density, which equals $\rho s - \rho o$
$\rho s, \rho o$ = density of floc and water, respectively
d_f = floc diameter

The researchers verified the relationship empirically. No significant influence on the effect of agitation or of a flocculation aid was found. Another size–density relationship for hydroxide flocs is the empirical relationship of Tambo and Watanabe (1967, 1979a). In their experiments aluminum hydroxide flocs with kaolinite and colored organics were measured (Figure 7-3). The derived relationship is:

$$\rho e = \frac{a}{(d_f/1)^{K_\rho}} \qquad \text{(Eq 7-3)}$$

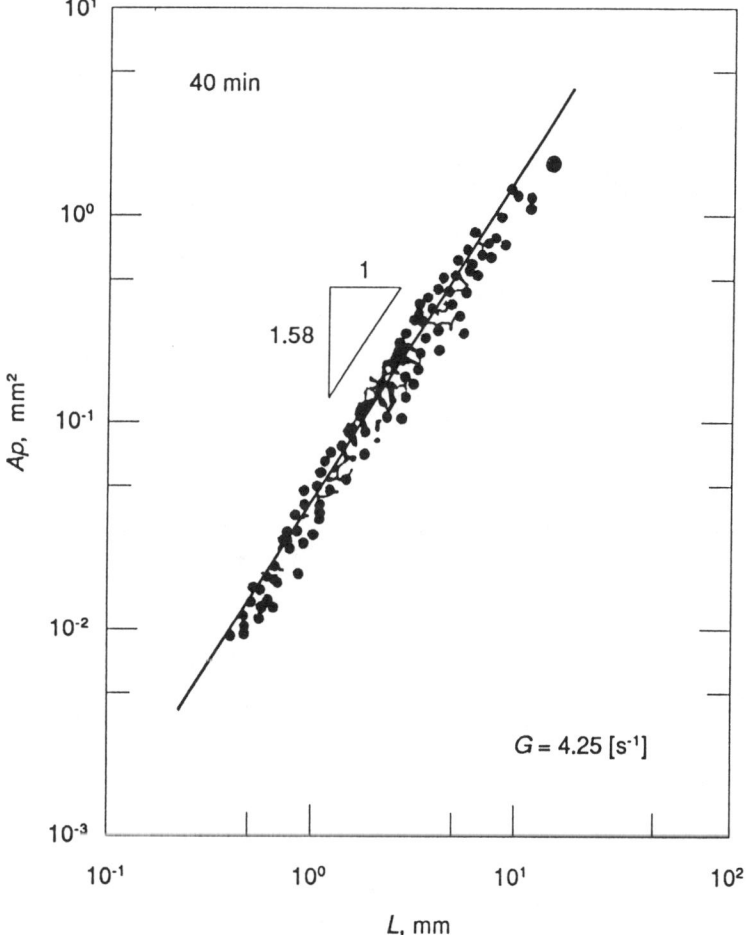

Source: Aratani, T. et al. 1988. Evaluation of Floc Form by Fractal Dimension. Kagaku Kogaku Ronbunshu (Jpn. Soc. Chem. Engrg.), *14:3:395.*

Figure 7-2 Floc area versus floc perimeter (fractal dimension of floc).

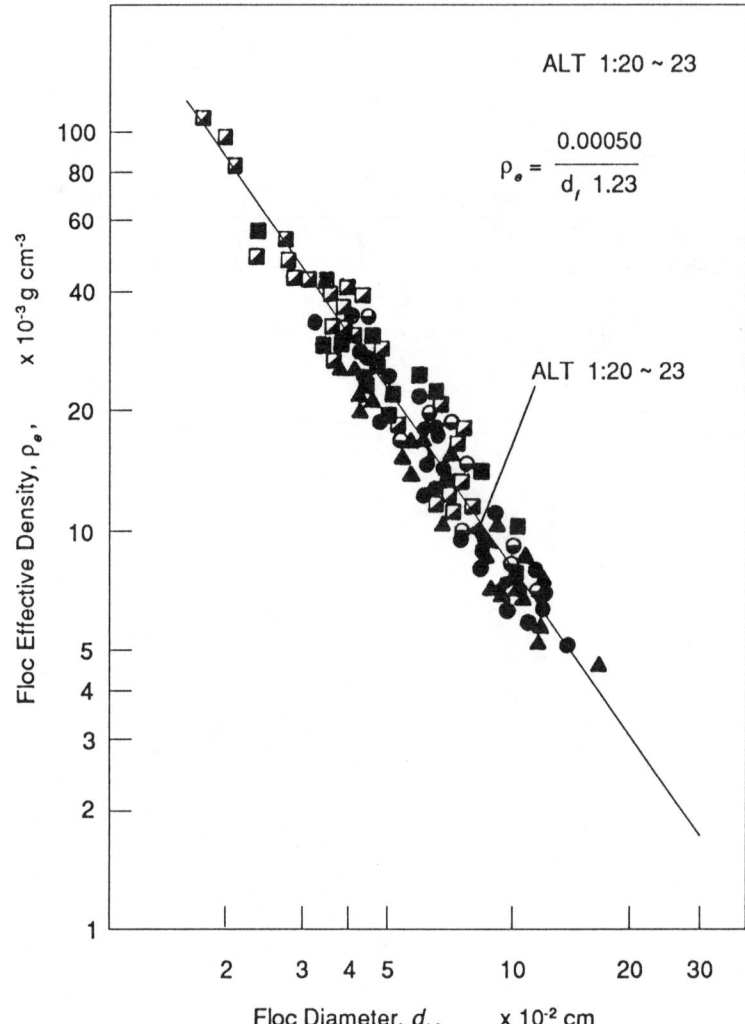

Source: Reprinted with permission from Water Research, 13:5:409. N. Tambo & Y. Watanabe, Physical Characteristics of Flocs, I: The Floc Density Function and Aluminum Floc. Copyright 1979, Pergamon Press PLC.

Figure 7-3 Plot of floc density function (coagulant is alum).

The floc porosity by consequence is:

$$1 - \varepsilon f = \frac{a}{(\rho m - \rho w)(d_f/1)^{K_\rho}} \qquad \text{(Eq 7-4)}$$

The constants a and K_ρ are functions of the ALT ratio (aluminum concentration to suspended particle concentration) as shown in Figure 7-4. In this manner the influence of the coagulant dosage on the floc density is taken into account.

No significant influence of the pH, the alkalinity, or the addition of a flocculation aid could be noted in the relationship between floc size and its density under ordinary conditions. The exponent value K_ρ is about 1.0 or a little larger in ordinary aluminum kaolinite flocs. However, a colored floc with aluminum hydroxide shows a much higher K_ρ value. More intense agitation produced denser flocs. Computer

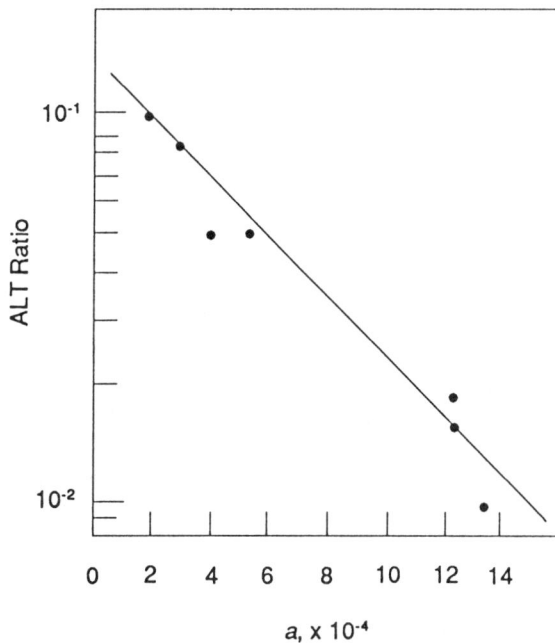 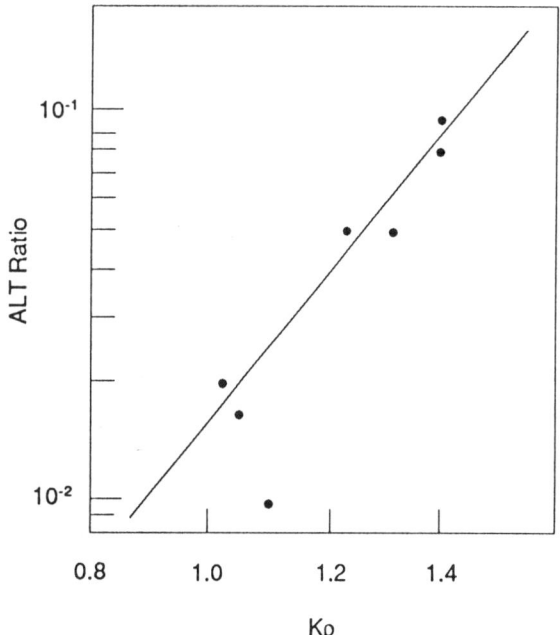

Source: Reprinted with permission from Water Research, *13:5:409. N. Tambo & Y. Watanabe, Physical Characteristics of Flocs, I: The Floc Density Function and Aluminum Floc. Copyright 1979, Pergamon Press PLC.*

Figure 7-4 Relationship between ALT ratio and constants at the neutral pH.

simulation of the floc formation process had been performed with random collisions of various size clusters in the course of agglomeration by Watanabe and Tambo (1971), and the above-mentioned experimental results were confirmed by the model.

Clark et al. (1988) noted the influence of the aluminum concentration and the pH on the density of aluminum–humic acid flocs. They applied a short breakup step after 30 min of flocculation, following which the broken flocs could recover again. The floc density after recovery did not vary monotonically with increasing breakup mixing intensity. Two optimal $\overline{GT}$ ranges for floc densification were found.

Domasch et al. (1986) derived, on the basis of theoretical considerations, a relationship between floc size and floc porosity.

$$1 - \varepsilon f = c'' MN^{-\Omega} = c''^{\,a} \left(\frac{d_f}{d_{f\max}}\right)^{K_\rho} \tag{Eq 7-5}$$

Where:

$a \quad = 1/(\Omega + 1)$
$K_\rho \quad = 3.\Omega/(\Omega + 1)$

The relationship was tested experimentally by flocculation of glass spheres or quartz particles with partially hydrolyzed polyacrylamide. The value found for Ω was 0.47 and was independent from the shape and size of the primary particles.

Klimpel and Hogg (1986) fitted their experimentally measured floc size–floc porosity data with the following equation:

$$\log(1 - \varepsilon_f) = \frac{\log(1 - \varepsilon_{f,l})}{1 - (d_f/x_c)^{-c''}} \quad \text{(Eq 7-6)}$$

Where:

$\varepsilon_{f,l} = 1 - \Phi_f$

In the experiments, for 1 min a nonionic polyacrylamide was continuously fed to a suspension of quartz particles. The polymer concentration and the agitation intensity did not change the measured floc porosity. With an increase in mixing time, the flocs became more compact. The influence of the primary particle size is important. The porosity decreases if the diameter of the primary particles increases. This is due to a different number of primary particles in the flocs. For flocs with the same number of primary particles, those formed from particles with smaller diameter are more compact than those formed from larger primary particles. More concentrated suspensions yield less porous flocs.

The above-mentioned relationship between floc size and density is observed with flocs generated in a conventional flocculator. Hozumi and Tambo (1974) revealed that in a solids contact clarifier, in which contact flocculation is occurring, the effective floc density of a floc in a floc blanket is about twice as high as that of the same size floc generated in a conventional flocculator. This may occur partially because of an aging effect of the flocs during an elapsed time in a floc blanket (see "Aging of Flocs," below). However, another important factor might be a mode of flocculation with one-by-one attachment of small flocs onto a suspended large floc. By an introduction of a suitable polyelectrolyte to the flocculation process prior to introducing a suspension into an agitated floc blanket, Suzuki et al. (1980) succeeded in generating pellet-like higher-density flocs.

Recently, Tambo and Matsui (1989) generated a high-density floc of almost constant density with diameter through a strict one-by-one attachment-mode flocculation of microflocs onto a grown floc surface in a fluidized bed. This procedure involves the introduction of polyelectrolyte directly at the bottom inlet of the fluidized bed, through which kaolinite microflocs flow into the bed. Without any prior flocculation, the microflocs with polyelectrolyte attached to their surface flow into the fluidized floc bed and adhere to the surface of the grown flocs. A suitable intensity of agitation is given to the fluidized floc bed in order to restrict irregular growth of the attached microflocs on the grown floc surface in the bed. Through this operation, a dense spherical floc of uniformly laminated microflocs is generated with the highest density attainable and is described as a pellet floc. The effective density of the pellet floc is as high as 0.2 g/cm^3, as shown in Figure 7-5. Assuming 1-mm floc, the density of the pellet floc is 10 times as high as that of a conventional floc.

FLOC STRUCTURE

Many investigations have been performed with different techniques in order to study the structure of flocs. Table 7-1 summarizes the work executed on the study of the floc structure.

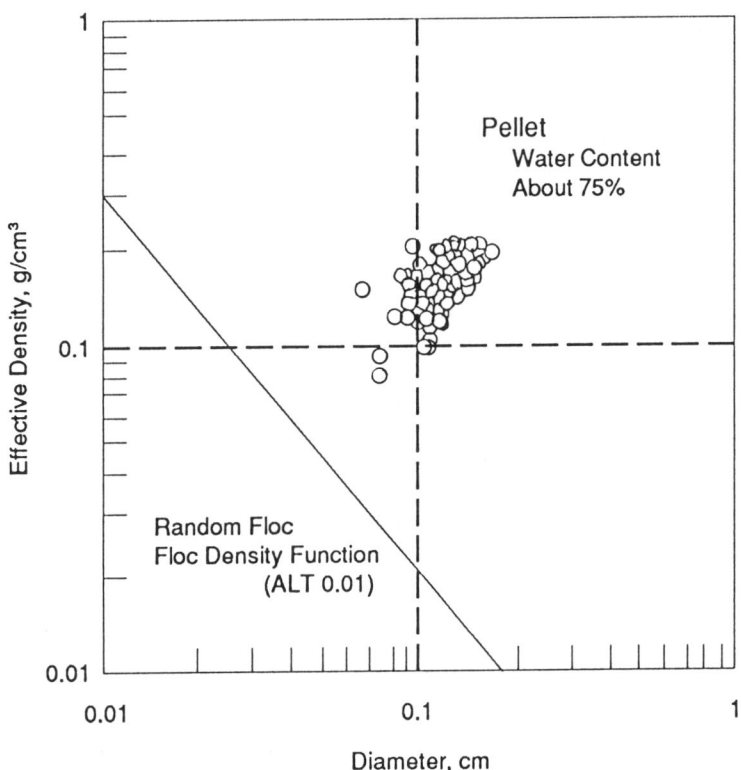

Source: Tambo, N. & Matsui, Y. 1989. Performance of Fluidized Pellet Bed Separator for High Concentration Suspension Removal. Jour. Wtr. Supply Res. & Tech. (AQUA), 38:1:16.

Figure 7-5 Density of pellet and floc.

Table 7-1 Investigations of Floc Structure

Authors	Type of Suspension	Type of Data Used
Vold (1963)	Silica sols in organic solvents	Computer simulation
Sutherland (1967, 1971)	Carbon black suspensions	Computer simulation
Michaels and Bolger (1962a, 1962b)	Kaolinite suspensions	Sedimentation behavior and plastic flow behavior
Tambo and Watanabe (1967, 1971, 1979a)	Aluminum hydroxide kaolinite flocs	Floc density measurements and computer simulation
Lagvankar and Gemmell (1968)	Ferric hydroxide kaolinite flocs	Floc density measurements
Firth and Hunter (1976) and van de Ven and Hunter (1977)	Different suspensions in water and organics	Energy dissipation
Weitz and Huang (1984)	Gold colloids	Kinetics
François and Van Haute (1984, 1985b)	Aluminum hydroxide kaolinite flocs	Floc strength measurements
Klimpel and Hogg (1986)	Quartz particles, polyelectrolyte flocs	Floc density measurements
Tambo and Matsui (1989)	Aluminum hydroxide flocs with polyelectrolyte	Fluidized pellet bed's floc density measurement
François (1988)	Aluminum hydroxide kaolinite flocs	Addition of polyelectrolyte during floc buildup

Through computer simulation Vold (1963) obtained a rigid floc structure on three levels. Flocs were formed by successive random addition of individual spherical particles. No internal rearrangements occurred. The particles not included in the core formed projecting tentacles that gave a rough surface to the floc. Those tentacles could entangle to form weak aggregates. Vold (1963) compared this shape with the shape of flocculated unstable silica colloids in organic solvents.

Sutherland (1966) demonstrated that in the Vold model the random addition of the primary particles was not truly random and, hence, the Vold model was unable to fit the experimental findings. The predicted floc density did not decrease for flocs larger than 500-fold. Using Smoluchowski's rate equation for perikinetic flocculation, Sutherland (1967) treated flocculation as a series of random collisions between primary particles and particle clusters. This mechanism leads to an open network since the bonds are considered as rigid. The previous "Single Smoluchowski Model" was modified by Sutherland and Goodarz-Nia (1971). However, they did not derive the size–density function. They modified the random rotation of the clusters and chose the collision sequence in a slightly different manner. Instead of using the same collision rate for all the particles, they altered it for each pair of aggregates. The different cluster models were all in qualitative agreement with experiments on carbon black suspensions. Michaels and Bolger (1962a, 1962b) used a multilayer floc model for fitting experimental data of sedimentation and plastic flow of very concentrated kaolinite suspensions. The observed phenomena were explained satisfactorily.

More direct experimental indications for a multilevel floc structure are found in the density–size relationship of a floc as measured by Tambo and Watanabe (1967, 1979a) and Lagvankar and Gemmell (1968). Watanabe and Tambo (1971) and Tambo and Watanabe (1979a) performed a computer simulation study that reflects the actual flocculation phenomenon. They proved their experimental results of the floc size–density relationship on the computer by a random collision model that takes into account various size cluster collisions in the course of floc growth. They derived the following equation for very low ALT ratio floc.

$$\rho e = (d_f/1)^{-0.9} \qquad \text{(Eq 7-7)}$$

The exponent of –0.9 is much higher than the value of –0.676 proposed by Lagvankar and Gemmell (1968) based on the Vold model. An exponent value of about –1.0 is more practical, and the multisize-cluster floc collision model for floc growth is more natural.

Similarly to Michaels and Bolger (1962b), Firth and Hunter (1976) used a Bingham model to describe the flow of an electrically charged colloidal sol. They compared the single-particle model, the hard-floc model, and the elastic-floc model with experimental measurements. Only the elastic-floc model could give a satisfactory flow pattern. Therefore, the unit of flow cannot be a single sphere or a hard nondeformable floc. A detailed calculation of the energy dissipation in a flowing sol exhibiting plastic behavior showed again the validity of the previous statement. Van de Ven and Hunter (1977) improved previous calculations. They also defined an elaborate four-level floc structure.

The previous models are mainly continuously growing models. However, some models are constructed as multiple-level models, which contain several different level agglomerates in their structure.

Weitz and Huang (1984) could determine in the growth kinetics of gold colloids two different regimes, which are limited by different mechanisms. They had to use different power functions in order to fit the two regimes. They also used the fractal concept developed by Mandelbrot (1982). The clusters formed from single particles had a fractal dimension Df = 2.49; the aggregates formed from clusters had a fractal dimension Df = 1.75. This is also strong evidence for a multilevel floc structure.

François and Van Haute (1984, 1985b) demonstrated the existence of a four-level floc structure for hydroxide flocs (i.e., primary particle, flocculi, floc, and floc aggregates). Freshly formed flocs were ruptured with different $\overline{G}$-values. The relationship between the average floc size after rupture and the $\overline{G}$-value during rupture displayed the existence of two regions. Stripping experiments were designed in order to study the floc structure in more detail. The resulting floc structure model agrees with the model of Firth and Hunter (1976) and van de Ven and Hunter (1977). Klimpel and Hogg (1986) analyzed the size–porosity relationships of quartz particles flocculated with a nonionic polyacrylamide. The relationship could be divided into three regions. The low size region contains small, low-porosity flocs. These small compact flocs may then act as the basic unit for the subsequent growth flocs. The flocs in the third region are large, highly porous aggregates. The porosity depends on the number of small flocs contained, the densities and sizes of these smaller flocs, and the degree of compaction and breakage. The differences in floc density might be explained as a result of response of the floc to turbulence, leading to compaction in the case of small-size flocs and to breakage in the case of larger flocs.

In general, the evidence for a multilevel elastic floc structure is abundant. It is very likely that the concept for a four-level structure of floc aggregates describes the reality. Densely packed flocs consist of a number of very dense and strong flocculi, which are formed at the highest rate of shear to which the system was ever subjected. Porous, weak floc aggregates are composed of those flocs. The bonds between the different parts are elastic.

François (1989a) researched the influence of many flocculation conditions on the four-level floc structure. There exists a logarithmic correlation between flocculus diameter and the rate constant in a first-order flocculation equation with various experimental conditions in coagulant dose or in time of rapid mixing.

The flocculi, which are the building stones of the structure, are formed in the first moment after destabilization. Thus, the $\overline{G}$-value during slow mixing has no influence on the floc structure; only the size of the aggregates is influenced. It is the $\overline{G}T$-value during rapid mixing that determines the structure of the flocculi. A too small $\overline{G}T$-value yields incomplete flocculi; too large $\overline{G}T$-values will also limit the flocculus size. Optimal $\overline{G}T$-values in the rapid mixing process were found to be around 20,000. If the $\overline{G}$-value during rapid mixing is low, the shape of the flocculi will be elongated. If the $\overline{G}$-value is high, the shape is spherical. The flocculus surface is tortuous if intermediate $\overline{G}$-values (500 s^{-1}) are used. A lower pH value during flocculation results in more porous flocs. The higher the concentration of suspended solids in the water, the larger the formed flocculi, but the smaller the formed aggregates. At the same time, flocs with a higher density are formed. The influence of a change in concentration of solids is more important in less-concentrated suspensions than it is in concentrated suspensions. The influence on floc structure with the change of temperature in an installation can be translated as the influence of the change in $\overline{G}$-values. In a given installation, the $\overline{G}$-values will decrease if the temperature decreases. Thus, a lower temperature corresponds with a lower $\overline{G}$-value and needs a longer rapid mixing time in order to keep an optimum $\overline{G}T$-value in the given installation. The influence of salts cannot be generalized. Some salts ($NaHCO_3$) have an enormous impact on flocculus size, others ($CaCl_2$) have no significant influence. However, the presence of salts in the water always yields more porous structures.

FLOC STRENGTH AND BREAKUP

An important characteristic of flocs in flocculation processes is their binding force, which resists external forces that act to break them down. Two approaches might be necessary to handle this phenomenon. One exploits the static nature of floc particles related to the binding force of flocs. Another involves a dynamic expression of floc strength, which is connected to the rate of breakdown under a fluid regime. Most flocculation processes are carried out under turbulent regimes. Hence, strength evaluations of flocs are usually discussed in terms of behaviors of floc particles under turbulent agitation conditions.

Floc Strength and Maximum Floc Size

Floc strength can be defined directly or indirectly in many ways. Direct measurement is given as the binding strength expressed in Newtons per square centimetre, or N. However, for practical purposes the maximum or mean floc diameter of an equilibrium particle size distribution of flocs under an agitation condition is often used as an indirect measurement for the floc strength.

A useful example is a derivation of Tambo and co-workers. Tambo et al. (1970a) and Tambo and Hozumi (1979) tried to establish a relationship between the maximum floc size and the agitation intensity in a flocculator to evaluate relative floc strength. They focused their attention on a well-known phenomenon that the maximum floc size under an agitation condition decreases with the increase of the agitation intensity. Formulation of the above relationship was performed as follows (with a correction given).

The intensity of the agitation in a turbulent condition can be evaluated by the rate of energy dissipation in a unit volume effective for the flocculation, ε_o [$J/cm^{-3}s$], after Kolmogoroff's locally isotropic turbulence theory (Heintz 1959; Batchelor 1958; Levich 1962). The maximum floc diameter under a given agitation condition (a dissipation rate of turbulent energy) is controlled by the two counteracting forces, the binding force of the floc and the break force exerted from fluid motion.

The binding force $B(N)$ is proportional to the net sectional area A_n (cm^2) at the plane of rupture multiplied by specific floc strength (N cm^{-2}) and is shown as Eq 7-8 by the introduction of Tambo and Watanabe's floc density function:

$$B = \sigma A_n \propto d_f^2 (1 - e) \propto \sigma d_f^{2-k_p} \, (N) \qquad \text{(Eq 7-8)}$$

where σ is a constant to show mean binding strength with respect to net sectional area of floc (N cm^{-2}).

On the other hand, the maximum external breakup force acting on a floc by microturbulent eddies can be calculated under the assumption of Kolmogoroff's locally isotropic turbulence theory (Levich 1962). The difference of the dynamic force Δf exerted on the unit area of opposite side of floc with a diameter d_f is:

$$\Delta f \propto \rho w \frac{|\overline{v_1' - v_2'}|^2}{2} \qquad \text{(Eq 7-9)}$$

Where:

v_1', v_2' = the fluctuation velocities of the water at points separated from each other by a distance of d_f

$|\overline{v_1'-v_2'}|$ = the absolute value of the time mean of the difference of the above two velocities

The value of $|\overline{v_1'-v_2'}|$ can be expressed through Kolmogoroff's theory as:

$$v_{\lambda=d} = \alpha(\varepsilon_o d_f/\rho w)^{1/3}, d_f \gg \lambda_o \text{ (inertia subrange)} \quad \text{(Eq 7-10)}$$

$$v_{\lambda=d} = \beta\sqrt{\varepsilon_o/\mu} \; d_f, \; d_f \ll \lambda_o \text{ (viscous subrange)} \quad \text{(Eq 7-11)}$$

$$\lambda_o = (v^3 \rho_w/\rho_o)^{1/4} \text{ (microscale)} \quad \text{(Eq 7-12)}$$

Where:

ε_o = the mean rate of energy dissipation per unit volume in a basin effective for flocculation (J/cm^{-3}s^{-1})

μ = the absolute viscosity ($P_a s$)

v = the kinematic viscosity (cm s^{-1})

λ_o = the microscale of the turbulent eddies (cm)

α and β are constants, and the values are about unity and $1/\sqrt{15}$, respectively (Fuchs 1964).

Substituting Eq 7-10 and 7-11 into Eq 7-9, and multiplying surface area of the floc, the total breakup force ΔF can be shown as

$$\Delta F \propto \rho_w \varepsilon_o^{2/3} d_f^{8/3}, d_f \gg o \quad \text{(Eq 7-13)}$$

$$\Delta F \propto \rho_w (\varepsilon_o/\mu) d_f^4, d_f \ll o \quad \text{(Eq 7-14)}$$

The critical condition of breakup of floc is considered to be the breakup force equal to the binding force. Therefore, equating Eq 7-8 and 7-13 or 7-14 with suitable coefficients gives the following equations for maximum floc diameter under a given flocculation condition:

$$d_{f\max} \propto (\sigma/\rho_w)^{3/(2+3k_\rho)} \cdot \varepsilon_o^{-2/(2+3k_\rho)}$$

$$= K* \varepsilon_o^{-2/(2+3k_\rho)}, \qquad d_f \gg \lambda_o \quad \text{(Eq 7-15)}$$

$$d_{f\max} \propto (\sigma/\rho_w)^{1/(2+k_\rho)} \cdot \varepsilon_o/\mu^{-1/(2+k_\rho)}$$

$$= K \varepsilon_o^{-1/(2+k_\rho)}, \qquad d_f \ll \lambda_o \quad \text{(Eq 7-16)}$$

where K and $K*$ are the constants that contain the strength of binding force.

The effective energy dissipation rate ε_o is proportional to the third power of the rate of flocculator rotation, Nr (s^{-1}), under the turbulent agitation conditions. The value ε_o is also proportional to the second power of $\overline{G}$-value. Thus, Eq 7-15 and 7-16 can be rewritten for practical application:

$$d_{f\max} = K* \varepsilon_o^{-2/(2+3k_\rho)} \propto G^{-4/(2+3k_\rho)} \propto Nr^{-6/(2+3k_\rho)}, d_f \gg \lambda_o \quad \text{(Eq 7-17)}$$

$$d_{f\max} = K* \varepsilon_o^{-1/(2+k_\rho)} \propto G^{-2/(2+k_\rho)} \propto Nr^{-3/(2+k_\rho)}, d_f \ll \lambda_o \quad \text{(Eq 7-18)}$$

As Tambo and Watanabe (1967, 1979a) reported, a constant of the floc density function K usually takes the value in the range of 1.0 to 1.5. Thus, the maximum size of flocs may change with agitation intensity as the following numerical relationships:

$$d_{f\max} = K* \varepsilon_o^{-(0.4-0.3)} \propto G^{-(0.8-0.6)} \propto Nr^{-(1.2-0.9)}, d_f \gg \lambda_o \quad \text{(Eq 7-19)}$$

$$d_{f\max} = K \varepsilon_o^{-(0.33-0.28)} \propto G^{-(0.66-0.56)} \propto Nr^{-(1.0-0.86)}, d_f \ll \lambda_o \quad \text{(Eq 7-20)}$$

This relationship between agitation intensity and floc diameter was verified with clay–aluminum flocs. The measured data of maximum floc diameter versus the rate of agitation are plotted on the logarithmic paper as shown in Figures 7-6 and 7-7. As shown in the figures, the maximum floc diameter decreases linearly on the plot as the rate of agitation, Nr (rpm), increases. The slopes of these plots are always in the range of –(1.0–1.1). Figure 7-8 shows the same relationship on the logarithmic plot of the data between the maximum floc size, $d_{f\max}$, and the mean effective rate of energy dissipation ε_o. In the paddle-blade mixer used for the measurement, the effective rate of mean energy dissipation ε_o is calculated from the measured overall rate of energy dissipation ε on the relationship of $\varepsilon_o \doteq 0.1\ \varepsilon\ [\text{J/cm}^{-3}\text{s}]$ (Tambo et al. 1970a; Tambo and Hozumi 1979).

François (1985) revealed that the significant $\overline{G}$-value in determining the maximum floc size is not the mean value but the value in the impeller-affected zone.

The microscale of a turbulent eddy is a little larger than the observed maximum floc size but in the same order. The maximum floc size and the microscale are in the same order. Hence, in practice, the equation of the maximum floc size in the viscous subrange can be used in calculating floc growth kinetics in order to keep consistency of the turbulent fluid regime covering the scale from the primary particles to the maximum size flocs.

The linear relationship between the specific rate of energy dissipation ε or the $\overline{G}$-value, and the maximum or average floc diameter has been proposed by many researchers. Table 7-2 shows several values of the exponent m in the equation of $D_{f\max} \propto \overline{G}^{-m}$. In introducing the theoretical formula the authors assumed miscellaneous breakup modes. These will be explained in the following section, "Breakup of Flocs."

Many researchers have tried to establish a method to evaluate the floc strength itself, but even now no standard method is proposed. In the method of evaluating strength σ by Tambo and Hozumi (1979), the breakup phenomenon is defined in the viscous subrange, and the floc binding strength, σ (N cm^{-2}) can be derived from Eq 7-8, 7-14, and 7-3 as:

$$\sigma = \kappa\,(\varepsilon/\nu)\,d_{f\max}(2+k_\rho)\cdot(\rho e/a\,1^{k_\rho})\ (\text{N cm}^{-2}) \quad \text{(Eq 7-21)}$$

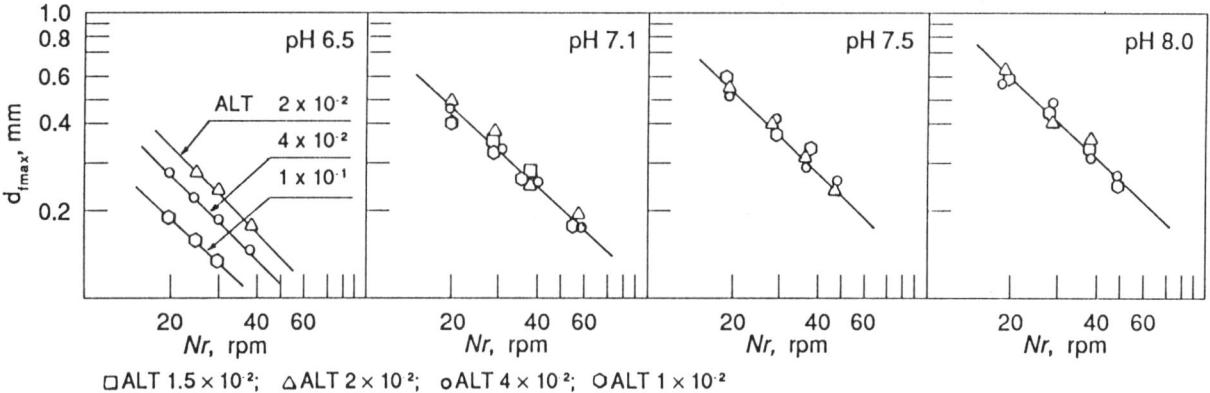

Source: Reprinted with permission from Water Research, 13:4:421. Norihito Tambo & H. Hozumi, Physical Characteristics of Flocs, II: Strength of Floc. Copyright 1979, Pergamon Press PLC.

Figure 7-6 Relationship between maximum floc diameter and the rate of agitation rotation of clay–aluminum flocs with variable pH.

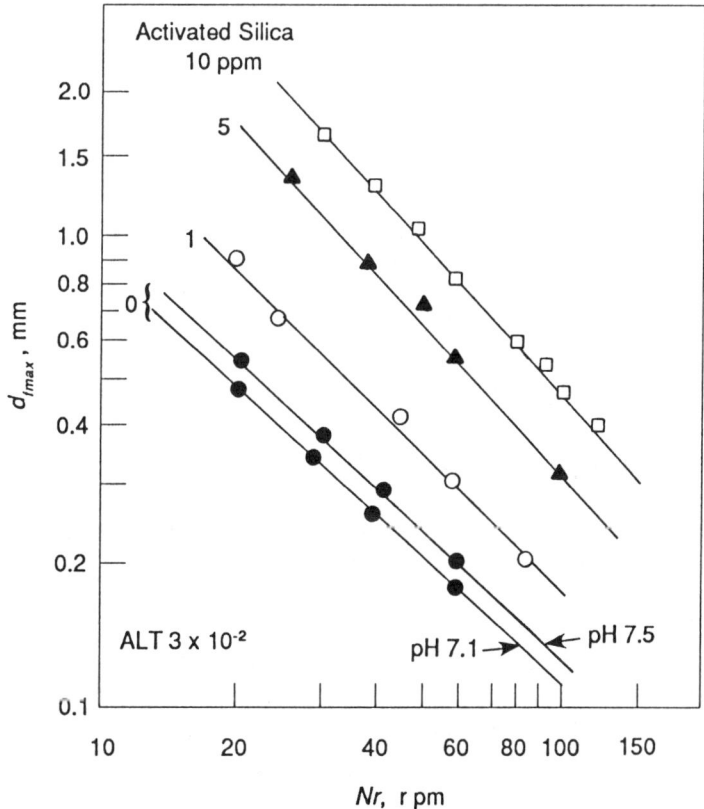

Source: Reprinted with permission from Water Research, 13:4:421. Norihito Tambo & H. Hozumi, Physical Characteristics of Flocs, II: Strength of Floc. Copyright 1979, Pergamon Press PLC.

Figure 7-7 Relationship between d_{fmax} and Nr of clay–aluminum flocs with variable activated silica dosage.

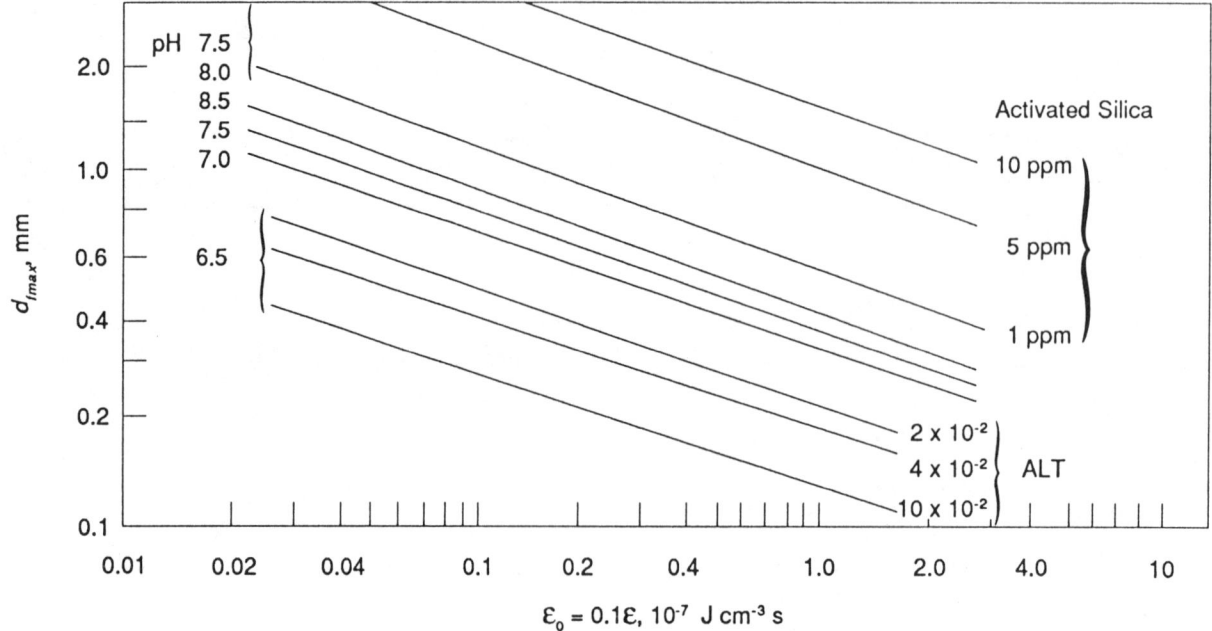

Source: Reprinted with permission from Water Research, *13:4:421. Norihito Tambo & H. Hozumi, Physical Characteristics of Flocs, II: Strength of Floc. Copyright 1979, Pergamon Press PLC.*

Figure 7-8 Relationship between maximum floc diameter and the effective rate of energy dissipation of clay–aluminum flocs.

Where:

κ = a constant (–)
a = a constant of floc density function (cm)
ρe = the effective density (g cm^{-3})
σ = floc binding strength (N cm^{-2})

From the equation, the binding strength of floc can be estimated when the constant is evaluated. If maximum floc diameters of the same type of flocs under different agitation conditions are measured, the constant κ of Eq 7-21 can be determined. After Tambo and Hozumi, through many measurements Glasgow and Jyh-Ping Hsu (1982) gave evaluations to calculate the bindings force as $B = 13.56d^{2.40}$ for ferric hydroxide kaolinite flocs of pH 9.5, $B = 8.54d^{2.4}$ for the ferric hydroxide kaolinite floc of pH 7.4, $B = 24.8d^{2.45}$ for aluminum kaolinite flocs, and so on, with B in dynes; d in centimetres. Bache and Al-Ani (1989) measured the equilibrium floc diameters in a vertical vibrating water column in which ξ changes linearly, and calculated the binding strength after the equation of Tambo and Hozumi. They proposed $B = 25d^{2.5}$, with respect to aluminum sulfate kaolinite floc of ALT ratio = 1/10.

Breakup of Flocs

Floc breakup mechanisms under turbulent agitation conditions are very complicated. Although an equilibrium floc size under an agitation intensity gives a universal expression of $d_{f\max} \propto \overline{G}^{-m}$, the mechanisms to support the relationship are various. Table 7-2 shows several examples.

Thomas (1964) suggested that under turbulent flow conditions the principle mechanism leading to floc rupture is pressure differences on opposite sides of the floc, which cause bulgy deformation and rupture; and the breakup of the floc is resisted by the yield stress, which increases with the energy dissipation per unit mass of fluid. He introduced the concept of floc disruption frequency by assuming it was equal to the eddy frequency defined by Levich (1962).

Tambo and co-workers (Tambo et al. 1970a; Tambo and Hozumi 1979) simply assumed the balance of the dynamic force exerted on the opposite side of a floc and derived the relationship between $d_{f\max}$ and $\overline{G}$ or ε under the assumption of Kolmogoroff's locally isotropic turbulence (see Eq 7-19 and 7-20). They did not further discuss the details of the mechanisms but took the density function into account.

Argaman and Kaufman (1970) postulated that rapid changes in velocity and the inertia of the floc particles caused disruptions. They did not clearly describe the exact breakup mechanism and the size of the breakup fragments under an agitation condition. However, they recognized from the experimental results that the average size of flocs is closely related to the scale of mean square velocity fluctuation. As the possible breakup model, the stripping of individual primary particles from the floc surface was considered to be a credible model.

Table 7-2 Values of the exponents m for the function $d_{f\max} \propto \overline{G}^{-m}$

Authors	m	Regime of Turbulence	Kind of flocs	Type of Data Used
Thomas (1964)	5 1	Inertia subrange Viscous subrange	Modeling	Theoretical formula Theoretical formula (checked with experimental data)
Tambo et al. (1970a, 1979)	0.6 ~ 0.8 0.56 ~ 0.66	Inertia subrange Viscous subrange ($\overline{G}$ = 5 ~ 45/s)	Model floc and aluminum hydroxide kaolinite floc	Theoretical formula and experimental study

Parker et al. (1971, 1972)	2
	1
	
	0.5
	0.5
	0.17 ~ 0.37
		Surface shear:
		  Inertia subrange
		  Viscous subrange
		  ($\overline{G}$ = 10 ~ 160/s)
		Filament flucture:
		  Inertia subrange
		  Viscous subrange
		  ($\overline{G}$ = 10 ~ 160/s)
			Model floc
			Ferric and aluminum floc
			
			Model floc
			Model floc
			Activated sludge floc
				Theoretical formula and experimental study
				
				Theoretical formula
				
				Experimental study

Tomi et al. (1978)	2 0 1	Inertia subrange Viscous subrange Intermediate	Modeling	Theoretical formula
Leentvaar et al. (1983)	1.0 0.6	Inertia subrange Viscous subrange	Ferric chloride floc	Experimental study with polymer
François (1987b)	0.7 ~ 1.5 0.3 ~ 0.5	Inertia subrange Viscous subrange	Aluminum floc	Experimental study

Parker et al. (1971, 1972) discussed the relationship on the same basis as the preceding authors. They first proposed and verified the occurrence of two distinct breakup mechanisms. One is surface erosion of primary particles (fine particle erosion), and another is bulgy deformation–floc splitting (large-scale splitting). In the preceding paper Argaman and Kaufman (1971) suggested that the principle mode of breakup of metal coagulants floc was surface erosion of floc by turbulent drag. However, for biological flocs having rigid filamentous network, filament breakage by tensile failure to yield two floc fragments was considered to be a significant form of breakup.

Kusuda (1973) discussed a relationship between floc structure and the agitation conditions. He suggested that the floc is a Bingham body with a basic structure that is regarding to the $\overline{G}$-value. On the plastic behaviors of flocs, he suggested an expression of breakup.

Pandya and Spielman (1982) cited an experimental study of floc breakup in a four-roller extensional flow (i.e., flow field surrounded by four rotating rollers in parallel arrangement). The flow visualization with photographic observation of ferric hydroxide kaolinite flocs in the simple extensional flow field revealed two distinct mechanisms for their breakage. They recognized occurrence of splitting into a relatively small number of daughter fragments whose size was comparable to that of the parent flocs and continual disintegration by erosion to produce extremely fine particles from the extremities of the parent floc. Although this experiment was performed in a simple laminar-flow field, the direct visualized observation of the two mechanisms is worth remembering. Gross splitting into a few daughter fragments caused simultaneous streaming of many fine particles from the extremities of the larger separating fragments. This resulted in the dual phenomena of gross splitting and continuous erosion owing to microparticle removal by shear. The mode of split may vary with shape and internal morphology of the aggregate. If the internal morphology is uniform, microparticle erosion would dominate.

François (1987b) discussed the influence of an increase of the velocity gradient on the floc dimensions. He concluded that flocs with dimensions of the order of magnitude of the turbulence scale for inertial conversion will be ruptured to large fragments. Those subjected to viscous forces will be ruptured by erosion of small particles from the floc surface. Those results coincide with the proposal of Parker et al. (1971, 1972). François proposed his strength factor and regrowth recovery factor through measurement of floc diameters in a successive variable-intensity flocculation agitation with slow, high, and medium steps. From the relative diameter of the slow- and high-level agitation he tried to define floc strength.

In another article, François (1989b) fitted the relationships between $d_{f\max}$ and $\overline{G}$-value for undisturbed flocs and for regrown flocs. For both types of flocs, two relationships were found: one for the strength (or regrowth ability) of the links between flocs in an aggregate, another for the strength (or regrowth ability) of the links between the flocculi forming a floc. Numerical data for the exponent values m are deduced and summarized in Table 7-3.

The formation of very large structures for a well-defined coagulant dose called the reference coagulant dosage (François 1985; François and Van Haute 1986) is due to the very high strength of the flocs composing the aggregate. If no rapid mixing step is applied during coagulation, then flocs with very small strength and regrowth coefficients are formed. The floc strength and regrowth coefficients prove again the validity of the theory of the critical mixing time (François and Van Haute 1984). A pH between 7.0 and 7.5, which results in a card-house structure (François 1989a),

Table 7-3 Range of Exponent Values m in the Relationship Between Diameter and $\overline{G}$-Value

Independent Variable	Exponent, m
Strength of aggregates	0.110 to 0.457
Strength of flocs	0.314 to 1.441
Regrowth ability of aggregate	0.016 to 0.348
Regrowth ability of flocs	0.205 to 0.553

yields the strongest flocs. An increased salt content always increases the strength of flocs.

Tomi and Bagster (1975) discussed rupture of flocs reinforced with a polymer network under laminar flow conditions in which only simple shear field is counted as the breakup force. They proposed idealized linking structures of floc with the polymer network and analyzed the behavior of the link under laminar shear force from fluid movement. Matsuo and Unno (1981) discussed the phenomenon and concluded that the floc breakup is caused by surface shear brought about mainly through the difference in deformability of floc and water.

In the preceding discussions, every researcher assumed that there was a stable maximum or mean diameter that derived from a relationship between the intensity of agitation and the floc strength. However, it is also known that an equilibrium state is generated at a balancing point of two counteracting reactions. In the case of flocculation, an equilibrium particle-size distribution is generated on a balance of floc growth and breakage.

If an examination if based entirely on the dynamic balance of growth and breakage of flocs in a turbulent field, two kinetic constants to characterize the growth and breakage rate on a population balance model will determine the maximum diameter or mean diameter on an equilibrium floc size distribution.

An early work that treats the phenomenon on a dynamic balance, Argaman and Kaufman's paper (1970), assumed that the stripping of individual primary particles from the floc surface is one of the more credible mechanisms of the breakup. By the model, the rate with which primary particles are released depends on the surface shear, the floc size, and the size of primary particles. Assuming that the shearing stress depends on mean square velocity of turbulent fluctuation, and the average size of floc is inversely proportional to the mean square velocity fluctuation, they proposed a rate equation of breakup. However, they did not handle any floc size distribution. Hence, no direct information of an equilibrium floc diameter was derived.

Broadway (1978) discussed the dynamic balance of floc size on an assumed Gaussian cumulative particle size distribution under laminar flow condition. He proposed a function to describe the mean ultimate floc diameter in laminar flow field.

Pandya and Spielman (1982) proposed a basic dynamic expression to describe floc size distribution, taking the two breakup mechanisms of splitting and erosion into account. If this kind of approach matures, the strength of flocs will be defined as a dynamic balance. In that case, the rate of breakage function would be composed of the effective frequency of turbulent force and strength factor of the floc.

Up to the present, it has been popular to solve a dynamic population balance model under a condition that the maximum floc size is a priori under the given agitation condition. The papers of Fair and Gemmell (1964), Tambo and Watanabe (1979b), and Tambo et al. (1979) are typical of the approach.

AGING OF FLOCS

Aging includes all irreversible changes of texture and floc structure from the moment that the flocs are completely formed. In pellet floc reactors and in sludge-thickening installations, relatively long-aged flocs are present in the process. The characteristics of an old floc can differ considerably from those of a freshly formed floc. More than 30 years ago, attention had been drawn to the phenomenon of floc aging (Bale and Schmidt 1959; Feitknecht 1954; Fricke and Meyring 1933).

The Aging Mechanism

Three aging mechanisms can be distinguished: condensation–polymerization, aggregation–cementation, and crystallization. These processes have been studied by many investigators. With the exception of the work of Brown and Hem (1975) and of François (1987a), investigations usually concern pure hydroxide flocs without enmeshed solid particles.

Stumm and Morgan (1962) obtained different ultraviolet spectra for fresh and old aluminum solutions. Stumm and O'Melia (1968) interpreted aging as a change of the chemical structure of the aluminum complexes. Hem and Roberson (1967) measured different solubilities for fresh and old hydroxides. Dependent on pH, the solubility reaches a constant value after a few months to four years (Hem et al. 1973). Most of the investigators reported a decrease of pH with time. This decrease in pH is a result of the continuing polymerization. Dousma and DeBruyn (1978) demonstrated that during polymerization the atoms are linked with an increasing number of hydroxide groups.

The interplay over two years of different forms of aluminum complexes has been studied by Smith and Hem (1972). Apart from the monomeric complexes, some unstable polynuclear structures with 20 to 400 atoms per structure are formed during hydrolysis of the aluminum coagulant. Large microcrystalline $Al(OH)_{-3}$ particles are also formed.

Dependent on the ratio of OH^- to Al^{3+}, a quantity of monomers is obtained. This quantity remains constant with time. The polynuclear structures grow with time and are finally converted into microcrystalline particles.

Numerous investigators identified the different insoluble products. The results of the work of Aldcroft et al. (1969), Ginsberg et al. (1962), Hem and Roberson (1967), Hem et al. (1973), Hsu and Bates (1964), and Smith and Hem (1972) together lead to good insight into the crystallization process.

The formation of different aged products is dependent on the pH and the initial ratio of OH^- of Al^{3+}. As a first step, an amorphous boehmite gel is always formed. By condensation–polymerization between the different boehmite particles and in each particle itself, a pseudoboehmite with a low degree of crystallinity is formed. Due to the condensation–polymerization process a decrease in pH was always noted, independent from the initial ratio of OH^- to Al^{3+}. The crystallization during aging occurs only for ratios larger than 2 to 2.75. For OH^- to Al^{3+} ratios between 3 and 3.3 crystals of gibsite, bayerite, nordstrandite, or a mixture of them are formed. In nordstrandite some aluminum ions are substituted by other ions. It can be considered as a bayerite crystal with a distorted structure. A recrystallization of bayerite to gibsite can be started by foreign ions (Na^+, K^+) dissolved in the water. Sometimes, after recrystallization the gibsite is called hydrargillite. The inverse process has also been observed. Bayerite will be formed in a pH range between 7.5 and 9.5. For pH values smaller than 7.5, gibsite will be formed. When silicates are dissolved in the water, the separation between bayerite and gibsite occurs at pH 7.0.

During the aggregation–cementation process an increasing number of contact points are formed within the floc aggregate itself.

The Aging Process

The aging process of hydroxide flocs can be split into three periods (François 1987a). The initial shrinkage of the flocs is mainly due to cementation–aggregation. The average diameter of the suspension shifts to lower values. In 6–8 h the initial floc size is halved. The shrinkage of the flocs during aging is definitely not due to erosion of particles from the surface. The distribution of floc diameters decreases because the large aggregates are more susceptible to cementation than the smaller ones. The filterability of the flocs does not change. During the next three to four days, the average floc diameter decreases due to condensation–polymerization and crystallization. The drop in diameter is smaller than in the first period. The diameter distribution continues to decrease. At the same time the dewaterability of the flocs starts to increase. In the third period, the average diameter increases because of Oswald ripening of the crystals. The diameter distribution becomes very narrow. The filterability of the aged suspension is very high compared with the non-aged floc suspension.

During each period the change of the average diameter as a function of age can be quantified by:

$$d_{f_i} = d_\infty + d' \cdot t_i^{\pm \omega} \qquad \text{(Eq 7-22)}$$

Where:

d_{fi} = floc diameter on moment $t = t_i$
d_∞ = floc diameter after infinite aging
d' = a constant

d_∞, d', and shrinkage constant ω are dependent on the flocculation conditions (François 1987a). Figures 7-9 and 7-10 show the change in time of the volume median floc diameter and of the linearization. The 75-mg/L kaolin suspended in distilled water was flocculated with an Al_2SO_4 solution (4.02 mg Al/L) (pH: 7.0, 25°C, rapid mixing—$\overline{G} = 389\ s^{-1}$, $t = 60$ s; slow agitation—$\overline{G} = 34\ s^{-1}$).

Kinetics of the aging process. The kinetics of the crystallization process are of a first order. The formation of nuclei of the trihydroxide crystals is the rate-determining step (Aldcroft et al. 1969). An increase of pH (Stumm and Morgan 1962) or of the ratio of OH^- to Al^{3+} (Smith and Hem 1972) promotes the crystallization process. The shrinkage constant in function of pH reaches a maximum at pH 7.3 (François 1987a). The presence of clay particles accelerates the formation of microcrystalline material (Brown and Hem 1975; François 1987a). Increased temperature increases the crystallization process (Aldcroft et al. 1969). The presence of foreign ions, on the other hand, slows down the crystallization process (Gastuche and Herbillon 1962), and anions especially play an important role in this process (Bale and Schmidt 1959). The mixing parameters influence the shrinkage for the crystallization process (Koga et al. 1979; François 1987a). An increase of the rapid mixing time to the critical mixing time increases the rate of shrinkage. For mixing times longer than the critical mixing time the shrinkage due to crystallization remains constant. The mixing intensity during the rapid and the slow mixing steps also influences the rate of shrinkage due to crystallization. A higher $\overline{G}$-value during rapid mixing causes a speedup of the crystallization process. Very high $\overline{G}$-value (e.g.,

276 MIXING

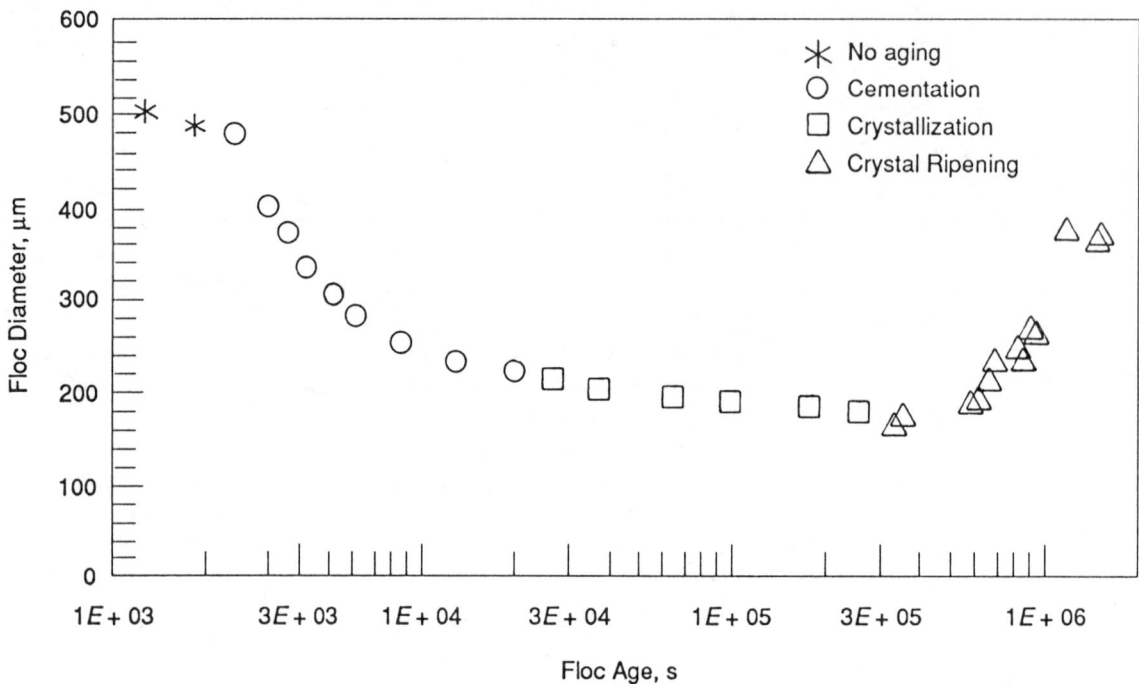

Figure 7-9 Evolution of floc size in function of floc age.

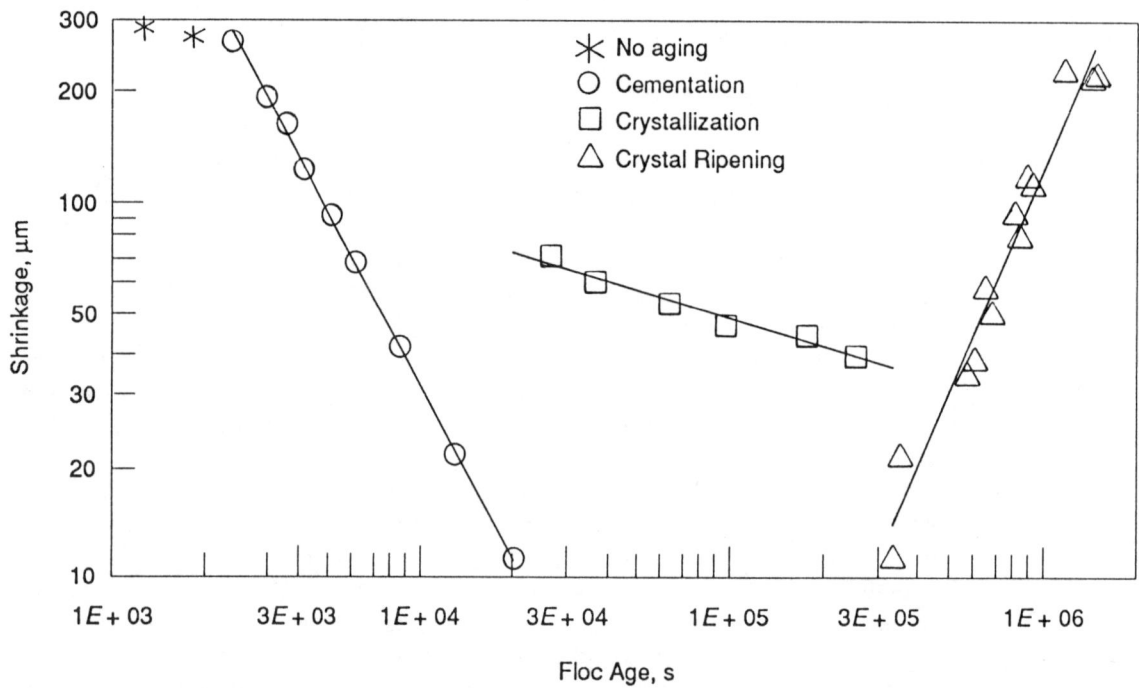

Figure 7-10 Evolution of the shrinkage in function of floc age.

1000 s^{-1}) strongly reduces the crystallization process. An increase of $\overline{G}$-value in a slow agitation, on the other hand, reduces dramatically the shrinkage due to crystallization.

Data on the shrinkage due to cementation are available (François 1987a). The pH, quantity of solids, $\overline{G}$-value in rapid and slow agitation all have an optimum for which the shrinkage is maximum; lower or higher values lead to a slowdown of the cementation process. The optima are pH 7.0, 50 to 75 mg/L solids, 400 s^{-1} and 30 to 40 s^{-1} $\overline{G}$-value.

Influence of Aging on Other Floc Characteristics

The age of a floc suspension influences not only the dimensions and filterability of the flocs but also their floc density, strength and regrowth ability, and electrophoretic mobility (François 1987a). Assuming that the shrinkage of the flocs is not due to erosion but to a reorganization of the floc structure, the effective density and porosity of a floc with age t_i can easily be expressed as a function of the diameter of the freshly formed floc and of the floc on that moment. The general expressions are:

$$\rho e | t = t_i = \frac{\alpha \cdot d_{f,o}^{3-k\rho}}{d_{f,i}^3} \qquad \text{(Eq 7-23)}$$

$$(1 - \varepsilon_f) | t = t_i = \frac{\alpha \cdot d_{f,o}^{3-k\rho}}{d_{f,i}^3 \cdot (\rho_m - \rho_w)} \qquad \text{(Eq 7-24)}$$

These expressions do not hold for the Oswald ripening period because the change in diameter is not only due to changes in floc structure. Figure 7-11 shows the effective density as it changes during the aging of the suspension. The experiment used in Figure 7-11 is the same as the one used in Figures 7-9 and 7-10.

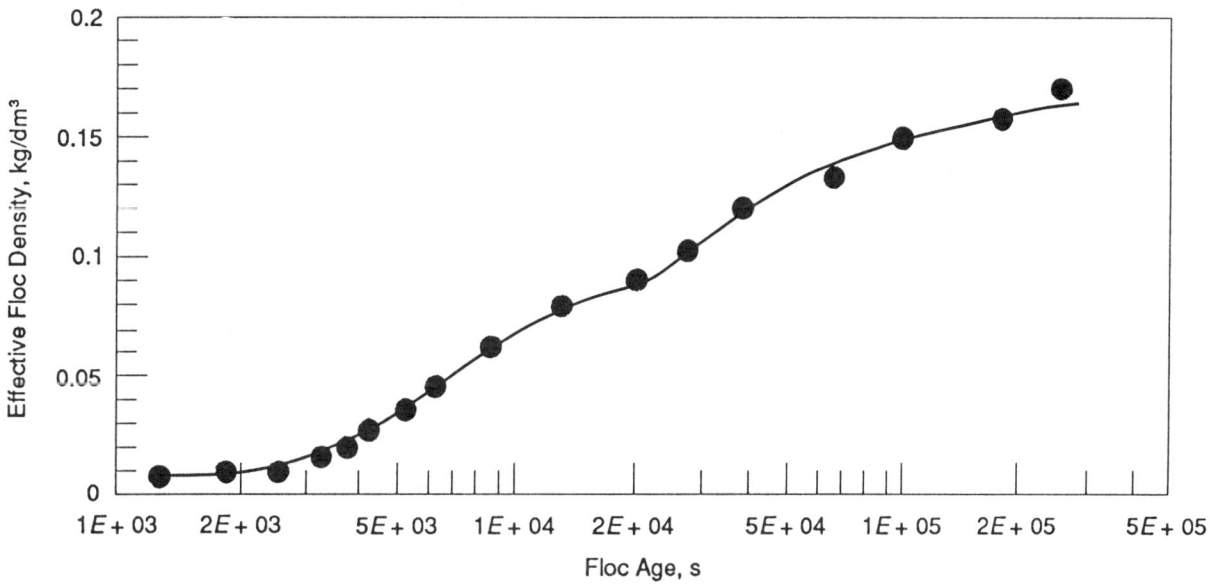

Figure 7-11 Evolution of the effective floc density in function of floc age.

278 MIXING

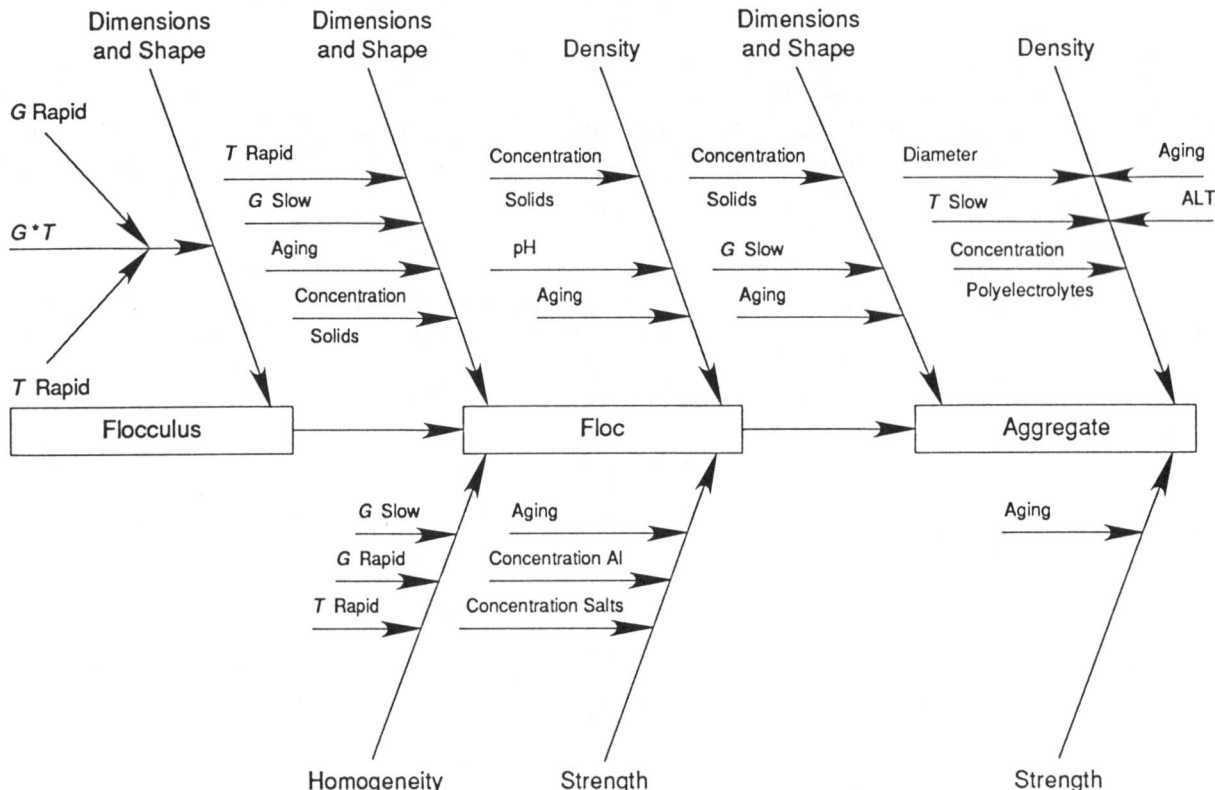

Figure 7-12 Most important influencing parameters.

Although the influence of shrinkage during crystallization on the average floc diameter is rather small, the increase in effective floc density—and thus the increase of the rate of sedimentation—is important.

It is evident that the strength and regrowth ability of the flocs are changed by the aging processes. This is clearly reflected in the strength and recovery coefficients and in the strength and recovery factors. During the aging process, the aggregates become less susceptible to shear forces. They are consolidated due to the cementation process during which the flocs form an increasing number of contact points. Also, the flocs in an aggregate consolidate during the cementation period. However, during the crystallization period the susceptibility of the flocs to a change in shear forces increases. This is due to the breakage of some internal bonds in the flocs. During the crystal ripening period, the crystallizing flocs become still more susceptible to rupture. The resistance of the total structure against rupture increases continuously except during the initial crystallization period. The ability of the total structure to recover after rupture increases continuously as a function of age. Very old flocs even regrow to agglomerates larger than the original structures.

The electrophoretic mobility of flocs in suspension remains constant for several hours. Then the electrophoretic mobility decreases rapidly because of the change of charge on the hydroxides during crystallization and because of the increasing influence of the electrophoretic mobility of the enmeshed clay particles. After about one week the electrophoretic mobility becomes constant. Because the crystallization starts from the surface of the enmeshed solids, the moment that the electrophoretic mobility starts to decrease is a measure of either or both the thickness of the hydroxide layer and the homogeneity of the hydroxides in suspension. Using this criterion, a very homogeneous distribution of hydroxides can be obtained with a rapid mixing during 60 to 120 s and a $\overline{G}$-value of rapid mixing of 400 s^{-1}. The higher the $\overline{G}$-value in a slow agitation, the more homogeneous the distribution of the hydroxides.

CONCLUSIONS AND RECOMMENDATIONS

For the rational design of flocculation and sedimentation processes, an understanding of the characteristic nature of flocs is very important. In this chapter, change of floc density with floc diameter and strength of floc, which is connected with floc size under an agitation condition, were discussed. Then details of floc structure were discussed, with multiple-level floc formation hypothesis, as was aging of floc for an extended agitation operation.

The parameters that influence the different characteristics of the different composing parts of aluminum hydroxide floc structures can be presented as a fishbone diagram (Figure 7-12). Several characteristics of floc nature can be predicted, using proposed mathematical models based on theoretical considerations or experimental measurements.

Because floc characteristics are determined by all interacting parameters (mixing conditions and other), more attention should be given to the difference between an average $\overline{G}$-value of the system and the local $\overline{G}$-value. It is also clear that a mixing system with a small spread of $\overline{G}$-value might result in a floc population with more homogeneous characteristics, which yields a better controllable process. To discuss those characteristics, dynamic analysis to describe floc size distribution with the maximum attainable floc size should be performed.

For the future, use of polyelectrolytes will be much more popular than today. Floc structures, strength and density of cationic polyelectrolyte flocs, and those of metal hydroxide flocs with anionic or nonionic polymers need to be measured.

References

ALDCROFT, D. ET AL. 1969. Crystallization Process in Aluminum Hydroxide Gels: IV, Factors Influencing the Formation of the Crystalline Trihydroxides. *Jour. Appl. Chem.*, 19:6:167.

ARATANI, T. ET AL. 1988. Evaluation of Floc Form by Fractal Dimension. *Kagaku Kogaku Ronbunshu* (Jpn. Soc. Chem. Engrg.), 14:3:395.

ARGAMAN, Y. & KAUFMAN, W.J. 1970. Turbulence and Flocculation. *Jour. San. Engrg. Div.*, SA2:4:223.

BACHE, D.H. & AL-ANI, S.H. 1989. Development of System for Evaluating Floc Strength. *Wtr. Sci. Tech.*, 21:529.

BALE, H.D. & SCHMIDT, P.W. 1959. Small Angle X-Ray Scattering from Aluminum Hydroxide Gels. *Jour. Chem. Phys.*, 31:6:1612.

BATCHELOR, G.K. 1958. *The Theory of Homogeneous Turbulence*. Cambridge University Press, London.

BROADWAY, J.D. 1978. Dynamics of Growth and Breakage of Alum Floc in Presence of Fluid Shear. *Jour. Envir. Engrg.*, 104:EE5:901.

BROWN, D.W. & HEM, J.D. 1975. Chemistry of Aluminum in Natural Water: Reaction of Aqueous Aluminum Species at Mineral Surface. U.S. Geological Survey, *Wtr. Supply Paper*, 1827-F.

CLARK, M. ET AL. 1988. Hydrodynamic Conditions of Aluminum–Humic Acid Floc. AWWA Annual Conf., Orlando, Fla.

DOMASCH, K. ET AL. 1986. Zur Physikalish Begrundeten Modellierung des Flockungs Prozessess Teil: III, Zum Zusammenhang Zwischen Flockengrosse und Flockenporositat. *Chem. Tech.*, 38:384.

DOUSMA, J. & DEBRUYN, P.L. 1978. Hydrolysis–Precipitation of Iron Solution. *Jour. Colloid Inter. Sci.*, 64:1:154.

FAIR, G.M. & GEMMELL, R.S. 1964. A Mathematical Model of Coagulation. *Jour. Colloid Sci.*, 19:4:360.

FEITKNECHT, W. 1954. Ordnungsvorgange bei Kolloiddispersen Hydroxyden und Hydroxysalzen. *Kolloid Z.*, 136:52.

FIRTH, B.A. & HUNTER, R.J. 1976. Flow Properties of Coagulated Colloidal Suspensions, *Jour. Colloid Inter. Sci.*, 57:2:248.

FRANÇOIS, R.J. 1985. "The Influence of Mixing Parameters and Water Quality on the Flocculation of Kaolinite With Aluminum Sulphate." In *Chemistry for the Protection of the Environment II*. Elsevier, Amsterdam.

———. 1987a. Aging of Aluminum Hydroxide Flocs. *Wtr. Res.*, 21:5:523.

———. 1987b. Strength of Aluminum Hydroxide Flocs, *Wtr. Res.*, 21:9:1023.

———. 1988. Growth Kinetics of Hydroxide Flocs. *Jour. AWWA*, 80:6:92.

———. 1989a. Influence of the Flocculation Process on Hydroxide Floc Structure. Unpublished.

———. 1989b. Influence of the Flocculation Process on Floc Strength and Floc Regrowth Ability. *Jour. AWWA*. In review.

FRANÇOIS, R.J. & VAN HAUTE, A.A. 1984. "Floc Strength Measurements Giving Experimental Support for a Four Level Hydroxide Floc." In *Studies in Environmental Science 23: Chemistry for Protection of the Environment*. Elsevier, Amsterdam.

———. 1985a. The Role of Rapid Mixing Time on a Flocculation Process. *Wtr. Sci. Tech.*, 17:7:1091.

———. 1985b. Structure on Hydroxide Flocs, *Wtr. Res.*, 19:10:1249.

———. 1986. The Influence of the $Al_2(SO_4)_3$ Coagulation Dosage on the Coagulation–Flocculation Process and on the Floc Characteristics. *Wtr. Supply*, 4:473.

FRICKE, R. & MEYRING, K. 1933. Zur Alterung Junger Aluminiumhydroxydgele. *Z. Anorg. Allg. Chem.*, 214:269.

FUCHS, N.A. 1964. *The Mechanism of Aerosol*. Pergamon Press, Oxford, England.

GASTUCHE, M.C. & HERBILLON, A. 1962. Etude de Gels Dalumine: Cristallisation en Milieu Desionise. Societe Chemique.

GINSBERG, H. ET AL. 1962. Uber die Bildung von Kristallinem $Al(OH)_3$ und Unwandlung von Bayerit in Hydrargillit. *Z. Anorg. Allg. Chem.*, 318:239.

GLASGOW, L.A. & JYH-PING HSU. 1982. An Example Study of Floc Strength. *AIChE Jour.*, 28:5:779.

HEINTZ, J.O. 1959. *Turbulence: An Introduction to Its Mechanism and Theory*. McGraw-Hill, New York.

HEM, J.D. & ROBERSON, C.E. 1967. Chemistry of Aluminum in Natural Water: Form and Stability of Aluminum Hydroxide Complexes in Dilute Solution. US Geological Survey. *Wtr. Supply Paper*, 1827-A.

HEM, J.D. ET AL. 1973. Chemistry of Aluminum in Natural Water: Chemical Interaction of Aluminum With Aqueous Silica at 25°C. US Geological Survey, *Wtr. Supply Paper*, 1827-E.

HOZUMI, H. & TAMBO, N. 1974. A Rational Design of Suspended Solids Contact Clarifier. *Kogyo Yosui* (Jour. Jpn. Ind. Wtr. Supply), 193:10:23.

HSU, P.H. & BATES, T.F. 1964. Formation of X-ray Amorphous and Crystalline Aluminum Hydroxides. *Min. Mag.*, 33:749.

KLIMPEL, R.C. & HOGG, R. 1986. Effects of Flocculation Conditions on Agglomerate Structure. *Jour. Colloid Inter. Sci.*, 113:1:121.

KOGA, K. ET AL. 1979. Relationship Between Basic Characteristics of Flocs and Stirring Condition. *Jour. Jpn. Wtr. Wks. Assn.*, 535:4:39.

KUSUDA, T. 1973. Effect on Floc Properties by Condition of Floc Formation. *Proc. Jpn. SCE*, 217:9:33.

LAGVANKAR, A.L. & GEMMELL, R.S. 1968. A Size–Density Relationship for Flocs. *Jour. AWWA*, 60:9:1040.

LEVICH, V.G. 1962. *Physicochemical Hydrodynamics*. Prentice-Hall, Englewood Cliffs, N.J.

MANDELBROT, B.B. 1982. *The Fractal Geometry of Nature*. Freeman, San Francisco, Calif.

MATSUO, T. & UNNO, H. 1981. Forces Acting on Floc and Strength of Floc. *Jour. Envir. Engrg.*, 108:EE3:527.

MICHAELS, A.S. & BOLGER, J.C. 1962a. Settling Rates and Sediment Volumes of Flocculated Kaolin Suspensions. *Ind. Engrg. Chem. Fund.*, 1:1:24.

———. 1962b. The Plastic Flow Behavior of Flocculated Kaolin Suspensions. *Ind. Engrg. Chem. Fund.*, 1:3:153.

PANDYA, J.D. & SPIELMAN, L.A. 1982. Floc Breakage in Agitated Suspensions: Theory and Data Processing Strategy. *Jour. Colloid. Inter. Sci.*, 90:2:517.

PARKER, D.S. ET AL. 1971. Physical Conditioning of Activated Sludge Floc. *Jour. Wtr. Poll. Cont. Fed.*, 43:9:1817.

———. 1972. Floc Breakup in Turbulent Flocculation Process. *Jour. San. Engrg. Div.*, SA1:2:79.

SMITH, R.W. & HEM, J.D. 1972. Chemistry of Aluminum in Natural Water: Effect of Aging on Aluminum Hydroxide Complexes in Dilute Aqueous Solution. US Geological Survey, *Wtr. Supply Paper*, 1827-D.

STUMM, W. & MORGAN, J. 1962. Chemical Aspect on Coagulation. *Jour. AWWA*, 54:8:971.

STUMM, W. & O'MELIA, C.R. 1968. Stoichiometry of Coagulation. *Jour. AWWA*, 60:5:514.

SUTHERLAND, D.N. 1966. Comments on Vold's Simulation of Floc Formation. *Jour. Colloid Inter. Sci.*, 22:3:300.

———. 1967. A Theoretical Model of Floc Structure. *Jour. Colloid Inter. Sci.*, 25:3:373.

SUTHERLAND, D.N. & GOODARZ-NIA. 1971. The Effect of Collision Sequence. *Chem. Engrg. Sci.*, 26:12:2071.

SUZUKI, K. ET AL. 1980. New Clarifier Is Small, Economic and Energy Efficient. *Pulp & Paper, Can.*, 81:6:1.

TAMBO, N. ET AL. 1970a. The Strength of Floc Particles. *Jour. Jpn. Wtr. Wks. Assn.*, 427:4:4.

———. 1970b. Rational Design of Flocculator (I). *Jour. Jpn. Wtr. Wks. Assn.*, 431:8:21.

———. 1979. A Mathematical Model of Turbulent Flocculation Process. *Memories of the Faculty of Engineering*, Hokkaido Univ., 15:2:163.

TAMBO, N. & HOZUMI, H. 1979. Physical Characteristics of Flocs, II: Strength of Floc. *Wtr. Res.*, 13:4:421.

TAMBO, N. & MATSUI, Y. 1989. Performance of Fluidized Pellet Bed Separator for High Concentration Suspension Removal. *Jour. Wtr. Supply Res. & Tech. (AQUA)*, 38:1:16.

TAMBO, N. & WATANABE, Y. 1967. A Study on the Aluminum Floc Density. *Jour. Jpn. Wtr. Wks. Assn.*, 392:10:2.

———. 1979a. Physical Characteristics of Flocs, I: The Floc Density Function and Aluminum Floc. *Wtr. Res.*, 13:5:409.

———. 1979b. Physical Aspect of Flocculation, I: Fundamental Treatise. *Wtr. Res.*, 13:5:429.

THOMAS, D.G. 1964. Turbulent Disruption of Flocs in Small Size Suspension. *AIChE Jour.*, 10:4:517.

TOMI, D.T. & BAGSTER, D.F. 1975. A Model of Floc Strength Under Hydrodynamic Forces. *Chem. Engrg. Sci.*, 30:3:269.

———. 1978. The Behaviour of Aggregates in Stirred Vessels, I: Theoretical Consideration on the Effects of Agitation. *Trans. Inst. Chem. Engrs.*, 56:1:1.

VAN DE VEN, T.G. & HUNTER, R.J. 1977. The Energy Dissipation in Sheared Coagulated Sol. *Rheol. Acta*, 16:534.

VOLD, M.J. 1963. Computer Simulation of Floc Formation in a Colloidal Suspension. *Jour. Colloid Inter. Sci.*, 18:7:684.

WATANABE, Y. & TAMBO, N. 1971. A Study on the Aluminum Floc Density, III. *Jour. Jpn. Wtr. Wks. Assn.*, 445:10:2.

WADELL, H. 1932. Volume, Shape, and Roundness of Rock Particles. *Jour. Geol.*, 40:443.

———. 1933. Sphericity and Roundness of Rock Particles. *Jour. Geol.*, 41:310.

WEITZ, D.A. & HUANG, J.S. 1984. Self-Similar Structures and the Kinetics of Aggregation of Gold Colloids. In *Kinetics of Aggregation and Gelation* by F. Family. Elsevier Science Publishers, New York.

8

Mixing and Scale-up

Mark M. Clark, Ph.D., Assistant Professor, Department of Civil Engineering, University of Illinois at Urbana–Champaign, Urbana, Ill. USA

François Fiessinger, Vice President, Zenon Environmental, Inc., Burlington, Ontario, Canada

INTRODUCTION

This chapter examines some traditional scale-up relationships, and reviews what is known about the sensitivity of rapid mixing and flocculation to scale. For example, if there is successful operation on the laboratory scale (good floc formation, settling of turbidity, and/or organics removal), should the same performance be expected on the prototype or plant scale? If the laboratory test is a batch test, but the plant is continuous, should that be a cause for worry? Although it does not give complete answers to all these questions, this chapter should at least develop a framework for discussion and future investigation. (Chapter 10 provides an excellent discussion of some of the more practical and managerial aspects of flocculation piloting.)

In the water treatment field, there are no quantitative scale-up procedures such as those sometimes used in the chemical and pharmaceutical industries (e.g., Bisio and Kabel 1985). In fact, most models in the water treatment field contain no scale-sensitive features at all. Is it really necessary to know anything about scale-related effects and scale-up in rapid mix and flocculation? From the traditional design standpoint, the answer is *no*. Design of full-scale plants can continue to be based on conservative standards developed through experience, and coagulant doses can still be picked by using the jar test or by trial and error. Nevertheless, as regulation of chlorinated organics, turbidity, and other compounds becomes more stringent, there may well develop a need for better understanding of scale-related phenomena, and the more subtle differences between the performance of lab and pilot or full-scale plants.

To the authors' knowledge there has never been a thorough test of flocculation model predictions in systems of different size. Models based on the Smoluchowski equations and average $\overline{G}$-values will predict the same performance in small and large tanks if the residence time distribution and $\overline{G}$-values are the same on both scales. A good example of the failure of a model to stand up to a scale test was the case of the early droplet breakup models. Droplet breakup models were originally developed assuming isotropic turbulence, which for equal residence times amounted to specifying a single parameter, the $\overline{G}$-value. Tests of these models failed when breakup results from large- and small-scale experiments were compared for the same average $\overline{G}$-values. Even though the $\overline{G}$-values were the same, droplet size distributions on the two scales were different. In other words, some scale-related effect was not captured in the breakup models (Konno, Aoki, and Saiti 1983). As discussed below, floc breakup is probably also sensitive to scale.

The rest of this chapter will examine traditional scale-up problems and laws in chemical processing, discuss likely scale-related effects in rapid mix and flocculation, and present some initial experimental work specifically aimed at uncovering sensitivity to scale. An attempt will also always be made to specify what is being scaled up. For example, does "scale-up of flocculation" mean small- and large-scale similarity of the particle size distribution function, the same removal of turbidity during settling, or the same sludge dewatering characteristics?

SCALE-RELATED EFFECTS IN NATURE AND CHEMICAL PROCESSING

The influence of scale in natural and engineered systems can be seen everywhere (Schmidt-Nielsen 1984). For example, the thickness of support beams in tall masonry buildings must increase faster than the linear scale change. This is because the weight of the building (and hence the required carrying capacity of the support beams) increases more than proportionately with the support beam cross section. At some limiting scale, the "foundation" fills the entire first floor. Of course, the development of high-strength steel freed modern architects from requiring massive load-bearing first floors. Shapes of airplanes and birds are found to change as their scale increases. For geometrically similar scale-up of a small airplane or bird, the surface area of the wing (and therefore the lifting ability) does not increase in proportion to the mass; hence, staring out a jetliner window, one may note that special modification to the wing shape is required to increase lift during takeoff and landing. It also seems that ducks and geese struggle more than sparrows during their takeoffs and landings.

Closer to the present objectives are the experiences of chemical engineers in trying to design full-scale installations based on laboratory or pilot-scale tests. It has long been known that for simple geometric scale-up of batch and continuous-flow reactors, the internal mixing effectiveness on the two scales may differ if the normalized (unit mass or volume) mixing power is just maintained constant. Scale-up deliberations are made more difficult because it is not always clear which characteristic of the mixing is most important for determining the desired "product" quality. In some cases, maintaining similarity in the residence time distribution might be most significant, while in other cases the mixing in the immediate region of the impeller or reactant injector might control product quality. In still other operations, the ability of the mixing device to maintain solids in suspension might be most critical for ensuring large-scale success. It is important to observe that in geometrically similar scale-up, required power input changes differently depending on what characteristic of the mixing must be held constant.

TRADITIONAL SCALE-UP LAWS

One of the most pervasive scale-up laws is constant average unit mass energy dissipation rate, $\bar{\varepsilon}$. Constancy of the unit mass energy dissipation rate is defined as:

$$\bar{\varepsilon}_l = \bar{\varepsilon}_s \qquad \text{(Eq 8-1)}$$

where l and s stand for large and small scales. $\bar{\varepsilon}$ can be defined in the following way:

$$\bar{\varepsilon} = \frac{P}{V\rho} \qquad \text{(Eq 8-2)}$$

Where:

P = the power input to the fluid
V = the volume of fluid in the mixer
ρ = the fluid density

It is important to note that P refers not to the power rating of the mixer but to the actual power used to mix the fluid. The familiar Camp and Stein (1943) $\bar{G}$-value is

$$\bar{G} = \left(\frac{P}{V\mu}\right)^{1/2} \qquad \text{(Eq 8-3)}$$

Therefore, for constant temperature, $\bar{\varepsilon}_l = \bar{\varepsilon}_s$ implies that $\bar{G}_l = \bar{G}_s$. Since in the flocculation context, $\bar{G}$ refers to small-scale particle transport, it is perhaps not surprising that certain micromixing-sensitive chemical reactions appear to observe the constant $\bar{\varepsilon}$ scale-up law (see following section and chap. 5). Nevertheless, shouldn't reactions like flocculation and coagulant precipitation also be sensitive to features of the larger-scale mixing? One characteristic of this macromixing is the circulation or turnover time

$$\tau_c = \frac{V}{q} \qquad \text{(Eq 8-4)}$$

where q is the impeller pumping rate. For the traditional backmix reactor, q has been related to impeller geometry and speed (Uhl and Gray 1966):

$$q = N_q N D^3 \qquad \text{(Eq 8-5)}$$

Where:

N = the impeller speed
D = the impeller diameter
N_q = the impeller pumping number

Then for the classic batch scale-up "law" of constant turnover time, τ_c (small scale) = τ_c (large scale), there is the relation

$$\frac{V_l}{N_q N_l D_l^3} = \frac{V_s}{N_q N_s D_s^3} \quad \text{(Eq 8-6)}$$

Assuming that the flow in the mixing vessel is fully turbulent, one can also assume that N_q is constant.* Canceling N_q and noting for geometrically similar scale-up that $V_s = k_v D_s^3$ and $V_l = k_v D_l^3$ produces the simple requirement that

$$N_l = N_s \quad \text{(Eq 8-7)}$$

In other words, for geometrically similar vessels in which ratios of dimensions describing the geometrical features are held constant, constant turnover time implies constant mixer speed. To derive an expression for the power requirement implied by this scale-up law, there is the correlation (Uhl and Gray 1966)

$$P = P_0 N^3 D^5 \quad \text{(Eq 8-8)}$$

Assuming fully developed turbulence on the large and small scales (hence P_0 = constant), Eq 8-7 and 8-8 can be used to arrive at

$$\frac{P_l}{P_s} = \frac{D_l^5}{D_s^5} \quad \text{(Eq 8-9)}$$

Normalizing P_l and P_s by $\rho k_v D_l^3$ and $\rho k_v D_s^3$, respectively, yields

$$\bar{\varepsilon}_l = \frac{\bar{\varepsilon}_s}{D_s^2} D_l^2 \quad \text{(Eq 8-10)}$$

Finally, since one can consider $\bar{\varepsilon}_s$ a constant for fixed laboratory conditions, Eq 8-10 implies that

$$\bar{\varepsilon}_l \propto \left(\frac{D_l^2}{D_s^2}\right) = S^2 \quad \text{(Eq 8-11)}$$

where S is the linear scale-up factor. Also note that for geometrically similar vessels the tank diameter will always be proportional to the impeller diameter; hence

$$\frac{D_l}{D_s} = \frac{D_{vl}}{D_{vs}} = S \quad \text{(Eq 8-12)}$$

If $\log \bar{\varepsilon}_l$ is plotted against $\log S$, Eq 8-11 plots as a straight line with slope of 2. The implications of this scale-up law should be noted: *unit mass* energy input increases as the square of the scale factor, and P increases as S^5. The same scale-up law

---

*In comparing different-sized vessels at the same unit mass energy dissipation rate or same $\bar{G}$, it is sometimes found that while the flow in the larger vessel will be fully developed turbulence, the flow in the smaller vessel might be in transition. This would violate the assumption of constancy of N_q. See "Scale-up Experiments in Flocculation," p. 297.

(Eq 8-11) results for continuous-flow reactors if the mean residence time $\bar{t}$ and the ratio r_t of the throughput flow rate Q and the impeller pumping rate are held constant (Nienow et al. 1974):

$$\bar{t} = \frac{V}{Q} \qquad \text{(Eq 8-13)}$$

$$r_t = \frac{q}{Q} = \frac{\bar{t}}{\tau_c} \qquad \text{(Eq 8-14)}$$

Beresford et al. (1970) have suggested that for continuous-flow reactors, the scale-up law of Eq 8-11 is useful in maintaining residence time distribution (RTD) similarity.

Another classic scale-up law is derived by specifying constant impeller Reynolds number, where

$$\text{Re}_p = \frac{ND^2}{\nu} \qquad \text{(Eq 8-15)}$$

Holding Re_p constant on small and large scales results in the relation

$$\frac{N_l}{N_s} = \left(\frac{1}{S}\right)^2 \qquad \text{(Eq 8-16)}$$

Using Eq 8-8, one can eliminate N in Eq 8-16 and, with a little algebra, arrive at

$$\bar{\varepsilon}_l \propto \left(\frac{1}{S}\right)^4 \qquad \text{(Eq 8-17)}$$

Notice the stark contrast between the dependence on S in Eq 8-11 and 8-17.

Constant impeller tip speed has been suggested as a possible scale-up law for maintaining solids in suspension. This suggestion presumably derives from the observation that point velocities in geometrically similar vessels are roughly constant if normalized by the impeller tip speed ND. Then holding ND constant yields the relation

$$\frac{N_l}{N_s} = \frac{1}{S} \qquad \text{(Eq 8-18)}$$

Again using Eq 8-8 to eliminate N as above produces the scale-up law

$$\bar{\varepsilon}_l \propto \frac{1}{S} \qquad \text{(Eq 8-19)}$$

which is intermediate between Eq 8-11 and 8-17.

The final scale-up law results from holding constant the micromixing time (Corrsin 1957)

$$t_m = \text{const.}\left(\frac{L^2}{\bar{\varepsilon}}\right)^{1/3} \qquad \text{(Eq 8-20)}$$

where L is the concentration macroscale. Now, holding t_m constant on small and large scales, while also assuming that L is proportional to the impeller diameter (and for geometrically similar vessels, the tank diameter), yields the scale-up relation

$$\bar{\varepsilon}_l \propto S^2 \qquad \text{(Eq 8-21)}$$

which is, of course, the same relation (Eq 8-11) obtained by assuming constant turnover time. It is very interesting that laws related to the small-scale and large-scale fluid mechanics are the same.

The "Penney" diagram has been used to illustrate how unit power input to a reactor must change on scale-up (Penney 1971), and it is used here to illustrate four scale-up laws (Figure 8-1). Note that, depending on the mixing requirement, drastically different scale-up is suggested.

INITIAL MIXING

It is accepted that the initial dispersion of destabilizing chemicals in the huge raw water flows characteristic of municipal water treatment facilities must be done in relatively short-detention-time, often high-intensity mixing units (rapid or initial mixers). The traditional rationale for this unit process has been to ensure chemical uniformity or homogeneity at the very beginning of the treatment system. There also seems to be an increasing awareness that intense initial mixing may be required due to the speed of the hydrolysis and precipitation reactions that occur when metal salts such as alum or ferric chloride are added to a raw water. The

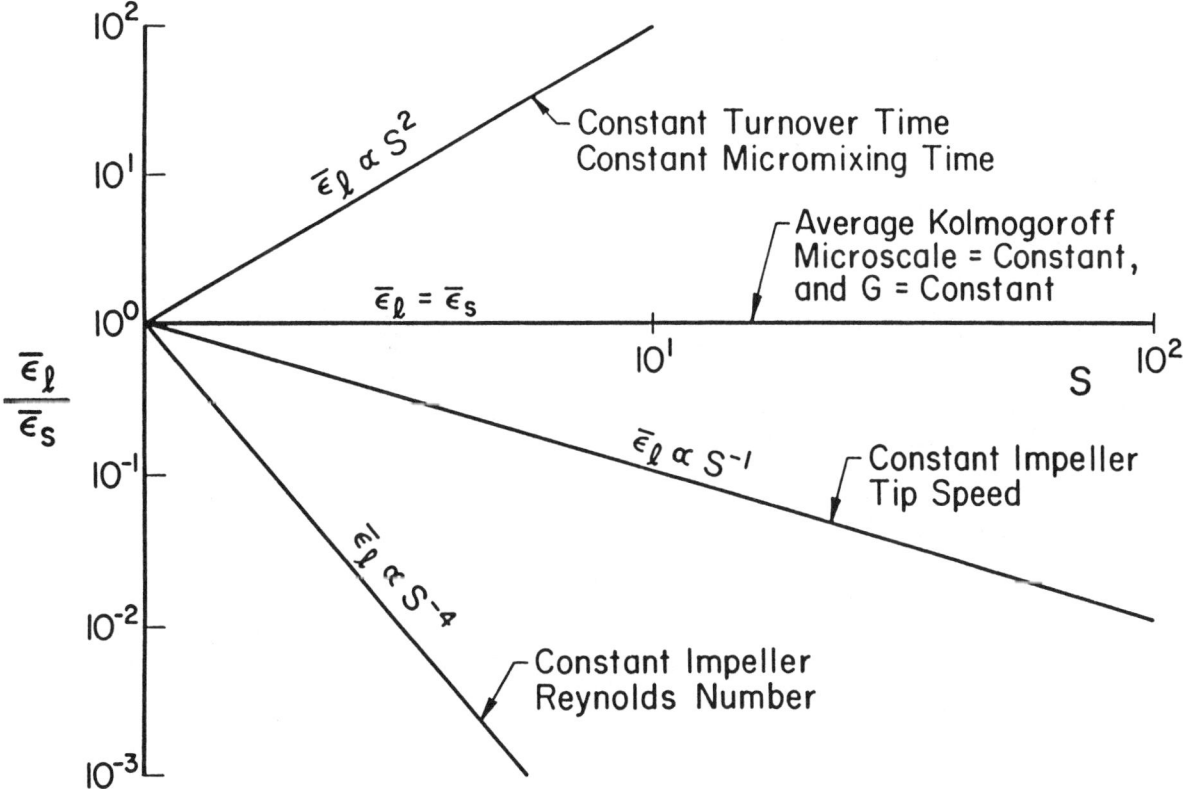

Figure 8-1 Traditional scale-up laws.

importance of reaction speed in aluminum and iron precipitation was pointed out in chap. 5. When complex reactions are much faster than the mixing, then the reaction can become transport limited and very sensitive to the details of the small-scale mixing.

Data from Charles et al. (1987) support the contention of the importance of initial mixing in coagulation and flocculation. In these experiments, a 10-mg/L solution of humic acid was flocculated with preneutralized aluminum chloride (OH/Al ratio = 2.0). Flocculation performance was monitored during modified jar tests, in which different types of initial mixing were investigated. Using a syringe assembly and small-bore glass tube, coagulant was added at either of two locations—1 cm below the surface or at the impeller level. Two initial mix $\overline{G}$-values were investigated for the different injection locations, $\overline{G}$ = 80 s^{-1} and 1440 s^{-1}. During the standard flocculation period of 30 min, the $\overline{G}$-value was 50 s^{-1}. Therefore, four different initial mixing conditions were evaluated. The most intense mixing should correspond to paddle-level injection at $\overline{G}$ = 1440, while the least intense mixing would correspond to surface-level injection at $\overline{G}$ = 80. Flocculation performance was evaluated by monitoring turbidity and organic removal during a 60-min settling period. Some of the results of these experiments are shown in Figure 8-2. Surprisingly, the least intense mixing results in the best treatment, while the most intense

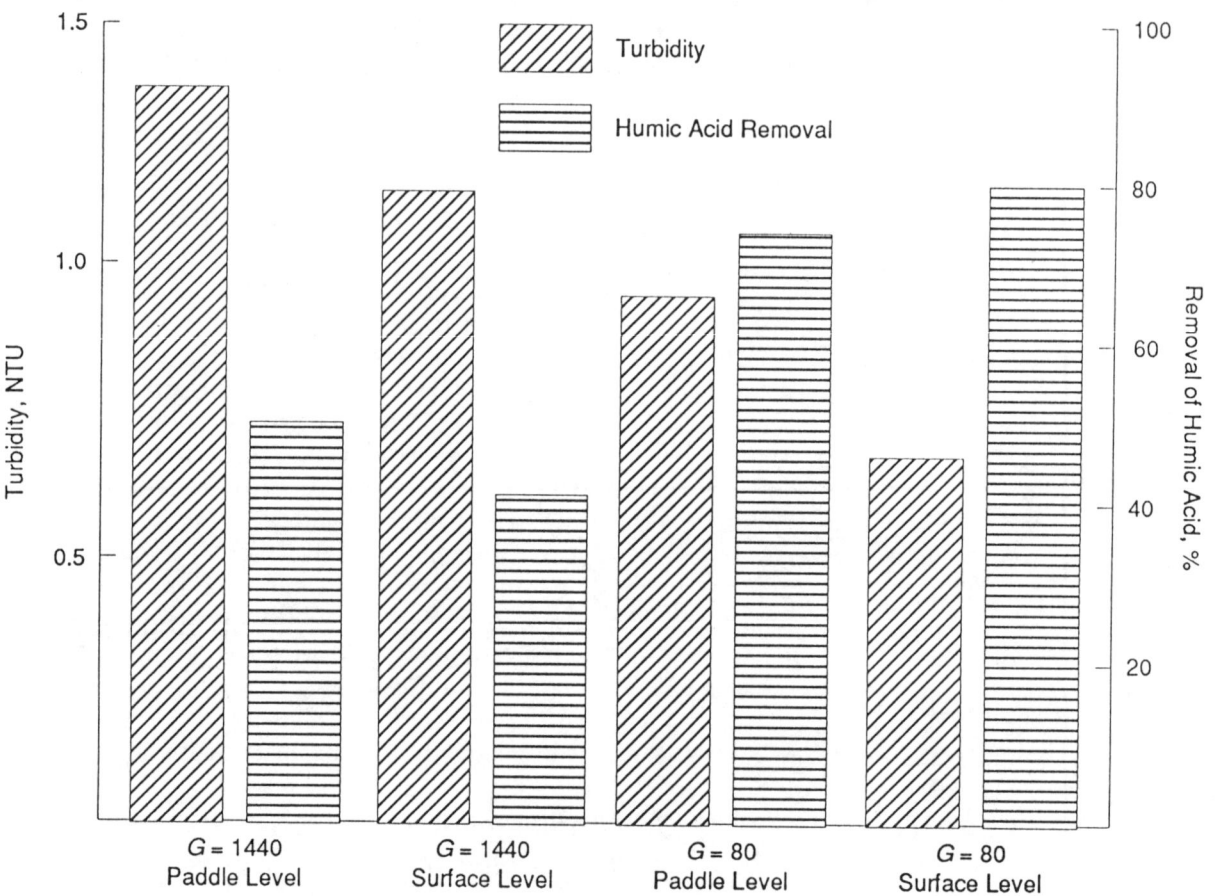

Source: Charles, P. et al. 1987. Influence des Conditions de Melange Initial sur la Coagulation. AGHTM Conf., Nice, France.

Figure 8-2 Influence of initial mix on flocculation settling.

mixing results in the poorest performance. This result seems to correspond to anecdotal reports from the field, where treatment plants sometimes seem to work just as well when the initial mix turbines are turned off. Of course, one could argue that in these plants the mixing was so bad to begin with that turning off the mixer made little difference. Nevertheless, an open mind should be maintained about the specific type of mixing required for particular coagulant–contaminant combinations.

Needless to say, there seems to be much that is not understood about initial mixing and coagulant precipitation. A look at well-studied reactions from the chemical engineering literature may provide clues to unraveling the mysteries of initial mixing. The competitive–consecutive reaction has been studied extensively by chemical engineers:

$$A + B \xrightarrow{k^1} R \qquad \text{(Eq 8-22)}$$

$$R + B \xrightarrow{k^2} R_s \qquad \text{(Eq 8-23)}$$

The intermediate product R is produced by reaction of A and B; afterward A and R can be thought of as competitors for the common reactant B. Such reactions are thought to occur in a wide variety of polymerization reactions, as well as the aluminum hydrolysis reactions given in chap. 5. When A and B are added to a mixed reactor, the speed of the reaction will determine the extent of the region in which A and B are actually reacting. For slow reactions, the concentration of A and B can be homogenized down to the molecular scale before much reaction has taken place; hence, the reaction occurs throughout the mixing vessel. For infinitely fast reactions, however, since the reaction is controlled by molecular diffusion and simple viscous stretching (micromixing), the reaction zone shrinks to the plane separating A- and B-rich zones.

For reactions such as Eq 8-22 and 8-23, conversion can depend on details of both the micro- and macromixing. A localized picture of the situation for the fast-reaction case is shown in Figure 8-3. Here, a thin reaction zone separates globs of solution A and B. Figure 8-3 suggests that the reaction and mixing have been selected for R_s, the secondary product. Levenspiel (1972) cites a probability calculation with which it can be shown that as the reaction speed increases, intermediate R has a decreasing chance of escaping B-rich zones and thus being converted to R_s; hence, poor micromixing favors R_s, while perfect mixing favors R.

Selectivity in consecutive–competitive reactions also depends on the method of contacting the reactants. A simple thought experiment shows that when A is slowly added to a B-rich solution, the formation of R_s is favored: the R formed in Eq 8-22 is immediately converted to R_s (Eq 8-23) due to the high concentration of B. However, when B is slowly added to an A-rich solution, all of the B is initially consumed in the first reaction. Not until a sufficient amount of R is created can R successfully compete with A for B. In this method, selectivity for the intermediate R can be optimized.

Bourne and co-workers (Bourne, Kozicki, and Rys 1981; Bourne et al. 1981; Belevi, Bourne, and Rys 1981) studied mixing and fast chemical reaction for the consecutive–competitive reaction, Eq 8-22 and 8-23. The reaction studied was the coupling of 1-naphthol (A) with diazotized sulfanilic acid (B) in a dilute alkaline solution. The reaction products, a monazo dyestuff (R) and a bisazo dyestuff (R_s), were measured spectrophotometrically. A and B were injected simultaneously into a conventional turbine mixer operated in both the continuous and semicontinuous modes, and the relative product distribution of R and R_s was determined. This distribution was expressed using the segregation index

$$X = \frac{2[R_s]}{[R] + 2[R_s]} \quad \text{(Eq 8-24)}$$

Therefore, $X = 0$ for perfect mixing and $X = 1$ for perfect segregation. Consistent with the above qualitative discussion, Bourne and co-workers found that decreases in segregation occurred for higher impeller speeds (greater $\bar{\varepsilon}$), lower concentration of B, and higher stoichiometric ratio of A to B. In modeling this process, Bourne and co-workers imagined that small spherical B-rich globs were created initially into which A would diffuse during the lifetime of the glob. Hence, they solved the diffusion equation for A in a spherical volume, and product distribution was determined according to Eq 8-22 and 8-23 and appropriate equilibrium relationships. This resulted in the theoretical prediction that X would be a function of the lifetime of the spherical glob, the concentration and feed rate of the reactants, k_1/k_2, and the second Damköhler number

$$\text{Da}_{\text{II}} = \frac{k_2 {R'}^2 [B]_O}{D_m} \quad \text{(Eq 8-25)}$$

Where:

k_2 is from Eq 8-23
R' = the initial radius of the B-rich spheres
$[B]_O$ = the initial concentration of B

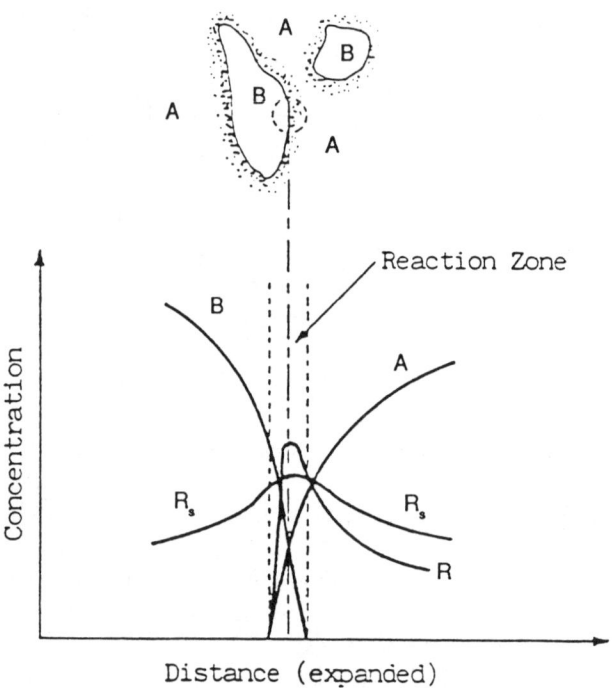

Source: Levenspiel, O. Chemical Reaction Engineering. John Wiley and Sons, New York, Copyright 1972.

Figure 8-3 Concentration profiles for fast competitive–consecutive reaction.

Segregation data (Eq 8-24) from their experiments could be fairly well fit when R' was set equal to about half the spatial average Kolmogoroff microscale

$$\eta = \left(\frac{\nu^3}{\bar{\varepsilon}}\right)^{1/4} \qquad \text{(Eq 8-26)}$$

Bourne and co-workers found that when A and B solutions were injected near the impellers in different-sized but geometrically similar vessels, a particular segregation X could be scaled-up by maintaining an approximately constant, average unit mass energy input rate $\bar{\varepsilon}$. This result might have been anticipated, since Eq 8-26 will be constant for constant $\bar{\varepsilon}$ (and temperature). However, when reactant B was injected near the surface of the reactor, a higher unit power input was required on scale-up, which was explained by invoking the importance of convection or turnover time (i.e., the gross flow patterns). The approximate scale-up law for the "bad" reactant introduction was $\bar{\varepsilon}_l \propto S$. Notice that this scale-up relation is intermediate between $\bar{\varepsilon}_l$ = constant and Eq 8-11. It is probable that the gross flow patterns become more important as the reactant injection point is moved farther from the impeller.

One of the limitations of the modeling of Bourne and co-workers was in not capturing the effect of the reactant injection location. In the typical Rushton turbine, the local Kolmogoroff microscale can vary spatially by a factor of 10^2 due to the large spatial variations in ε. Also, for finite reaction times, the initial B-rich glob would experience a continuous variation in mixing intensity as it is convected from the injection point to other regions of the vessel. For unpremixed reactants, it would, therefore, be useful to describe both the macro- and micromixing phenomena. In chap. 5, in the section on "Mixing and Aluminum Precipitation," a different mixing and chemical reaction model has been developed for aluminum neutralization with base. This is a more complex model than that of Bourne and co-workers, because it models both micro- and macromixing. Because of this, the model has more time scales, hence more possible scale-up relationships. For example, in addition to micromixing and turnover time scales (Eq 8-21 and 8-11), the model includes the vortex incorporation time

$$t_{\text{inc}} = 12.2 \left(\frac{\nu}{\bar{\varepsilon}}\right)^{1/2} \qquad \text{(Eq 8-27)}$$

as well as the turbulent cloud diffusivity

$$D_T = C_{DT} D^2 N \qquad \text{(Eq 8-28)}$$

Equation 8-27 leads, of course, to a scale-up law of $\bar{\varepsilon}_l$ = constant. Using the same methods developed in the previous section, Eq 8-28 leads to the scale-up law $\bar{\varepsilon}_l \propto S^{-4}$ (Eq 8-17). Hence, a model for a seemingly simple process like the base neutralization of aluminum can have a number of apparently conflicting scale-up laws. Conflicts in scale-up laws are, however, not new to engineers. For example, those who construct physical models of fluid mechanical systems such as channels, rivers, and bays invariably find a conflict between the requirement of constant Reynolds and Froude numbers. Depending on the goal of the scale-up, certain scale-up laws may exert more influence than others. Multi-time scale models such as those described above, as well as future models of more realistic coagulant precipitation and speciation, will ultimately rely on experiments for elucidation of the most influential scale-up law.

In summary, little is known about rapid mix, much less any sensitivity to scale. However, the models and data reviewed suggest the need to be on the lookout for certain effects. From what is presently known, it can be speculated that since coagulant precipitation is sensitive to both micro- and macromixing, scale-up must consider not only energy dissipation rate but also the reactant injection point and the contacting method. The numerous time scales presented suggest numerous theoretical scale-up relations. Well-conducted experiments should uncover the controlling mechanism(s).

FLOCCULATION

In flocculation, it is of interest to characterize the reaction speed relative to the mixing time scales. The classic expression for the initial decay rate of a monodispersion is

$$\frac{dn}{dt} = \frac{-2\alpha d^3}{3} \overline{G} n^2 \qquad \text{(Eq 8-29)}$$

Where:

n = the concentration of the monosized particles
α = the collision efficiency factor ($0 < \alpha < 1$)
d = the particle diameter
$\overline{G}$ = characteristic velocity gradient (Eq 8-3)

The solution for the second-order equation (Eq 8-29) is

$$n^{2nd} = \frac{n_0}{1 + K_c \overline{G} t} \qquad \text{(Eq 8-30)}$$

where K_c is the dimensionless reaction constant

$$K_c = 2/3 \, \alpha d^3 n_0 \qquad \text{(Eq 8-31)}$$

From Eq 8-30, the flocculation half time (i.e., $n/n_0 = 0.5$) is

$$t_{1/2} = K_c^{-1} \overline{G}^{-1} \qquad \text{(Eq 8-32)}$$

For all typical values for $\overline{G}$, d, n_0, and α, flocculation half times are quite long, on the order of hours and even days. Comparing these half times with the mixing time scales in Figure 5-19 (chap. 5), it is readily evident that typical turnover, residence, and mixing times are generally orders of magnitude less than flocculation half times. Therefore, it can be concluded as a general result of the principle of time scale separation (e.g., Mercier 1984) that flocculation rate as modeled by Eq 8-29 should be independent of the subtleties of micromixing. Rather, average detention times, the RTD, average concentration gradients, and average power input would be expected to be more important in scale-up. This is one of the major contrasts between initial mix and flocculation.

Since slow reactions are often approximated as pseudo first order, any non-linearity can be investigated further in Eq 8-29. For example, Eq 8-29 can be linearized as in Weber (1972):

$$\frac{dn}{dt} = -K_c \overline{G} n \qquad \text{(Eq 8-33)}$$

the solution being

$$n^{1st} = n_0 \exp(-K_c \overline{G} t) \qquad \text{(Eq 8-34)}$$

In Figure 8-4, the ratio of Eq 8-34 to Eq 8-30 is plotted for 1-μm particles and $\alpha = 0.5$, and for a wide range of n_0 and $\overline{G}t$. Here the bimolecular or second-order kinetics are well-approximated by linear kinetics. There is a simple mathematical explanation for this, demonstrated by expanding the exponential in Eq 8-34:

$$\exp(-K_c \overline{G} t) = (1 + K_c \overline{G} t + 1/2! \, K_c^2 \overline{G}^2 t^2 + \ldots)^{-1} \qquad \text{(Eq 8-35)}$$

Note that for sufficiently small values of $K_c \overline{G} t$, the exponential is sufficiently approximated by the first two terms in the expansion. Hence, accepting the limitations implied in the use of these equations (notably the strict applicability to only the initial flocculation rate of a monodispersion), it can be concluded that flocculation rate has strong linear tendencies.* As pointed out in chap. 4 and 5, the yield from linear reaction kinetics is *uniquely* determined by the RTD and reaction kinetics. This is another reason why flocculation *rate* would be insensitive to micromixing.

This linearity should have some implications in scale-up (Clark 1986). For the simple kinetic scheme used above, it might be sufficient for scale-up to specify RTD similarity. This makes scale-up of batch flocculation relatively simple, since residence time similarity is achieved simply by maintaining the same processing or flocculation time, and reaction rate similarity is obtained simply by starting with the same solution and using the same $\overline{\varepsilon}$- or $\overline{G}$-value. For scale-up of continuous-flow flocculators, a somewhat different story emerges, since a rather severe scale-up of $\overline{\varepsilon}$ may be required (Eq 8-11). However, since $\overline{G}$ is involved both in the macromixing and the reaction rate, a compromise scale-up law is evidently involved. In the absence of breakup effects (see below), it is likely then that $\overline{G}$ = constant for scale-up of batch flocculation, while $\overline{\varepsilon} \propto S^b$ for continuous-flow flocculation, with $0 < b < 2$.

The above discussion concerns the ideal case of irreversible particle aggregation, or flocculation. In reality, aggregates and flocs are also deaggregating and breaking up due to forces exerted by the surrounding fluid and flow boundaries. Therefore, it is important to be aware that during "flocculation" in water treatment,

---

*Typically in the water treatment literature, the transformation from Eq 8-30 to 8-34 is stated to derive from either the assumption that floc volume concentration is constant during flocculation, or that volume is conserved when two floc particles aggregate. In reality, these statements appear to be excuses for the transformation. As shown above, the linear transformation can work *mathematically* only if the reaction is relatively slow and does not go to completion. It would be coincidental if floc volume is conserved or if the "coalescing spheres" assumption holds. On the other hand, there may be other physical explanations for first-order kinetics (for example, see Friedlander [1977]).

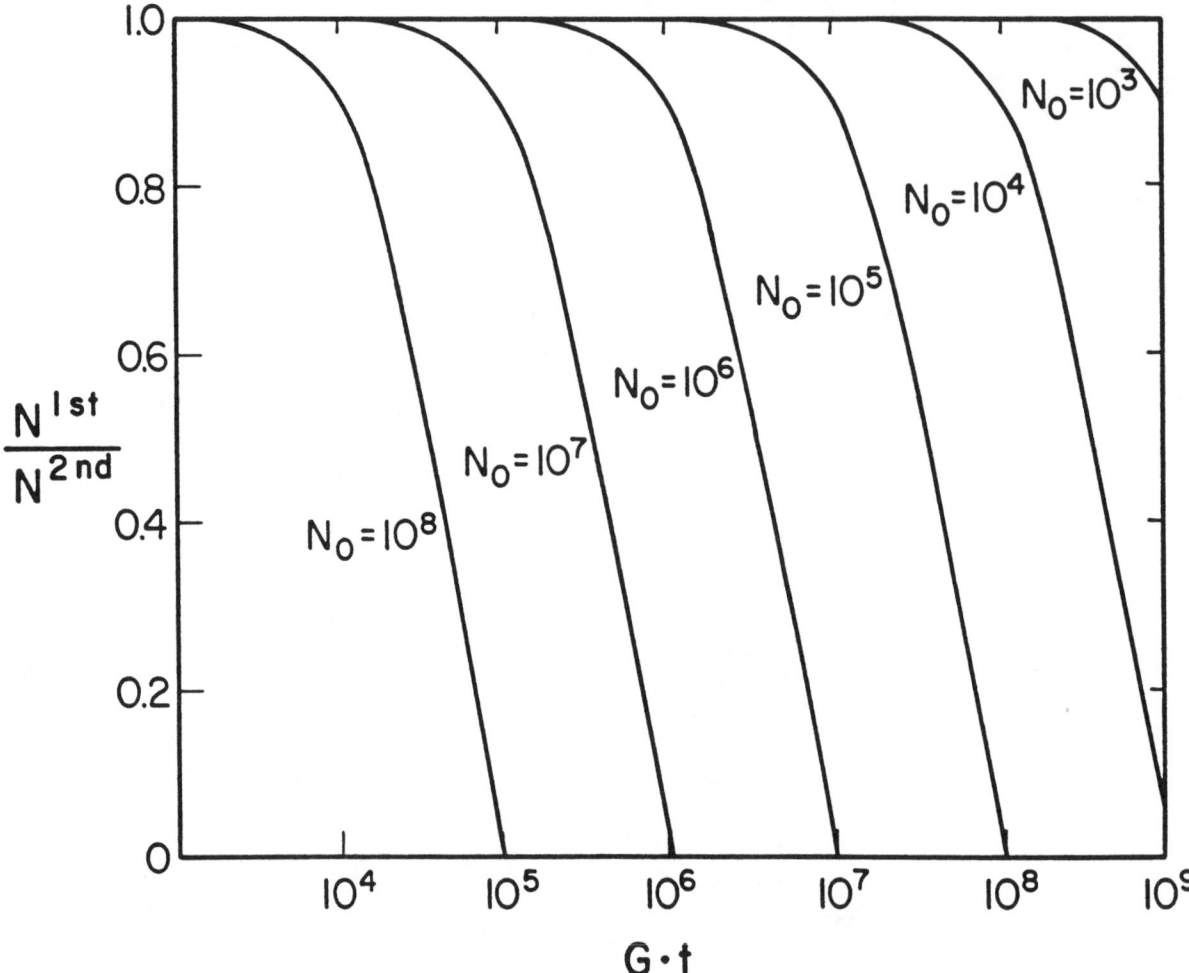

Figure 8-4 Comparison of first- and second-order coagulation kinetics.

both aggregation and deaggregation are occurring simultaneously. In a sense then, flocculation might be thought of as a reversible reaction

$$\text{small floc and particles} \underset{\text{breakup}}{\overset{\text{flocculation}}{\rightleftarrows}} \text{larger flocs and particles} \quad \text{(Eq 8-36)}$$

It is interesting that the forces that bring particles together and tear them apart are both generated by the stirring of the vessel. This implies some optimum mixing intensity for successful flocculation.

It is also believed that there may be an approximate maximum stable size for the floc, suggesting that flocs that grow above this size are likely to be broken up. For example, the following relation

$$d_{\max} = \frac{c}{G^e} \quad \text{(Eq 8-37)}$$

is observed experimentally and can be derived assuming isotropic turbulence (Parker, Kaufman, and Jenkins 1972). Here c is related to the floc strength and $\overline{G}$ is related to the disruptive forces in the surrounding fluid. However, as opposed to flocculation, the breakup rate law is not easy to postulate because of difficulties in representing the solid mechanics of the floc and uncertainty in how fluid forces are exerted on the floc (e.g., through viscous or inertial fluid forces?). In addition, the individual floc structures are known to be nonhomogeneous, and this structure varies randomly from floc to floc. This also implies that breakup fragments are not uniform, which has been observed experimentally. Parker and co-workers (1972) modeled breakup in flocculators by assuming isotropic turbulence, simple erosion and splitting mechanisms, and very simplified fragment size distributions. Pandya and Spielman (1982) modeled breakup with more general rate and "daughter" size expressions, but ended up with numerous unknown parameters.

As opposed to coagulation, there is a quickly growing consensus that breakup has less to do with average turbulence parameters, such as the average energy dissipation rate, and more to do with extremes in flow. Müller (1982) found that plots of maximum floc size versus average energy dissipation rate for different D/D_v ratios would collapse to the same curve if replotted as maximum floc size versus energy dissipation in the impeller region. Clark (1988) provided photographic evidence that during emulsification, oil droplets broke up exclusively in the region directly behind a special grid-type impeller. This study also showed that droplet breakup was probably due to the large pressure fluctuations experienced by the droplets in the impeller region. Konno, Aoki, and Saito (1983) also showed striking photographic evidence that during emulsification in the conventional turbine mixed-flow reactor, a large proportion of droplets was broken directly behind the impeller blade. The breaking drops seemed to follow the outward flow behind and along the impeller blade, taking a roughly helical path. Further, they showed that the dynamics of the droplet size distribution function could not be modeled assuming isotropic turbulence; they found that the drop breakup rate expression had to include two terms

$$k_b(d) = k_{bB}(d) + k_{bA}(d) \qquad \text{(Eq 8-38)}$$

Where:

$k_{bB}(d)$ = the breakup rate in the roughly isotropic region
$k_{bA}(d)$ = the breakup rate in the nonisotropic region behind the impeller

and

$$K_{bA}(d) = C_1 N \int_{u_{\text{crit}}}^{\infty} P[u(d)] \, d[u(d)] \qquad \text{(Eq 8-39)}$$

Where:

C_1 = a constant for geometrically similar vessels
P = the probability density for the relative velocity $u(d)$
u_{crit} = the "critical" breakup velocity

Konno, Aoki, and Saito (1983) also compared droplet distribution dynamics for geometrically similar batch-operated vessels, and found that for equal average energy dissipation rate $\bar{\varepsilon}$, breakup occurred faster as the vessel size was increased.

This suggests that energy input scales with some inverse power of the characteristic vessel dimension.

Konno, Aoki, and Saito (1983) and others have noted that the assumption of isotropy of the turbulent fluctuations always leads to the scale-up law $\bar{\varepsilon}_l$ = constant (Eq 8-1). This can be seen in a simple way by noting in Eq 8-37 that d_{max} is a constant for $\bar{\varepsilon}$ = constant. Accepting Eq 8-38 and 8-39 and the observation that droplet size in geometrically similar batch reactors is reduced faster on scale-up with $\bar{\varepsilon}$ = constant, the following observations can be made:

- Since $k_b(d)$ increases and $k_{bB}(d)$ remains constant, then $k_{bA}(d)$ must increase with increasing scale.

- Using the approach presented in the section "Traditional Scale-up Laws," N in Eq 8-39 decreases as $D_v^{-2/3}$. Since C_1 is a constant, this implies that the integral is increasing greater than in proportion to $D_v^{2/3}$.

These observations indicate that even though the circulation rate through the impeller region *decreases* with increasing scale (and $\bar{\varepsilon}$ = constant), therefore indicating a lower chance (on average) of breakup, this effect is more than offset by a breakup mechanism of *increasing* intensity. A closer look at the probability density of the nonisotropic turbulence fluctuations proposed by Konno, Aoki, and Saito (1983) helps explain this last observation. $P[u(d)]$ was assumed to be a normal distribution

$$P[u(d)] = \sqrt{\frac{\overline{u^2(d)}}{2\pi}} \exp\left\{-\frac{u^2(d)}{2\overline{u^2(d)}}\right\} \qquad \text{(Eq 8-40)}$$

Since the mean square value, $\overline{u^2}(d)$, is the parameter of the distribution, scale-up should follow from similarity of $\overline{u^2}(d)$. In mixed reactors, $\overline{u^2}(d)$ is usually given by

$$\overline{u^2}(d) \propto (NT)^2 \qquad \text{(Eq 8-41)}$$

Using the techniques of traditional scale-up laws, Eq 8-41 yields the same scale-up law (Eq 8-19) as constant tip speed, $\bar{\varepsilon}_l \propto S^{-1}$. Apparently, this scale-up law holds more sway in the breakup of liquid droplets in batch reactors, because S in the correct scale-up law obviously has a negative exponent.

Van't Riet and Smith (1975) have shown that behind the conventional turbine impeller, a pair of vortices is developed (Figure 8-5). It is interesting that a particle in the vortex would apparently have the same motion as the breaking drops observed by Konno, Aoki, and Saito (1983). Further, van't Riet and Smith have shown that within the vortex, nondimensional velocity, pressure, centrifugal acceleration, and shear rate are related to the paddle Reynolds number. These results are consistent with the scale-up behavior noted by Konno, Aoki, and Saito, since the breakup mechanism is at least partially related to the fluid mechanics in the impeller region.

It is likely that the discussion of droplet breakup is also relevant to the breakup of flocs. It is reasonable to expect that if the breakup of a dispersed oil phase is controlled by the fluid mechanics in the impeller region, the same would hold for the breakup of a dispersed (and fragile) floc phase. Nevertheless, in actual flocculation reactors, it is not known whether scale phenomena related to flocculation or breakup are more important. Lacking a calibrated aggregation–deag-

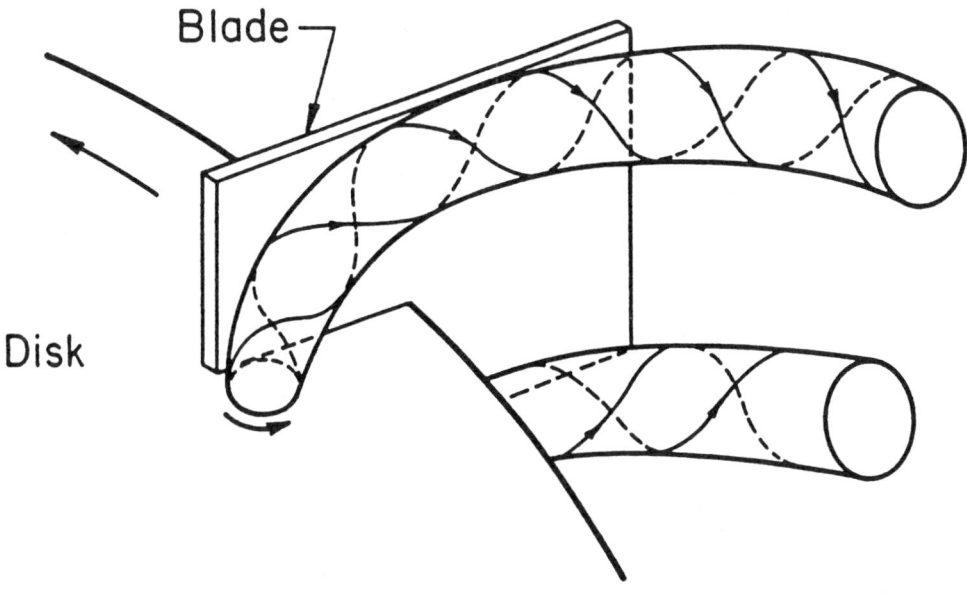

Source: Van't Riet, K. & Smith, J.M. 1975. The Trailing Vortex System Produced by Rushton Turbine Agitators. Chem. Engrg. Sci., *3:1093.*

Figure 8-5 Trailing vortex system behind the disk turbine impeller blade.

gregation model with well-modeled hydrodynamics, the only choice is to run some scale-up experiments.

SCALE-UP EXPERIMENTS IN FLOCCULATION

Little information is available in the flocculation literature on scale effects. Before the existing data on scale-up are presented, the design of scale-up experiments should be discussed. It should be clear from the preceding material that geometric similarity allows enormous simplifications. One need only look at standard texts on the dimensional analysis of power consumption in reactors to see that the list of possible influential dimensionless groups drops precipitously once one assumes geometric similarity (e.g., Uhl and Gray 1966). Geometric similarity has some important implications if, by "scale-up," one means using a 1-L vessel to simulate a 10-mgd compartmentalized flocculator. The lab mixing vessel might look a little like an ice-cube tray if strict geometric similarity were maintained (see "Summary," p. 302). On the other hand, if the intent is to precisely quantify laboratory-observable scale effects in flocculation, then strict geometric similarity should be maintained.

It should be evident from the preceding development that an equally important requirement in scale-up experiments is to maintain similarity in either batch or continuous-flow operation. This approach seems to reject the traditional batch "jar test" as an effective scale-up tool, since almost all full-scale plants will for the time being operate on a continuous-flow basis. This viewpoint is a little too extreme (see "Summary"), but it is recommended here for more unequivocal laboratory quantification of scale-related effects.

(see "Summary"), but it is recommended here for more unequivocal laboratory quantification of scale-related effects.

The final important design feature of good scale-up experiments is to ensure that initial conditions for the various scale units are the same. For a flocculation experiment, this requirement is most simply satisfied by having the same source of water + coagulant for the various reactors. If one is specifically studying flocculation (as opposed to rapid mix or rapid mix and flocculation), then the source should be a rapid-mix unit that is large enough to supply all reactors being tested. This assumption seems to throw out another possible influential dimensionless group, the ratio of particle size to reactor diameter. It is hoped that this group is of secondary importance.

It is also necessary to be precise in specifying what performance aspect is to be scaled-up. For example, since particle and floc size is such an important variable in downstream performance, the particle size distribution (PSD) of flocs leaving the flocculator would be a viable scale-up parameter. In the authors' opinion, this might turn out to be a less-than-optimal choice for a scale-up parameter. Since the shape of the PSD can be expected to be influenced by both particle aggregation and breakup, a particular PSD may be unique at a given scale. Nevertheless, if one has the capability of measuring the PSD it would be a very interesting scale-up parameter. Two more easily measured parameters are turbidity removal during settling and, in the case of organic-rich waters, organic removal during settling. These are both measured with fairly standard laboratory equipment. However, when measurements are made during settling, it is an integrated flocculation–settling process that is being investigated—not flocculation alone.

To the authors' knowledge, the only published experiments aimed specifically at finding scale effects in flocculation are those of Oldshue and Mady (1978). They took water from the effluent of the rapid-mix stage of a 10-mgd surface water treatment plant for testing in two small-scale batch mixers of 4.6- and 7.6-cm diameter. Therefore, the recommendation of common source water was satisfied. Geometrically similar Rushton-type mixers were used with a variety of impellers. The flocculation time was 10 min, and turbidity values were measured several times during a 60-min settling period following flocculation. Oldshue and Mady chose to report the influence of flocculation $\overline{G}$-value on the residual turbidity at 60 min in the two tanks. Using these data, the apparent b values in the general scale-up relation, $\overline{\varepsilon}_l \propto S^b$ have been calculated for the condition of *minimum residual turbidity* at the end of 60 min (Table 8-1). These scale-up exponents are both negative and of great magnitude. The value for the Lightnin* A211 impeller ($D/D_v = 0.3$) is even more severe than the constant Reynolds number law discussed above. Here is clear evidence for the importance of breakup in flocculation. More finely broken floc settle less well and result in higher residual turbidity during settling.

Due to the relatively small-scale difference for the two reactors of Oldshue and Mady ($S = 1.65$) and the relatively weak (and noisy) minima in the 60-min turbidity data; Oldshue and Mady did not recommend extrapolating their results to full-scale design. In addition, the severe scale-up implied by some of these results is really unimaginable. For example, even for the smallest magnitude exponent in the table ($b = -1.0$), full-scale $\overline{G}$-values are predicted to be 24 and 8 for scale-up factors of $S = 10$ and $S = 100$, respectively. Yet the general conclusion that $\overline{\varepsilon}$ is proportional to some inverse power of S seems inescapable.

---

*Lightnin Mixing Equipment Company, Rochester, N.Y.

Table 8-1 Apparent Scale-up Relations for Experiments by Oldshue and Mady (1978)

Impeller Type	$\overline{G}$-Value at Minimum Turbidity		b-Value
	4.6-cm-Diameter Impeller	7.6-cm-Diameter Impeller	
Rake ($D/D_v = 0.8$)	148	112	−1.11
Lightnin* A212			
$\quad D/D_v = 0.3$	110	34	−4.68
$\quad D/D_v = 0.2$	58	27	−3.05
Lightnin A200 ($D/D_v = 0.2$)	77	60	−0.99
Rushton ($D/D_v = 0.2$)	116	59	−2.69

*Lightnin, Rochester, N.Y.

It is possible that scale-up exponents determined in two or more small-scale experiments are not applicable to larger-scale vessels. This would be a general conclusion for multi-time scale models, and, as shown above, several time scales are important in flocculation and breakup. This is illustrated schematically in Figure 8-6, where it is presumed that the constant turnover time, constant $\overline{\varepsilon}$, and constant tip speed laws are most important in flocculation scale-up. For small S-values a hypothetical scale-up law with decreasing slope has been drawn in, suggested by the data of Oldshue and Mady (1978). From the previous discussion, the negative slope seems consistent with an influential breakup mechanism. However, as scale increases at this slope, the average $\overline{G}$-value is decreasing rather severely. At some point, the $\overline{G}$-value is simply not high enough for effective particle contacts. Therefore, the slope of the scale-up line must increase in order to achieve similarity in final flocculator performance. One might argue that even with a flattening of the slope, the relative intensity of the breakup mechanism would be increasing. Nevertheless, the severely increasing turnover time (lower probability of passing through the impeller zone) would offset this effect.

Bradley (1988) investigated batch scale-up in two fully baffled Rushton mixers of 1- and 20-L, which represents a scale-up factor of 2.72. Solutions of 10 mg/L humic acid were flocculated with aluminum sulfate, and turbidity and humic acid removal were monitored during the course of a 60-min settling period. The aluminum concentration was $10^{-3.5}$, and the pH was 7.0. Power consumption in the vessels was determined using the classic correlation discussed above, Eq 8-8, and published values for the power number P_o as a function of paddle Reynolds number (Eq 8-15) (Holland and Chapman 1966). As mentioned above, it may not always be reasonable to assume N_q is constant if the smaller vessel is too small or is operated at too low a Reynolds number. Contrasting P_o values in the two vessels is then of some interest, since P_o and N_q are usually assumed to behave similarly as a function of Re_p. In Figure 8-7, while the larger vessel reaches the fully turbulent P_o value of 6 at $\overline{G} \simeq 100$, the fully turbulent P_o value is not reached in the smaller vessel until about $\overline{G} = 400$. Nevertheless, the effect of this divergence in P_o is minimized when the scale-up relation is not too nonlinear in S. For example, for the condition $\overline{\varepsilon}$ (or $\overline{G}$) = constant, it is easy to show using the methods of traditional scale-up laws that $N_1/N_2 = S^{-2/3}$. Using Bradley's data, this relation is maintained even at small-scale $\overline{G}$-values below 50.

Bradley (1988) performed two basic experiments. In the first experiment, a common rapid-mix stage was used for each vessel. This was achieved simply by beginning rapid mix and coagulant addition in the larger vessel, which initially contained 21 L of humic acid. After coagulant addition and 60 s of rapid mixing at a standard $\overline{G}$-value of 1000 s^{-1}, 1 L was dipped out into the smaller vessel and both vessels were agitated for 30 min using separate mixers operating at the same calibrated $\overline{G}$-values. In the second experiment, each vessel was operated independently, having its own 60-s rapid-mix stage. Two flocculation $\overline{G}$-values were compared in these experiments, $\overline{G} = 40$ and 100. By the previous argument, the first experiment isolated the coupled flocculation settling process, whereas the second experiment included rapid-mix effects. While not showing high sensitivity to scale, Bradley's results yield the same conclusion as Oldshue and Mady (1978). Table 8-2

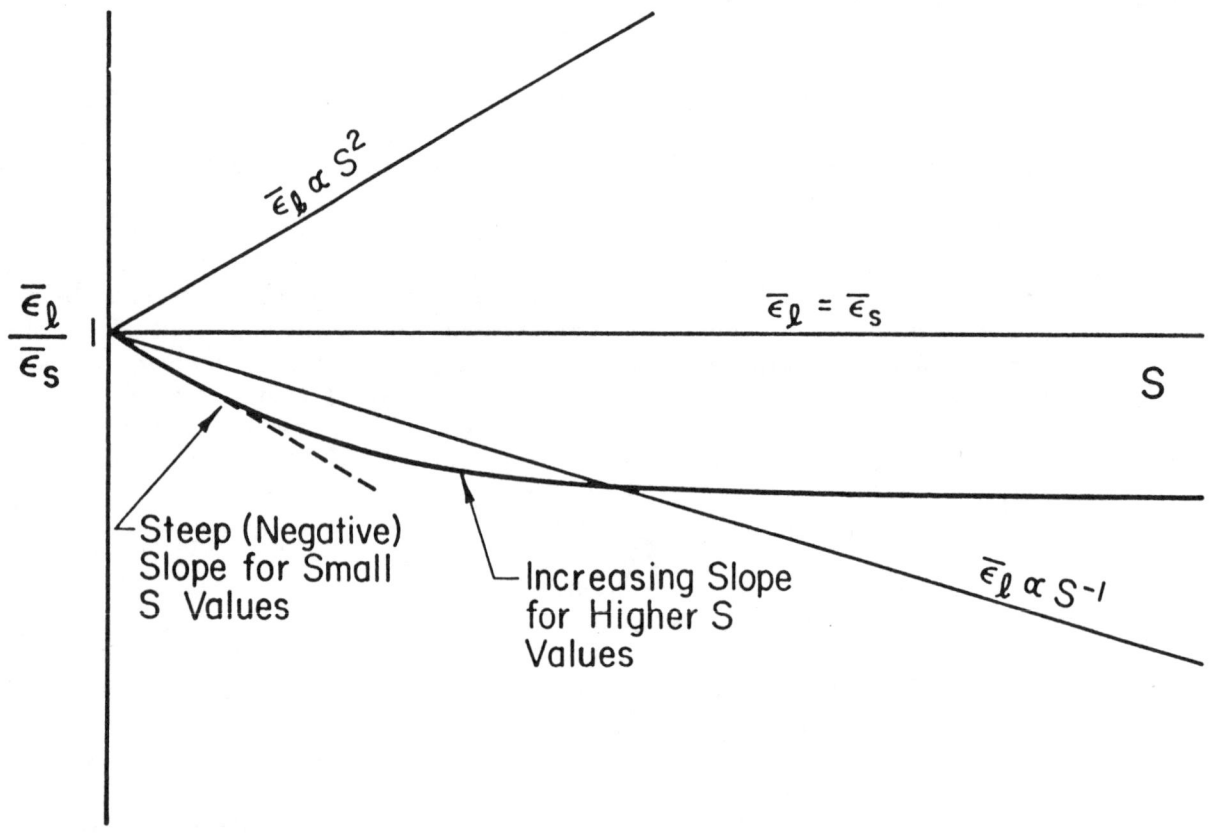

Figure 8-6 Hypothetical change of scale-up law with scale factor.

Table 8-2 Data for Turbidity and Humic Acid Removal After 60 min of Settling (Bradley 1988)*

$\overline{G}$-Value		Same Rapid Mix		Separate Rapid Mix	
		1 L	20 L	1 L	20 L
40	Turbidity, *NTU*	0.40	0.50	0.58	0.63
100		—	—	0.55	0.66
40	Organic removal, %	78	72	86	82
100		—	—	88	80

*System was 10-mg/L humic acid flocculated with aluminum sulfate.

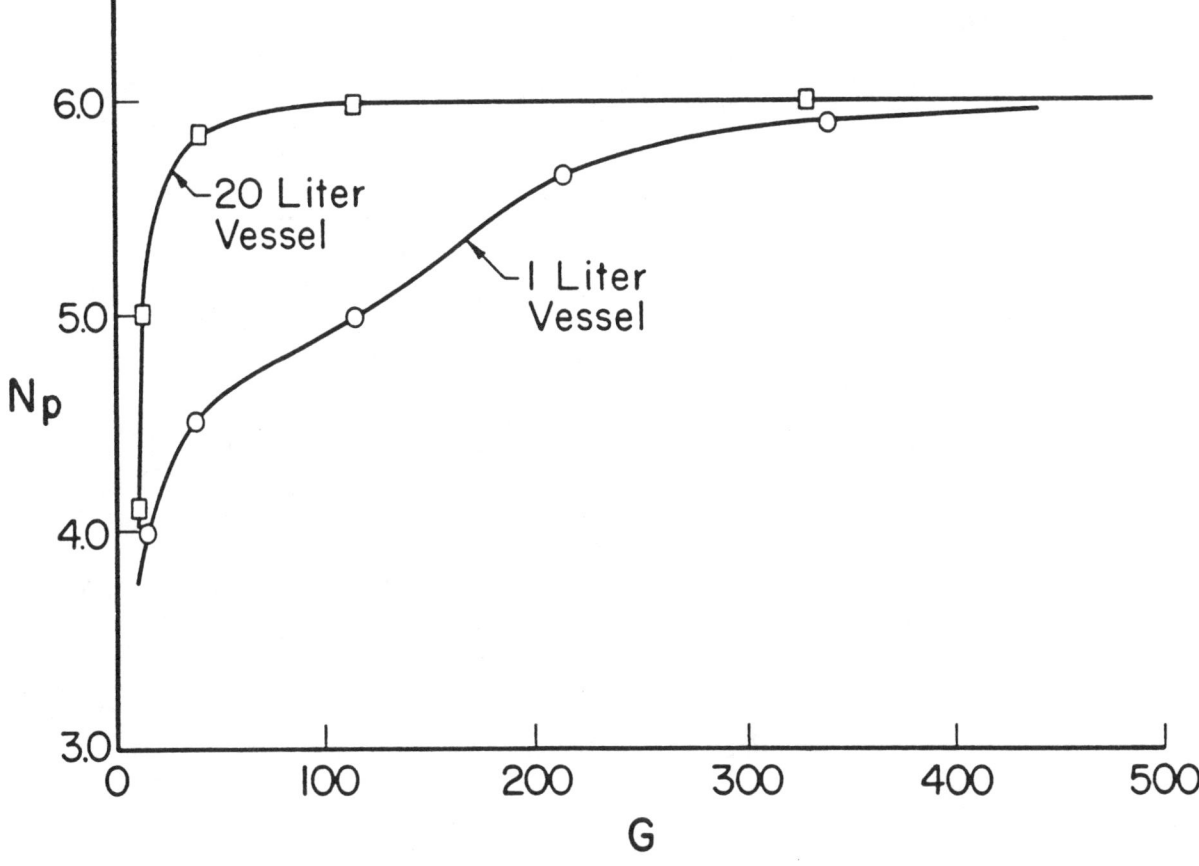

Figure 8-7 Variation of power number with $\overline{G}$ for 1- and 20-L Rushton vessels.

shows that for both experiments, there was higher residual turbidity and lower humic acid removal in the larger vessel when it was operated at the same $\overline{G}$-value.

Although not a strong tendency, some sensitivity to the $\overline{G}$-value is also noted. The divergence between small- and large-scale performance seems to increase as $\overline{G}$ increases from 40 to 100, at least for the case of separate rapid mixing. This suggests that flocculation could also be sensitive to the scale of the rapid-mix device.

It is concluded in this section that the limited data for batch flocculation do indicate scale-related sensitivity. In the next section data aimed at showing sensitivity to the mode of operation—batch or continuous—are reviewed.

COMPARISON OF BATCH AND CONTINUOUS FLOCCULATION

Leprise (1987) began with the simple hypothesis that if the jar test could be legitimately extrapolated to larger-scale continuous-flow flocculation, then a necessary condition (and an interesting experimental check) would be to have similarity in performance between a 1-L jar test and a 1-L continuous-flow flocculator. The system utilized in his work was a 10-mg/L humic acid solution coagulated with a preneutralized aluminum coagulant sold under the trade name WAC.* The same chemical system was investigated by Wiesner, Lahousine-Turcard, and Fiessinger (1986) in their 1-L batch jar tests.

Figure 8-8 is a schematic of the apparatus used by Leprise (1987). Initial mixing in this system occurs following coagulant addition and pH adjustment as the solution flows to the vessel. Tube length and flow rate are adjusted so that the batch initial mix and flocculation times of 60 s and 30 min of Wiesner, Lahousine-Turcard, and Fiessinger (1986) are duplicated. The $\overline{G}$-value in the feed tube was estimated to be 100 s^{-1}, while the flocculator $\overline{G}$-value was held at 40 s^{-1} for similarity with batch flocculation tests of Wiesner, Lahousine-Turcard, and Fiessinger. The batch initial mix $\overline{G}$-value in the experiments of Wiesner, Lahousine-Turcard, and Fiessinger was around 1500, so that the two experiments are not completely similar in $\overline{G}$-value.

The RTD was measured with a conductivity meter. Pulse injection of sodium chloride was done with a syringe in the feed tube just upstream of the vessel. The NaCl conductivity was recorded continuously on a strip-chart recorder. This signal was properly normalized as discussed in chap. 4, and the resulting residence time density function was found to yield a mean residence time quite close to the theoretical value, V/Q. The residence time density function was approximately exponential, even at relatively low $\overline{G}$-values.

In Figure 8-9, Leprise's results are superimposed on the results of Wiesner, Lahousine-Turcard, and Fiessinger (1986). Here is plotted the pH–aluminum concentration region of effective removal of humic acid through flocculation and settling. The region is considerably smaller in the case of continuous-flow flocculation. Of course, the a priori expectation is that for positive reaction orders, the batch reactor should outperform the continuous reactor for the same average detention time (see chap. 4).

These data raise some questions about the use of the standard batch jar test to simulate performance of continuous-flow facilities. If the full-scale residence time density function tends toward the exponential well-stirred case, important differences between small- and large-scale units could be found due simply to the residence time density function—regardless of any scale-related effects. On the other hand, as the full-scale residence time density function becomes more similar to plug or piston flow (as it should for compartmentalized flocculators), then more similar performance should be found between the jar test and full-scale facilities.

SUMMARY, CONCLUSIONS, AND RESEARCH RECOMMENDATIONS

This chapter has discussed the possible influence of scale on the performance of rapid mix and flocculation. Traditionally, scale-up has not been an active area of research in sanitary and environmental engineering. Perhaps as the modeling of water treatment processes has relied more and more on computer simulation, it has been assumed implicitly that any scale sensitivity would be accounted for. In fact, this would be true if the model accurately treated both "reaction" *and* mixing. It is likely that many of these mathematical and simulation models would not stand up to a rigorous test in different scale systems. A model that passed such a test would probably be a better model.

---

*AUTOCHEM, Groupe Elf-Aquitaine, Paris la Défense, France.

Since it is already *known* that effective water treatment $\overline{G}$-values are in the range of 20 to 60, why worry about scale-up? This is probably a good argument for conventional large-scale water treatment plants that are not operated at too high a rate and that do not have to meet too high treatment standards. After all, even if flocculation performance is not quite what was expected, can't it always be made up for in the filtration stage? The answer is, "yes, probably."

Nevertheless, as treatment standards get tighter, as smaller and higher-rate processes are developed, and as new processes for solid–liquid separation (e.g., direct filtration and membranes) are developed, it will be of considerable advantage to understand the apparent subtleties of scale-related effects. In all these cases, the flocculation "product" may have more relative impact on the efficiency of downstream processes.

The jar test has been the de facto scale-up tool in rapid mix and flocculation. It is difficult to name a test that has been such an effective means worldwide of determining the treatability of natural waters. In fact, the jar test should probably not be changed—it works too well now! On the other hand, those interested in scale-up

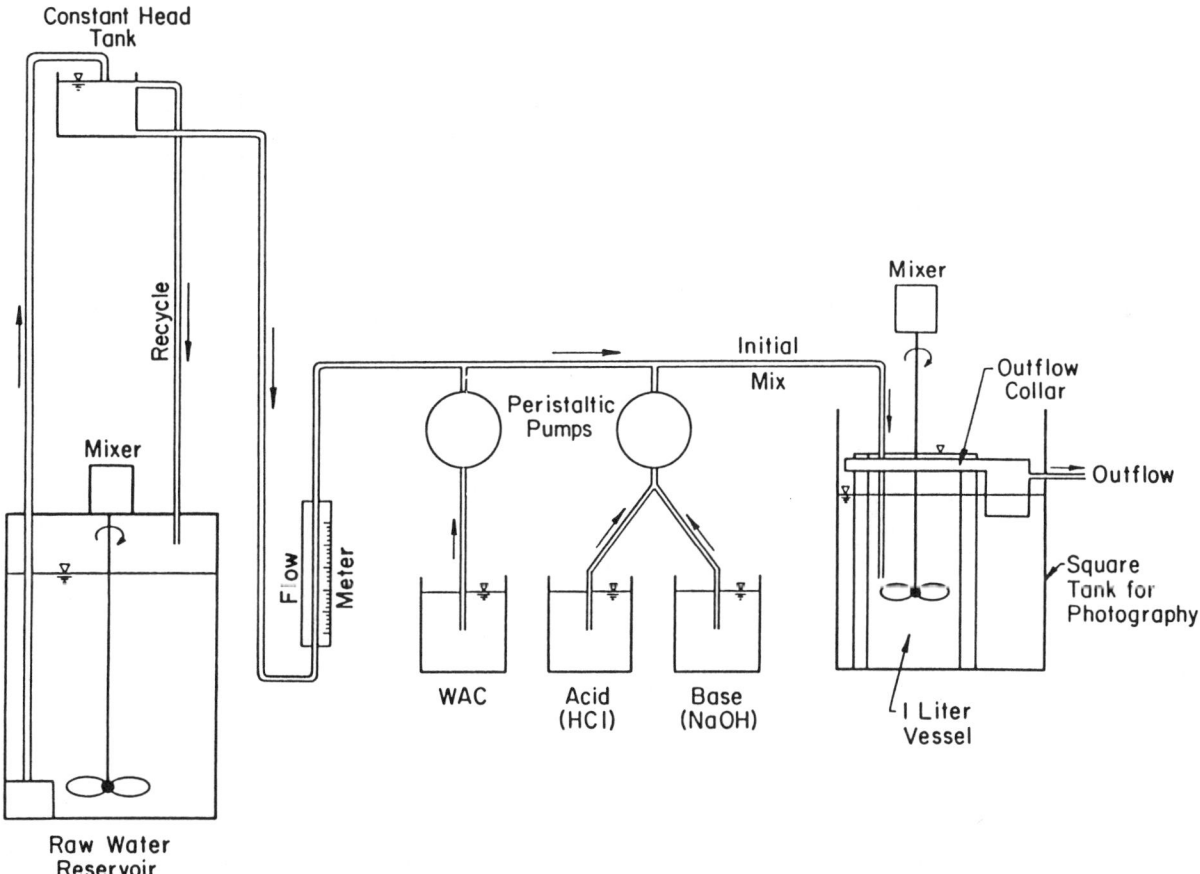

Source: Leprise, P. 1987. Floculation en Continue. D.E.A. Report, Societe Lyonnaise des Eaux, Le Pecq, France.

Figure 8-8 Continuous-flow flocculation pilot used by Leprise.

304 MIXING

need to be aware of some of the limitations of the jar test. To isolate scale-related effects, the following are the most important experimental requirements:

- In rapid-mix tests, in addition to geometric similarity, coagulant contacting order, coagulant injection point, and operation mode (batch or continuous) are the most important similarity variables.
- In flocculation, similarity in starting solutions and RTD should be most important. The latter restriction implies that if the tests are to be relevant to continuous-flow large-scale plants, the small-scale tests should also be continuous flow.
- Geometric similarity is also important to isolate scale-related effects in flocculation. In order to avoid the "ice-cube tray" effect mentioned earlier, larger-volume small-scale tests might be considered.

Regarding geometric similarity in flocculation tests, there is anecdotal evidence that it is difficult to design inlet and outlet features of small-scale, continuous-flow flocculators. For example, Leprise (1987) found for certain chemical conditions it was very difficult to achieve steady-state floc concentration in his 1-L flocculator. Apparently, more dense floc will preferentially remain in the reactor, leading to continual solids buildup. Special designs might be required for outlets to increase

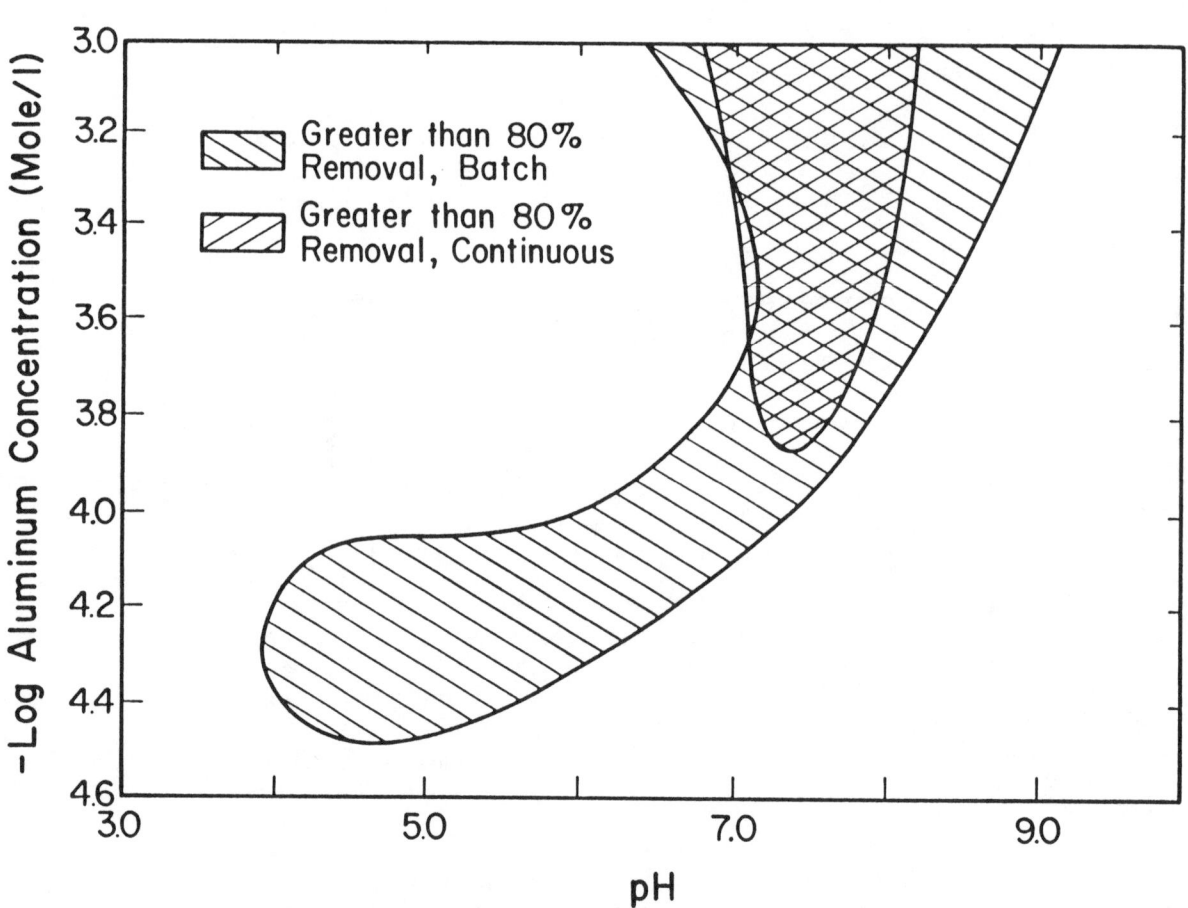

Figure 8-9 Humic acid removal after 1 h of settling, batch versus continuous flow.

upflow velocities. This partial geometric nonsimilarity might be a necessary compromise.

The Camp and Stein (1943) characteristic velocity gradient, $\bar{G}$, has been frequently discussed and criticized (Argaman and Kaufman 1970; Oldshue and Mady 1978; Cleasby 1984; Clark 1985). It is difficult to name a parameter of more obvious relevance in rapid mixing and flocculation. $\bar{G}$ must be important since it reflects the power input to the mixer and is a measure of the intensity of mixing. Nevertheless, a fairly complete consensus now exists about certain theoretical and observed limitations of $\bar{G}$. The present chapter has added to that list of qualifications, and this can be best illustrated by referring to Figure 8-1. Here for $\bar{G}$ = constant (the traditional scale-up assumption), all other mixing time scales are changing with scale*—and they are changing in different ways. Stated in another way, constancy of the $\bar{G}$-value does not ensure constancy of mixing quality as scale changes.

Finally, most of the results discussed in this chapter were developed for the standard configuration Rushton turbine. This is because chemical engineers have made the most thorough studies of mixing, and the standard Rushton turbine is most frequently studied by chemical engineers. Although the authors hope that new mixing studies will be done in vessels more typical of those used in water treatment, the above studies should still capture the most important effects in the general mixing situation in which energy is imparted to turbulence through a mechanical impeller.

References

AMITHARAJAH, A. 1981. Initial Mixing. Proc. AWWA Seminar on Coagulation and Filtration: Back to the Basics, St. Louis, Mo.

ARGAMAN, Y. & KAUFMAN, W.J. 1970. Turbulence and Flocculation. *Jour. San. Engrg.*, SA2:4:223.

BATCHELOR, G.K. 1967. *An Introduction to Fluid Dynamics*. Cambridge University Press, Cambridge, England.

BELEVI, H.; BOURNE, J.R.; & RYS, P. 1981. Mixing and Fast Chemical Reaction, II: Diffusion–Reaction Model for the CSTR. *Chem. Engrg. Sci.*, 36:10:1649.

BERESFORD, H.I. ET AL. 1970. *Trans. Instn. Chem. Engrs.*, 48:T21.

BISIO, A. & KABEL, R.L. 1985. *Scaleup of Chemical Processes*. Wiley-Interscience, New York.

BOURNE, J.R. ET AL. 1981. Mixing and Fast Chemical Reaction, III: Model–Experiment Comparisons. *Chem. Engrg. Sci.*, 36:10:1655.

BOURNE, J.R.; KOZICKI, F.; & RYS, P. 1981. Mixing and Fast Chemical Reaction, I: Test Reactions to Determine Segregation. *Chem. Engrg. Sci.*, 36:10:1643.

BRADLEY, J. 1988. Scale-up of the Flocculation Phase in Reservoir Water Treatment Processes. Independent Study in Civil Engineering, University of Illinois, Urbana, Ill.

CAMP, T.R. & STEIN, P.C. 1943. Velocity Gradients and Internal Work in Fluid Motion. *Jour. Boston Soc. Civ. Engrs.*, 30:219.

CHARLES. P. ET AL. 1987. Influence des Conditions de Melange Initial sur la Coagulation. AGHTM Conf., Nice, France.

CLARK, M.M. 1985. Critique of Camp and Steins, RMS Velocity Gradient. *Jour. Envir. Engrg.*, 111:6:741.

———. 1986. Scale-up of Laboratory Flocculation Results. Proc. AWWA Annual Conf., Denver, Colo.

———. 1988. Drop Breakup in a Turbulent Flow, II: Experiments in a Small Mixing Vessel. *Chem. Engrg. Sci.*, 43:3:681.

---

*Recall that the Reynolds number can be considered the ratio of viscous to inertial time scales—(e.g., Batchelor 1967).

CLEASBY, J.L. 1984. Is Velocity Gradient a Valid Turbulent Flocculation Parameter? *Jour. Envir. Engrg.*, 110:5:875.

CORRSIN, S. 1957. Simple Theory of an Idealized Turbulent Mixer. *AIChE Jour.*, 3:3:329.

FRIEDLANDER, S.K. 1977. *Smoke, Duct and Haze: Fundamentals of Aerosol Behavior.* Wiley-Interscience, New York.

GOLDSTIEN, A.M. 1973. The Application of Simulation Vessel Flow Studies to an Industrial Chemical Reactor Mixing Problem. *Chem. Engrg. Sci.*, 28:1021.

HILL, J.C. 1976. Homogeneous Turbulent Mixing With Chemical Reaction. *Ann. Rev. Fluid Mech.*, 135.

HOLLAND, F.A. & CHAPMAN, F.S. 1966. *Liquid Mixing and Processing in Stirred Tanks.* Reinhold Publishing, New York.

KONNO, M.; AOKI, M.; & SAITO, S. 1983. Scale Effect on Breakup Processes in Liquid–Liquid Agitated Tanks. *Jour. Chem. Engrg. Jpn.*, 16:4:312.

LEPRISE, P. 1987. Flocculation en Continue. D.E.A. thesis (French Master's thesis), Ecole Nationale des Ponts et Chaussees, Paris.

LEVENSPIEL, O. 1972. *Chemical Reaction Engineering.* John Wiley and Sons, New York.

MERCIER, R.S. 1984. The Reactive Transport of Suspended Particles: Mechanisms and Modeling. Ph.D. thesis, Massachusetts Institute of Technology and the Woods Hole Oceanographic Institution, Cambridge, Mass.

MULLER, H.W. 1982. Untersuchungen zum Transportprozess bei der Flockung in Turbulenten Strmungen. *Chem. Tech.*, 34:10:517.

NIENOW, A.W. ET AL. 1974. Constant Turnover Time as a Scale-up Criterion for Agitated Tanks. *Chem. Engrg. Sci.*, 29:1043.

OLDSHUE, J.Y. & MADY, O.B. 1978. Flocculation Performance of Mixing Impellers. *Chem. Engrg. Prog.*, 74:103.

PANDYA, J.D. & SPIELMAN, L.A. 1982. Kinetics of Floc Breakage in Agitated Liquid Suspensions. *Jour. Colloid Inter. Sci.*, 90:2:517.

PARKER, D.S.; KAUFMAN, W.J.; & JENKINS, D. 1972. Floc Breakup in Turbulent Flocculation Processes. *Jour. San. Engrg. Div.*, SA1:2:79.

PENNEY, W.R. 1971. *Chem. Engrg.*, 78:6:86.

SCHMIDT-NIELSEN, K. 1984. *Scaling: Why Is Animal Size So Important?* Cambridge University Press, Cambridge, England.

UHL, V.W. & GRAY, J.B., eds. 1966. *Mixing: Theory and Practice.* Academic Press, New York.

VAN'T RIET, K. & SMITH, J.M. 1975. The Trailing Vortex System Produced by Rushton Turbine Agitators. *Chem. Engrg. Sci.*, 30:1093.

WEBER, W. 1972. *Physicochemical Processes for Water Quality Control.* Wiley-Interscience, New York.

WIESNER, M.R.; LAHOUSINE-TURCAUD, V.; & FIESSINGER, F. 1986. Organic Removal and Particle Formation Using a Partially Neutralized $AlCl_3$. Proc. AWWA Annual Conf., Denver, Colo.

Design and Operation

9

Design of Impellers for Mixing

James Y. Oldshue, Ph.D., P.E., Vice President—Mixing Technology, Lightnin, Rochester, N.Y. USA

R. Rhodes Trussell, Ph.D., P.E., Senior Vice President, James M. Montgomery Consulting Engineers, Inc., Pasadena, Calif. USA

IMPELLERS

In the two typical process operations, commonly called "rapid mixing" and "flocculation," usually followed by some sort of settling or filtration, just about every impeller type ever imagined has been used at one time or another. A constant that governs the results of all impellers is that no mechanical action of the impeller blades on the particles is a major source of phenomena in the process. The impeller blades are designed to set up a certain fluid regime, and the impeller design required is based on fluid mechanics.

In order to classify the types of impellers used and proposed, a review is needed of some concepts of the fluid motion in a mixing vessel. All the power applied by the mixer produces a pumping capacity Q and a velocity head $V^2/2g$, H. This is similar to evaluating any pumping device, but is somewhat less quantitative than is the case with a pump and piping system. One problem is that the discharge of a mixing impeller occurs across an area somewhat larger than the actual impeller peripheral discharge area. The discharge also normally has a velocity with one component perpendicular to an arbitrarily chosen discharge plane, and a tangential component with the tendency to swirl the fluid around the impeller rotation centerline.

Depending on the shape of the tank and the power level applied to the fluid, baffles either at the tank wall or around the impeller in the form of a stator ring are often used to increase the normal flow component and decrease the tangential flow component.

For any process, if the impeller Reynolds number is approximately 1000 or higher, the flow is normally turbulent, and a velocity probe placed in the fluid would normally exhibit a pattern as shown in Figure 9-1.

Using the average velocity, it is possible to calculate tangential, axial, and/or radial components, and, with the proper integration, the impeller flow through the arbitrarily defined discharge area.

Different impeller designs produce different velocity profiles. Radial-flow impellers have a typical profile as shown in Figure 9-2A, while the new fluidfoil impellers have a profile like that shown in Figure 9-2B. The so-called axial-flow turbines have a flow pattern as shown in Figure 9-2C. This is a function of the blade angle.

In mixing technology, the velocity gradient at a point is called a shear rate. It has the units of reciprocal time, in this chapter, reciprocal seconds. The only way particles can meet each other in a mixing vessel is by having shear rates present. Figure 9-3 shows the definition of shear rate if particle A and particle B have different velocities. With no shear rate, A and B would follow each other around in a frozen circle just like two dots on a rotating disk.

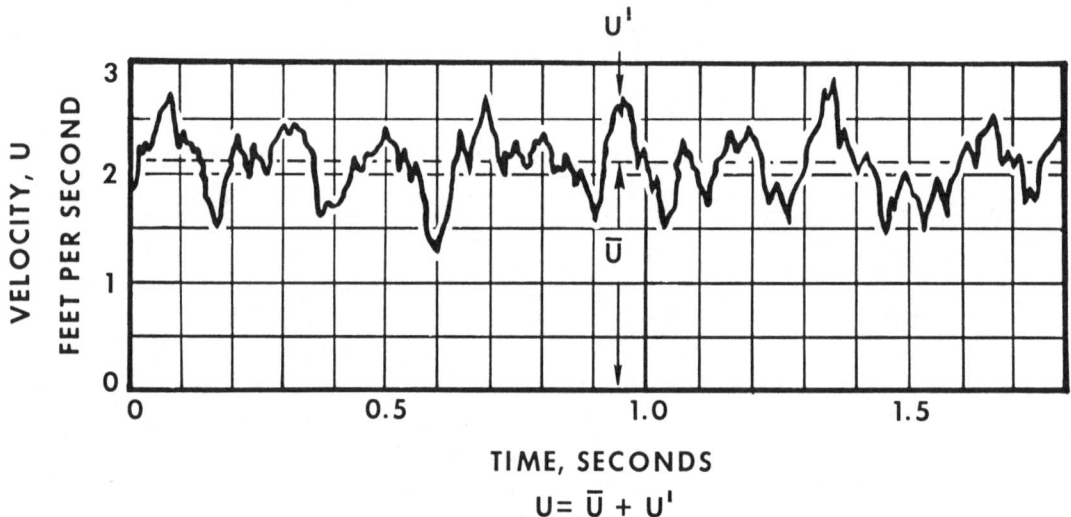

Courtesy of Lightnin, A Unit of General Signal, Rochester, N.Y.

Figure 9-1 Typical pattern of velocity with time near an operating impeller.

For any shear rate there is an accompanying shear stress, expressed in this chapter in terms of pounds per square inch. This shear stress is a product of the viscosity times the shear rate. If the fluids are non-Newtonian, then the viscosity used must be the actual viscosity at the shear rate at that point. With non-Newtonian fluids, viscosity is a function of time and/or shear rate. There are also fluids with viscoelastic properties, but these do not often occur in flocculation systems for water and waste treatment.

Fluidfoil impellers were developed to minimize the high, localized shear rate in the impeller zone, therefore providing a high ratio of flow compared to velocity head

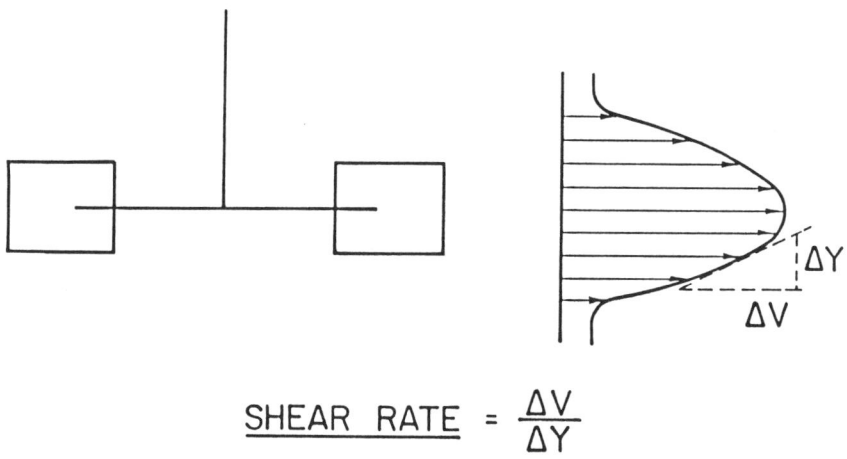

A. Radial flow

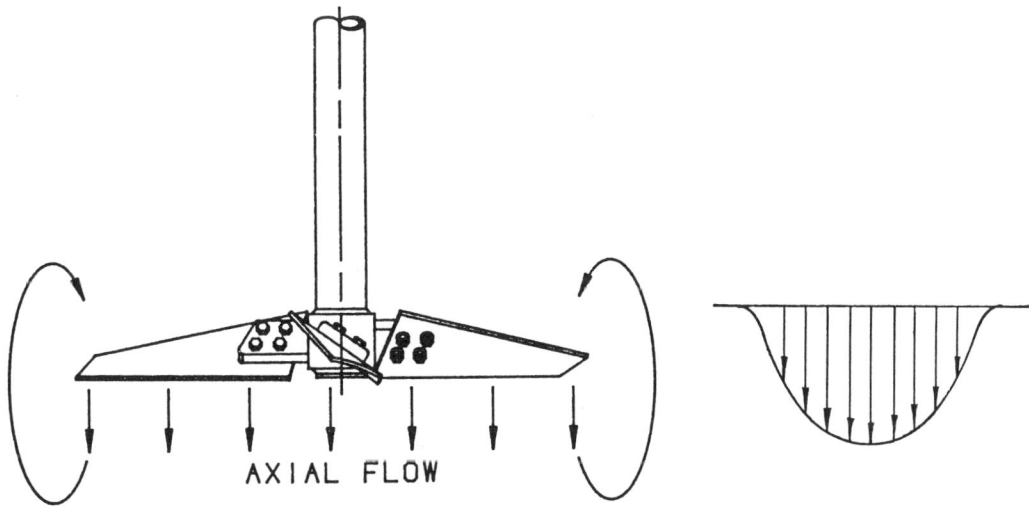

B. Fluidfoil

Courtesy of Lightnin, A Unit of General Signal, Rochester, N.Y.

Figure 9-2 Typical velocity profiles from common impeller types.

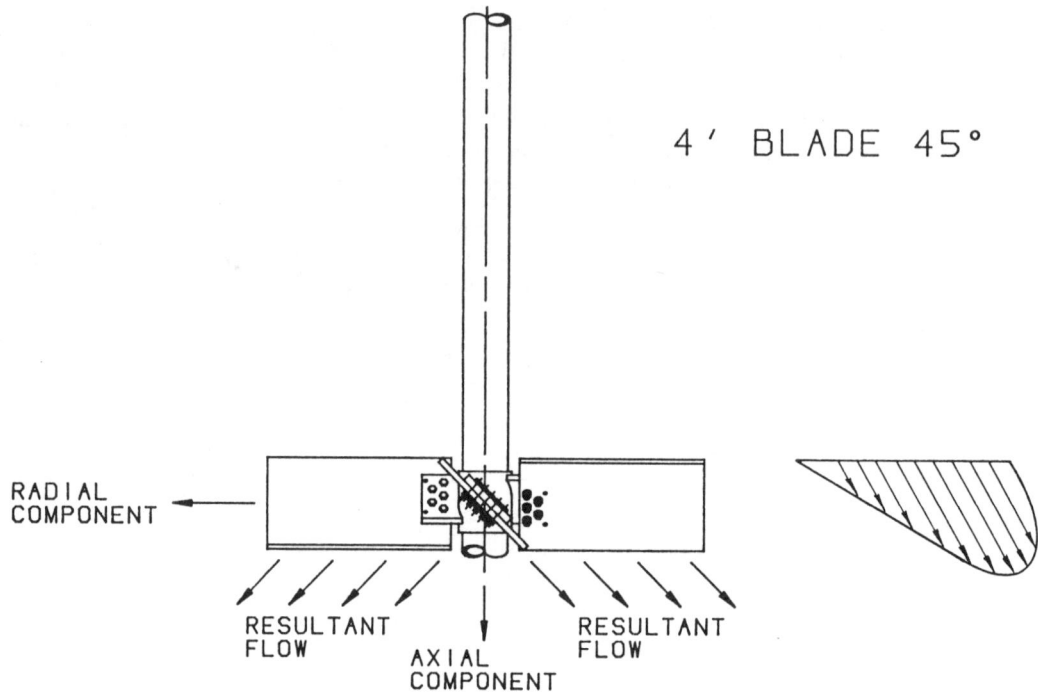

c. Axial flow

Courtesy of Lightnin, A Unit of General Signal, Rochester, N.Y.

Figure 9-2 Typical velocity profiles from common impeller types (continued).

for a given power level. For waterlike-viscosity materials, the most efficient impeller that can be made from a single thickness of metal is typified by the A310 impeller* shown in Figure 9-4A. Other impellers in this family are called by other names, such as high efficiency, hydrofoil, and many other flow-descriptive terms. A diagram of the flow pattern from the A310 impeller is shown in Figure 9-4B. Table 9-1 shows the relationships among several important variables in radial-flow, axial-flow, and 45° axial-flow impellers.

Another characteristic of the various impellers is the velocity fluctuations, and for the A310 the velocity fluctuations at a particular speed and diameter are also shown in both the axial and radial direction (Figures 9-4C and 9-4D).

Macroscale/Microscale

All the power that is put into a fluid eventually appears as heat. In a mixer, all the power is converted to heat via viscous shear. Viscous shear is present in turbulent flow at the macroscale level. Under the mixing conditions used in water treatment, the dividing line between the microscale and macroscale is between 100 and 1000 µm. Particles in the macroscale are only affected by the average velocity.

---

*All of the impellers referred to in this chapter are produced by Lightnin, Rochester, N.Y., and identified by a proprietary number. Other manufacturers make similar equipment, but no universal nomenclature is available.

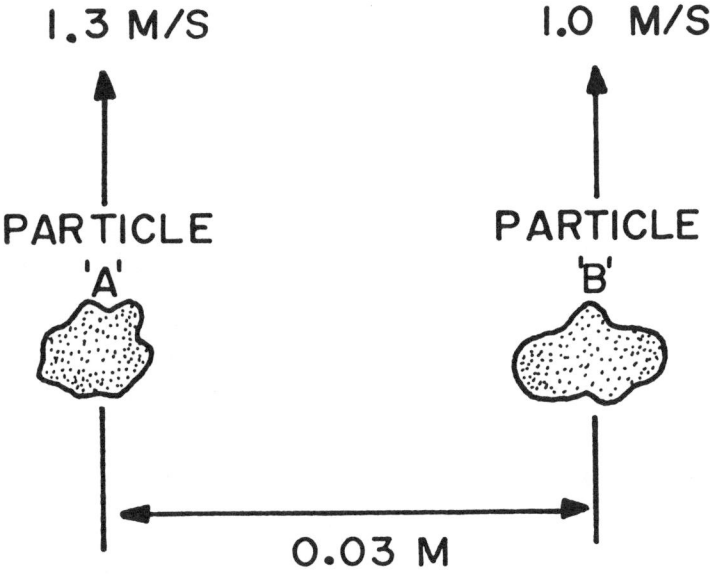

Courtesy of Lightnin, A Unit of General Signal, Rochester, N.Y.

Figure 9-3 Relationship between shear rate and particles suspended in a fluid system.

Table 9-1 Relative Comparison of Impeller Designs

Impeller Type	Diameter	Flow	hp	rpm	Torque	Fluid Force
True axial flow	1.0	1.0	1.0	1.0	1.0	1.0
45° four-blade aft	1.0	1.0	1.6	0.7	2.3	1.7
True radial flow	1.0	1.0	6.0	0.7	8.6	3.4

In contrast, particles that are less than 100 μm see the effect of the fluctuating velocities. The ultimate minimum length scale L of the velocity fluctuation for the mixer at the microscale level is a function of the viscosity and the power per unit volume, $L \propto (v^3/\varepsilon)^{1/4}$. Turbulence is much more complicated than power per unit volume, particularly the average power per unit volume over the entire tank. Nevertheless, power per unit volume is the major determinant of microscale activity. At an eddy size of about 1 μm the fluid does not know where the energy came from; therefore, where particles of this size are concerned, the type of impeller used is immaterial.

314   MIXING

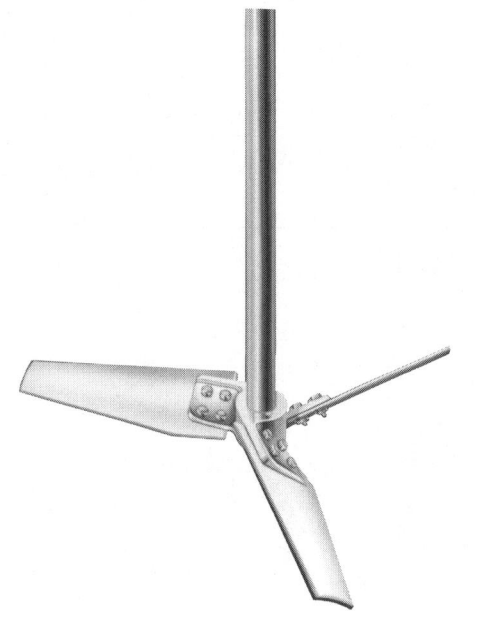

A. The Lightnin A310 fluidfoil impeller.

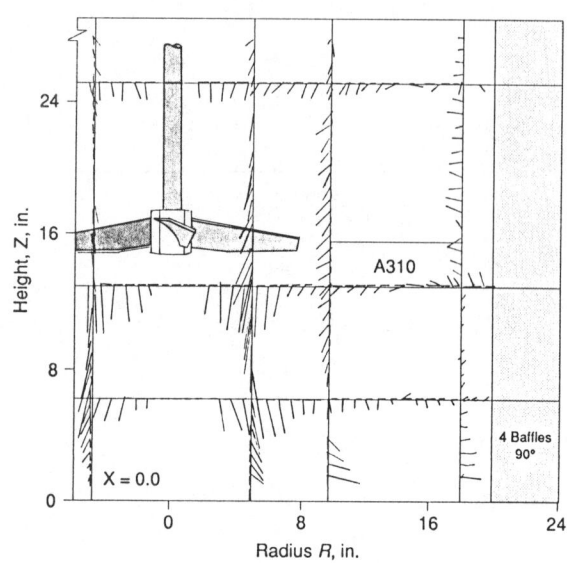

B. Velocity vectors near a fluidfoil.

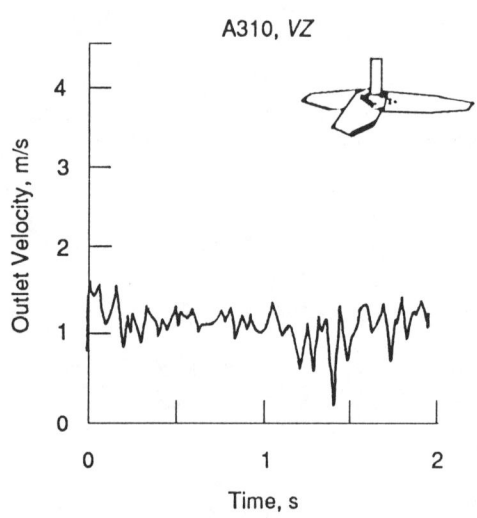

C. Vertical velocity profile.

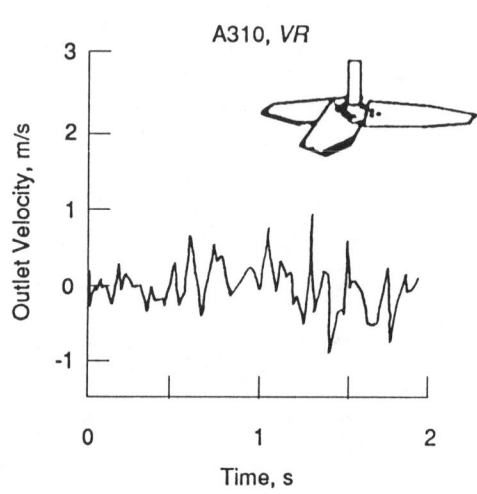

D. Radial velocity profile.

NOTE: The Lightnin A310 impeller shown above is a good example of an impeller design that minimizes local impeller shear and maximizes pumping. Velocity profiles in the radial and vertical direction illustrate its behavior.

Courtesy of Lightnin, A Unit of General Signal, Rochester, N.Y.

Figure 9-4 The Lightnin A310 impeller.

An axial-flow turbine (A200), typified by a 45° angle impeller, is shown in Figure 9-5A. By examining a velocity flow pattern from the laser velocimeter (Figure 9-5B) it is evident that the direction of flow makes an angle of approximately 45° from the vertical. There actually is some up-flow in the center of the impeller.

To demonstrate the microscale effects and the impeller zone shear rate effects, Figure 9-5C shows the velocity fluctuation in the Z (axial) direction. Figure 9-5D shows the velocity fluctuation in the R (radial) direction. Figures 9-5C and 9-5D are at the same power levels as Figures 9-4C and 9-4D. It is visually apparent that the microscale mixing is considerably higher. The velocity head, therefore shear rates, are higher than those of the A310, indicating why the power number of the A200 is 1.25 and of the A310 is 0.3. The effect of blade angle on power number and flow is shown in Table 9-2.

When impellers of the axial-flow type are placed in draft tubes, there is a significant pressure drop within the draft tube. To prevent projected hub-area backflow, impellers normally have a very large hub, similar to that shown in Figure 9-6. Impellers operating in open tanks do not encounter a significant system resistance, so there is not much backflow in the center of the impeller. Therefore, the use of a large-diameter hub (on the order of one-fourth the impeller diameter) is not indicated in impellers operating in the low-viscosity range in large tanks.

Another extreme in design is the radial-flow impeller, or the so-called "Rushton Turbine." This impeller is designed to promote radial flow and it results in considerable impeller shear. Figure 9-7 shows such an impeller, its velocity vectors, and velocity profiles. The power number of the R100 is approximately 5.0 in contrast to the A310's 0.3. This is due to the much greater shear rate of the impeller.

Figure 9-7 refers to a radial-flow turbine with a disk. A radial-flow turbine without a disk is shown in Figure 9-8. The disk was used with the R100 impeller many years ago only as a mechanical means of attaching vertical blades to the impeller shaft. However, the disk did have an effect on the fluid mechanics, and tended to give a more uniform pressure drop across both sides of the radial-flow turbine, allowing it to pump in a radial direction. When a disk is not present, the pressure drop across the impeller is not balanced, and the impeller pumps with an axial component either up or down, depending on the location of the impeller, the proximity of the tank bottom, and the proximity to other impellers.

At the macroscale level, Figure 9-9A shows that the maximum shear rate for an R100 impeller is proportional to tip speed, which is the product of impeller rotational speed and diameter. Tip speed is often specified by water plant designers in an effort to control maximum shear. Figure 9-9B shows that the radial-flow turbine has an average shear rate proportional to the impeller rotational speed N, and independent of impeller diameter. Therefore, the impeller rotational speed, and not

Table 9-2 Effect of Blade Angle, Axial-Flow Impeller

Tip Angle, degrees	N_P	N_Q
34	1.0	1.0
25	0.58	0.835
17	0.29	0.663
10	0.11	0.479
8	0.07	0.418

316 MIXING

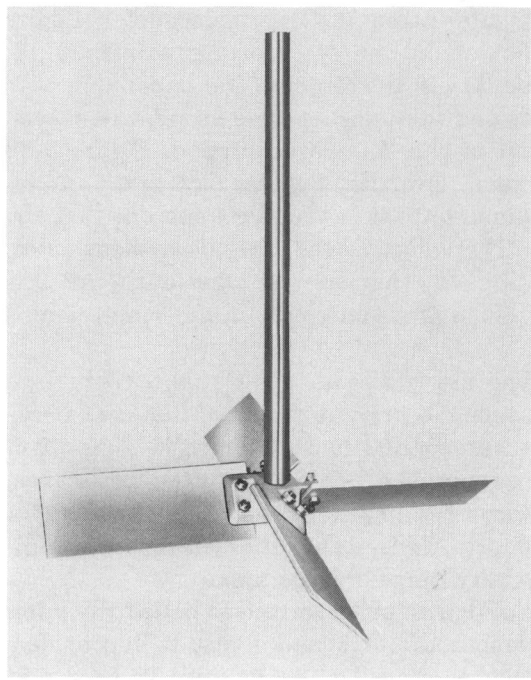

N = 162.7 rpm
Power = 277 W
Flow = .140 m³/s
Total Flow = 0.311 m³/s

Shear Gradients
Max = 39 s⁻¹
Ave = 16 s⁻¹
N_P = 1.25
N_Q = 0.77

A. The Lightnin A200: a 45° pitched-blade turbine.

B. Velocity vectors surrounding a 45° pitched-blade turbine.

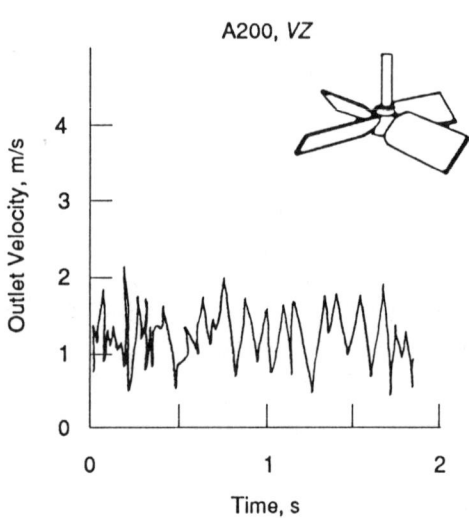

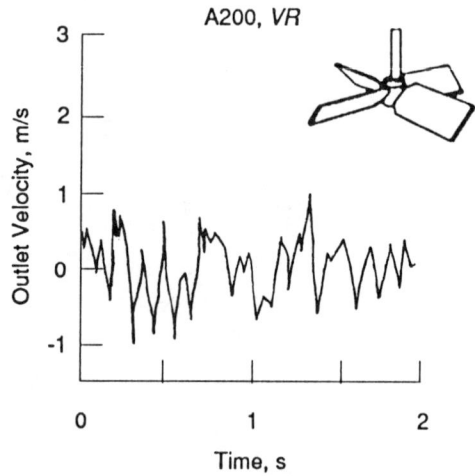

C. Vertical velocity profile.

D. Radial velocity profile.

NOTE: The A200 is a good example of the 45° pitched-blade turbine, which has been commonly used for flocculation for the past two decades. Note that, in contrast to the A310, this impeller pumps in the radial direction.

Courtesy of Lightnin, A Unit of General Signal, Rochester, N.Y.

Figure 9-5 The A200 axial-flow turbine.

NOTE: This impeller is designed with a large hub area. Such designs are necessary to reduce backflow when operating in draft tubes.

Courtesy of Lightnin, A Unit of General Signal, Rochester, N.Y.

Figure 9-6 Modified axial-flow impeller.

the tip speed, is the most appropriate means for controlling average shear with these impellers.

Studies with axial-flow turbines, such as the A200 (Figure 9-10), show that the maximum shear rate is apparently a function of impeller speed rather than impeller tip speed for these impellers as well. Further work is going on, and it may be that the proper location for the shear rate regime has not yet been determined, since impeller tip speed is four times higher than the velocity leaving the impeller.

Figure 9-10 also shows that the maximum shear rate for the A310 impeller is a function of impeller speed. This impeller requires a tip speed about five times higher than the fluid velocity, and again it may be that not all the measurements have been made that would settle the question of speed or tip speed in the maximum shear rate measurement with axial-flow impellers.

These differences in the performance of different impeller designs should be kept in mind, especially where the plan is to substitute one where the other has been used in the past.

Camp's G

At a given point in a mixing tank, the product of shear rate times shear stress is power per unit volume. Since shear stress is the product of the shear rate and the viscosity, power per unit volume is equivalent to the product of the square of the shear rate squared and the fluid viscosity. Thus, at a given point in the tank, the

318   MIXING

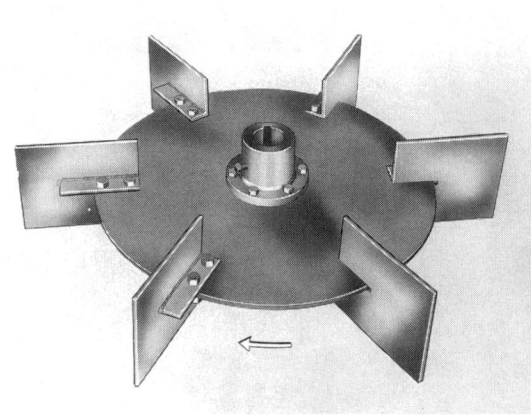

A. The Lightnin R100: a radial-flow turbine.

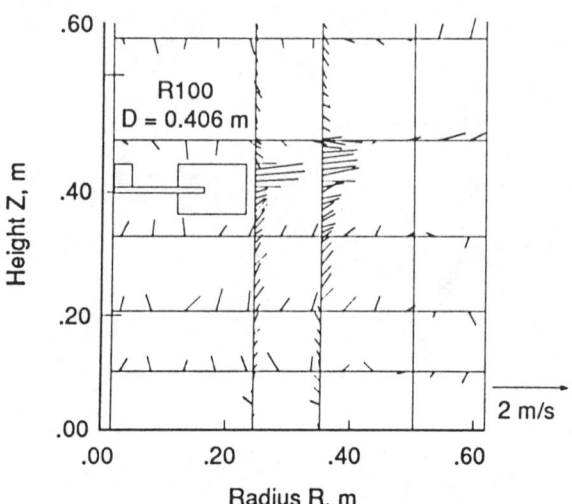

Speed = 101.8 rpm
Power = 334 W
Flow = .082 m³/s
Total Flow = .313 m³/s

Shear Gradients
Max = 61 s⁻¹
Ave = 20 s⁻¹
N_P = 5.18
N_Q = 0.72

B. Velocity vectors surrounding a radial-flow impeller.

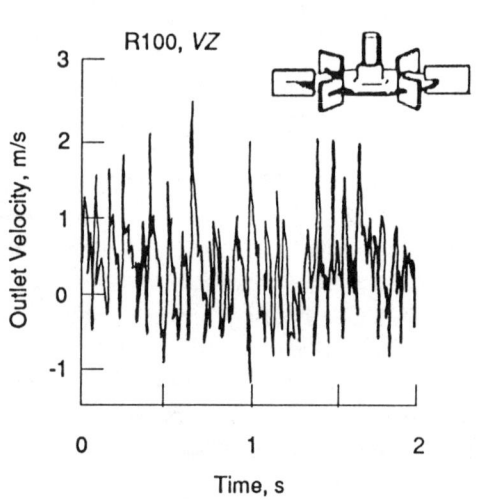

C. Axial velocity profile.

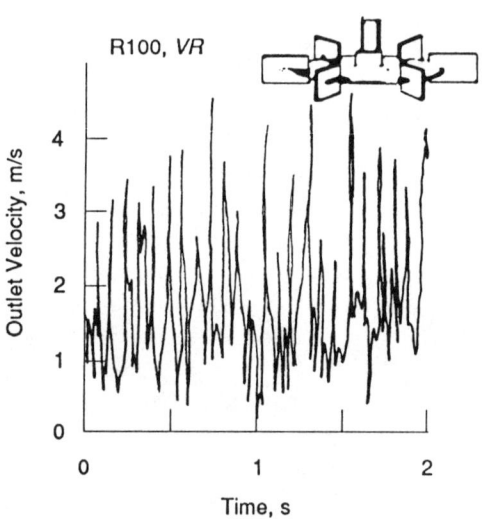

D. Radial velocity profile.

NOTE: The R100 is an example of a radial-flow turbine, another impeller that has often been used in flocculation. Note that the R100 produces radial flow and shear even more intense than the 45° pitched-blade turbine.

Courtesy of Lightnin, A Unit of General Signal, Rochester, N.Y.

Figure 9-7 A radial-flow impeller.

DESIGN OF IMPELLERS 319

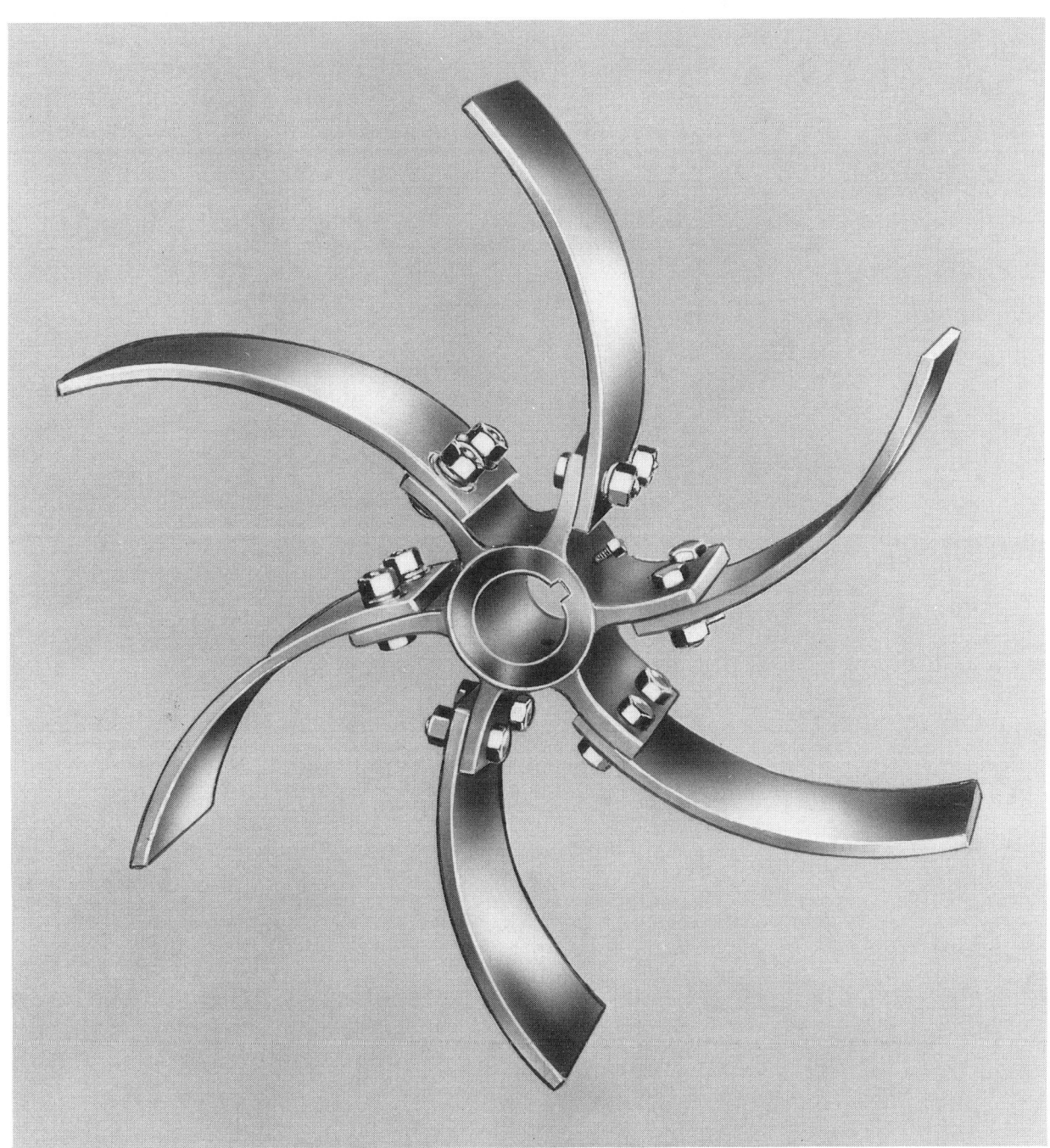

Courtesy of Lightnin, A Unit of General Signal, Rochester, N.Y.

Figure 9-8 Radial-flow turbine without disk.

320 MIXING

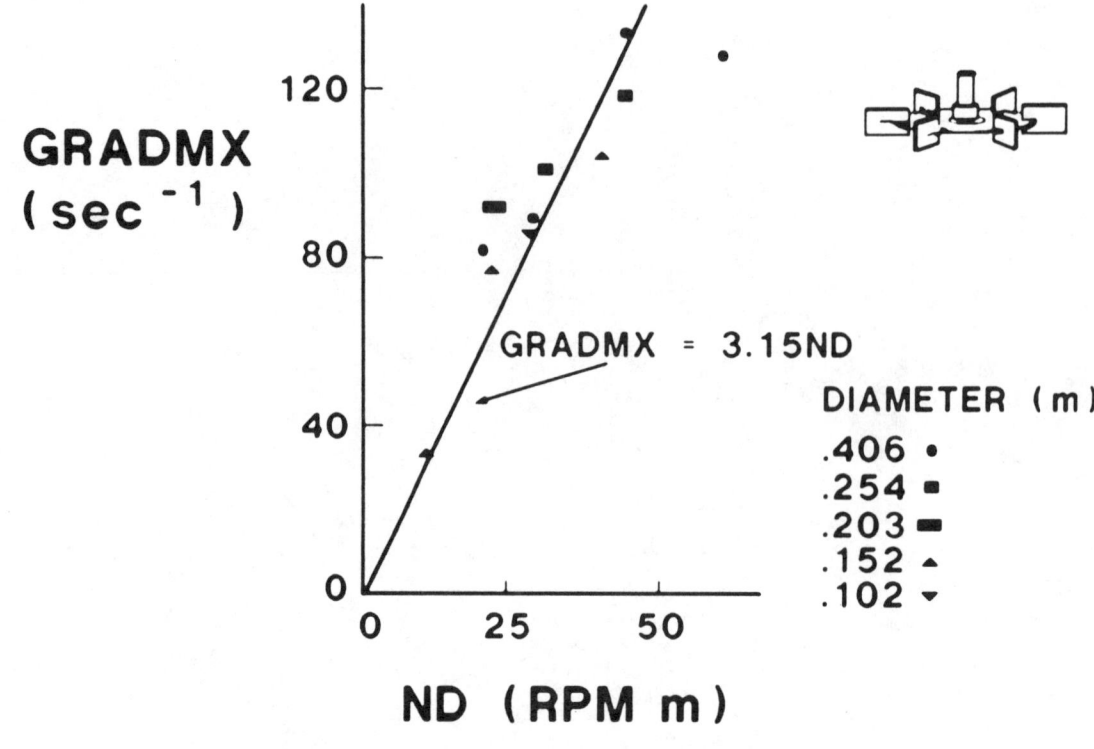

A.

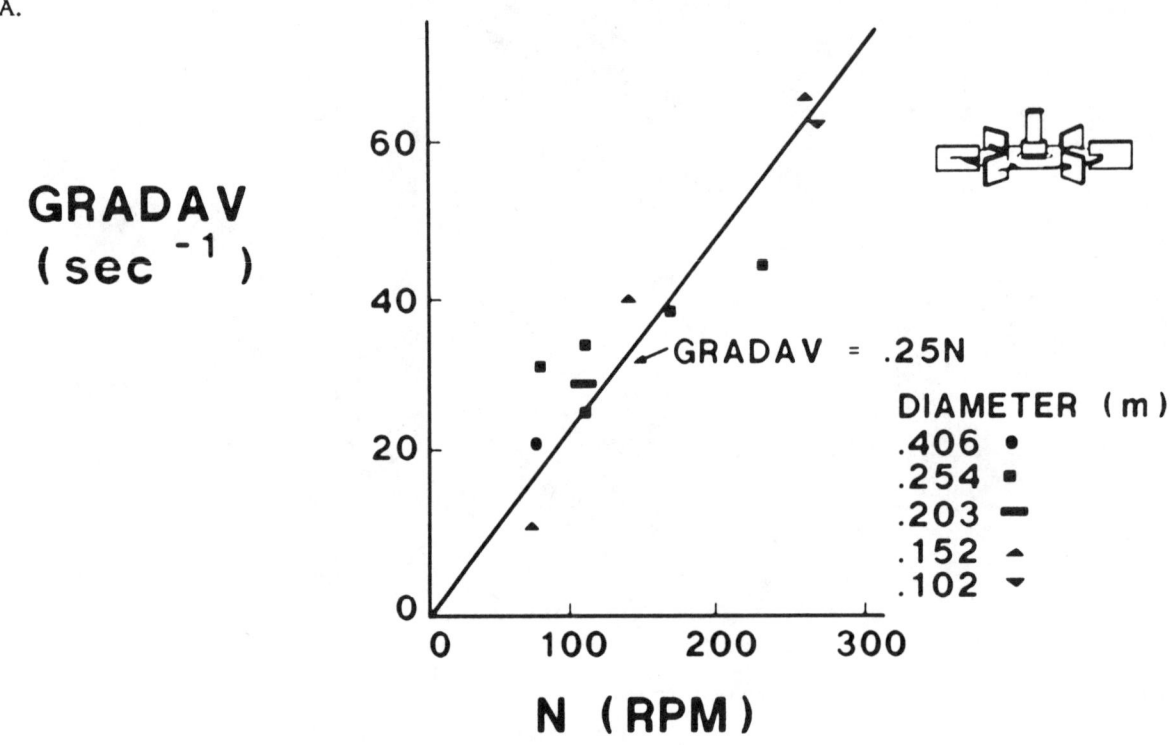

B.

NOTE: For a radial-flow impeller (R100) the maximum shear rate is a function of the tip speed, and the average shear rate is determined by the impeller's rotational speed alone.

Courtesy of Lightnin, A Unit of General Signal, Rochester, N.Y.

Figure 9-9 Maximum shear rate for an R100 impeller.

DESIGN OF IMPELLERS 321

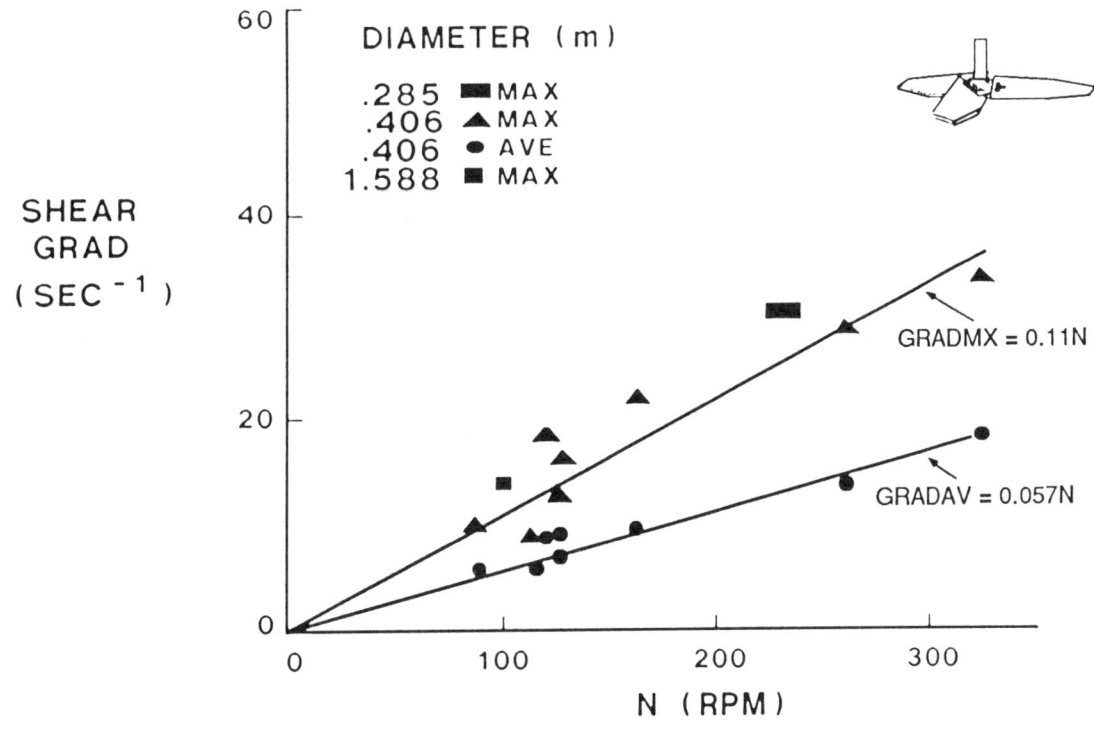

A.

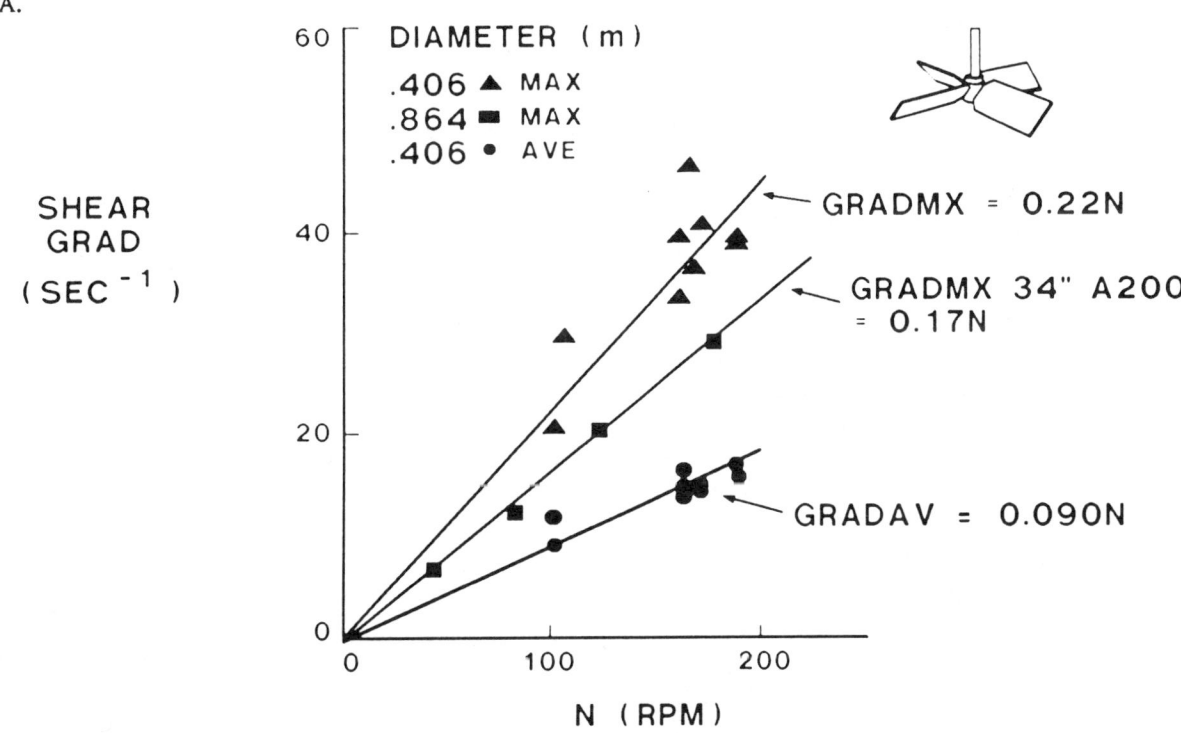

B.

NOTE: Like the radial-flow turbine, the pitched-blade turbine shows a maximum shear that increases with rotational speed and impeller diameter. The maximum shear of the fluidfoil impeller is relatively independent of impeller diameter.

Courtesy of Lightnin, A Unit of General Signal, Rochester, N.Y.

Figure 9-10 Shear rates for the fluidfoil (A310) and pitched-blade turbine (A200).

shear stress is equal to the square root of the power per unit volume divided by the viscosity. Since point values are not readily obtainable, the concept has been extended to the average value for the entire tank (Camp's "G" concept). This is not rigorous, but it has been widely employed by the water works industry with some success. The concept is more satisfactory in describing microscale effects (<1 μm) than macroscale effects (>1000 μm). Microscale shear is a function of the average power per unit volume. Macroscale shear is a function of power per unit volume at a particular point in the tank.

The power per unit volume near the impeller is approximately 100 times higher than the average tank value. As a result, G is different at various points in the tank. That is one major difficulty of comparing performance of impellers in flocculation. At the same G, the individual values of shear rate and shear stress and their distribution around the tank will be quite different for different impellers. The performance of different impeller types should be based on flocculation testing. The G for one impeller should be compared to the G for another impeller at the same level of flocculation performance.

GT

The profile of shear rate values around a mixing vessel does not tell the entire story of the performance of the tank in flocculation. Camp (1956) also proposed that the product of G and the mixing basin detention time (GT) be used to characterize performance. Multiplying the average G value in the tank by the residence time gives an indication of the total work applied to the process. The term "total shear flow" will be used here to describe the "GT" phenomenon.

In a normal multiphase system in which one of the phases is a liquid droplet, a gas bubble, or a solid aggregate, there is a combination of dispersion and coalescence. In some processes, once a particle size, a bubble size, or droplet size is achieved, there is no further coalescence, and the ultimate particle size that results from the high shear rate with sufficient time gives an ultimate particle, bubble, or droplet size that remains constant with further time of exposure within the vessel. On the other hand, for many processes there is a coalescence or flocculation going on as well as a dispersion, so those processes arrive at a dynamic equilibrium that eventually will be steady state with time. This has been shown to be the case in flocculation in water treatment where the maximum floc size (for a given impeller) decreases as G increases (Argaman and Kaufman 1968).

The use of the parameter GT is very appropriate for talking about the same kind of impeller system and the same type of chemical system, either organic or inorganic, but with different impellers the same flocculation may not be given by the same value of GT.

SCALE-UP PRINCIPLES

To show some of the problems in scaling-up an impeller system, an example will be used in which the scale ratio between the prototype and the model is 7:1. The corresponding volume ratio is 343:1. Table 9-3 illustrates the impact of scale on several variables. The 11 variables being examined are listed in the first column as "properties." They are power, power per unit volume, speed, diameter, impeller pumping flow, flow per unit volume, tip speed, Reynolds number, Froude number, and Weber number. The impact of scale-up from an 80-L pilot process to a plant-scale process of 27,400 L is shown in the four columns to the right. This entire table is based on geometric similarity of impeller and tank. As scale-up occurs, it is not possible to keep all the tank's properties constant. Hence, each of the four columns on the right shows the result of keeping 1 of the 11 properties constant as scale is increased.

In the third column, power per unit volume is held constant. This is standard US practice for design of flocculators and flash mixers. As scale increases, the speed decreases, the impeller pumping flow increases, but the flow per unit volume decreases. This is a key factor in scale-up: for a given energy input there will be a longer circulation time and a longer blend time on full scale than on pilot scale.

The tip speed goes up, which is normally related to maximum impeller zone fluid shear. Tip speed is usually associated with maximum shear rate and thus increases on scale-up. The average shear rate is more normally associated with operating speed, and that decreases. Thus, the two shear rates diverge and give a greater variation in larger tanks than in small tanks, as shown in Figure 9-11. Referring back to Table 9-3, it is seen that the Reynolds number also increases, so that larger tanks appear more turbulent than do their small-scale geometrically similar models.

For a rapid-mixing tank it might be desirable to keep the blending time constant as scale is increased (see fourth column). Holding blend time constant requires that the operating speed stay the same in the model and the prototype, thus requiring that the power per unit volume ratio increase with the square of the scale ratio. In this case, P/V is 49 times higher at plant scale than at pilot scale, a possible but not a practical scenario.

The fifth column keeps the impeller tip speed constant. This criterion is also occasionally used for the design of flocculation tanks. This approach decreases the power and flow per unit volume so that it is not a practical scale-up criterion.

The sixth column shows that in an attempt to obtain an equal Reynolds number, the total power in the large tank decreases. This does not seem a sensible mixing criterion either.

Thus, it is apparent that a big tank is considerably different from a small tank, and there will be major difficulties trying to model the process on small scale with the same sensitivity to the variables that are important on full scale.

Table 9-3 Properties of a Fluid Mixer on Scale-up

Property	Pilot Scale, 80 L	Plant Scale, 27,400 L			
P	1.0	343	16,000	49	0.14
P/Vol	1.0	1.0	49	0.14	0.0004
N	1.0	0.27	1.0	0.14	0.02
D	1.0	7.0	7.0	7.0	7.0
Q	1.0	93	343	48	7.0
Q/Area	1.0	—	7.0	1.0	0.14
Q/Vol	1.0	0.27	1.0	0.14	0.02
ND	1.0	1.9	7.0	1.0	0.14
$\dfrac{ND^2\rho}{\mu}$	1.0	13.2	49	7.0	1.0
$\dfrac{N^2 D}{g}$	1.0	0.4	7.0	3.0	0.003
$\dfrac{N^2 D^3 \rho}{6}$	1.0	8580	117,650	2300	47

P/V = Const. Q/V = Const. ND = Const. $ND^2\rho/\mu$ = Const.

324 MIXING

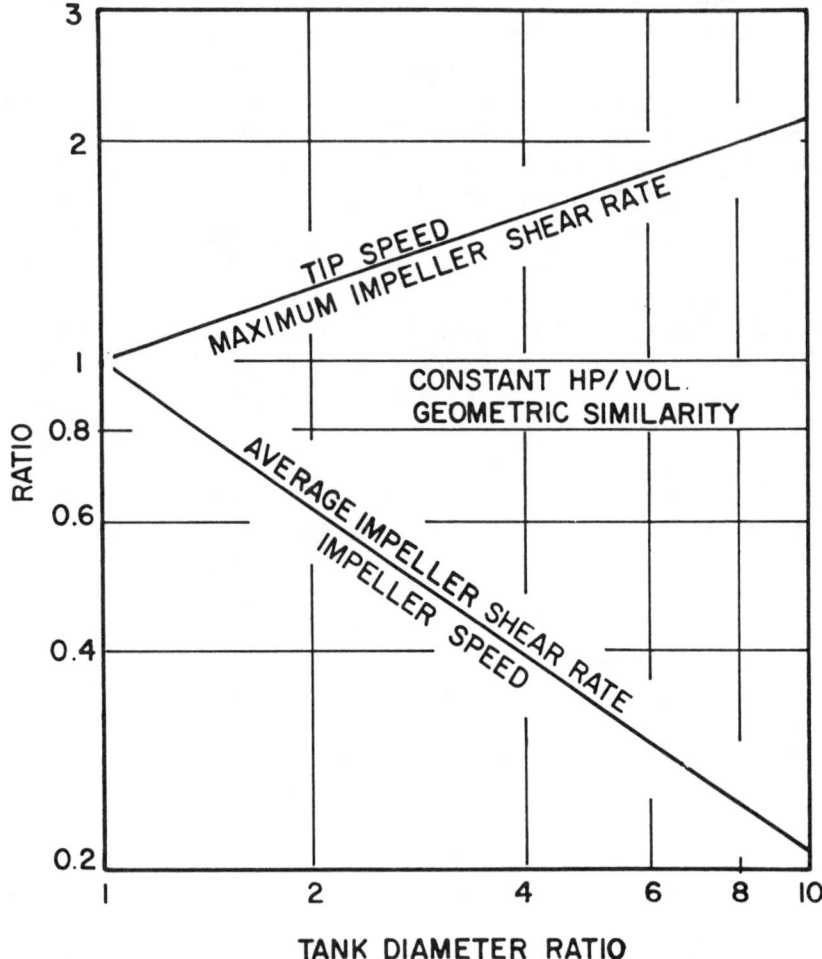

Courtesy of Lightnin, A Unit of General Signal, Rochester, N.Y.

Figure 9-11 Two shear rates diverging.

The typical problem in treating water, both in rapid mixing and flocculation, is the need to model these processes in a way that describes what is likely to happen when basic parameters are changed to full scale. This normally requires a nongeometric approach to small-scale equipment. The problem with a small-scale model can be summarized as follows:

- The blend time tends to be very short relative to full scale (rapid mixing is easier).
- The maximum shear rate tends to be lower than in full scale (less floc breakup).
- The average shear rate tends to be higher than in full scale (more floc buildup).
- There is a much more uniform distribution of shear rates in the small scale than there is in large scale.
- The small-scale tanks appear to be more viscous than full-scale tanks.

The geometry needed to modify the situation in the pilot plant is a smaller D/T ratio and a blade that is narrower than the full-scale design. Both of these items will decrease the blend time and increase the maximum shear rate. These sorts of variables affect the macroscale mixing situation, which is in the general range of 1000 μm and larger. This means that the impeller used in the pilot plant most often is of the A200 type, shown in Figure 9-10, since it is a relatively inefficient blending impeller compared to the A310, and also has a fabricated construction that is amenable to making the blades either normal width or much narrower.

Another possibility, which is not recommended at present, is to use a radial-flow turbine in the pilot plant in order to scale-up to an axial-flow impeller (particularly an A310) in full scale. However, there is marked difference in the performance of radial-flow impellers and axial-flow impellers, and the relationship between blending, tip speed, and maximum and average shear rate is not completely understood at present.

It turns out that the fluctuating components of the velocity that are in the area where the power is dissipated into heat are mainly a function of power per unit volume. Thus, power per unit volume is an important parameter to keep in mind in carrying out pilot experiments and comparing them to full-scale performance. In flocculation, concern normally focuses on all the various velocities and shear rates throughout the tank and, therefore, the use of a particular parameter, such as turbidity, that requires many different kinds of individual fluid-mechanic relationships to obtain a particular process result. However, one ratio that should be maintained constant is the thickness of the material used to make the impeller blades divided by the impeller diameter. This ratio on a typically full-scale unit is 1:24, and it does affect the pattern of turbulence in and around the impeller zone.

DYNAMIC SIMILARITY

There are four fluid forces that are important in mixing operations. As shown in Table 9-4, these four forces are inertia force put in by the mixer and three forces that are opposing the process result—viscosity, gravity, and surface tension.

It is not possible to keep all these ratios constant during scale-up. It turns out that any two can be maintained constant, but the others must be allowed to change. In many other examples of fluid mechanics where the result of the experiment is the force, drag coefficient, or other parameter of the fluid, different fluids are sometimes used in the model and in the prototype.

The mixing processes used in water treatment are at a higher level of complexity. In addition to physical forces, phenomena like chemical diffusion, turbulent diffusion, reaction kinetics, shear rates, microscale velocity fluctuations, etc. are all involved. It is difficult to imagine a dimensionless scale-up parameter for such

Table 9-4 Hydraulic Similitude

Geometric	$\dfrac{X_M}{X_P} = R_X$	
Dynamic	$\dfrac{(F_I)_M}{(F_I)_P} = \dfrac{(F_\mu)_M}{(F_\mu)_P} = \dfrac{(F_g)_M}{(F_g)_P} = \dfrac{(F_\sigma)_M}{(F_\sigma)_P} = R_F$	

NOTE: Requirements for dynamic similarity for fluid forces, inertia force, viscosity force, gravity force, surface-tension force in both model, M, and prototype, P.

complex processes that has the same quality as simple physical parameter like Reynolds number. If the process result can be defined in a very quantitative way, then dynamic similarity is a good possibility for scale-up relationships. For example, if the blend time is measured by a certain arbitrarily chosen but similar type of result in various experiments, both on small scale and large scale, blend time multiplied by the impeller speed often correlates with Reynolds number over a wide range of density, viscosity, diameter, and speed relationships. However, in the case of flocculation, no way of expressing flocculation performance quantitatively that lends itself to being a part of a dimensionless process group has been developed. Thus, there is not much that can be done about dynamic similarity.

A major variable in fluid mixing experiments is the impeller speed. Normally, the first thing to be done is to measure the effect of impeller speed on the process. In flocculation, the usual process variable is the resultant turbidity of the supernatant that remains after the flocculation slurry has been allowed to settle or has been filtered. Figure 9-12 shows that turbidity reaches an optimum, minimum value at a given impeller speed and is higher in value at either lower or higher speeds. The general interpretation of this type of result is that at low speed there is insufficient shear rate and/or microscale mixing to provide the proper floc agglomeration, while at higher speeds the increasing shear stress resulting from the impeller causes floc breakup.

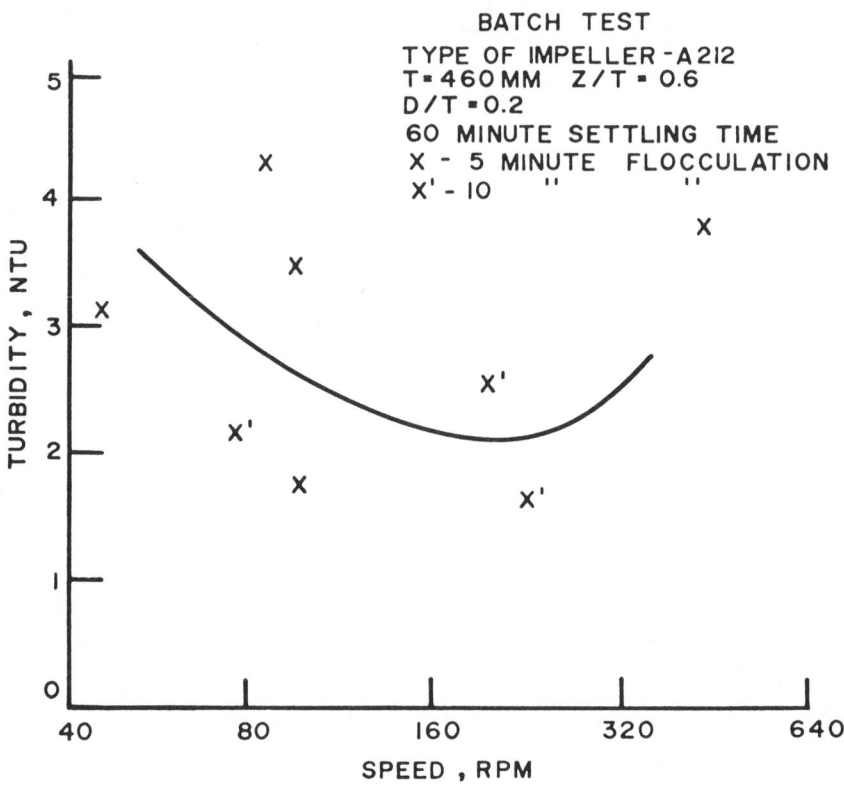

NOTE: At low speed, mixing is inadequate to cause floc agglomeration. At high speeds, impeller shear causes excessive particle breakup.

Courtesy of Lightnin, A Unit of General Signal, Rochester, N.Y.

Figure 9-12 Relationship between impeller speed and performance.

OTHER IMPELLER TYPES

The A310 (Figure 9-4) is about the best shape that can be produced from bending and forming a uniform-thickness material. More efficient airfoils can be designed by using more of a shaped structure, and the new composite materials allow this to be done quite economically.

In an open tank, there is the opportunity for flow to recirculate around the impeller, forming what is referred to as a tip vortex. The use of proplets, shown in Figure 9-13A, improves the flow by approximately 10 percent at a given power level compared to that obtained without the use of these proplets. Figure 9-13 also shows the velocity scans of this type of impeller.

At the other end of the impeller spectrum are the very slow-speed, dangling-plate impellers; gate-type impellers with oftentimes different blades and crossbeams; the reel-type impellers, normally with horizontal shaft; and the horizontal-turbine impellers. All types are shown in Figure 9-14. These impellers usually operate at very low speeds, do not set up any marked flow pattern, and tend to more or less "jiggle" the flow as the blades pass by at low speeds. The fluid mechanics and mixing properties of these impellers are not well-documented. Results obtained in experiments with some of these impellers will be documented in the section on experimental data comparing various impellers.

FLOW PATTERNS IN LARGE BLENDING TANKS

The actual circulation time in a large tank is similar to what one would expect from several smaller tanks in series. This observation has been demonstrated by experiments. Experimenters placed a radio transmitter in a plastic pill approximately 8 mils in diameter and made it neutrally buoyant. They then placed a stationary antenna around the periphery of a radial-flow impeller. As they analyzed the number of circulations that passed the antenna, which they could measure as a function of time by the radio signal as the particle came by the antenna, they found that small tanks behaved essentially as one would expect; the circulation time is calculated quite accurately by dividing the volume of the tank by the pumping capacity of the impeller.

The experimenters continued their tests in large tanks. As the tanks got larger, the number of apparent loops increased. The largest tank (1000 gal) had 5 to 10 loops depending on the impeller speed, demonstrating that in large systems the flow does not go entirely in a way that could be shown by drawing arrows and visualizing the major flow pattern. Rather, it takes a path with a series of loops and other flow paths that give an average circulation time longer than that predicted by the relationship V/Q. Thus, the blending time of a large tank is actually much longer and much less uniform than is predicted by the simple V/Q ratio shown in Table 9-1. Figure 9-15A shows the possible path of a particle exhibiting the circulation time in a series of tanks. Figure 9-15B shows how the experimental data were fitted to an equivalent number of tanks in series.

FLOCCULATION EXPERIMENTS

The problem with removing suspended solids in a water or waste stream is that often the solids all have the same charge so they tend to repel each other and stay in suspension. The idea of flocculation is to use chemicals to neutralize the charge on the particles such that aggregation and flocculation can occur if velocity gradients are present. Mixing in flocculation normally takes place in two steps: first

328   MIXING

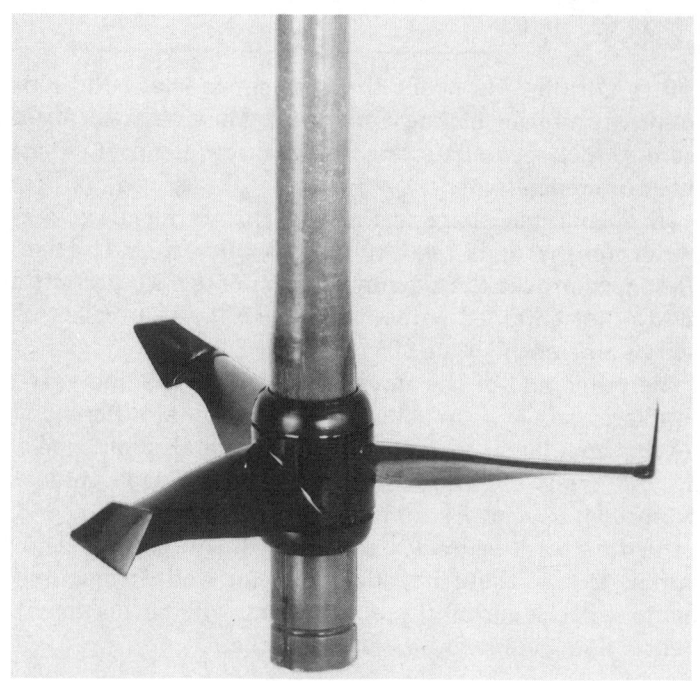

A. Lightnin fluidfoil with proplets.

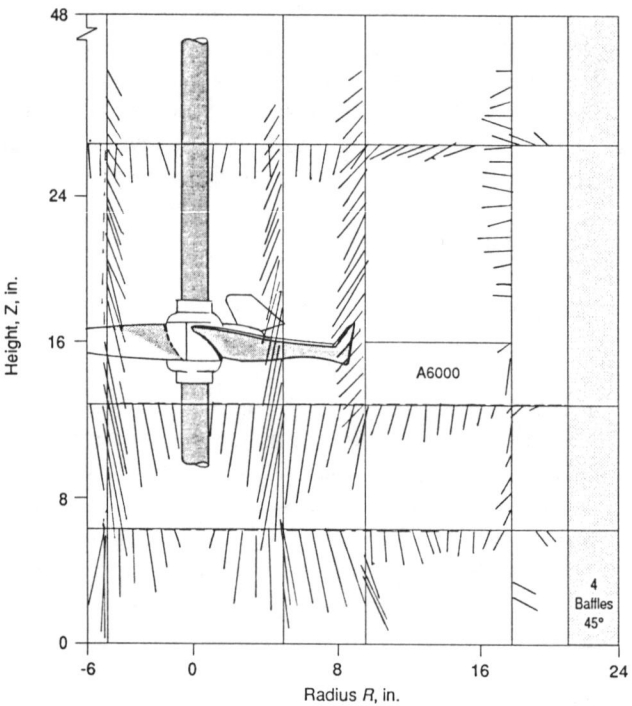

B. Fluidfoil with proplets velocity vectors.

NOTE: The addition of proplets to a fluidfoil design reduces the tip vortex, increasing pumping.

Courtesy of Lightnin, A Unit of General Signal, Rochester, N.Y.

Figure 9-13 Composite fluidfoil with proplets (Lightnin A6800).

DESIGN OF IMPELLERS 329

a rapid mixing step designed to blend treatment chemicals with the bulk fluid, and second a slow stirring step designed to build increasingly larger aggregates that will settle out.

When inorganic chemicals are used, it does not seem that there is any maximum power level for rapid mixing. Rapid mixing is typically designed on the basis of horsepower per thousand gallons, and it is difficult to predict any performance decrease or increase by using different types of impellers. Studies on the rapid-mix part of the process have been relatively inconclusive, and most experiments and observations have focused on the performance of flocculation following rapid mixing.

A. Dangling-plate flocculator.

B. Gate-type flocculator.

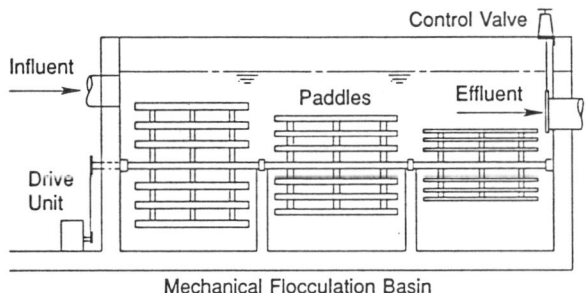

C. Horizontal-shaft paddle.

D. Horizontal-shaft turbine.

NOTE: The dangling-plate, gate, horizontal-shaft paddle, and horizontal-shaft turbine are all alternatives used in flocculator design. The hydraulics of these impellers are poorly understood, and their repair often requires a plant shut down. Nevertheless, they remain in use for traditional reasons and because of concerns deriving from misapplication of vertical turbines.

Courtesy of Lightnin, A Unit of General Signal, Rochester, N.Y.

Figure 9-14 Other types of impellers used in flocculation.

330 MIXING

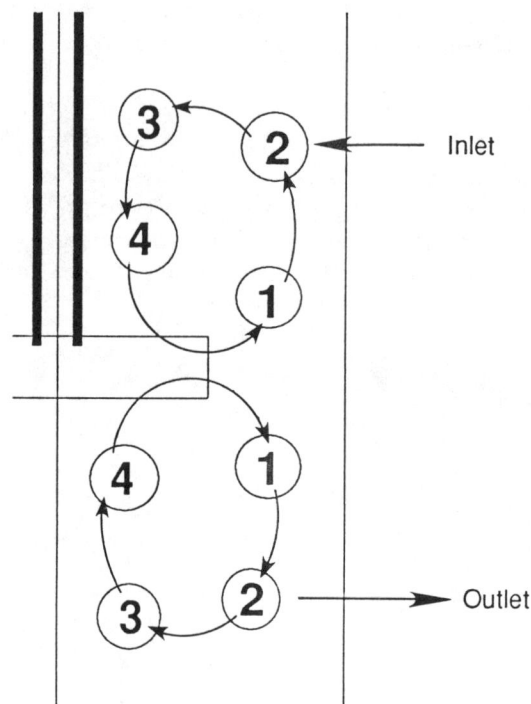

A. Particle path in mixer.

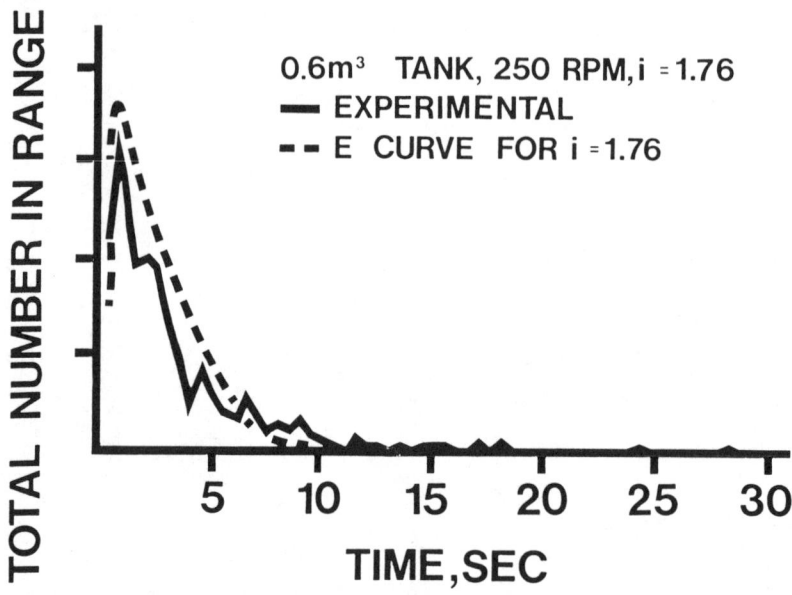

B. Dye curve from mixer.

NOTE: Note that the impeller creates two recirculation regimes that operate with some independence. A tank's performance can be modeled by assuming it is made up of a certain number of smaller, well-mixed tanks. The experimental results from a dye curve in a real mixed tank can be simulated by assuming an arbitrary number of tanks in series. In the case shown above (B), assuming 1.76 tanks produced reasonable results.

Courtesy of Lightnin, A Unit of General Signal, Rochester, N.Y.

Figure 9-15 Blending: Possible paths for particles in a mixing tank (A). Dye curve from mixer (B).

Bench and Pilot-Scale Tests

The jar test is almost universally used to model rapid mixing and flocculation on a bench scale. A jar test mixer would normally have to be calibrated in place to find out how to run the test to give the same flocculation performance as observed in the plant. The jar test normally has a particular tank size and impeller diameter, and the fluid mechanics obtained from the system do not necessarily relate to any given full-scale impeller/tank combination where mixing parameters are concerned.

As a general rule, a pilot test vessel should be on the order of 30 in. in diameter to allow accurate modeling of the impeller and also to allow other ratios to be used in the pilot plant to more closely duplicate the mixing and fluid mechanics obtained in a particular plant-size unit.

Experimental data comparing various impellers. A batch study made in pilot scale—two different-size tanks (18 and 30 in. in diameter)—as well as full-scale measurements in tanks that were 25 ft square were conducted over a period of three months, using water passing through full-scale rapid-mix tanks. A variety of different impeller types were used, including the R100 (Figure 9-7), the A200 (Figure 9-6), the A212 (Figure 9-16A), a marine A100 propeller (Figure 9-16B), and a gate-type impeller used as typical of that class of impeller types (Figure 9-14B).

The material from the full-scale rapid mixer (Figure 9-14) was mixed in the batch tanks. A schematic of the water treatment plant where the tests were conducted is shown in Figure 9-17. The tank was then allowed to settle for various lengths of time and the turbidity measured (in nephelometric turbidity units [NTUs]). Figure 9-18 shows the general relationship between turbidity and settling time. A 60-min settling time was used for most of the comparisons, although Figure 9-18 can be used to evaluate other settling times. In addition, the color was measured. It turned out that the source of water, Lake Massabesic, N.H., was primarily ice- and snow-covered and the entering-water turbidity and color were consistent during all these experiments. The various impellers were used with various combinations of D/T ratio. A typical result from these experiments was shown in Figure 9-12, and the results from all the experiments are summarized in Table 9-5.

In all cases, the impeller speed for minimum turbidity at a 60-min settling times was used to correlate the effect of various mixing parameters. The principle conclusions were that, for any tank size, the larger the D/T ratio with a given impeller type, the higher was the G at optimum conditions, and the optimum process result was improved with the higher D/T (Tables 9-6 and 9-7). For the open impellers, defined as any impeller less than 0.6 D/T ratio, the higher the flow and the lower the shear rate compared to another impeller type, the better the performance at the optimum point. The higher-flow impellers allowed the introduction of a higher G factor at the optimum point, apparently leading to a better process performance.

The runs comparing the rake, gate, and dangling plate (Table 9-5) show less sensitivity in regard to the optimum speed. Nevertheless, at any particular speed used to achieve a comparable turbidity G was higher by something on the order of 2 to 10 (Tables 9-6 and 9-7) when compared to the optimum G for an open impeller. In addition, the power numbers for these impellers are very high. This means that the speed for a given G factor will be relatively low, which in turn means that the torque required to turn the impeller (P/N) will be higher for these units than for the open impellers. The cost of the mixer speed reducer is largely a function of the torque required, so this phenomenon of increased power and reduced speed makes the capital cost of the mixer drive for these slow-speed impellers quite high.

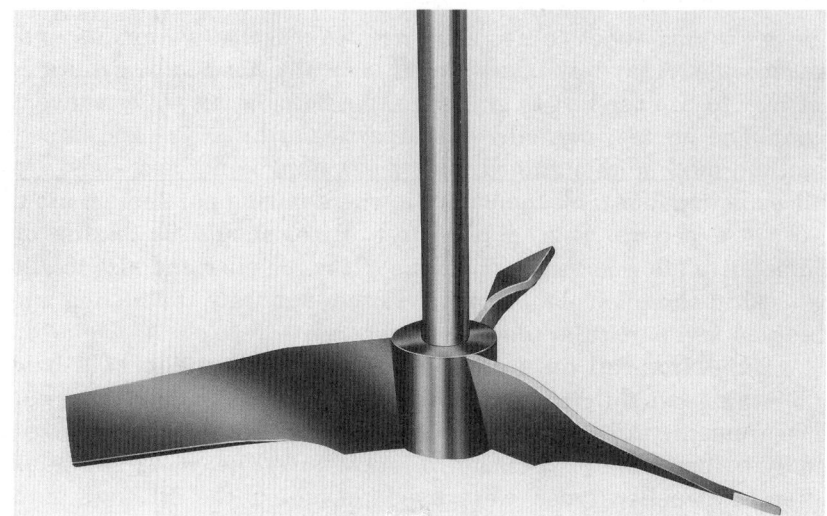

A. Lightnin A212: an earlier design of current fluidfoil impeller.

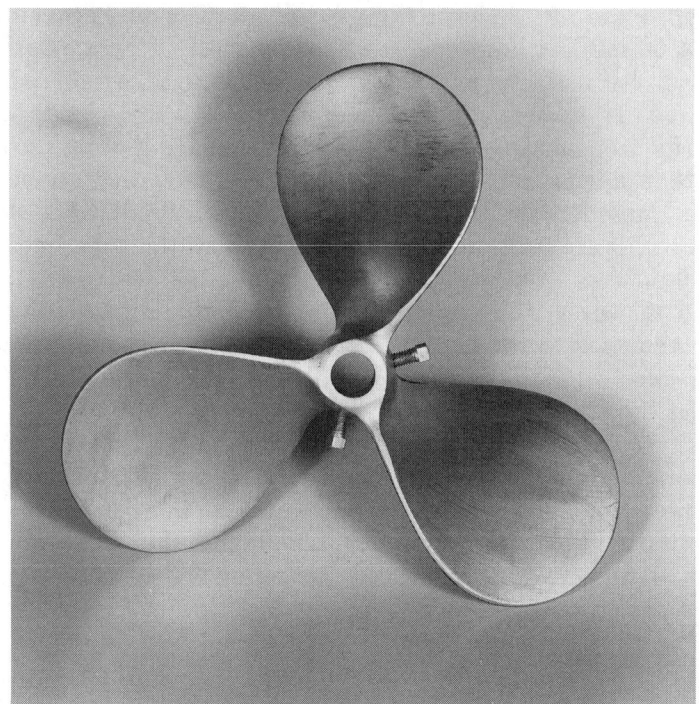

B. Lightnin A100: A marine axial-flow impeller.

Courtesy of Lightnin, A Unit of General Signal, Rochester, N.Y.

Figure 9-16 Alternative impellers used in flocculation experiments.

DESIGN OF IMPELLERS   333

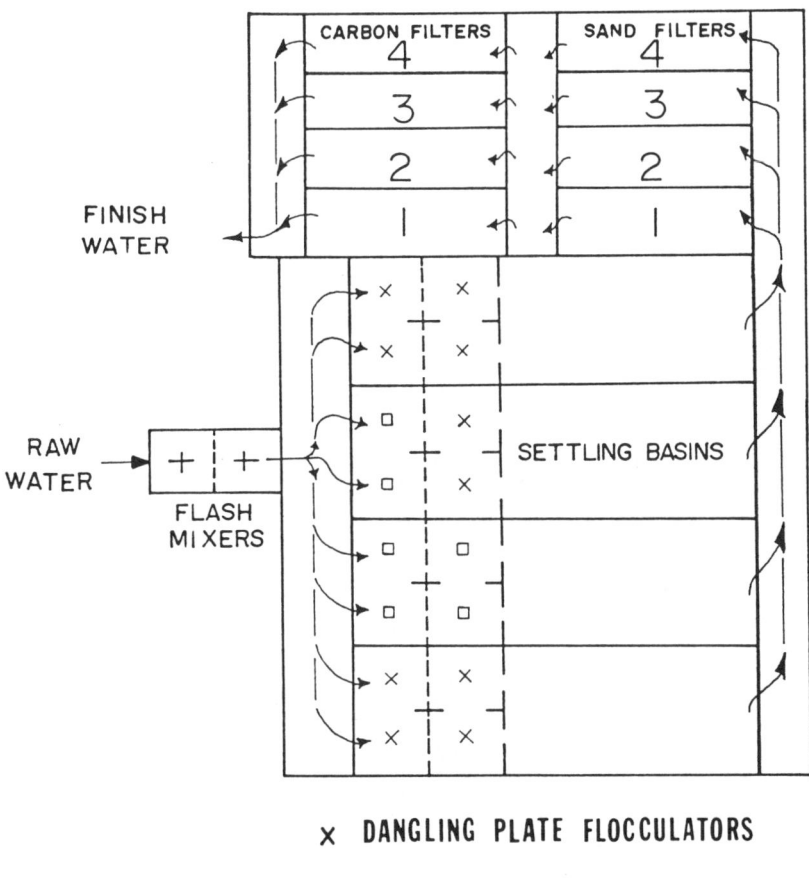

× DANGLING PLATE FLOCCULATORS

☐ LIGHTNIN A212 FLOCCULATORS

NOTE: Full-scale flocculation tests were conducted at the Manchester, Mass., water treatment plant. Shown above is the overall flow scheme of the plant along with the location of various impellers compared during the tests.
Courtesy of Lightnin, A Unit of General Signal, Rochester, N.Y.

Figure 9-17 Treatment plant schematic.

Comparing the results from the two different scales (Tables 9-6 and 9-7), it is seen that the G for the larger tank is considerably lower than it was in the small tank for the same geometry at the optimum speed (Figure 9-19). An interpretation of this phenomenon is the fact that at the same maximum impeller zone shear rate, the power level is considerably less as tank size increases, and this apparently is adequate to give excellent performance on the larger systems. There is no way really to predict this outcome from basic fluid mechanics since the process steps in the formation of floc particles and their eventual settling are too complex to allow any assurance as to what the controlling factor would be. Table 9-5 gives the data of various geometries tested, and also gives some indication of the scatter and variability of these kinds of results.

As shown in Figure 9-17, the plant unit was equipped with four separate parallel trains of flocculators, each having two stages in a series. There were two trains with dangling-plate impellers in both stages in the series. One train had turbine impellers in both stages; the other had turbine impellers in stage one and a dangling plate in stage two. Of course, there is a difference between batch pilot plants

334 MIXING

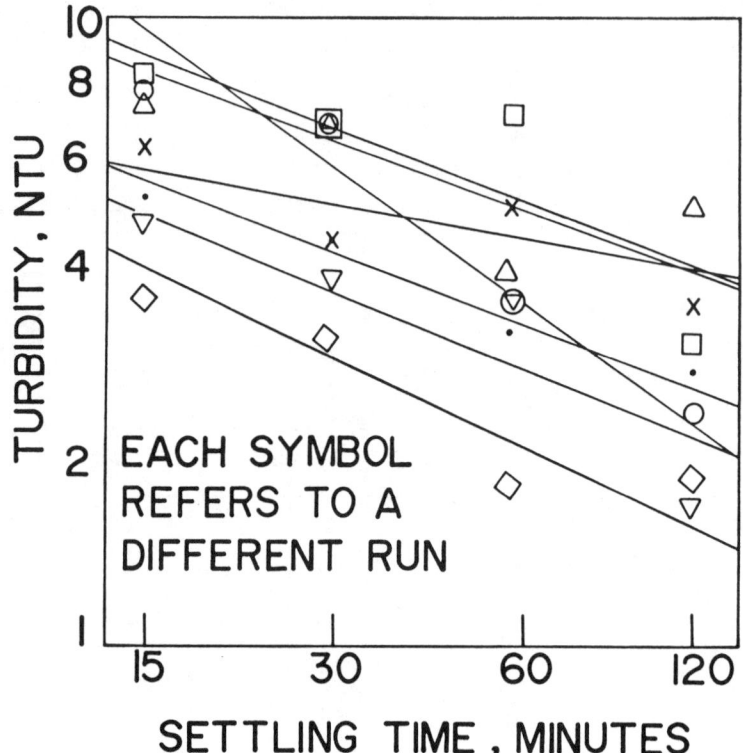

NOTE: Turbidity removal improved with settling time.

Courtesy of Lightnin, A Unit of General Signal, Rochester, N.Y.

Figure 9-18 Batch flocculation tests: settling time.

and continuous-flowing full-scale plants, so the comparative results cannot be considered to be under exactly the same mixing conditions at small and full scale. Table 9-8 summarizes the results of these full-scale experiments. Unfortunately these results were run over a limited period of time and flocculation experiments are not highly reproducible. As a result, comparisons with batch-scale data are difficult.

Summary

Current thinking about flocculation equipment is that there is a significant process difference between the large-diameter (D/T of 0.95 to 1.0) impellers (rakes, horizontal paddles, dangling plates, etc.) and the newer fluidfoil impellers of the A310 type. The power level can be considerably lower with the new fluidfoil impellers, and the actual capital costs involved with the size of the gear reducer required are normally much lower. Perhaps more important, vertical turbines can be repaired or maintained with fewer process interruptions. However, the fluidfoil impellers must be properly designed and also require relatively large D/T ratios. The experimental work compared A212s, which are a step beyond the A200. Comparable data of this type do not exist for the fluidfoil impellers, but there are several operating plants that indicate that the basic principles presented here are applicable.

Table 9-5 Summary of Batch Flocculation Experiments

460-mm Tank										
Impeller	D/T		Performance							
A200	0.2	N	81	84	186	365				
		NTU	2.1	4.0	3.3	3.5				
A212	0.2	N	45	75	85	95	95	185	228	436
		NTU	3.1	2.1*	4.3	3.4	1.8	2.5*	1.6*	3.8
A212	0.3	N	20	79	80	84	136	197		
		NTU	5.0	2.9	2.5	3.1	2.8	2.5		
A212	0.5	N	20	30	70	130				
		NTU	2.3	1.9	1.5	2.1				
R100	0.2	N	27	60	139	291				
		NTU	4.1	3.3	1.9	3.8				
RAKE	0.8	N	92	30	58	73				
		NTU	2.2	1.3	3.0	3.1				
760-mm Tank										
Impeller	D/T		Performance							
A212	0.13	N	72	134	303	481				
		NTU	3.0	2.7	2.3	3.2				
A212	0.2	N	38	42	48	58	72	142	181	202
		NTU	4.1*	3.4	4.7*	3.9	3.5	2.4*	3.2*	2.4
A212	0.3	N	20	23	52	60	132	164	234	
		NTU	3.5	5.2	2.9	2.6	3.4	4.7	3.9	
A212	0.4	N	22	40	73	115				
		NTU	2.7	1.9	1.1	3.3				
R100	0.2	N	20	50	84	180				
		NTU	5.8	3.2	2.7	4.0				
RAKE	0.8	N	8	14	25	54				
		NTU	1.8	1.3	0.9	2.7				

Source: Lightnin, A Unit of General Signal, Rochester, N.Y.
*10-min flocculation; all other 5 min. Settle time = 60 min in all experiments.

Table 9-6 Minimum Turbidity Values (NTU) at Different D/T

460-mm Tank					760-mm Tank			
R100	A200	RK	A212	D/T	A212	RK	A200	R100
		1.3		0.8		1.0		
			1.5	0.5				
				0.4	1.1			
				0.4	1.1			
			2.5	0.3	2.6			
2.5	3.0		2.0	0.2	2.4		4.0	3.0
				0.13	2.3			

Table 9-7 G at Minimum Turbidity (1/s) (NTU)

R100	460-mm A200	Tank RK	A212	D/T	A212	760-mm RK	Tank A200	R100
		148		0.8		112		
			92	0.5				
				0.4	106			
			110	0.3	34			
116	77		58	0.2	27			59
				0.13	31			

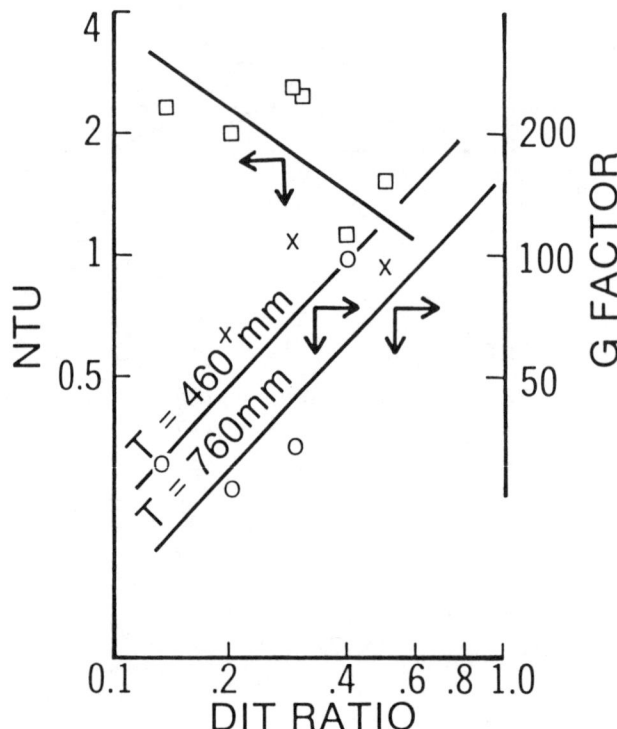

NOTE: This summary of the batch results shows the relation between D/T ratio, G, tank size, and performance. Generally, lower G's are required in larger tanks and larger D/T raios improve performance.

Courtesy of Lightnin, A Unit of General Signal, Rochester, N.Y.

Figure 9-19 Batch flocculation results.

Table 9-8 Summary of Results From Full-Scale Flocculation Experiments

A212 Same Speed Each Stage		A212 High Speed Second Stage		A212 Low Speed Second Stage		Dangling Plates	
N	NTU	N	NTU	N	NTU	N	NTU
3.3	2.3	3.3	2.9	5	2.7	0.6	3.3
3.4	5.1	3.3	3.2	5	2.3	0.6	2.6
4.8	2.5	3.5	3.9	5	1.9	0.7	1.9
5.9	3.1	7.8	3.5	6.2	3.8	0.9	1.9
7.5	3.4			7.2	2.6	1.0	2.3
8.0	2.4			8.1	3.0	1.0	2.2
11.5	2.9			8.5	2.1	1.0	2.0
11.8	3.2			8.5	1.9	1.1	1.9
				10.2	3.2	1.1	1.7
				10.2	2.3	1.3	2.6
				10.2	1.9	1.5	1.6
				10.5	4.2	1.5	2.6
				11.5	3.2		
				11.5	1.7		

Source: Lightnin, A Unit of General Signal, Rochester, N.Y.

POWER AND FLOW COMPARISONS FOR IMPELLERS

A process mixer is designed by specifying the power and the diameter of the impeller required for the particular impeller type and fluid properties of the process. This determines the process performance. The next step is the calculation of the impeller speed. For that purpose, it is necessary to use a Reynolds number/power number curve, shown in Figure 9-20, which must be available for the particular geometric design of the impeller and the tank.

In flocculation, most of the time the speed is out in the high Reynolds number range, where the curve is flat, as shown in Figure 9-20. The power number in this region is constant, and typical values are listed in Table 9-9.

Another parameter of interest is the pumping capacity of the impeller, often expressed as a flow number, N_Q, which is Q/ND^3. Table 9-9 gives values for N_Q as well, which can be used to estimate the pumping capacity of the impeller in the system. There is less published data comparing the flow number to the power number, and by their very nature flow number data are more approximate. In addition, there must be a definition of the discharge area of the impeller used in the flow number. For Table 9-9, this area is defined as the plane normal to the principle discharge direction of the impeller. For the radial-flow turbine, it is a circumferential cylinder of a height equal to the blade height. For the fluidfoil impellers, it is a disk in the plane of a discharge area at the diameter of the impeller. For an axial-flow turbine, A200, it is a conical surface corresponding to the discharge angle of the particular impeller used in the system.

In addition, there usually is entrainment as the jet stream from the impeller spreads out into the flow from the tank. This entrainment can vary from a few percentage points to as high as 1000 percent.

Field power measurements. The most accurate and reliable measurement of power in the field is obtained with a continuously recording wattmeter. It must be used in conjunction with a motor curve; a typical one is illustrated in Figure 9-21. This combination allows the estimation of motor output power based on the impeller

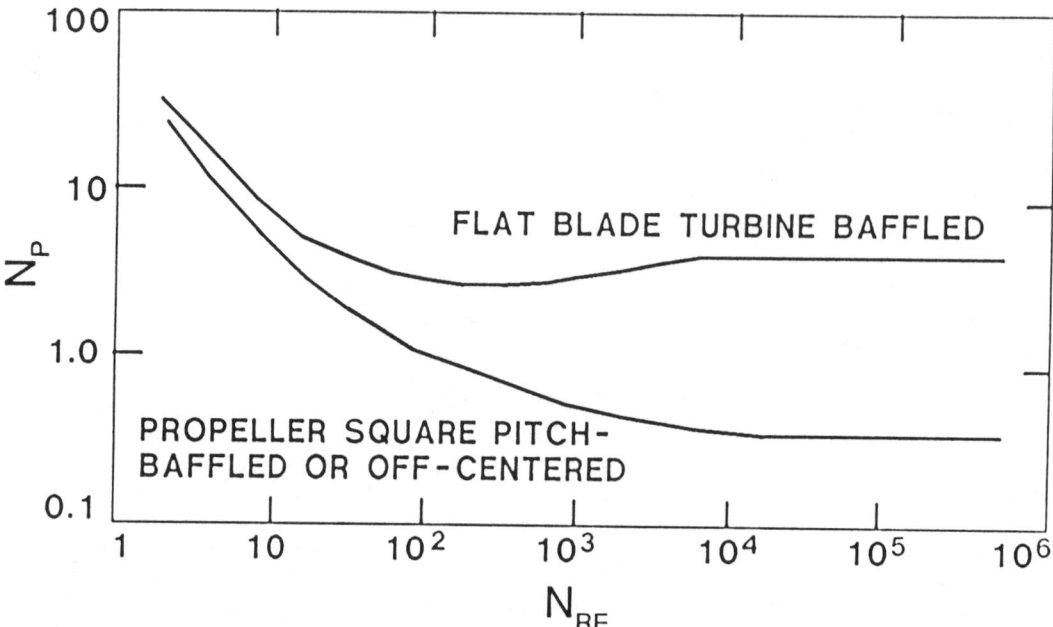

NOTE: The power number for a specific impeller design (N_p) decreases with increasing Reynolds number (N_{RE}) until the turbulent zone is reached. Above a Reynolds number of 10³ or so, N_p is usually constant. It is important to have a curve of N_p vs. N_{RE} for the particular basin configuration being compared.

Courtesy of Lightnin, A Unit of General Signal, Rochester, N.Y.

Figure 9-20 Power number versus Reynolds number.

Table 9-9 Power* and Flow Numbers for Various Lightnin Impellers

Impeller Type	N_P	N_Q	$(Q/P)_R$†	
A6000	0.23	0.59	2.30	
A310	0.30	0.56	1.51	Flow
A200	1.27	0.79	1.00	
C102	0.23	0.56	1.97	
A315	0.75	0.73	1.34	Pressure
C104	0.60	0.52	0.60	
R100	5.20	0.72	0.18	
R510	0.65			Shear
R500	0.45			

Source: Lightnin, A Unit of General Signal, Rochester, N.Y.
*Power number is $Pg/[\rho N^3 D^5]$.
†Ratio $(Q/P) / (Q/P)$ at constant Q,D.

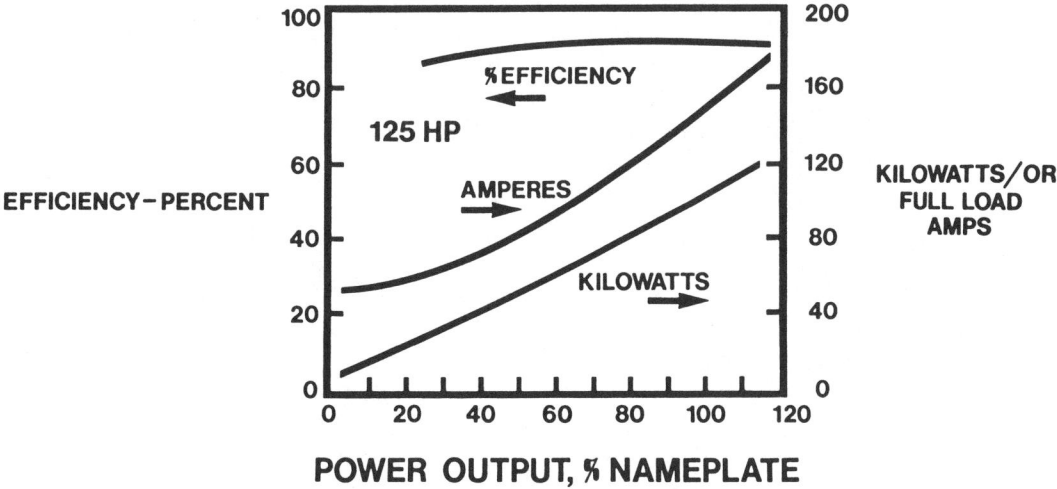

NOTE: This type of curve must be obtained for the mixing motor in order to do field calibration of flocculators.

Courtesy of Lightnin, A Unit of General Signal, Rochester, N.Y.

Figure 9-21 A typical motor curve.

input electrical readings. The curve in Figure 9-21 is for a large motor, and it is seen that the curve of output horsepower and input watts is relatively proportional down to about 10 percent of motor output rating.

Estimation then must be made of the power lost in the speed-reduction mechanism. For helical and spiral bevel sets, there is normally a loss of about 2 percent of speed-reducer rated power at a given speed for each set of reductions, so a double reduction unit would have 4 percent losses and triple reduction would have 6 percent losses. If a more accurate determination is required, the speed-reducer manufacturer should be contacted for more detailed data on mechanical losses.

The use of ammeter readings is often adequate, but there is less linearity of amps versus power at the low motor loadings. Ammeter readings must normally be adjusted for actual voltage, which introduces another estimation in the procedure.

Small motors, on the order of 1 hp or less, and particularly fractionless power motors have curves of watts and amps that are much less linear with output horsepower. Thus, it is very difficult to obtain appropriate readings to use with small motors, particularly in laboratory or pilot-plant equipment. In these cases, one can obtain the power number for the particular impeller being used. Then an estimation can be made of the effect of different geometries, and power can be calculated from the measurement of the speed. One caution in measuring speed of laboratory or pilot-plant equipment: sometimes the torque required for the tachometer is high enough to reduce the speed of the motor during the measurement of the speed from what it was during the actual experiment. Experiments have also been conducted using a torquemeter to measure torque directly and using strobe lights to measure high rotational speeds.

Measuring flow in mixing vessels. In the laboratory, flow can be measured by several different techniques ranging from pitot tubes to photographs, hot wire/hot film velocimeter, and the laser–Doppler velocimeter. Measuring flow in an actual installation is much more difficult. Most often propeller meters are used, placed in an array on a rod that allows the velocity profile across an entire discharge line to be measured simultaneously. Propeller meters are sensitive to the angle of flow coming into them to some degree, and they should be oriented properly to the actual flow being measured. Pilot tubes or other dynamic flowmeters can also be considered for flow measurement on full scale. The physical installation and holding of the probes in a known location are complicated mechanical questions to be addressed in the actual field measurement.

If flow is the basic parameter required for the process, then a flow specification and a measurement technique can be valuable. On the other hand, if it is merely thought that flow will give an indication of the performance expected from the mixer, then some judgment must be exercised as to whether the expense and time to measure flow will yield an answer that is commensurate with the accurate prediction of process performance.

Measurement of mixing effectiveness on continuous flow. A continuously flowing stream can have two extreme residence time distributions. The stream flowing in a long pipeline under turbulent flow will have what is termed "plug flow." This means that all portions of the fluid are in the conduit for approximately the same length of time. The other extreme is a well-mixed tank with a random, statistical distribution of residence times.

There are two types of tests considered for measurement of these properties in continuous flow. One is a pulse test, in which a short pulse of a soluble tracer chemical is introduced and concentrations made of the output concentration. The other is a step change. These techniques are discussed in chap. 5. These techniques can also be used to estimate whether the tank is approximated by one well-mixed stage or is better approximated by several well-mixed stages in a series.

Geometrical considerations. The least power with a single impeller is required for a Z/T ratio 0.6. Any other Z/T ratio with a single impeller will require more power for the same volume of fluid.

Obviously, the economic optimization is dependent on factors other than just the power of the mixer. Using tanks with higher Z/T ratios may require more mixer horsepower, but may result in a more economical installation overall. As shown in Figure 9-22, the use of a second, third, or subsequent impeller is often indicated when tanks have very high Z/T ratios. This, however, neglects the fact that, particularly with axial-flow turbines (A200), there is a definite zoning action with each individual impeller because of the 45° angle of the flow coming off the axial-flow impeller. This action can result in temperature or concentration differences, but they usually are not serious in typical rapid-mix or flocculation systems.

The general principal is that when a Z/T ratio is over 1.0 and up to 1.6, serious consideration should be given to the use of a second impeller. For every incremental Z/T increase of 0.8, subsequent impellers are considered for installation.

If it is necessary to mix the last volume of fluid as it leaves the tank, then placing the lower impeller one-sixth diameter off the tank bottom is often considered. This is not the optimum location for an impeller for overall power requirements, but it is the only possibility when mixing the last volume of fluid during draw-off is required.

It is typical to use three impeller blades with the fluidfoil, axial-flow type of impellers. With the axial-flow turbine, it is typical to use four or six blades; with the radial-flow turbine (R100), four to eight blades are often used.

Steady (bottom) shaft bearings. In the actual design of a mixer, the question arises: "Should there be a steady bearing in the bottom of the tank?" Steady bearings can be designed to have a wearing sleeve on the shaft and a wearing sleeve inside the steady bearing holder, so that maintenance and replacement of the

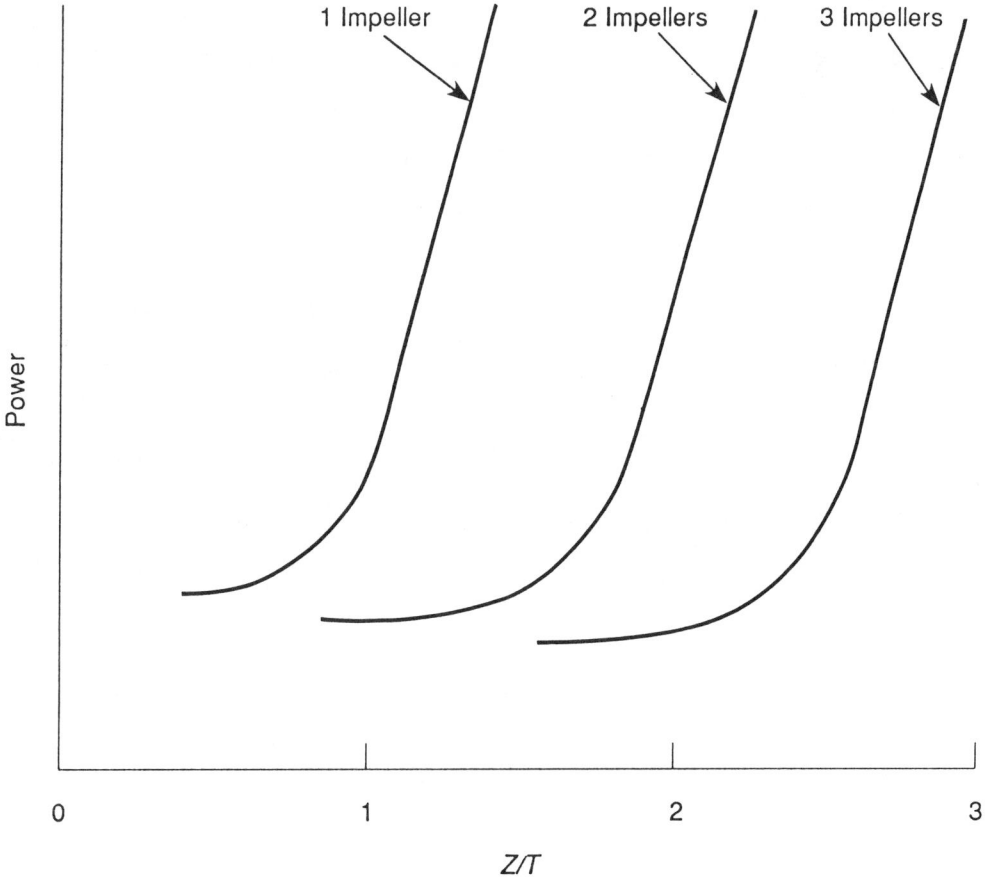

Power Required for Suspension Versus Z/T

NOTE: The ideal position for a single impeller is such that the ratio of the distance above the tank bottom (Z) and the equivalent tank diameter is 0.6. For deeper tanks multiple impellers should be considered.

Courtesy of Lightnin, A Unit of General Signal, Rochester, N.Y.

Figure 9-22 Power requirement versus impeller position.

sleeves can be done relatively inexpensively. Nevertheless, some maintenance will always be necessary with a steady bearing.

In principle, it is always possible to design an overhung shaft for any given length, speed, and impeller diameter by making the shaft a large enough diameter. That diameter may be uneconomical or impractical, but it could be done. The steady bearing, in general, allows the use of a smaller-diameter shaft than with an overhung shaft. Its use is strictly a question of economics as to whether savings in shaft diameter are balanced by the increased maintenance expense.

References

AMERICAN WATER WORKS ASSOCIATION. 1969. *Water Treatment Plant Design.* McGraw Hill, New York.

ARGAMAN, Y. & KAUFMAN, W.J. 1968. Turbulence in Orthokinetic Flocculation. Rept. No. 68-5, Sanit. Engrg. Res. Lab., College of Engrg., Univ. of Calif.—Berkeley, Berkeley, Calif. (July).

BRIJ,; MOUGILAND, M.; & SOMASUNDARAN, P. 1986. *Flocculation, Sedimentation and Consolidation.* United Engineering Trustees, Inc.

CAMP, T.R. 1956. Flocculation and Flocculation Basins. *Trans. Amer. Soc. Civ. Engrs.*, 120:1.

CAMP, T.R. & STEIN, P.C. 1943. Velocity Gradients and Internal Work in Fluid Motion. *Jour. Boston Soc. Civ. Engrs.*, 30:10:217.

J.M. MONTGOMERY. 1985. *Water Treatment Principles and Design.* Wiley-Interscience, New York.

OLDSHUE, J.Y. 1966. Fermentation Mixing Scale-up Techniques. *Biotechnol. & Bioengrg.*, 6H:3.

———. 1970. The Spectrum of Fluid Shear in a Mixing Vessel. In *Chemeca 70 Conference.* Butterworths, London.

OLDSHUE, J.Y. & MADY, O.B. 1978. Flocculation Performance of Mixing Impellers. *Chem. Engrg. Prog.*, 74:103.

———. 1979. Flocculation Impellers: A Comparison. *Chem. Engrg. Prog.*, 75:72.

10

Pilot-Plant Studies for Design and Operation

Dale D. Newkirk, P.E., Regional Operations Manager,
 Metropolitan Water District of Southern California,
 Los Angeles, Calif. USA

R. Rhodes Trussell, Ph.D., P.E., Senior Vice President,
 James M. Montgomery Consulting Engineers, Inc.,
 Pasadena, Calif. USA

INTRODUCTION

This chapter provides a state-of-the-art review of the design and use of pilot-plant studies to determine the design, operation, modification, and evaluation of mixing processes in water treatment. It discusses considerations necessary to properly design a pilot testing program to meet objectives and highlights the strengths and weaknesses of various methods currently available to properly assess mixing processes in water treatment.

The primary mixing processes reviewed in depth include the rapid mixing process, where chemical coagulants are added to the treatment process to enhance coagulation of particulates, and the flocculation process, which is used to form agglomerates of particles (floc) for subsequent removal in the sedimentation and filtration processes. The mixing process is used in many other instances in the overall water treatment process such as for chemical addition (i.e., chlorine, ammonia, hydrogen peroxide, potassium permanganate, caustic soda, lime, sodium bicarbonate, chlorine dioxide, sodium silicate, carbon dioxide, acid, and a host of other chemical additives). Other mixing processes in water treatment include degassification, ozone addition, filter surface wash for physical removal of particulates in filters during backwash, and a host of other applications not discussed here. The rate of these reactions is a direct function of the quantity of energy applied, in contrast to the coagulation and flocculation processes, which are more sensitive to the rate and uniformity of energy application. The use of ozone as a coagulant aid is discussed to some extent in this chapter, given its significance in enhancing coagulation and flocculation. Other authors (JM Montgomery 1985; Sanks 1980) provide further reading on mixing during ozone addition and other treatment processes not discussed here.

The significance of mixing during coagulant addition (rapid mixing) and flocculation cannot be overstated and should never be underestimated during the design or modification of full-scale water treatment facilities. Many preliminary design studies that include pilot testing have a tendency to focus on filtration design, with perhaps a cursory review of chemical coagulant types and dosages at best. It is the authors' experience as well as others' (Tanaka et al. 1987) that more than a 100 percent improvement in overall treatment process particulate removal can be achieved by enhancement of rapid mixing and flocculation. In the real world, this improvement may mean the difference in meeting state and federal regulations, which are becoming more stringent (i.e., Surface Water Treatment Rule). This may also translate to significant cost savings when alternatives such as full expansion of a treatment facility are weighed against an upgrade of the rapid mixing or flocculation process alone.

The definition of pilot-scale testing is important in reviewing this chapter. Three scales are generally referred to in the water industry: bench-, pilot-, and full-scale. Bench-scale testing refers to small batch reactors, and is commonly conducted in the laboratory. Pilot scale refers to miniature continuous-flow systems designed to model the full-scale treatment process or prototype system. Generally speaking, the pilot-scale system is designed at no less than 1 gpm (0.06 L/s) and as high as 10 gpm (0.6 L/s) for particulate removal in a single treatment process train; however, special systems require larger scales. Full scale is the prototype system referring to an actual treatment plant. It is not uncommon that all three scales are conducted at the same time during a pilot testing program when the full-scale facility is available. Bench-scale testing is always conducted simultaneously with pilot-scale testing to facilitate determinations on chemical coagulant dosages and to meet other requirements of the program.

This chapter is divided into three primary areas concerning pilot-plant studies, starting with elements for proper design of the program including specific factors for the design of the pilot plant. Three sections, following discussions on program design, highlight aspects of rapid mixing and flocculation that provide actual data and analysis with examples of pilot testing conducted recently. The final sections of the chapter are devoted to pilot-scale testing as compared to actual side-by-side comparison with full-scale testing.

A recurring theme of this chapter is that in pilot-plant work it is critical to include all treatment unit processes for full conventional treatment rather than to focus on a single process. Each individual unit process is dependent on the next to provide the overall treatment system performance. Pilot-plant testing is a valuable tool used in the area of research on rapid mixing and flocculation and is sometimes the only means to provide essential process data for the design of full-scale treatment facilities.

DESIGN OF PILOT STUDIES

In the increasingly complex world of emerging treatment technologies, more stringent state and federal regulations, and increasing numbers of regulated contaminants in drinking water, it is sometimes unclear what the specific objectives of pilot studies should be. This section will focus on the design of pilot study programs and pilot plants with special emphasis on rapid mixing and flocculation. It cannot be overemphasized, however, that pilot studies on rapid mixing and flocculation should not be conducted without consideration of subsequent unit processes designed for physical removal of floc (agglomerates of particulate matter) such as sedimentation, filtration, or microstraining. The reason for this is simple, since the best measure of optimization of rapid mixing and flocculation currently available is to test the performance and efficiency of removal in the physical removal processes. The physical removal processes, such as filtration, are the ultimate test of how well the characteristics of floc have been prepared for final physical removal (i.e., floc size, shape, and strength are put to the test). For this reason, this section includes discussion of the overall treatment process.

Pilot studies are normally performed to develop process design criteria for a new full-scale treatment facility, to avoid interruption of a full-scale facility that is on line to meet water quality and quantity needs, or to test improvements too costly to experiment on the full scale. Four general criteria can be used to identify whether pilot testing is needed: technical feasibility, economic feasibility, less expensive alternative, and process refinement (Tate and Trussell 1982). The size and complexity of the pilot testing program should be based on budgetary constraints, the level of information existing on the water to be tested, and the treatment alternatives to be tested, which can include innovative technologies. Optimization of the rapid mixing and flocculation processes to achieve better particulate removal should be one objective or alternative in all pilot testing programs oriented toward the goal of meeting state and federal regulations. The phrase "particulate removal" in this section is intended to be more broadly defined than turbidity to include pathogens, color, asbestos, and other drinking water contaminants of solid form. However, it is recognized that limited soluble organics and other contaminants are removed by adsorption to particulate matter and floc. The following is a more detailed list of some of the considerations used to assess whether pilot testing for particulate removal may be warranted:

- to develop process design criteria for a new source water supply including perhaps a new blend of source waters;
- to upgrade an existing treatment facility to meet more stringent state and federal regulations for particulate removal (i.e., turbidity, color, or asbestos) or specific pathogens (i.e., *Giardia lamblia, Cryptosporidium*, or virus removal);
- to optimize removal performance and cost effectiveness for an existing treatment facility by testing different chemical coagulants and injection sequences without interruption of the full-scale facility;
- to upgrade an existing treatment facility by testing new rapid mixing and flocculation methods that would be more costly for full-scale testing;

- to test new rapid mixing and flocculation technologies as part of research and development;
- to provide a state approval agency with proof that a new innovative technology will meet regulatory requirements;
- to troubleshoot a full-scale facility experiencing a problem water treatment condition.

The design of pilot studies requires special attention to the overall program objectives in order to remain on track. This is a critical consideration, since most pilot testing programs are expensive and time consuming. The key to the success of a pilot testing program begins with a well-written scope of work and technical approach. There is probably no such thing as overdefining a program; however, some flexibility is also necessary to accommodate the unexpected.

Scoping Pilot Studies

The primary goal of any pilot testing program is the proper selection of a water treatment process that will provide the necessary degree of removal to reliably meet the water quality requirements, at the lowest overall cost (capital plus operation and maintenance). The pilot testing program is usually tailored to meet the specific needs of a particular agency or water purveyor, which can vary substantially. In fact, no two pilot testing programs are ever identical in specific details of scoping. However, this chapter deals in the details that share common ground among pilot testing programs. The following considerations all influence the selection of a treatment process scheme (JM Montgomery 1985):

- cost-effectiveness of the system both in terms of capital and operations and maintenance (O&M) costs including off-site requirements;
- overall system reliability;
- flexibility and simplicity of operation (user friendly to plant operations staff);
- ability to meet water quality objectives;
- adaptability of process to both seasonal and long-term changes in raw water quality;
- capability of process to be upgraded in cases where water quality or drinking water regulations are changed (e.g., if a direct filtration plant is designed, provisions for addition of future sedimentation basins should be included);
- capability of process to meet both hydraulic peaks and quality excursions (reserve capacity);
- availability of skilled O&M personnel;
- post-installation service and chemical delivery;
- state and federal requirements; and
- ease of construction of facilities.

To meet its objectives, the pilot testing program may need to incorporate trade-offs between unit process design. For example, it may be more cost effective to consider expanding the pretreatment process (i.e., rapid mixing and flocculation) than expanding the treatment plant or expanding filters. Other unit process trade-offs may be to add tube settlers in an existing sedimentation basin and expand filters while leaving the rapid mix and flocculation processes fixed at higher hydraulic loadings. These trade-offs are usually commonly considered for upgrade of existing facilities to meet greater hydraulic loadings or more stringent regulatory requirements. New facilities usually do not require these types of process design criteria trade-offs; rather, they focus more on innovative or alternative technologies such as direct filtration, two-stage filtration, ozone pretreatment for coagulation enhancement, and many other emerging treatment technologies.

The first primary step to developing a pilot testing program scope is to develop a list of alternatives that require further testing at the pilot scale to provide the process criteria needed for final evaluation and design. Chapters 9 and 11 include further detail on existing alternatives for the rapid mixing and flocculation process in water treatment. The focus of pilot testing in the area of rapid mixing and flocculation should be oriented more toward the sensitivity of the particular source water quality with regard to detention times, energy application (G), and need for uniformity of energy application and range of energy needed. Pilot testing should be capable of giving insights into such things as the level of G appropriate for mixing; variable-speed turbine flocculators versus horizontal paddle type; or, in the case of rapid mixing, pumped injection versus static mixers, which are general classifications of equipment types. Process design criteria such as detention time are also possible through pilot testing, in addition to treatment enhancements through use of various coagulants and coagulant aids. The following is a list of requirements that must be addressed in preparing a scope for pilot testing:

- Review existing data. The scope of a pilot testing program may be greatly reduced or fine-tuned by careful consideration of existing data sources. In the case of a new source water, existing data may be available from other treatment facilities currently on line, or for an existing facility in the general area treating a similar source water quality. In the case of an upgrade for an existing facility, careful review and analysis of historical operations records can be used to help identify process criteria such as presented in Figure 10-1. Figure 10-1 is a statistical summary of plant effluent turbidity over a span of five years at different flow ranges. This type of data is essential to evaluate plant performance. It is likely that the historical records will not be in the useful and proper format seen in Figure 10-1, so this analysis should be included in the scope.

- Perform a literature review. A brief literature review for the past five years should always be performed to capture available information pertaining to a specific pilot testing program. Information on pilot testing that may help support findings or facilitate scale-up is growing. Given the costly and time-consuming nature of pilot testing, it is worthwhile to streamline the program without "reinventing the wheel." However, pilot testing is also very sensitive to site-specific source water quality conditions, making it crucial to be selective in reviewing existing data sources.

- List viable alternatives. Alternative selection is a critical function in developing a scope. Testing programs for process upgrade should always consider optimization of the rapid mixing and flocculation process through use of optional chemical coagulants and coagulant aids (i.e., polyelectrolytes, bentonite, sodium silicate, iron and aluminum salts, as well as ozone). Ozone will become more common for enhancing coagulation in those cases where it is required for disinfection or trihalomethane control. Modification of rapid mixing and flocculation is likely to be less costly than full treatment plant expansion or expansion of filtration systems for particulate removal.

- Consider parallel testing at pilot and full scale. Bench-scale testing should always be conducted parallel to pilot-scale work to help optimize treatment

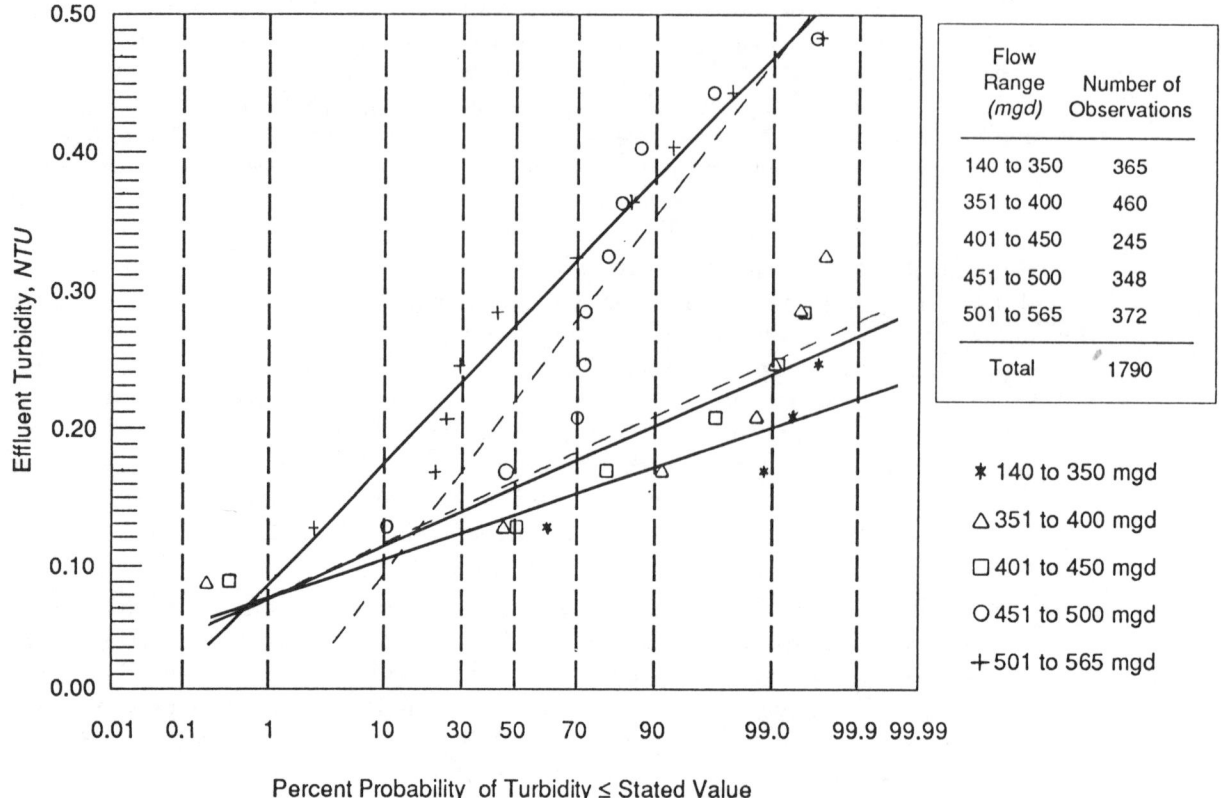

Source: Tanaka, T.S. et al. 1988. Conceptual Design Report for Joseph Jensen Plant Expansion. Metropolitan Water District of Southern California, R&D Report.

Figure 10-1 Summary of plant effluent turbidity at different flow ranges for month of June over five-year span.

chemical dosages. Bench-scale testing is also appropriate to provide statistical information on process parameters such as flocculation detention time. Figure 10-2 illustrates how a statistical analysis was performed on 51 separate tests over a detention time range of 1 to 30 min followed by batch filtration. These tests were conducted over a period of one year during a pilot testing program, providing good historical data. To perform this same function with pilot-scale testing would have been prohibitively expensive. Full-scale testing can only be considered in cases where upgrade of an existing facility is under study. One example is to conduct a study that matches both the pilot plant and the full-scale plant under the same process conditions as a control to determine how well the pilot facility models the full scale. Tracer studies on both the pilot and full scale should be considered for scale-up, depending on budgetary constraints. Lithium chloride has been used with success in meeting public-health acceptability (Tanaka et al. 1987).

- Provide a schedule. Scheduling is often critical in meeting treatment plant design and modification milestones. The schedule is usually a trade-off between

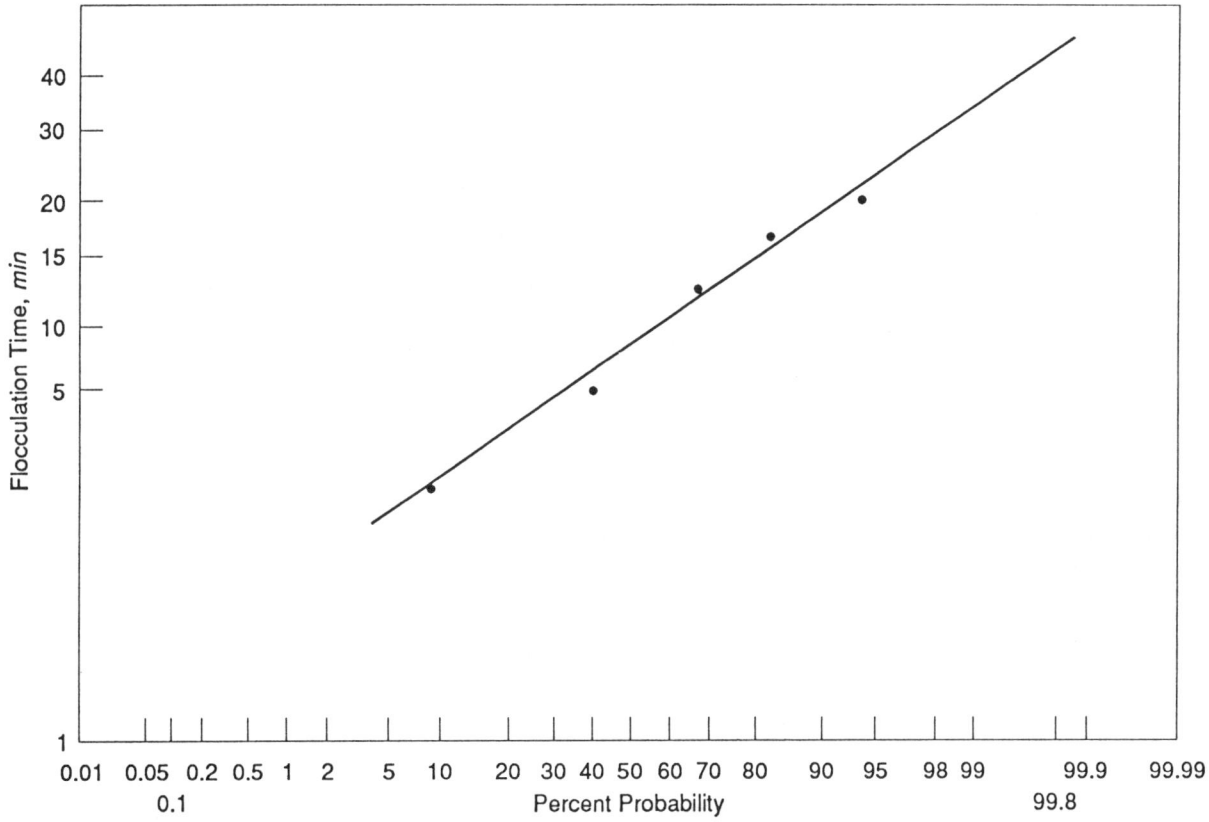

NOTE: Percent probability of a jar test effluent turbidity being less than or equal to 0.1 NTU at stated flocculation time.

Source: Tanaka, T.S. et al. 1988. Conceptual Design Report for Joseph Jensen Plant Expansion. Metropolitan Water District of Southern California, R&D Report.

Figure 10-2 Statistical analysis of 51 separate tests over detention time range of 1 to 30 min, followed by batch filtration.

the time needed to do accurate testing, cost, and the time allotted to implement. The schedule should be defined at the task level.

- Consider program organization and logistics. A program organization table is essential to expedite the testing without confusion and to avoid repeat testing as a result of communication breakdowns. Logistics are critical, since testing laboratories and pilot testing facilities should be in near proximity. Rapid data collection and analysis are also important to spot problem areas early, so that program adjustments can be made without valuable loss of time and resources.

- Provide a concise testing and monitoring plan. The frequency and type of data acquisition should be clearly understood and outlined prior to testing. Tables 10-1 and 10-2 illustrate how this plan can be developed for a full treatment process pilot study. It should be expected that this plan will require some modification during testing, since research rarely follows a predictable path. For example, a decision may be made to reduce the number of test runs in an area that is producing a consistent result and expanding another area that is not.

Source Water Quality

The water quality characteristics of the source water under consideration for water treatment are some of the most critical factors affecting the need and design of pilot-plant testing. Water treatment plant design practice has been well-established through decades of experience, making it sometimes difficult to perceive the need for testing at all. In reality, every source water is different as a result of outside influences on the drainage course including animal, vegetable, and mineral contact with the water. These outside influences vary continually over periods of an hour to 1000 years or more. Differences in source water can also result from very local influences. For example, the same water can pass through two separate reservoirs,

Table 10-1 Frequency of Analysis for Turbidity, Total Suspended Solids (TSS), and Particle Count for Bench-Scale, Pilot-Scale, and Full-Scale Tests

Parameter	0	1	2	3	4	5	6	7	8	9	10	11	12	13	14	15	16
Bench-scale*																	
Turb./Infl.			X														
Turb./Filt.																	X
Part. Count					X												
TSS/Infl.			X														
TSS/Filt.			X														
Pilot-scale†																	
Turb./Infl.																	X
Turb./Floc.																	X
Turb./Sed.																	X
Turb./Filt.																	X
Part. Count									X								
TSS/Infl.			X														
TSS/Floc.			X														
TSS/Sed.			X														
TSS/Filt.			X														
C12/Infl.													X				
C12/Effl.													X				
Bac. sample			X														
Full-scale sedimentation‡																	
Turb./Infl.					X												
Turb./Floc.					X												
Turb./Sed.					X												
Turb./Filt.													X				
Part.Count									X								
TSS/Infl.			X														
TSS/Floc.			X														
TSS/Sed.			X														
TSS/Filt.			X														

Source: Tanaka, T.S.et al. 1988. Conceptual Design Report for Joseph Jensen Plant Expansion. Metropolitan Water District of Southern California, R&D Report.

*Bench-scale tests—routine tests to be conducted during the pilot- and full-scale tests.
†Pilot-scale tests—approximately 6 months of intensive pilot-scale tests (24 h/day, 5 days/week).
‡Full-scale tests—approximately 2 months of full-scale tests (24 h/day, 5 days/week).

causing significant differences in treatability and source water characteristics. Table 10-3 compares two source waters originating from the same upstream source to Lake Castaic and Lake Silverwood at the Metropolitan Water District of Southern California. These variable changes in source water quality act to confound the would-be pilot tester, who, driven by budgetary constraints, is often asked to determine treatment process criteria in three to six months in a pilot plant. The methods available to characterize the source water and duplicate the source water conditions during pilot testing are a considerable challenge. The purpose of this section is to provide insight into how this task can best be done.

Characterization of source water. Source water monitoring data are becoming more prevalent today than some years ago. This is particularly true if an existing treatment facility is on line with the source water of interest. Table 10-4 summarizes the water quality constituents that play an important role in the treatment process. Water quality variables such as turbidity can be cyclic, heavily influenced by storm weather patterns. At least 10 years of monitoring data are

Table 10-2 Tasks and Process Parameters To Be Varied For Determining Existing Plant Capacity, Pilot Tests

Task		Process Parameters								
		Influent Turbidity, NTU	Flash G, $1/s$	Mix T, s	Flocculation G, $1/s$	T, min	Baffles	Sedimentation T, min	Q, mgd	Filtration Q/A, gpm/ft^2
Determine existing plant capacity with existing chemicals	A	0.5	TBD*	TBD	TBD	TBD	Wooden	120	200	3.0
	B	"	"	"	"	"	with	90	300	4.5
	C	"	"	"	"	"	slotted	60	400	6.0
	D	"	"	"	"	"	openings	45	550	8.5
	A	1.2†	TBD	TBD	TBD	TBD	Wooden	120	200	3.0
	B	"	"	"	"	"	with	90	300	4.5
	C	"	"	"	"	"	slotted	60	400	6.0
	D	"	"	"	"	"	openings	45	550	8.5
	A	10	TBD	TBD	TBD	TBD	Wooden	120	200	3.0
	B	"	"	"	"	"	with	90	300	4.5
	C	"	"	"	"	"	slotted	60	400	6.0
	D	"	"	"	"	"	openings	45	550	8.5
	A	25	TBD	TBD	TBD	TBD	Wooden	120	200	3.0
	B	"	"	"	"	"	with	90	300	4.5
	C	"	"	"	"	"	slotted	60	400	6.0
	D	"	"	"	"	"	openings	45	550	8.5
	A	45	TBD	TBD	TBD	TBD	Wooden	120	200	3.0
	B	"	"	"	"	"	with	90	300	4.5
	C	"	"	"	"	"	slotted	60	400	6.0
	D	"	"	"	"	"	openings	45	550	8.5
	A	65	TBD	TBD	TBD	TBD	Wooden	120	200	3.0
	B	"	"	"	"	"	with	90	300	4.5
	C	"	"	"	"	"	slotted	60	400	6.0
	D	"	"	"	"	"	openings	45	550	8.5

Source: Tanaka, T.S. et al. 1988. Conceptual Design Report for Joseph Jensen Plant Expansion. Metropolitan Water District of Southern California, R&D Report.

*TBD—to be determined in troubleshooting tests of pilot plant to match the existing Jensen plant.
†1.2—the 1.2 NTU influent level is used to denote the existing Jensen influent turbidity at the time of testing.

extremely helpful to perform statistical analyses on key water quality parameters. It is likely that not all constituents listed in Table 10-4 will be available over a 10-year period. Some spot testing of parameters is advisable in the absence of historical information. Figure 10-3 illustrates examples of two data-analysis schemes, showing daily average turbidity over a nine-year period. Both data analyses yield useful information. The statistical treatment is useful since it shows the likelihood of a given turbidity event occurring, which can be ultimately translated to a feel for the reliability of a treatment process (Tanaka et al. 1987). The historical treatment of the data gives the reviewer a good visual representation of the variability in addition to the maximum and minimum values. Knowing the historical characteristics of water quality parameters is essential to making judgments on scale-up, pilot testing program design, and ultimately the treatment process design criteria. In fact, the

Table 10-3 Lake Silverwood and Lake Castaic Effluent Water Quality Characteristics

Parameter	Silverwood	Castaic
Silica, mg/L	14	19
Calcium, mg/L	22	30
Magnesium, mg/L	17	15
Sodium, mg/L	77	64
Potassium, mg/L	4	4
Carbonate, mg/L	0	1
Bicarbonate, mg/L	100	122
Sulfate, mg/L	43	55
Chloride, mg/L	121	83
Nitrate, mg/L	2.2	1.5
Fluoride, mg/L	0.1	0.3
Total dissolved solids, mg/L	350	334
Total hardness, mg/L	123	136
Total alkalinity, mg/L	82	101
Carbon dioxide, mg/L	1.4	2.3
pH	8.1	8.0
Specific conductivity, $\mu mho/cm$	648	591
Turbidity, NTU	2.1	1.3

Source: Tanaka et al. 1987. Pilot Scale and Full Scale Investigations for the Design of Additional Direct Filtration Modules for an Existing 320 mgd Treatment Plant. AWWA Annual Conf., Kansas City, Mo.

Table 10-4 Important Source Water Parameters

pH
Turbidity
Total dissolved solids
Alkalinity
Dissolved oxygen
Temperature
Conductivity
Total suspended solids
Total organic carbon
Total hardness

Source: Tanaka et al. 1987. Pilot Scale and Full Scale Investigations for the Design of Additional Direct Filtration Modules for an Existing 320 mgd Treatment Plant. AWWA Annual Conf., Kansas City, Mo.

PILOT-PLANT STUDIES 353

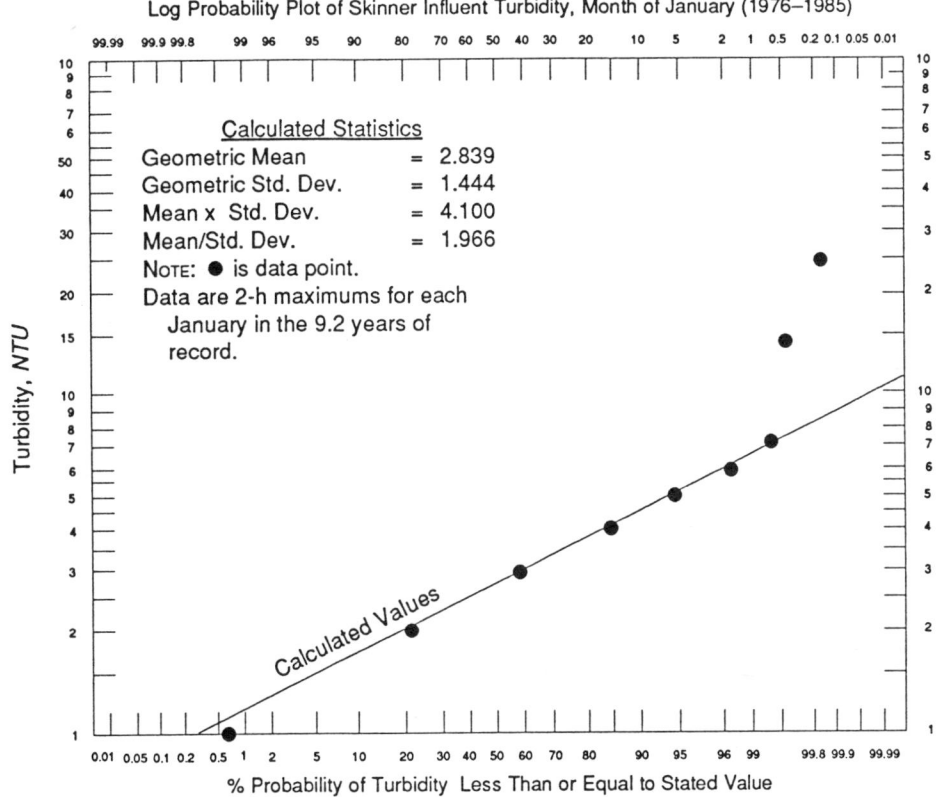

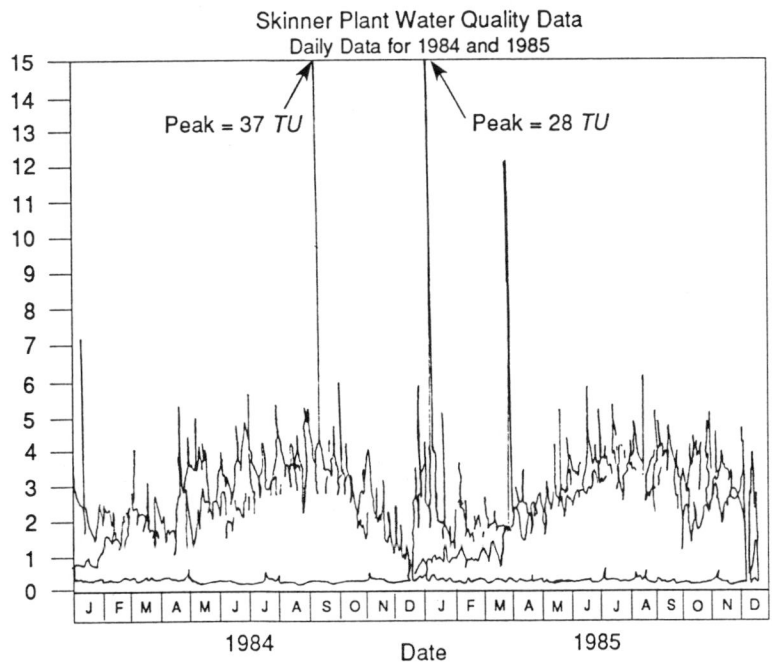

Source: Tanaka et al. 1987. Pilot Scale and Full Scale Investigations for the Design of Additional Direct Filtration Modules for an Existing 320 mgd Treatment Plant. AWWA Annual Conf., Kansas City, Mo.

Figure 10-3 Two data analysis schemes showing daily average turbidity.

source water characteristics will help define the overall treatment process configuration and use of innovative technologies, as illustrated in Table 10-5 (Tate and Trussell 1982).

Methods for duplicating source water variability. Most pilot-plant testing programs utilize the existing source water available at the time of study. This can present significant logistical problems in view of usual source water quality variations. Two primary forms of particulate matter, suspended solids (turbidity) and algae, have perhaps the greatest impact on pilot testing for particulate removal. This section highlights several methods used to artificially induce higher levels of particulate matter during pilot testing, which can greatly accelerate the time necessary for the testing program. The problem with duplicating turbidity excursions is in the limitations of the methods to precisely identify and duplicate the surface properties on particles in the source water and in the experiments. Given the many problems associated with duplicating higher turbidity levels, it has still provided a useful tool to the pilot tester (Tanaka et al. 1987; Tanaka et al. 1988).

Table 10-5 Process Selection Guideline

Water Quality Parameters	Process and Components	Applicability	Comments
Turbidity	In-line filtration Coagulation Filtration	Low turbidity, low color	Greater operator attention required; shorter filter run than with direct filtration and conventional treatment; additional sludge-handling facilities may be required; pilot-plant studies may be required; lower capital and O&M costs
	Direct filtration Coagulation Flocculation Filtration	Low to moderate turbidity, low to moderate color	Greater operator attention required; greater sludge-handling facilities may be required; pilot-plant studies may be required; lower capital and O&M costs; better filter run time than in-line filtration but shorter than conventional treatment
	Conventional Coagulation Flocculation Sedimentation Filtration	Moderate to high turbidity, moderate color	Detention time in sedimentation basins allows for adequate contact time for taste and odor and color removal chemicals; more operational flexibility and less operator attention required
	Microscreening	Removal of gross particulate matter (e.g., algae)	Process relies on straining mechanism; process could not meet water quality objective if used alone

Source: JM Montgomery. 1985. Water Treatment Principles and Design. Wiley-Interscience, New York.

The best way to artificially seed a source water with higher turbidity for pilot testing is to collect naturally occurring sediments from the submerged shoreline of the reservoir or natural drainage course under study. This material has already been exposed to the natural conditions existing in the source water. The material should be placed in a tank of sufficient size, mixed with source water, and allowed to settle for a period of one to three days. The settling time should be based on the usual time after a heavy rainfall that turbidity notably rises in the source water reservoir effluent. In the absence of this information, a best guess based on the size of the closest source water reservoir and inlet/outlet configuration will suffice. This method will produce a supply of fine particulate matter, which can then be diluted with source water to the desired turbidity for testing. It will also be necessary to check the artificial source water solution to ensure that the resulting water quality characteristics have not changed vastly. A check of total dissolved solids, hardness, alkalinity, and pH will help identify the match quickly. Some settling of material will continue to occur in the source water storage vessel, so a mixer is needed. Focus on this is not essential, since it is anticipated that turbidity fluctuations would occur in the natural condition anyway. It is important to mix the tank for only brief periods to avoid natural flocculation. Figure 10-4 shows actual data on filter run time generated using artificial turbidity where rapid mixing and flocculation evaluations were under way (Tanaka et al. 1987). It should be noted that comparison with a full-scale facility was also possible in this experiment, which very closely matched the pilot testing.

A second method for modeling high-turbidity events requires use of natural clay deposits. This method is appropriate in situations where existing shoreline deposits are difficult to obtain. The primary problem with this method is the lack of preseasoning of the particulate surface properties by naturally occurring minerals and organics. Without preseasoning, this approach can give an overly optimistic picture of process design criteria, similar to laboratory studies using kaolinite and bentonite clay minerals. Preseasoning can be accomplished to some extent by adding an organic acid such as humic acid. Figure 10-5 illustrates a bench-scale study using humic acid to preseason a naturally occurring clay in the vicinity of the source water reservoir under investigation. Results of the testing compared well with historical information on chemical dosage requirements during an actual high-turbidity incident at the full-scale treatment plant.

Perhaps the greatest menace to particulate-removal efficiency and filter run length is the presence of algae blooms in surface water reservoirs. Pilot-plant testing would ideally be accomplished during an algae-bloom episode; however, the two rarely coincide. Alternate methods are available to stress test the process testing in a pilot plant to simulate such a condition. The best approach is to perform the pilot tests in the upper surface water of the reservoir where the algae counts are higher, which may require moving the pilot test facility to the reservoir site and pumping source water in situ. This procedure, if performed near the outlet tower, can also afford the opportunity to test water at different depths, giving a variety of source water conditions. Transport of surface water in 5000-gal (19,000-L) tank trailers to the pilot-plant site has also been tried with success (Tanaka et al. 1988). In the absence of pilot-plant testing with nuisance algae conditions, the final analysis of process design criteria must take this into account, using experience and professional judgment. Where seasonal algae problems occur, testing should be scheduled during the algae season.

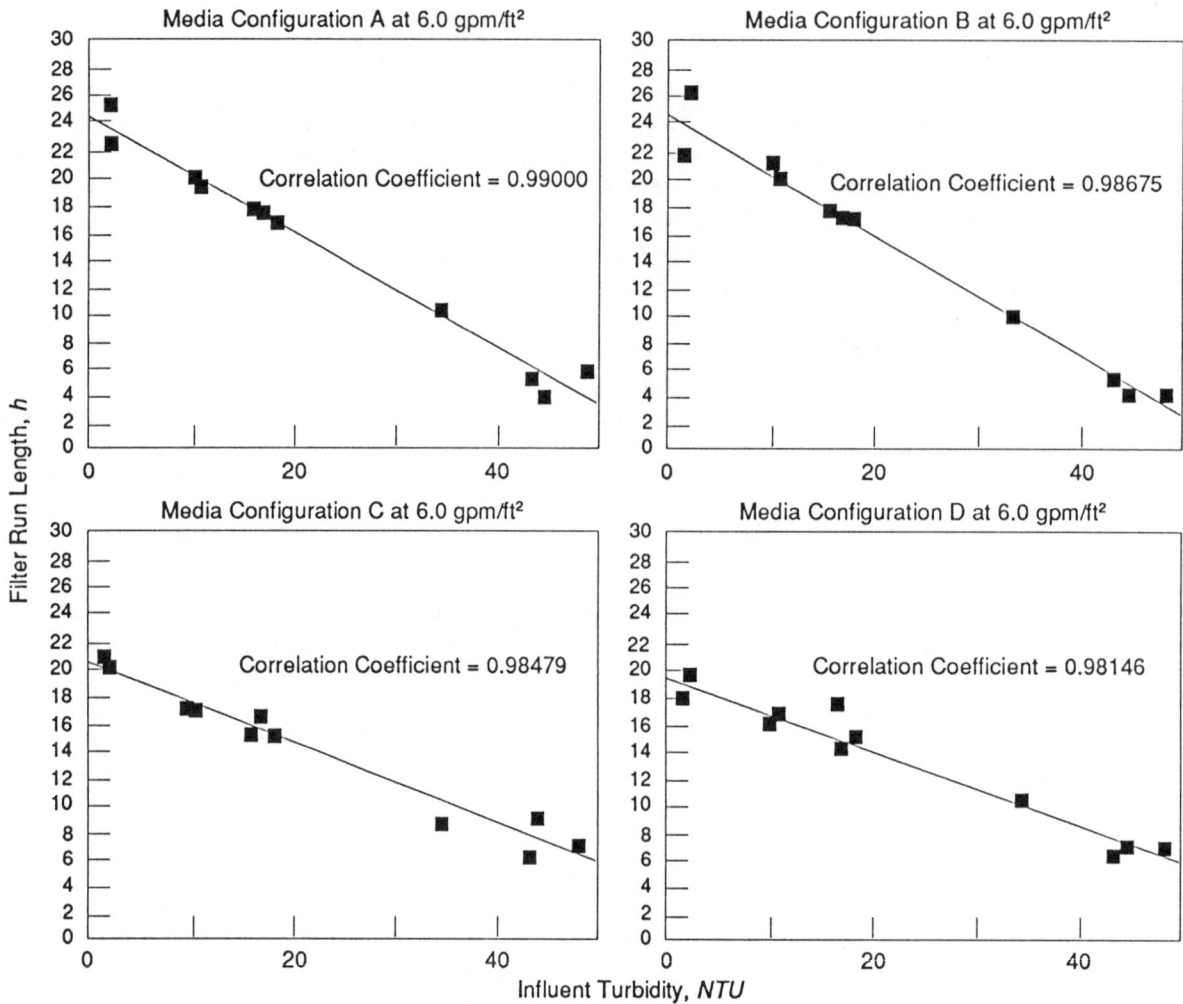

Source: Tanaka et al. 1987. Pilot Scale and Full Scale Investigations for the Design of Additional Direct Filtration Modules for an Existing 320 mgd Treatment Plant. AWWA Annual Conf., Kansas City, Mo.

Figure 10-4 Effect of influent turbidity on filter run length at a filtration rate of 6 gpm/ft^2.

Scaling Pilot Studies

Rapid mixing and flocculation processes can be effectively scaled in continuous-flow pilot-plant facilities in the range of 1 to 10 gpm (0.06 to 0.6 L/s) without serious difficulty (Tate and Trussell 1982). In fact, process design parameters such as flocculation detention time can be tested in batch jar testing apparatus (Tate and Trussell 1982; Hudson 1981; Tanaka et al. 1988). In general, the larger the pilot test facility, the better in terms of scale-up. Budgetary constraints are the usual culprit in dictating pilot-plant sizing. Certainly, it would not be worth spending $500,000 for pilot-plant construction when the treatment facility improvements are in the same cost neighborhood. In all cases, full-scale testing should always be considered in addition to pilot testing, providing the overall program costs are not prohibitive. The Sunol Treatment Facility located in San Francisco is already equipped with

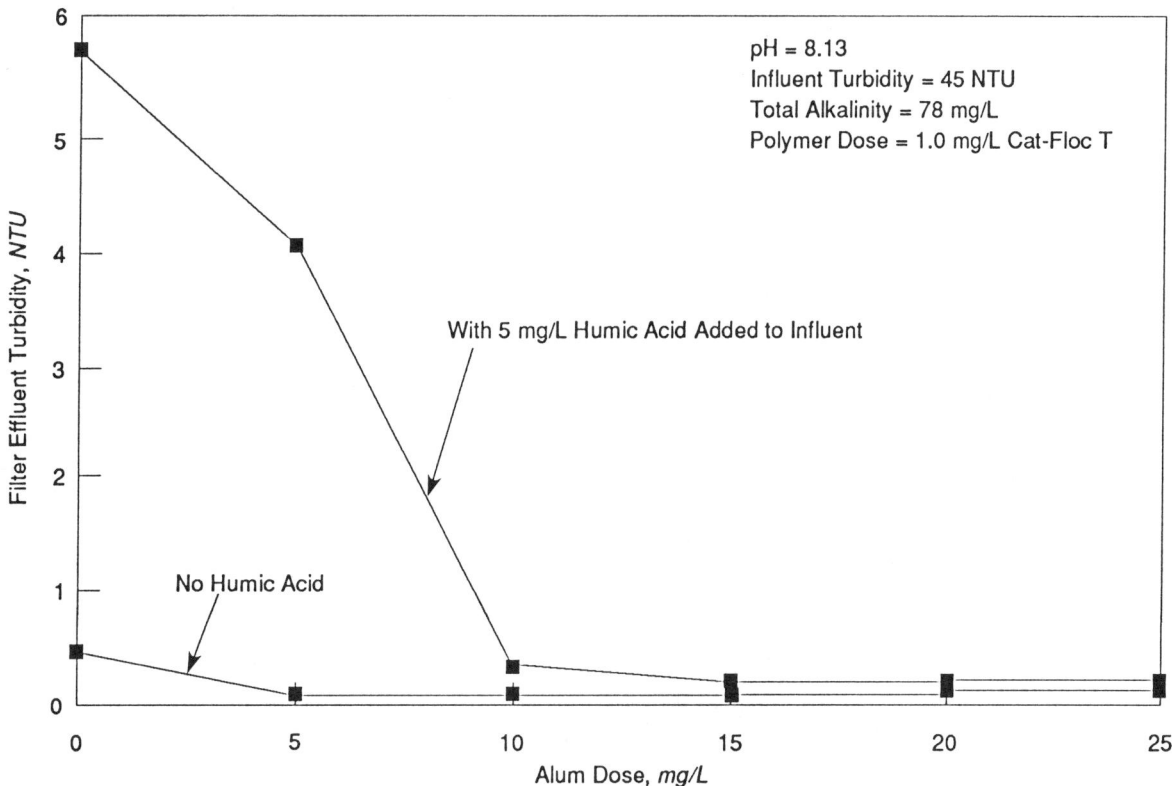

Source: Tanaka, T.S. et al. 1988. Conceptual Design Report for Joseph Jensen Plant Expansion. Metropolitan Water District of Southern California, R&D Report.

Figure 10-5 Effect of humic acid on filter effluent turbidity and required alum dose.

vertical turbine mixing equipment side by side with horizontal paddle-type flocculators. Full-scale opportunities such as these should not be overlooked prior to implementing a pilot test program, as long as source water quality conditions are similar. In fact, an urgent need exists to conduct a state-of-the-art full-scale treatment plant survey to take advantage of side-by-side treatment process comparisons such as the Sunol Treatment Facility to provide this important full-scale information. Chapter 8 provides further reading concerning scale-up considerations for pilot testing.

Pilot-Plant Design

Pilot-plant design varies considerably, depending on the objectives of the testing program. This section provides relatively general guidelines for the design of a pilot plant intended to determine the treatability of a source water for particulate removal by conventional treatment processes including rapid mixing, flocculation, sedimentation, and filtration. As mentioned earlier, studies on rapid mixing and flocculation would not be complete without including the physical removal processes to test floc settleability and resistance to shearing action in the filters among other considerations. The ozone unit process is also briefly considered here as a coagulant aid, given its recent popularity for trihalomethane control. Studies designed to model the full scale would be incomplete without precise duplication of all intended chemical injection points for the full scale. Injection of chlorine and other oxidants at the front of the process train should also not be overlooked in view of their significance in influencing particulate removal as well.

Process and instrumentation diagram. The first step to designing a pilot plant after the project scope has been developed is the same as for a full-scale treatment plant. This step involves the development of a process and instrumentation diagram, commonly referred to as a P&ID. Figure 10-6 shows an example of a P&ID used in the development of a pilot study at the Metropolitan Water District of Southern California. This step is usually preceded by development of a PFD (process flow diagram); however, in the case of a pilot plant, both functions can be performed essentially at the same time. The purpose of a P&ID is to define clearly the unit process steps, sample-tap locations, instrumentation and control equipment, instrument transducer locations, layout of all major hardware in schematic format, valves, chemical injection points, and local and remote control functions. The P&ID is perhaps one of the single most critical steps to designing a pilot plant, since it provides a blueprint for the physical hardware and layout. Such a diagram will prevent the confusion created by building a pilot plant "as-you-go" and finding out in the eleventh hour that some component has been overlooked or the hardware requirements have been underestimated. In fact, it is nearly impossible to provide a reasonable cost estimate or schedule without having such a diagram. The following key considerations must be reviewed when a P&ID is prepared:

- The overall project scope must be considered in determining the process layout and process criteria to be tested. The number of parallel process treatment trains is crucial to the system hardware and layout. Many functions may require manifolding, if a number of process treatment trains are considered. Four parallel process treatment trains will usually provide sufficient system flexibility to accomplish most pilot testing tasks. Multiple process treatment trains are essential to aid in the elimination of variables caused by variable source water conditions.

- In today's world of computers, instrumentation and control functions are becoming more affordable to the pilot tester. For example, automated pressure readings, turbidity monitoring, motor control, and alarmed weir overflow or motor failure are all possibilities for the modern-day pilot tester. In selection of these gadgets, it is important to keep in mind the cost and that hand calibration will still be necessary. Instrumentation and control functions will, however, play a greater role in pilot testing due to their savings in operator time and increased data-gathering potential.

- The choice of variables to be considered and testing scheme should be well-known before the P&ID is developed. Selection of instrumentation, sample point locations, and frequency of testing will influence hardware layout and selection.

- It is important to be cognizant of the flow rates throughout the pilot plant including sample size requirements. Large, high-flow samples in low-flow pilot plants of 1 gpm (0.06 L/s) could shock the steady-state condition of the treatment process. For example, a 1000-mL sample drawn off a flocculator compartment at a rate of 10 gpm (0.6 L/s) should not be considered in a pilot plant designed at 1 gpm (0.06 L/s).

PILOT-PLANT STUDIES 359

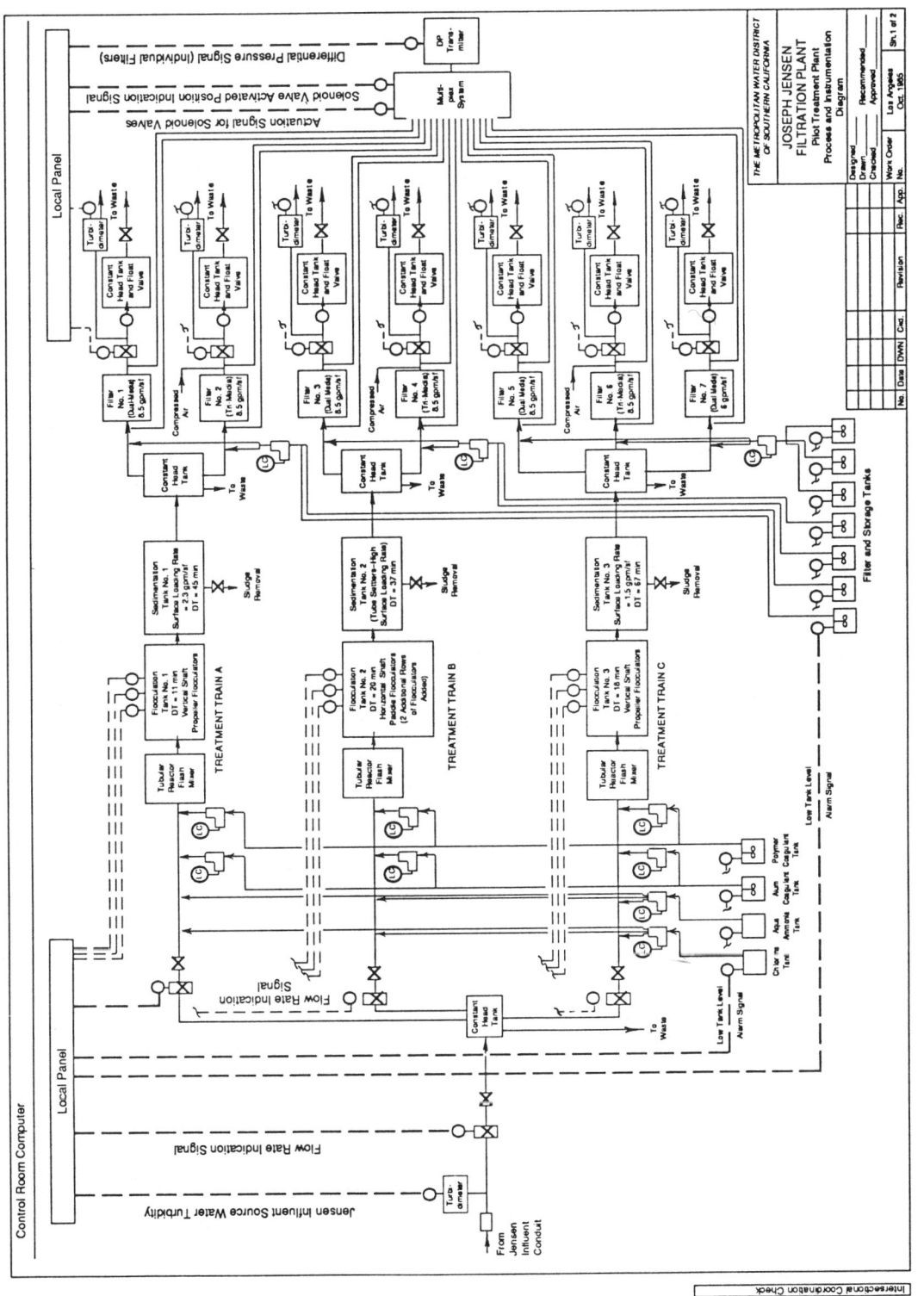

Figure 10-6 P&ID used in development of pilot study at Metropolitan Water District of Southern California.

- The schedule is important to consider, since shorter time frames can require more treatment process trains to expedite data collection. It is important, however, not to accelerate a program too fast, since something will inevitably be missed in the source water variability.
- Artificial injection of turbidity should be included in the P&ID since some equipment and controls will most likely be required.
- Variable-speed drives will be required on nearly every motor control to give flexibility over process parameter variability.
- Versatility in piping layouts, injection points, and sampling is vital in a pilot plant. The P&ID should include provision for adequate manifolding, bypass capability around unit processes, and in-depth sampling of filtration systems.

Hardware development. Hardware development is initially determined once a P&ID is available for the pilot plant. Selection of pilot-plant hardware should be based on the cost and intended longevity of the equipment. In some cases, after full-scale plant construction, the pilot plant is permanently stationed at the treatment plant for later use and process troubleshooting such as at the Los Angeles Aqueduct Water Filtration Plant. Hardware should be rugged and as reliable as the budget will allow, an investment that is often paid back many times over by the elimination of lost time from plant failures. When plant failure occurs in the middle of a test series, test runs may have to be repeated or an opportunity to test during an unusual source water condition may be missed. Hardware development should also be user friendly to the operator, who will be continually observing and evaluating data as it is generated.

Pilot-plant hardware will require some design and fabrication during the construction phase. Subsequent sections will focus on specific aspects of the rapid-mix and flocculation unit design. The following considerations relate to general hardware development:

- All motors and drives should be sized very conservatively. Motors should generally be sized 10 to 20 times larger than needed based on theoretical torque calculations. Some systems, such as chemical feed pumps and motors, come prepackaged and can be used as delivered.
- All pumps should be positive-displacement type to minimize shearing action. Chemical feed pumps are generally the peristaltic type often used for medical research. Other pumps used to transport process flow can be progressing-cavity type. Many positive-displacement pumps have a pulsing effect that can induce a shearing action on particulate matter. Minimizing shear is particularly important at all points beyond the flocculation process. Gravity feed over weirs and between unit processes is preferred, since it more closely models the full-scale treatment process and provides uniform distribution.
- A number of flowmeter types are available. Rotometers provide dependable service, but do not induce a signal for control purposes. Many magnetic flowmeters are on the market, but are much more expensive. These units provide dependable service and can be interfaced for control purposes. Chemical feeds are generally regulated by revolutions per minute (rpm), and many peristaltic units can be interfaced for control function. Other feed systems utilize plunger or hypodermic-style pumping actions often used for medical research.
- Clear plastic tubing or piping is generally preferred to determine settling of floc in lines or air entrainment problems.

- A modular design for treatment unit processes and pump systems greatly facilitates flexibility. Each unit process should be self-contained, and pumps and other motor-operated systems should be readily removable for ease of repair and quick alteration of hardware layout (Lang 1982).
- Mounting a pilot plant on a trailer provides the decided advantage of greater mobility for testing different source waters. Metropolitan Water District of Southern California as well as many others has developed trailer-mounted units. Other considerations come into play for shock mount resistance and flexible couplings needed for vibration control during transport.

Pilot-plant operation and shakedown. Operation of a pilot plant is as much an art as a science. Considerable experience and knowledge are needed to accurately model and scale-up to a full-scale facility. One of the most critical elements of operating a pilot plant is to establish good controls. This problem is made somewhat simpler if a full-scale treatment facility is available to operate in parallel with the pilot plant. Pilot-plant operation is always conducted by holding all process variables constant with the exception of one.

To test all process variables over the full range of values under investigation is a formidable and lengthy task. The usual approach is to set up a matrix of variables and explore it only to the extent necessary to arrive at a conclusion. It is not as necessary to explore regions of the matrix that are yielding constant results as those regions where parameters vary more significantly in overall treatment performance. Another key consideration is the number of repeat runs necessary to come to an accurate conclusion while avoiding problematic test runs. Only spot-check repeat test runs are necessary usually to verify that a false or problematic run has occurred. A good place to start is normally with those runs yielding somewhat "questionable" results based on experience and theoretical considerations. It is important that the source water quality not vary significantly when repeat runs are conducted.

One control should always be established while pilot testing is being conducted with parallel treatment process trains. This control requires that all treatment process trains are operated simultaneously. This indicates whether other variables are coming into play as a result of differences in hydraulics, chemical feed consistency, and chemical feed solution consistency among a host of other considerations.

One common problem is the uniformity of chemical feeds and mixing conditions between treatment trains. A good method to deal with this problem, for example, is to cross connect unit processes not being tested for parameter variation. For instance, if one wished to determine the effect of various G-values in the flocculator, the best way to rule out variables between differences in chemical feed application would be to pass the entire process stream through a single rapid-mix unit prior to subsequent treatment in the flocculation unit process. The same can be done for other unit processes if the pilot plant is so designed. Another useful control is to perform a tracer dye test through the pilot-plant unit process and compare it with those performed on full-scale facilities.

Staffing a pilot plant is another important function of its operation. Usually three to four operators are necessary to conduct seven-day, 24-h operation of a system. Continuous operation of a system is advantageous, since steady-state conditions are well-established. Turning a pilot plant off every day and restarting the system at the operator's convenience can lead to erroneous results. Continuous operation can also save time, depending on the nature of the testing. Operators can work in 12-h shifts, although they may make more mistakes, owing to fatigue. Operators should be sufficiently skilled at pilot-plant operation and well-grounded in theoretical knowledge of unit process fundamentals. Never staff a program with operators who have no experience in pilot-plant operation. One or two trainees may be appropriate if supervised by well-experienced individuals during day shifts. If time constraints are a problem, then the higher level of skilled operators (master's degree level) are appropriate to avoid oversights and mistakes requiring repeat of pilot test runs.

Outside electrical service can cause problems with digital equipment and microprocessor-controlled circuitry. Power surges and losses should be dampened with suitable electronic devices.

Laboratory testing should be conducted as quickly as possible for particle counting, turbidity, electrophoretic mobility, and other test parameters subject to rapid change with time. A small process laboratory in the vicinity of the pilot plant is advisable.

Rapid-mix design considerations. Rapid-mixer design is relatively straightforward. Rapid mixing can be accomplished through any number of commercially available in-line static mixing devices (JM Montgomery 1985). However, studies oriented toward testing a range of G energy values will usually require construction of a continuously stirred tank reactor (CSTR). The CSTR design is normally composed of a rectangular clear plastic container with a rotor and stator combination. Several inlet and outlet ports will facilitate the ability to study differences in detention time. High-speed, low-volume rapid mixers (Amirtharajah and Mills 1980) have been used to study very high rates of energy application with low volume. In general, G does not apply well to rapid mixing, owing to complications presented by microturbulence and hydraulic shearing at the solid–liquid interface (Amirtharajah 1982). Jet mixers are also possible at the pilot test scale (JM Montgomery 1988). The rapid mixer can be larger than the remainder of the pilot plant (oversized by hydraulic retention time), providing a portion of the stream is wasted prior to entry into the flocculation unit. This allows rapid-mixing pilot-plant designs of considerably larger scale. In fact, full-scale facilities can be linked with pilot scale downstream of rapid mixing and/or flocculation. Figure 10-7 shows several examples of pilot-scale rapid-mixing devices.

Flocculator design considerations. Flocculator design on the pilot scale has been well-documented (JM Montgomery 1985; Sanks 1980; Hudson and Wolfner 1967). This section highlights some of the well-known design considerations for flocculators on the pilot scale and updates this information with more recent designs.

The general design of a pilot-scale flocculator uses three-compartment, rectangular acrylic basins with slots provided at the baffle walls that allow replacement of different baffle configurations. The range of G should cover from 1 to 100 s^{-1} or higher. Flocculation basins on the pilot scale will require a higher G to model lower G-values in the full scale (JM Montgomery 1985). It is preferable to design the flocculator basin as an integral part of the sedimentation basin to provide a uniform transition for hydraulics (Tanaka et al. 1988). A bypass around the sedimentation basin is necessary to study direct filtration treatment. The size of the flocculation basin should allow detention times up to 30 min to be tested, so that the full range of detention times can be studied. Variation of detention time can be studied by providing higher initial process stream flows and wasting excess process stream after the flocculator or sedimentation basin if both basins are combined. Lighting is important around the flocculation basin to allow maximum visibility of floc formation. No rule of thumb is available to determine the best lighting configuration since it is site specific; it will require some experimentation.

The mixer design is important for flocculator basins. The most commonly used is a vertical flat-paddle type for simulating horizontal flocculator designs (chap. 11).

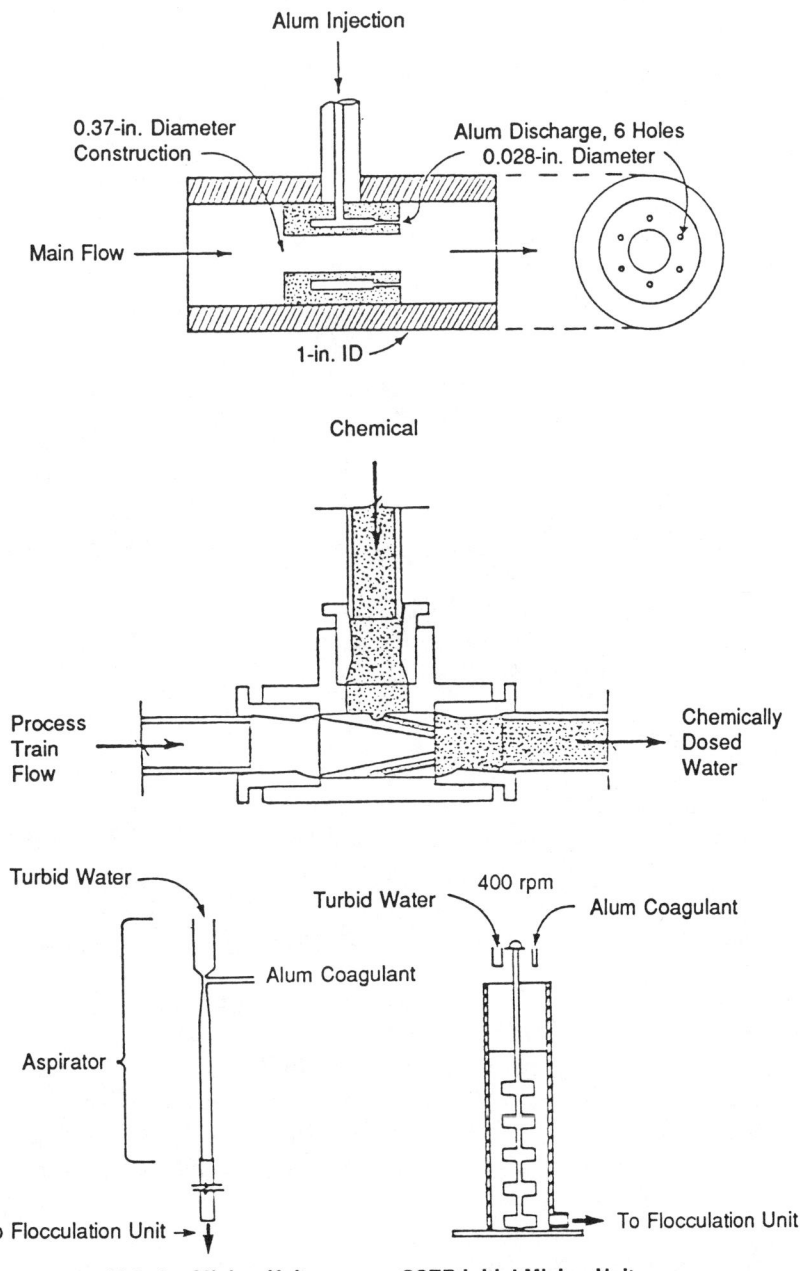

Source: Sanks, R.L. 1980. Water Treatment Plant Design for the Practicing Engineer. *Butterworth–Heinemann, Stoneham, Mass.*

Figure 10-7 Pilot-scale rapid-mixing devices.

Use of stators in the flocculation compartments is also possible. However, stators are not usually used in recent designs. Vertical propeller types are also possible to help simulate vertical turbine flocculators. It is difficult to fine-tune specific manufacturer designs in a pilot plant of 1 to 10 gpm (0.06 to 0.6 L/s). It should be the focus of the pilot-plant work to determine optimum ranges for G and the sensitivity of plant performance to variation of G. The mixer requires calibration by a torque meter, a stroboscope, and a tachometer. The energy of application formula for G is defined as:

$$G = \sqrt{\frac{P}{\mu V}} \; s^{-1} \tag{Eq 10-1}$$

Where:

P = power input (ft · lb/s)
μ = absolute viscosity (lb · s/ft^2)
V = volume of sample (ft^3)

The calibration curve, which consists of a relationship between motor rpm and the G-value, is handy to use during the pilot testing. The curve should appear uniform in appearance as a gently sloping c curve. As a rule of thumb, low ranges of G are appropriate if sedimentation is required, whereas higher ranges of G (100 s^{-1}) are more appropriate for direct filtration. Recent studies (Cleasby 1984) suggest that a better estimate of velocity gradient is related to the mean power input per unit mass to the two-thirds power as a more appropriate flocculation parameter than G for common waters.

Ozone contactor design considerations. Ozone contactor design on the pilot scale is briefly considered here because of the significance of ozone on particulate removal. Recent studies (Liang et al. 1989) demonstrated a 60 percent overall improvement in plant effluent turbidities, as indicated in Figure 10-8. It is believed that ozone reacts with organics on the surface of particles, thereby changing the surface properties and enhancing the effectiveness of coagulants. It is well-known that organics play a key role in the coagulation of particulate matter (Bales and Morgan 1984).

Ozone contactor design on the pilot scale has been well studied (JM Montgomery 1985; Stolarik 1981). The basic design requires use of clear glass, polyvinyl chloride, or acrylic columns with a countercurrent application of ozone. The detention time is usually 2 to 10 min and is varied by changing flow rate. Ozone transfer is achieved through a ceramic air stone and can be accompanied by a submerged turbine. Most piping will need to be stainless steel for ozone transport from one of many commercially available ozone generators. Ozone is highly toxic, so every safety precaution should be taken including alarm systems and training for operators concerning the hazard. A rotometer can be used to measure gas flow. A decision is necessary to use pure oxygen or air and should be guided by comparative cost considerations.

Monitoring Mixing Effectiveness

The methods available to monitor mixing effectiveness have already been briefly touched on. This section will provide further detail on the use of these techniques on the pilot scale. Other chapters, including chap. 11, also deal with some of these methods in their application at full scale. The techniques used to monitor mixing effectiveness that will be discussed here include the following:

- tracer studies,
- turbidity,
- particle counting,
- visual inspection and photography,
- zeta potential and electrophoretic mobility,
- velocity measurement,
- conductivity probe,
- streaming current detector, and
- visual dye testing.

The most commonly used techniques for monitoring mixing effectiveness at the pilot scale include tracer studies, turbidity, particle counting, and visual inspection. It should be noted at the outset that significant need exists in pilot testing to discover and apply new techniques at the research level to improve upon these time-honored methods.

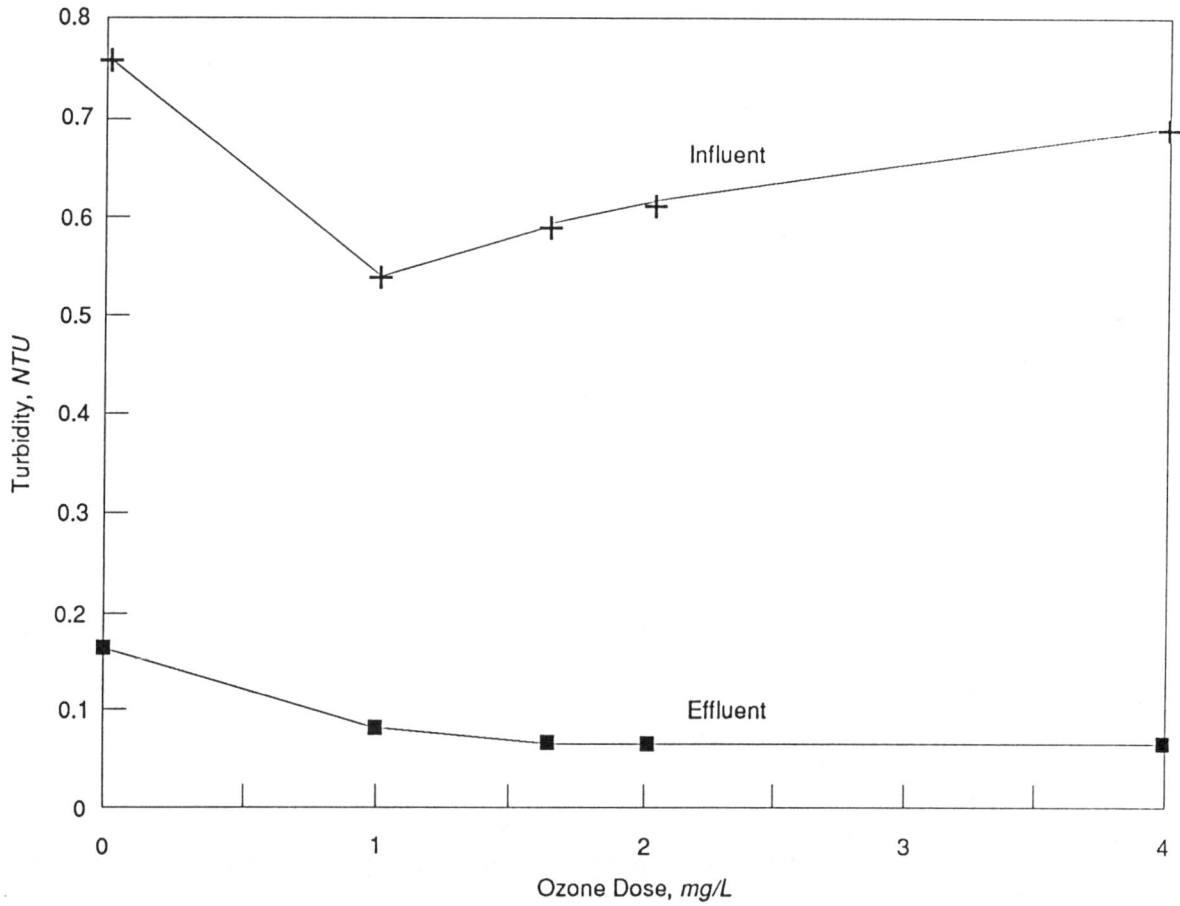

Source: Liang, S. et al. 1989. The Big Switch—Los Angeles Aqueduct Filtration Plant Treatment of State Project Water. 9th Ozone World Congress, Intl. Ozone Assn., New York.

Figure 10-8 Effect of ozone dose on turbidity.

Tracer studies. Tracer studies are an important indicator of mixing effectiveness. In fact, it is considered an essential part of any pilot testing work to first perform a tracer study to better understand the degree of short-circuiting and flow-through hydraulic characteristics in the various unit processes under consideration. Several methods exist to perform and analyze a tracer study (Hudson 1981). Tracer studies are performed by either spiking or continuous feed of a dye at the influent and testing for the concentration of dye at key locations at the influent and effluent points of all unit processes. Common dyes used for this purpose include Fluorescene and Rotamene B, which are analyzed in a colorimetric device that can provide a continuous dye concentration-time tracing. Lithium chloride is also used; however, testing is accomplished by discrete sampling and individual sample analysis using an atomic adsorption method for metals analysis. Use of lithium salts is popular at the full scale, since many of the organic dyes can be objectionable from a public health standpoint. Mixing effectiveness in dye testing is determined by the rate of change in dye concentration leaving the mixing basin or at key sample locations across the mixing vessel. Chapter 4 gives considerable detail on the use and methods of mathematically determining residence time distributions utilizing tracer study data.

Turbidity and particle counting. Both turbidity and particle counting are intended to measure the quantity of suspended particulate material. In actual practice, both measurements are limited in their ability to do this, and it is well for the pilot tester to fully understand the differences, complications, and interpretation of data prior to using these techniques. Both techniques remain, however, the most popular used in pilot testing, since turbidity or suspended particulate matter is regulated in drinking water. A detailed explanation of the instruments and measurement procedures is provided by *Standard Methods For the Examination of Water and Wastewater* (1989).

Turbidity is usually monitored on a continuous basis as a measurement of light reflectance at $90°$ from the light-beam path through particulate suspended matter. The measurement in itself will not directly measure mixing effectiveness.

Mixing effectiveness is determined by measurement at the clarified water or filter effluent. The degree of turbidity removal at these process-control points and overall treatment goals determine the degree of mixing effectiveness. These determinations are made on a comparative basis between parameters under study in parallel process treatment trains. Filter effluent turbidity should always be determined as a weighted average over the entire run length when scaling-up to full-scale facilities with many filters in service. A filter run length is determined by terminal head loss criteria and/or a terminal turbidity level goal such as 0.25 NTU.

Particle counting differs from turbidity measurement since the particle size and number are tested. Particle counters differ in their ability to do this, since several measurement techniques are available using principals of light reflectance, light omission, and electroconductivity. Particle counting is less popular for pilot testing, since it is difficult to measure, requiring individual discrete samples rather than continuous measurement. Particle counting should be performed, however, to provide further information on particle characteristics. Particle counting is essential for some testing where turbidity measurements can produce significant errors (i.e., while using powdered activated carbon).

Visual inspection, photography, and visible dye. Visual inspection of process water can provide valuable insight during rapid mixing and flocculation studies. Floc size and shape are readily viewed under good conditions of lighting, which should always be provided. Photography is a means of documenting floc development. Use of strobe lighting will allow motionless still photography. Microscopic examination of floc is also possible in situ or by using special viewing drop slides. Application of traces of visual dye can also reveal intricate detail of hydraulic patterns around mixing blades. Some form of visual inspection should always accompany pilot testing work, particularly in the area of visual inspection of floc development.

Zeta potential and electrophoretic mobility. Zeta potential and electrophoretic mobility are determined by the same procedure, which tests the degree of particle destabilization during the coagulation process. An electric current is passed through a cell equipped for microscopic examination. The velocity of particles is determined by timing the progress over a marked grid. Laser light is sometimes used in a dark-field microscopic analysis of extremely fine particles (Bales and Morgan 1984). The test methods are primarily oriented toward determination of surface properties on particles and can be used in concert with mixing tests. The technique is not intended to directly measure mixing effectiveness, but is provided here as a method of particular interest to studies on coagulation.

Streaming-current detector. The streaming-current detector is similar in concept to methods used to measure degree of particle destabilization (i.e., zeta potential). The detector is of recent origin and has its greatest application on the full scale. Pilot-scale applications are appropriate for process flows above 10 gpm (0.6 L/s). The instrument consists of a probe, which accepts a continuous sample stream and is located within 1 min after chemical injection into the rapid-mixing process. The method may have value for comparing various levels of mixing energies and chemical dosages. Chapter 11 includes additional information concerning full-scale use of streaming-current detectors.

Conductivity probe. The conductivity probe is a research tool (Lamb, Manning, and Wilhelm 1960) used to measure the effectiveness of mixing downstream of the mixing device. The method uses conductivity measurements over a spatial grid to determine concentration of a tracer salt such as sodium chloride. Sodium chloride is injected at the point of proposed chemical coagulant injection. The sodium chloride concentration can then be measured at various points in the downstream process stream to deduce the thoroughness of mixing. Stenquist and Kaufman (1972) provide a detailed application of the probe's use. The conductivity probe is usually used at the research level and has not been typically used for process studies at the pilot scale.

Velocity measurement. Velocity measurements are possible at key locations within the mixing vessel using various techniques. One technique uses induced magnetic fields similar to the magnetic flow-measuring meters in wide use. Thermal differences can also be related to fluid velocity using thermistors. Most of these methods have greatest application for the full scale and are not usually applied to process studies for pilot-plant work. Mention of the techniques here is in the interest of completeness and to offer possibilities for future application.

RAPID-MIXING EVALUATION

Rapid mixing, or initial mixing, as it is sometimes called, is a unit process whereby chemical coagulants (i.e., alum, ferric sulfate, ferric chloride, polyelectrolytes, etc.) are mixed with raw source water to facilitate coagulation of particulate matter. Coagulation is defined as the destabilization of the colloidal particles through the addition of coagulant to the water (Stenquist and Kaufman 1972). The significance of rapid mixing and coagulation is well-known (Stenquist and Kaufman 1972; Letterman, Quon, and Gemmell 1973; Argaman and Kaufman 1970). A 100 percent improvement can be achieved through efficient mixing versus uncontrolled backmixing chambers (Tanaka et al. 1987). Much theoretical and laboratory work has been done on rapid mixing and coagulation dating back to 1949 (Langelier and Ludwig 1949) and earlier. This section provides guidance on the data accumulation and evaluation for pilot-plant studies conducted for optimization of the rapid-mixing process for design and operation.

There are basically six major parameters that are traditionally used by the pilot tester to study rapid-mixing effectiveness. These include G, detention time, type of mixer, clarified-water turbidity, filter run length, and filter effluent turbidity. The value of G can vary substantially for pilot testing from 100 to 16,000 s^{-1} (Amirtharajah and Mills 1980). Detention times for backmix reactors are usually on the order of 1 s to 2 min or higher. The type and sequencing of chemical coagulants are also important to the overall treatment performance as measured by filter effluent turbidity and run length. A word of caution should be reiterated at this point concerning the direct use of G for full-scale design purposes. G is not an ideal parameter for rapid-mixing evaluations (Amirtharajah 1982) and differences can be found in actual scale-up. One of the primary problems is that the G-value requires a reactor volume in the calculation, which for some situations is difficult to assess, since energy fields vary in the form of turbulent eddies throughout the reactor volume. However, G is one of the easiest parameters to apply in pilot testing and serves as a reasonable indicator. Another word of caution is that filter effluent turbidity should be measured as a weighted average value over the entire run length to simulate a full-scale plant with a number of filter units in service. Measuring only the lowest turbidity value during the run will give an artificially low value.

Figure 10-9 shows two ways that G, detention time, and filter effluent turbidity can be shown for subsequent data analysis. In the absence of sufficient data to plot curves, a tabular format such as shown in Table 10-6 is appropriate for analysis. An interesting aspect of these two studies, using alum and polymer, is that low energies of 200 s^{-1} and short detention times yielded best overall performance. Prior to initiating pilot-scale studies, bench-scale jar tests are appropriate to identify proper chemical feed dosage requirements, as seen in Figure 10-10 (JM Montgomery 1977) in the form of a log–log isoturbidity topogram. These studies are best done at the bench scale in the interest of cost and time. More will be said on evaluation of rapid mixing effectiveness in the section on full-scale testing. Evaluation of rapid mixing is covered in greater detail in chap. 9 for vertical turbine flocculators and chap. 11 for horizontal-type flocculators.

FLOCCULATION EVALUATION

Flocculation is the unit process in water treatment where particles are made to combine into floc in preparation for physical removal in sedimentation or filtration. Formally, flocculation is defined as the collision and aggregation of the destabilized particles into relatively large aggregates known as flocs (Argaman and Kaufman 1970). The means of testing flocculation on the pilot scale is the same as described for rapid mixing. Pilot tests are usually conducted by holding one or the other process constant while varying the parameter under study. Again, G-value and detention time are two of the major parameters in flocculation; however, compartment number, baffle configuration, and mixer design also play a significant role.

Evaluation of pilot test data for flocculation is somewhat similar to that for rapid mixing. Figure 10-11 serves as a good illustration of the usefulness of G-value and detention time data. Tabular formats are also instructive, as seen in Tables 10-7 and 10-8. Performance of flocculation in conventional treatment facilities with

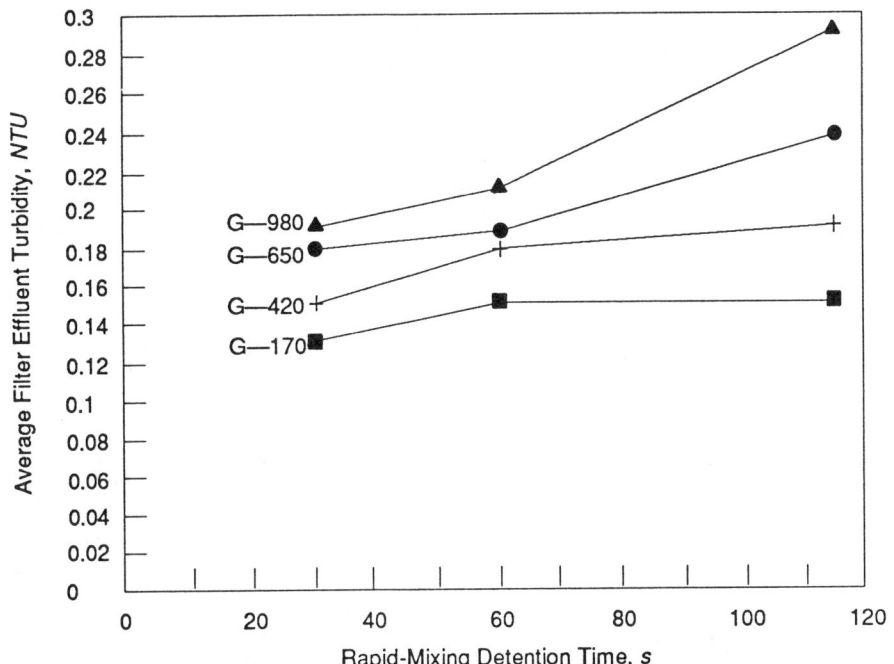

A. Filter effluent turbidity as a function of rapid-mixing detention time at four levels of mixing energy

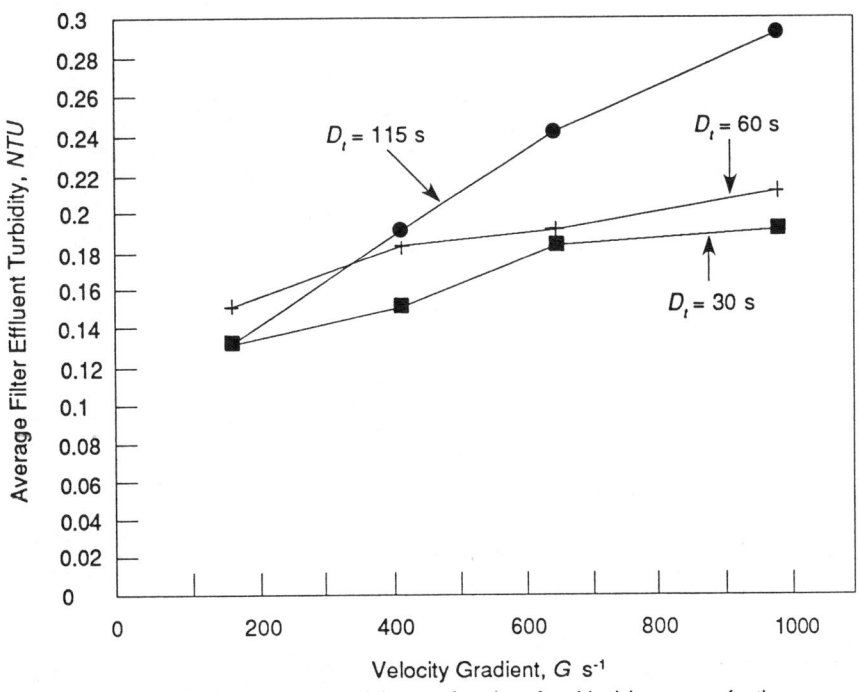

B. Filter effluent turbidity as a function of rapid-mixing energy for three detention times.

Source: Tanaka et al. 1987. Pilot Scale and Full Scale Investigations for the Design of Additional Direct Filtration Modules for an Existing 320 mgd Treatment Plant. AWWA Annual Conf., Kansas City, Mo.

Figure 10-9 Two ways to show detention time and filter effluent turbidity for subsequent data analysis.

370 MIXING

Table 10-6 Effect of Flash Mixing on Clarification Performance for High-Turbidity Water

Run No.	Train	Raw Water Turbidity, NTU	Pre-treat Flow, mgd	Coagulant Dose Alum, ppm	Coagulant Dose Polymer, ppm	Flocculation G, 1/s	Flocculation Detention Time, min	Flocculation Baffle Type	Flash Mix G, 1/s	Flash Mix Detention Time, s	Flash Mix Type*	Sedimentation Basin Effluent Turbidity, NTU
16	B	45.00	400	4.5	1.0	100/70/15	14.0	Slots	700	50	BM	9.6
	C	45.00	400	4.5	1.0	100/70/15	14.0	Slots	7200	2	IL	17.3
19	B	43.00	400	6.0	1.3	70/40/15	14.0	Slots	750	50	BM	5.5
	B	43.00	400	6.0	1.3	70/40/15	14.0	Slots	750	50	BM	7.3
	C	43.00	400	6.0	1.3	70/40/15	14.0	Slots	3500	2	IL	8.8
	C	43.00	400	6.0	1.3	70/40/15	14.0	Slots	3500	2	IL	8.8
20	B	43.00	400	18.0	1.5	40/30/10	14.0	Slots	500	50	BM	7.7
	C	43.00	400	18.0	1.5	40/30/10	14.0	Slots	2000	2	IL	11.8
21	B	43.00	400	12†	1.5	70/40/10	14.0	Slots	250	50	BM	4.8
	C	43.00	400	12†	1.5	70/40/10	14.0	Slots	2000	2	IL	11.0
15	B	28.00	400	4.5	1.0	100/70/35	14.0	Slots	700	50	BM	10.0
	C	28.00	400	4.5	1.0	100/70/35	14.0	Slots	7200	2	IL	14.0
58.8	A	65.00	300	25.0	2.0	50/50/50	18.0	O/U‡	1000	68	BM	9.7
	B	65.00	300	25.0	2.0	50/50/50	18.0	O/U	700	68	BM	8.7
	C	65.00	300	25.0	2.0	50/50/50	18.0	O/U	500	68	BM	6.2
	D	65.00	300	25.0	2.0	50/50/50	18.0	O/U	200	68	BM	4.9
58.10	A	65.00	300	25.0	2.0	50/50/50	18.0	O/U	200	68	BM	6.4
	B	65.00	300	25.0	2.0	50/50/50	9.0	O/U	200	2	IL	8.2
	C	65.00	300	25.0	2.0	50/50/50	18.0	O/U	200	2	IL	7.2
	D	65.00	300	25.0	2.0	50/50/50	9.0	O/U	200	68	BM	7.2
60	A	30.00	300	15.0	2.5	70/20/10	18.0	O/U	200	68	BM	4.2
	B	30.00	300	15.0	2.5	70/20/10	18.0	O/U	500	68	BM	5.5
	C	30.00	300	15.0	2.5	50/50/50	18.0	O/U	200	68	BM	5.1
	D	30.00	300	15.0	2.5	50/50/50	18.0	O/U	500	68	BM	5.3

*IL—indicates in-line type flash mixer, BM—indicates backmix type flash mixer.
†Indicates ferric chloride rather than alum.
‡Indicates over/under type baffle.

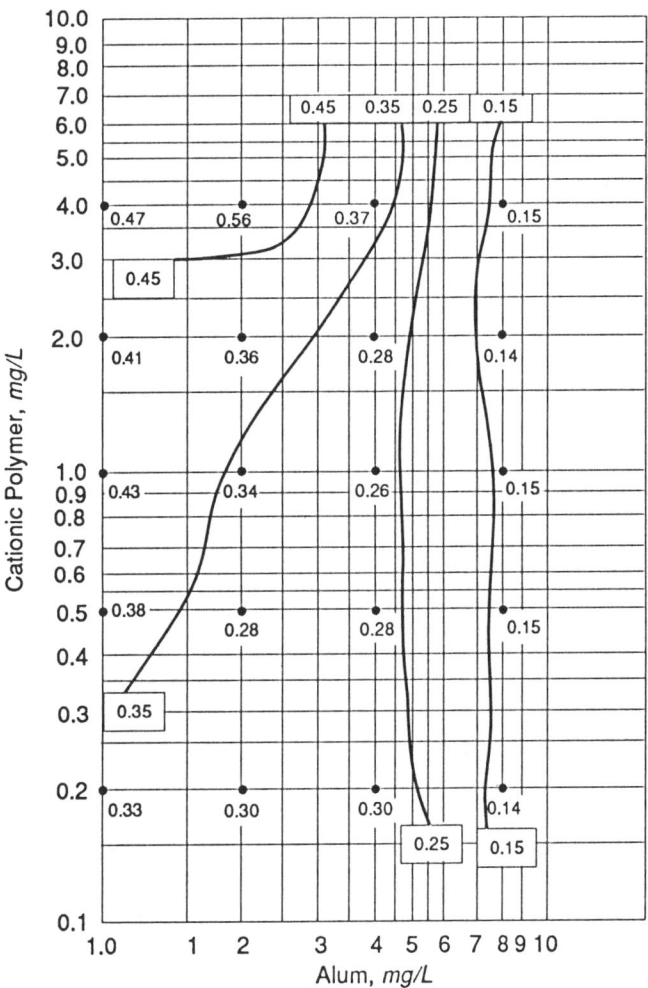

Source: JM Montgomery. 1985. Water Treatment Principles and Design. *Wiley-Interscience, New York.*

Figure 10-10 Log–log isoturbidity topogram.

sedimentation can be further analyzed by clarified effluent curves, as shown in Figure 10-12. Both of these studies utilized artificial variation of turbidity. The first study resulted in a find that low mixing energy and a detention time of about 20 min worked best in a direct filtration process. The second study, performed at the Jensen Filtration Plant (Metropolitan Water District of Southern California), concluded that low-turbidity water is relatively insensitive to the flocculation operational parameters studied. As a result, the high-turbidity flocculation results controlled flocculation requirements. The high-turbidity results indicated the most effective flocculation operation was as follows:

- approximately 20-min detention time minimum (longer detention times were not evaluated),
- over/under baffling, and

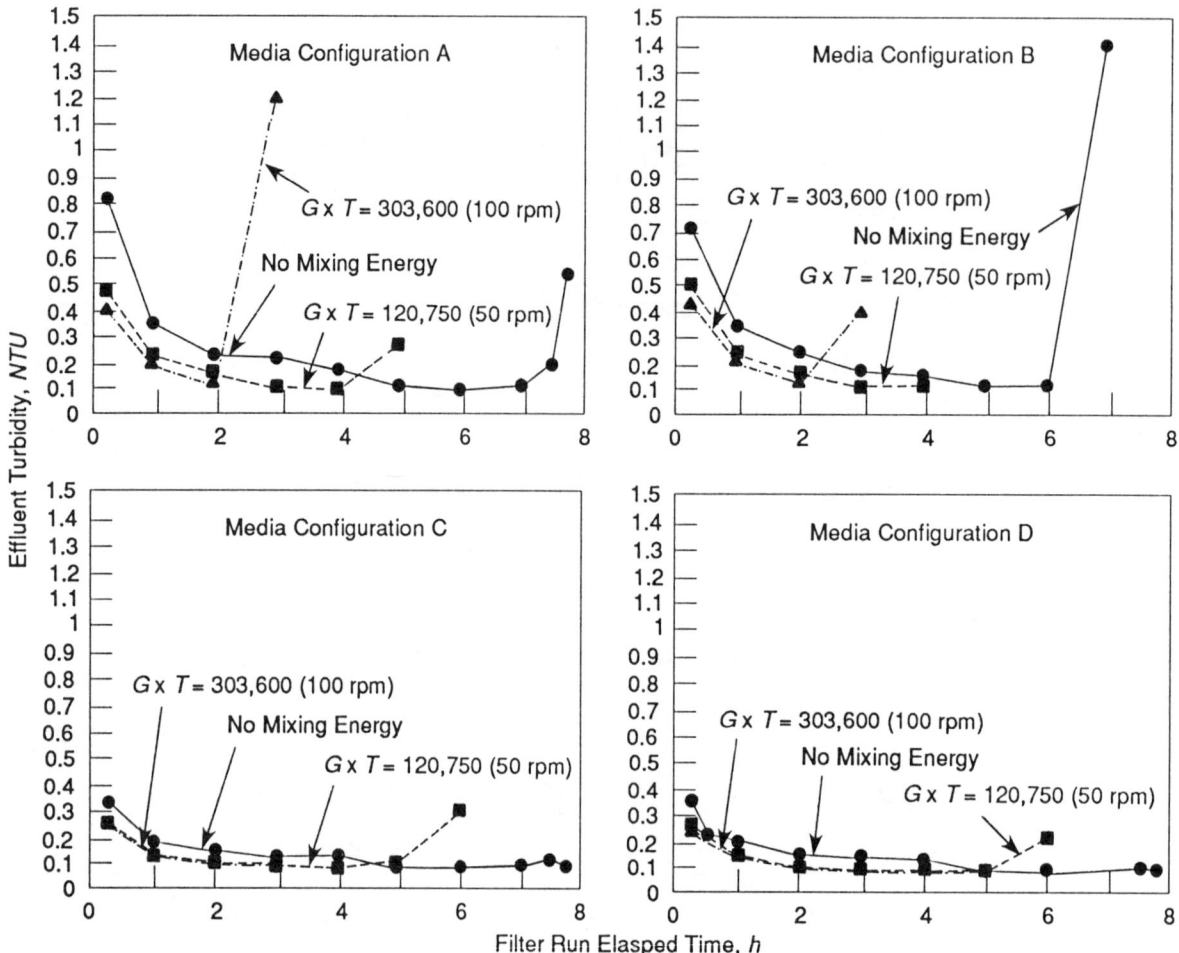

Source: Tanaka et al. 1987. Pilot Scale and Full Scale Investigations for the Design of Additional Direct Filtration Modules for an Existing 320 mgd Treatment Plant. AWWA Annual Conf., Kansas City, Mo.

Figure 10-11 Effect of flocculation mixing energy on pilot filter effluent turbidity.

- 70/20/10 and 50/50/50 energy input into the three-stage flocculation. These ratios provided equivalent performance. However, the 70/20/10 scheme requires 28 percent less total power input than the 50/50/50 configuration.

The first conclusion was further substantiated with historical plant records and bench-scale studies shown in Figure 10-2. Again, it should be restated that several pilot studies have determined that optimum G-values may be substantially overstated in pilot studies by as much as 100 percent (JM Montgomery 1985). The uncertainty of scale-up from pilot to full scale implies that variable-speed flocculators are advisable in situations where full-scale controls are not available.

Table 10-7 Effect of Flocculation on Clarification Performance Using High-Turbidity Water

Run No.	Train	Influent Turbidity, NTU	Pre-treat Flow, mgd	Coagulant Dose, mg/L Alum	Coagulant Dose, mg/L Polymer	Flash Mix G, 1/s	Flash Mix Detention Time, s	Flash Mix Type*	Flocculation G, 1/s	Flocculation Detention Time, min	Flocculation Baffles†	Sedimentation Basin Effluent Turbidity, NTU
58.1	A	65	300	25	2	200	68	BM	100/100/100	18	O/U	15.0
	B	65	300	25	2	200	68	BM	100/100/100	18	Slots	12.0
	C	65	300	25	2	200	68	BM	70/40/20	18	O/U	5.8
	D	65	300	25	2	200	68	BM	70/40/20	18	Slots	6.7
58.2	A	65	300	25	2	200	68	BM	70/40/20	18	O/U	6.8
	B	65	300	25	2	200	68	BM	70/40/20	18	Slots	6.3
	C	65	300	25	2	200	68	BM	100/100/100	18	O/U	21.0
	D	65	300	25	2	200	68	BM	100/100/100	18	Slots	17.0
58.3	A	65	300	25	2	200	68	BM	70/40/20	18	O/U	4.3
	B	65	300	25	2	200	68	BM	70/40/20	18	Slots	7.6
	C	65	300	25	2	200	68	BM	20/20/20	18	O/U	16.0
	D	65	300	25	2	200	68	BM	20/20/20	18	Slots	17.0
58.4	A	65	300	40	2	200	68	BM	70/40/20	18	O/U	5.5
	B	65	300	40	2	200	68	BM	70/40/20	18	Slots	6.4
	C	65	300	40	2	200	68	BM	20/20/20	18	O/U	15.0
	D	65	300	40	2	200	68	BM	20/20/20	18	Slots	16.0
58.5	A	65	300	25	2	200	68	BM	50/50/50	18	O/U	4.2
	B	65	300	25	2	200	68	BM	50/50/50	18	Slots	4.1
	C	65	300	25	2	200	68	BM	70/40/20	18	O/U	5.5
	D	65	300	25	2	200	68	BM	70/40/20	18	Slots	6.7
58.6	A	65	300	25	2	200	68	BM	20/20/20	18	O/U	7.3
	B	65	300	25	2	200	68	BM	20/20/20	18	Slots	8.4
	C	65	300	25	2	200	68	BM	70/40/20	18	O/U	6.4
	D	65	300	25	2	200	68	BM	70/40/20	18	Slots	7.2

*BM—indicates backmix-type flash mixer; IL—indicates inline-type flash mixer.
†O/U—indicates over/under-type flocculator.

Table continues next page

Table 10-7 Effect of Flocculation on Clarification Performance Using High-Turbidity Water, continued

Run No.	Train	Influent Turbidity, NTU	Pre-treat Flow, mgd	Coagulant Dose, mg/L Alum	Coagulant Dose, mg/L Polymer	Flash Mix G, 1/s	Flash Mix Detention Time, s	Flash Mix Type*	Flocculation G, 1/s	Flocculation Detention Time, min	Flocculation Baffles†	Sedimentation Basin Effluent Turbidity, NTU
58.7	A	65	300	25	2	200	68	BM	70/40/20	18	O/U	6.1
	B	65	300	25	2	200	68	BM	70/40/20	18	Slots	8.6
	C	65	300	25	2	200	68	BM	50/50/50	18	O/U	5.6
	D	65	300	25	2	200	68	BM	50/50/50	18	Slots	8.6
58.10	A	65	300	25	2	200	68	BM	50/50/50	18	O/U	6.4
	B	65	300	25	2	200	2	IL	50/50/50	9	O/U	8.2
	C	65	300	25	2	200	2	IL	50/50/50	18	O/U	7.2
	D	65	300	25	2	200	68	BM	50/50/50	9	O/U	7.2
59	A	66	300	25	2	200	68	BM	70/20/10	18	O/U	4.8
	B	66	300	25	2	500	68	BM	70/20/10	18	O/U	6.2
	C	66	300	25	2	200	68	BM	50/50/50	18	O/U	5.0
	D	66	300	25	2	500	68	BM	50/50/50	18	O/U	5.0
60	A	30	300	15	2.5	200	68	BM	70/20/10	18	O/U	4.2
	B	30	300	15	2.5	500	68	BM	70/20/10	18	O/U	5.5
	C	30	300	15	2.5	200	68	BM	50/50/50	18	O/U	5.1
	D	30	300	15	2.5	500	68	BM	50/50/50	18	O/U	5.3

*BM—indicates backmix-type flash mixer; IL—indicates inline-type flash mixer.
†O/U—indicates over/under-type flocculator.

Table 10-8 Effect of Flash Mixing on Clarification Performance for High-Turbidity Water

Run No.	Train	Raw Water Turbidity, NTU*	Flow, mgd Pre-treat	Flow, mgd Filter	Coagulant Dose, mg/L Alum†	Coagulant Dose, mg/L Polymer	Coagulant Dose, mg/L Filter Aid	Flash Mix G, 1/s	Flash Mix Detention Time, s	Flash Mix Type‡	Flocculation G, 1/s	Flocculation Detention Time, min	Flocculation Baffle Type§	Filter Run Time Below 0.1 NTU, h	Average Filtrate Turbidity, NTU	Reason for Termination
61	A	2.60*	300	400	50.0	3.0	0.00	200	68.0	BM	70/20/10	18.0	O/U	0.0	0.13	Turbidity
	C	2.60*	300	400	50.0	3.0	0.00	200	68.0	BM	50/50/50	18.0	O/U	0.0	0.16	Turbidity
	A	2.60*	300	400	50.0	3.0	0.02	200	68.0	BM	70/20/10	18.0	O/U	7.0	0.06	Head
	C	2.60*	300	400	50.0	3.0	0.02	200	68.0	BM	50/50/50	18.0	O/U	6.5	0.07	Head
62	A	2.00*	300	400	25†	3.0	0.02	200	68.0	BM	70/20/10	18.0	O/U	7.3	0.07	Head
	C	2.00*	300	400	25†	3.0	0.02	200	68.0	BM	50/50/50	18.0	O/U	8.0	0.09	Head
	A	2.00*	300	400	25†	3.0	0.05	200	68.0	BM	70/20/10	18.0	O/U	4.0	0.08	Head
	C	2.00*	300	400	25†	3.0	0.05	200	68.0	BM	50/50/50	18.0	O/U	0.5	0.11	Head
54	A	0.59	600	400	2.5	0.8	0.00	200	34.0	BM	10/10/10	9.0	O/U	30.5	0.08	Turbidity
	B	0.59	600	400	2.5	0.8	0.00	200	34.0	BM	10/10/10	9.0	Slots	32.0	0.08	Turbidity
	C	0.59	300	400	2.5	0.8	0.00	200	68.0	BM	10/10/10	18.0	O/U	32.0	0.09	Turbidity
	D	0.59	300	400	2.5	0.8	0.00	200	68.0	BM	10/10/10	18.0	Slots	35.0	0.08	Turbidity
55	A	0.56	600	400	2.5	0.8	0.00	200	34.0	BM	10/10/10	9.0	O/U	23.5	0.09	Turbidity
	B	0.56	300	400	2.5	0.8	0.00	200	68.0	BM	10/10/10	18.0	O/U	35.5	0.08	Turbidity
	C	0.56	600	400	2.5	0.8	0.00	200	34.0	BM	100/100/100	9.0	O/U	34.5	0.08	Turbidity
	D	0.56	300	400	2.5	0.8	0.00	200	68.0	BM	100/100/100	18.0	O/U	35.5	0.08	Turbidity
56	A	0.60	600	400	2.5	0.8	0.00	750	34.0	BM	100/100/100	9.0	O/U	37.0	0.08	Turbidity
	B	0.60	600	400	2.5	0.8	0.00	750	34.0	BM	100/100/100	9.0	Slots	32.0	0.09	Turbidity
	C	0.60	600	400	2.5	0.8	0.00	750	34.0	BM	70/40/10	9.0	O/U	37.5	0.09	Turbidity
	D	0.60	300	400	2.5	0.8	0.00	750	68.0	BM	70/40/10	18.0	O/U	56.5**	0.09	Time

NOTE: All filter runs performed with dual-media, constant-head filters.
*Indicates run was conducted using "shallow" lake water.
†Indicates ferric chloride rather than alum.
‡BM—indicates backmix-type flash mixer.
§O/U—indicates over/under-type flocculator.
**Indicates filter run time was projected.

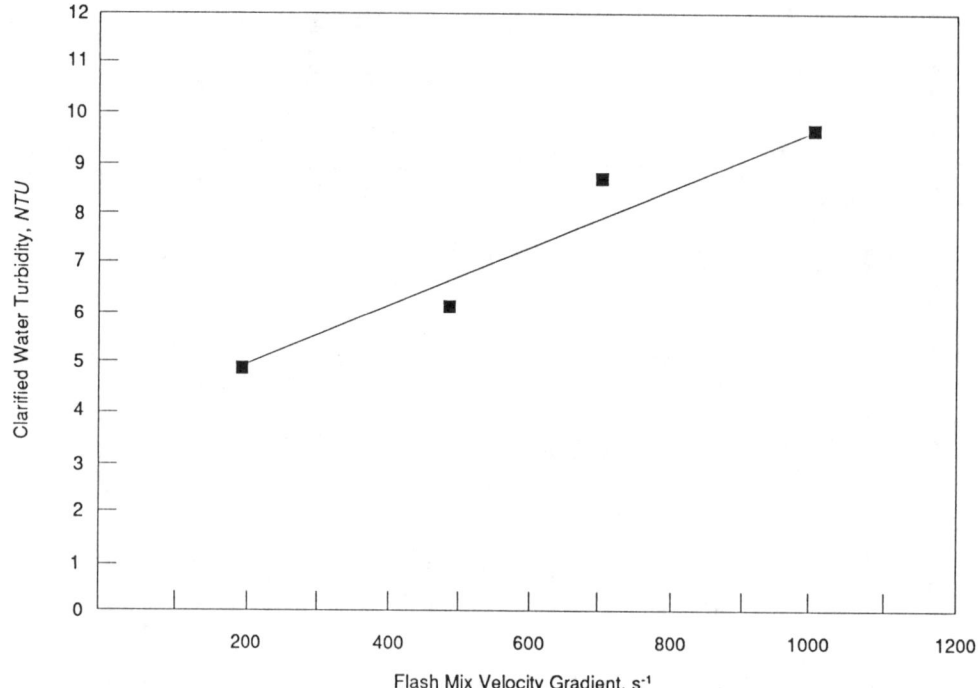

Source: Tanaka, T.S. et al. 1988. Conceptual Design Report for Joseph Jensen Plant Expansion. Metropolitan Water District of Southern California, R&D Report.

Figure 10-12 Clarifier performance versus flash mix energy.

FULL-SCALE TESTING

Actual pilot tests conducted in recent years can be directly compared to full-scale treatment facilities already in existence. Many more examples are available than presented here. However, this section provides insight into how well both pilot- and full-scale treatment facilities can track for rapid mixing and flocculation.

Studies conducted at the Metropolitan Water District of Southern California (MWDSC) demonstrated the excellent agreement between pilot- and full-scale process treatment evaluations. Figure 10-13 provides an illustration, comparing both head loss buildup and effluent turbidity from filter results between a direct filtration pilot plant and the R.A. Skinner Filtration Plant. In the same study, kaolin clay was introduced into the full-scale plant along with artificially seeded pilot-plant source water at a total suspended solids equivalent of 12 NTU with identical comparative results.

Similar testing performed at the MWDSC Joseph Jensen Filtration Plant again confirmed very close consistency between pilot- and full-scale evaluations. Tracer studies performed on the pilot test unit and the full-scale plant indicated excellent agreement, as shown in Figure 10-14. The figure compares a concentration time plot of tracer at the sedimentation basin effluent. This sample point was selected because both the pilot- and full-scale plants have flocculation basins integrated with sedimentation basins.

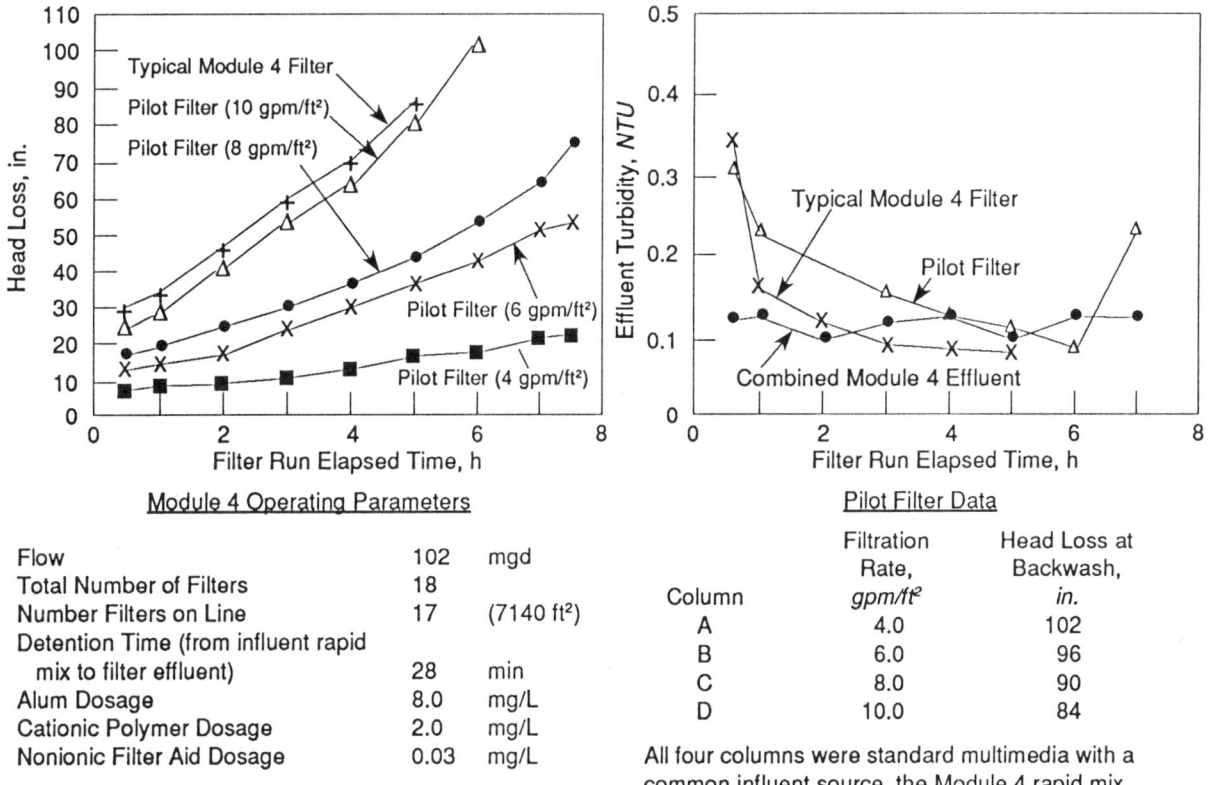

Source: Tanaka et al. 1987. *Pilot Scale and Full Scale Investigations for the Design of Additional Direct Filtration Modules for an Existing 320 mgd Treatment Plant. AWWA Annual Conf., Kansas City, Mo.*

Figure 10-13 Filter run head losses and effluent turbidities for various filter media configurations and filtration rates (influent turbidity = 30 NTU, koalin-induced).

The City of Los Angeles Department of Water and Power conducted extensive pilot tests on Owens Valley water supply. Results of the testing on particulate removal are in close agreement with the full-scale Los Angeles Aqueduct Filtration Plant, which, in fact, performs better than pilot-scale studies indicated. The filtration plant meets the 0.2 NTU goal determined during pilot testing and unit filter run volume greater than 5000 gal/ft^2 (2 mm/s).

CONCLUSIONS

The use of pilot-scale testing for design, operation, and modification of treatment plant design work has wide application and has proven effective. Considerable time, effort, and cost are often required for best results, depending on the scope of the evaluations. Much still remains to be understood in the area of scale-up from pilot to full scale for the rapid-mixing and flocculation processes. The next important step in

378   MIXING

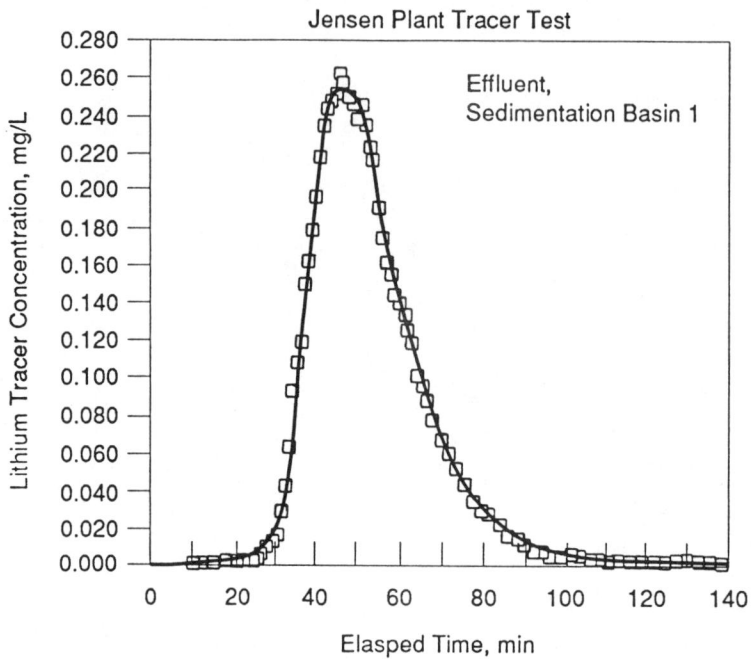

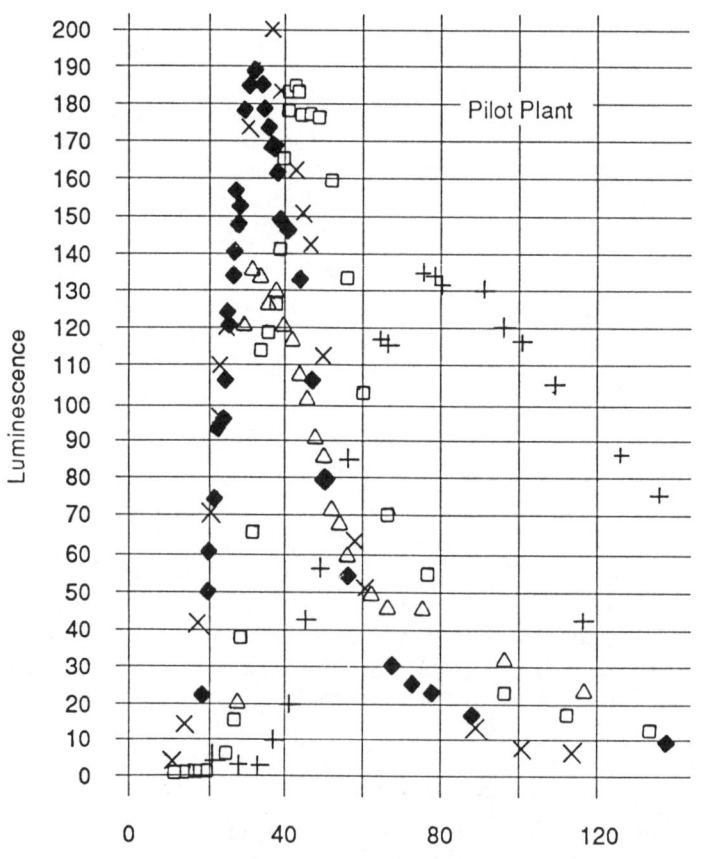

Source: Metropolitan Water District of Southern California.

Figure 10-14 Concentration time plot of tracer at sedimentation basin effluent.

the pilot testing field will be a compilation of major pilot- and full-scale testing work on various source water conditions. Close examination of this work to date will allow needs for future research to be properly assessed. At this point, it seems more research is needed in the area of scale-up for flash mixers and flocculator types, and more parallel tests need to be reviewed between full- and pilot-scale evaluations. As state and federal regulations become more stringent, and cost of facilities keeps increasing, more innovative treatment process technologies will emerge, with continued need of pilot testing as a valuable tool for process evaluations.

References

AMIRTHARAJAH, A. 1982. Design of Rapid Mix Units. In *Water Treatment Plant Design*, ed. by R.L. Sanks. Ann Arbor Science, Ann Arbor, Mich.

AMIRTHARAJAH, A. & MILLS, K.M. 1980. Rapid-Mix Design for Mechanisms of Alum Coagulation. AWWA Annual Conf., Atlanta, Ga.

ARGAMAN, Y. & KAUFMAN, W.J. 1970. Turbulence and Flocculation. *Jour. San. Engrg. Div.*, SA2:4:223.

BALES, R.C. & MORGAN, J.J. 1984. Surface Chemical and Physical Behavior of Chrysotile Asbestos in Natural Waters and Water Treatment. Cal. Tech., Report No. AC-8-84, Pasadena, Calif.

CLEASBY, J.L. 1984. Is Velocity Gradient a Valid Turbulent Flocculation Parameter. *Jour. Envir. Engrg.*, 110:5:875.

HUDSON, H. JR. 1981. *Water Clarification Processes: Practical Design and Evaluation.* Van Nostrand Reinhold, New York.

HUDSON, H. JR. & WOLFNER, J.P. 1967. Design of Mixing and Flocculating Basins. *Jour. AWWA*, 59:10:1257.

JM MONTGOMERY. 1977. Pilot Study on the Treatment of Water From Lake Casitas. Lake Casitas Metropolitan Water District, California.

———. 1985. *Water Treatment Principles and Design*. Wiley-Interscience, New York.

———. 1988. East County Prototype Study, for Contra Costa Water District. Calif.

KAWAMURA, S. 1973. Coagulation Considerations. *Jour. AWWA*, 65:6:417.

LAMB, D.E.; MANNING, F.S.; & WILHELM, R.H. 1960. Measurement of Concentration Fluctuations With an Electrical Conductivity Probe. *AIChE Jour.*, 6:682.

LANG, J. 1982. Selection of Pilot Plant Equipment. AWWA Annual Conf., Miami, Fla.

LANGELIER, W.F. & LUDWIG, H.F. 1949. Mechanisms of Flocculation in the Clarification of Turbid Waters. *Jour. AWWA*, 41:163.

LETTERMAN, R.; QUON, J.; & GEMMELL, R. 1973. Influence of Rapid-Mix Parameters on Flocculation. *Jour. AWWA*, 65:716.

LIANG, S. ET AL. 1989. The Big Switch—Los Angeles Aqueduct Filtration Plant Treatment of State Project Water. 9th Ozone World Congress, Intl. Ozone Assn., New York.

MOSER, R.; LEE, R.; & CREEL, S. 1987. Investigation of Rapid Mix Technology: Recommendations for the American Water System.

SANKS, R.L. 1980. *Water Treatment Plant Design for the Practicing Engineer*. Butterworth–Heinemann, Stoneham, Mass.

Standard Methods for the Examination of Water and Wastewater. 1989. APHA, AWWA, and WPCF. Washington, D.C. (17th ed.).

STENQUIST, R. & KAUFMAN, W. 1972. Initial Mixing in Coagulation Processes, S.E.R.L. Report No. 72-2, University of California, Berkeley, Calif.

STOLARIK, G.F. 1981. Ozonation and the Direct Filtration Process. AWWA Fall Conf., Palm Springs, Calif.

TANAKA, T.S. ET AL. 1987. Pilot Scale and Full Scale Investigations for the Design of Additional Direct Filtration Modules for an Existing 320 mgd Treatment Plant. AWWA Annual Conf., Kansas City, Mo.

———. 1988. Conceptual Design Report for Joseph Jensen Plant Expansion. Metropolitan Water District of Southern California, R&D Report.

TATE, C.H. & TRUSSELL, R.R. 1982. Pilot Plant Justification. Proc. 24th Annual Public Water Supply Engineer's Conf., Urbana, Ill.

11

Design of Mixers for Water Treatment Plants: Rapid Mixing and Flocculators

Robert D.G. Monk, P.E., Senior Vice President, Camp Dresser & McKee Inc., Walnut Creek, Calif. USA

R. Rhodes Trussell, Ph.D., P.E., Senior Vice President, James M. Montgomery Consulting Engineers, Inc., Pasadena, Calif. USA

INTRODUCTION

Mixing plays an important part in the treatment of water for potable use. Rapid mixing is often the first step in the treatment process train and provides both the mechanism for the dispersion of coagulants and time for chemical coagulation to occur.

The flocculation stage in water treatment follows coagulation or destabilization and provides a mechanism for the agglomeration or growth of the destabilized colloidal suspension in the raw water. It can also provide an environment for fine precipitates of the lime (or lime–soda ash) softening process to make contact with coagulant chemicals to maximize precipitation during settling.

Rapid mixing allows for the chemistry of the process to occur, while flocculation is the transportation step that leads to collisions between destabilized colloidal particles, thereby encouraging the buildup of flocs.

This chapter is designed to help the engineer identify, select, and design the rapid mixing and flocculation system appropriate for a specific location, water quality, and treatment process. The theoretical considerations of coagulation–flocculationare presented in chap. 1 and other chapters in this volume and are not addressed here. Chapter 9 outlines how installed mixing units can be tested for overall performance.

As Hudson (1981) points out, "... all processes used in water treatment are interrelated. A change in any one of these processes may require or enable a change in the design or management of one or more of the other. For example, an improvement in mixing and flocculation may make possible reduction in chemical dosages simultaneously with the allowance of higher settling basin loadings and filter rates and improved filtered water quality." The importance of the coagulation–flocculation process in the overall treatment process train must not be underrated. This chapter presents the practical aspects of the design of rapid mixers and flocculators that will lead to improvements in total treatment plant performance.

RAPID MIXERS

The purpose of rapid mixers in the water treatment process train is to obtain complete and uniform dispersion of the coagulant chemical throughout the total flow of raw water. Hydrolysis reactions and destabilization of the colloidal particles in the raw water are the primary objectives when the coagulant is introduced. Therefore, it is important to diffuse the coagulant as quickly as possible. In fact, the coagulating time is extremely short, less than a second, and best results are obtained if the colloids in the raw water are completely destabilized before part of the coagulant has begun to form a precipitate.

Ineffective coagulant dispersion permits local areas of high coagulant concentration and inappropriate pH levels, resulting in overdosing of coagulants. Further, prolonged mixing or backmixing with high energy dissipation will tend to break up the flocs that are forming. However, when sufficient metal ion salts must be added to form hydroxide precipitates of $Al(OH)_3(s)$ and $Fe(OH)_3(s)$ for sweep coagulation, then longer mixing time, of the order of 1 to 7 s, is needed, compared to only fractions of a second needed for hydrolysis and destabilization (see chap. 1).

Rapid Mixing Systems

To effectively disperse chemicals into the water requires high energy dissipation at the point of application. Energy dissipation can be accomplished by either hydraulic or mechanical means. In this text hydraulic mixing includes all systems in which there are no mechanical or moving parts; the energy for mixing is obtained through head loss in the water.

Hydraulic mixing. Any point where energy dissipation occurs in a water channel, pipeline, or chamber is a possible location for coagulant chemical mixing. Examples of such points of violent turbulence include:

- flow-measuring or flow-splitting weirs,
- hydraulic jumps,

382   MIXING

- excessive head dissipation devices,
- pipe discharges with relatively high discharge velocity, and
- static mixers.

Weirs. Weirs are often used in water treatment plants for either flow splitting or flow measuring. The latter include V-notch, rectangular, or trapezoidal weirs. The sudden free-fall in the hydraulic level over the weir induces the considerable turbulence in the water needed for chemical dispersion. The head loss should be 4 in. (102 mm) or greater. The height of the coagulant diffuser over the weir should be at least 1 ft (0.3 m), so that the speed of the falling coagulant solution can penetrate the nappe thickness (Schulz and Okun 1984). A small receiving chamber immediately below the weir with a 1- or 2-s detention time confines the turbulence resulting from the energy dissipation in the falling water (Hudson 1981). Figure 11-1 shows a section through a weir rapid mixer.

Hydraulic jumps. A "naturally" occurring hydraulic jump in a raw water channel or an induced jump at a Parshall or similar type flume is a good place for chemical injection. The advantage of the latter is that such an induced jump can be used for plant flow measurement as well as mixing. The coagulant can be added above the water level just upstream of the jump or through a diffuser located in a recess in the floor of the flume. Figure 11-2 shows a section through a Parshall flume used as a rapid mixer.

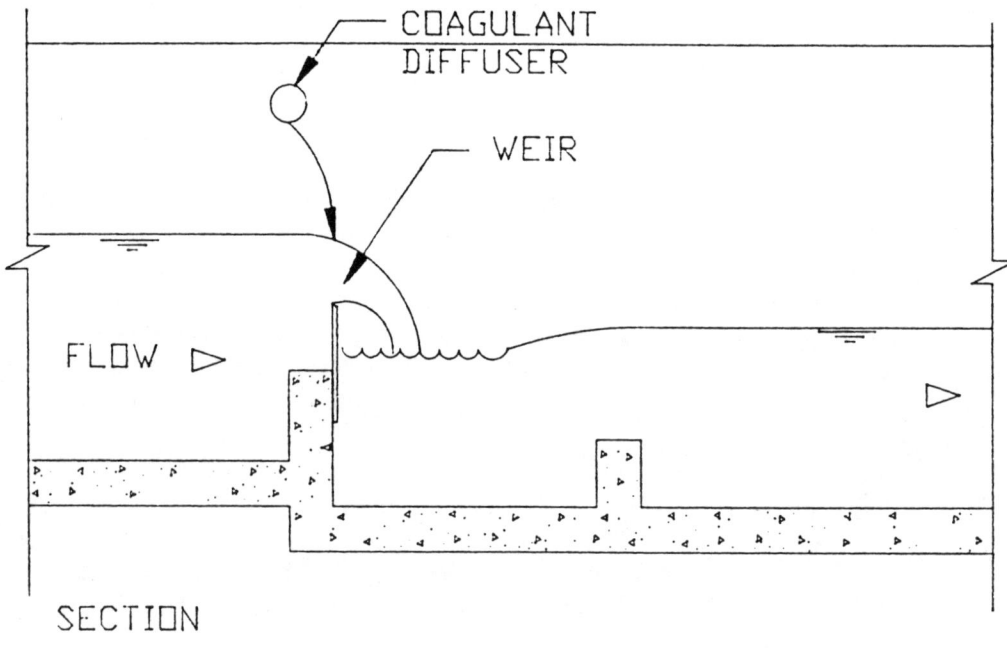

Figure 11-1 Weir rapid mixer.

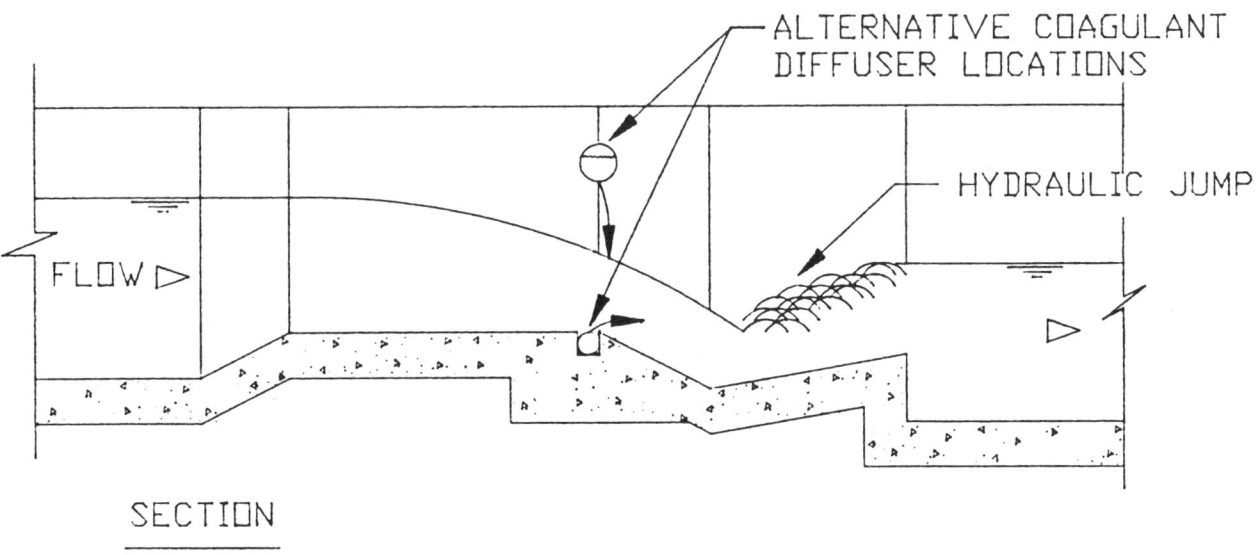

Figure 11-2 Parshall flume rapid mixer.

Energy dissipation. At many plants there is often excessive head in the raw water, which must be dissipated before the water enters the plant proper under open channel flow conditions. If this head is large, then the excess head or energy may be used to drive an electric generator.

However, often the head is insufficient for cost-effective energy recovery and therefore must be dissipated by means of a Bailey valve or similar device. A high degree of turbulence occurs in the chamber into which the valve discharges, and advantage can be taken of this energy to disperse chemicals. The Metropolitan Water District of Southern California (MWDSC) uses specially designed perforated slide gates to dissipate the excess raw water head and disperse the coagulant chemical (Bowers and Beard 1981). The use of this type of hydraulic mixer improved filtered-water turbidity by 50 percent without increasing coagulant chemical dosage. However, MWDSC has since replaced this type of hydraulic mixer.

Pipe discharges. Often the raw water entering a plant headworks enters with a relatively high velocity. Dissipation of the high-velocity head causes violent turbulence in the inlet chamber, which can be effectively used for initial chemical mixing. Alternatively, a dissipation plate can be positioned close to the discharging pipe and the coagulant chemical injected at the center of the plate and pipeline. An example of this type of mixer is shown in Figure 11-3.

Static mixers. Static rapid mixers (also called motionless or in line mixers) are specially designed to create intense turbulence through a fixed geometric design within a pipe. These produce, through three distinct mixing mechanisms, two-by-two flow division, radial mixing, and transient mixing. The mixing occurs over a specific length, and the mixing time can be calculated. Little or no backmixing with its associated energy loss occurs, and the head losses are relatively low. Static mixers are available for any diameter, from small-bore tubes up to pipe diameter in excess of 6 ft (2 m). Figure 11-4 shows details of a typical in-line static mixer. Static mixers for the mixing of chemicals and carrying water can be as simple as a series of 45° and 90° bends on the discharge line of a feed pump.

384 MIXING

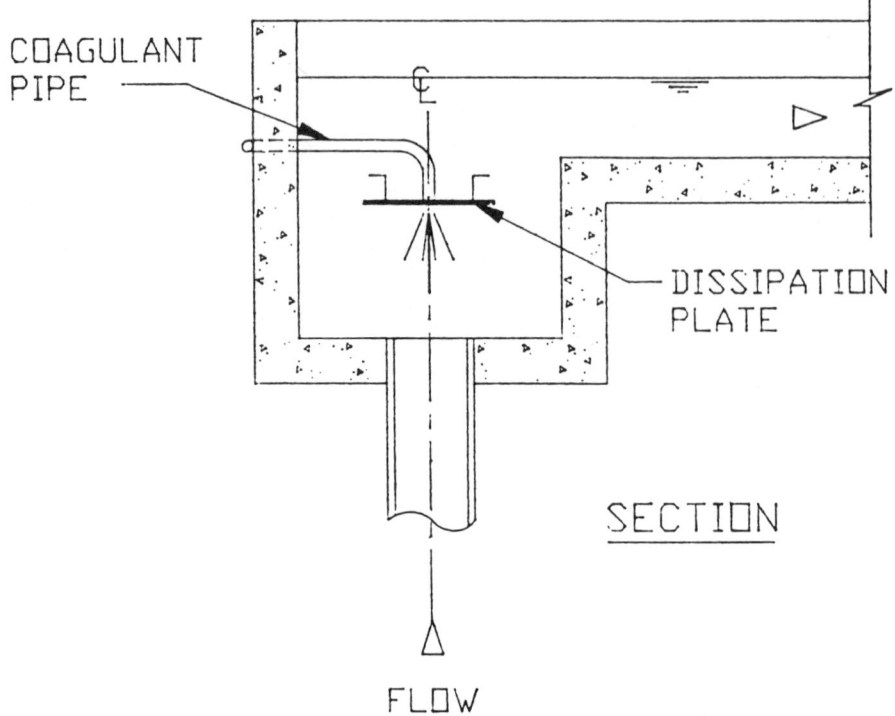

Figure 11-3 Energy-dissipation rapid mixer.

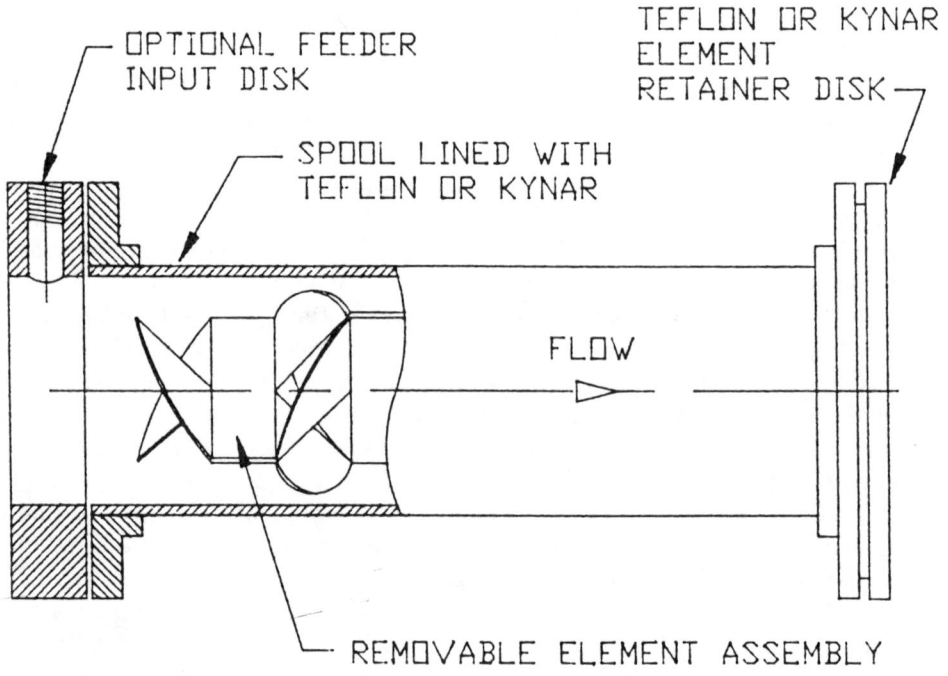

Courtesy of Komax Systems, Inc.

Figure 11-4 Motionless (static) mixer.

Mechanical mixing. Mechanical mixing systems in the context of this text include any mixing device that incorporates a mechanical device and moving parts. These include:

- completely mixed chamber,
- pump,
- pump jet diffusion,
- vertical turbine,
- horizontal turbine, and
- in-line mixers.

Completely mixed chamber. The mechanical mixer for this system consists of a mixing chamber of 10- to 60-s detention time in which a vertical shaft impeller is located; this method is the more conventional approach to rapid mixing. The entrance and exit locations will vary, depending on designer preference or plant configuration. Often the flow enters into the chamber at the bottom and exits at the top to one side. The impeller design may be a flat-blade turbine radial-flow type, pitched-blade axial-flow type, or hydrofoil. If the mixing box is deep, two blades may be required. Backmixing is a common feature of this type of mixer and is therefore appropriate for sweep floc coagulation when the mixing time is limited to seconds. Figure 11-5 shows the arrangement of a completely mixed rapid mixer.

Pump mixing. When low-lift raw water pumps are located at the treatment plant, these can be used for chemical dispersion, with chemicals injected into the suction pipe just upstream of the pump. Care must be taken that the chemicals used will not corrode the pump impeller, that the distance to the flocculation basin is not too great, or that turbulent conditions result in floc breakup.

Pump injection diffusion. In this application, a pump takes water from the raw water flow and pumps the water through an injection nozzle directly against the flow of the balance of the raw water confined in an inlet pipe. The coagulant chemical is injected into the pumped flow as close to the nozzle as possible (JM Montgomery 1985). Where energy recovery through a hydroelectric power generation system is practiced at the plant site, a high-pressure bypass line may be installed to provide the high-pressure water in lieu of the pump. A rate-of-flow controller can be included to control the rate of high-pressure flow to match changing raw water flows. Figures 11-6 and 11-7 show typical arrangements for this type of mixing system.

Shrouded radial turbine mixer. This type of mixer was designed to allow all of the mixing to occur within the confines of the turbine. This method of mixing prevents backmixing, thus confining the mixing time to fractions of a second. The mixing of the coagulant occurs within the volume of a shrouded radial-flow turbine. The turbine blades can be straight or curved. Figure 11-8 diagrammatically shows this type of mixer. The turbine mixer is appropriate for the adsorption–destabilization mechanism of coagulation because of the very short mixing time.

Selection Criteria

The type of mixing system used will depend upon a number of factors: the location of the plant; the characteristics of the raw water; whether the mixer is going into a new or a modified plant; the intended mixing operation; the treatment process; and if adsorption–destabilization or sweep floc coagulation is the major mode of coagulation.

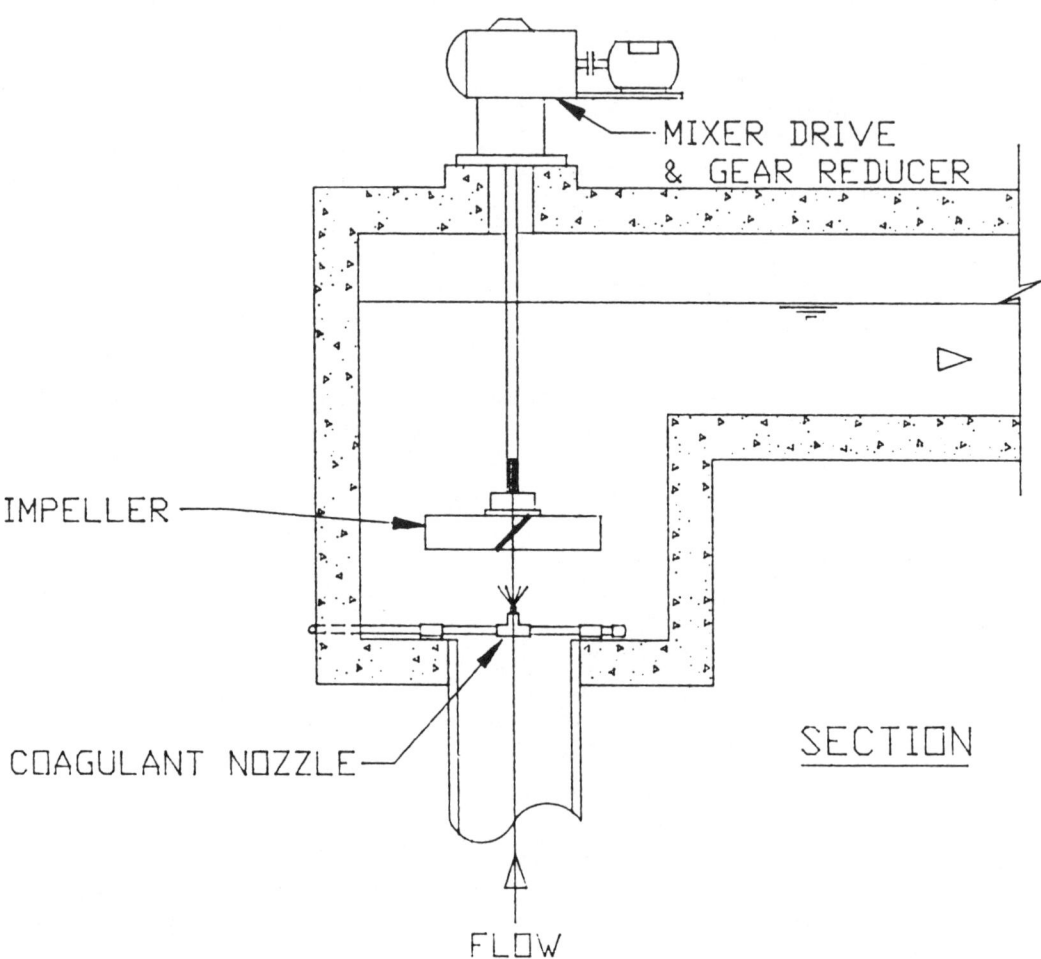

Figure 11-5 Completely mixed rapid mixer.

For example, in a developing country where the use of mechanical equipment should be minimized, a Parshall flume would effectively meet the joint function of plant flow measuring requirements and coagulant chemical mixing. This concept may also be appropriate for smaller plants where a simple operation is desired. Where there is excessive head at the headworks of a plant that must be dissipated, this excess energy should be considered for use in hydraulic mixing in lieu of a separate mechanical mixer. For plants where flow variations are not excessive or where the water characteristics are relatively stable, a motionless mixer should be evaluated.

When raw water coagulation jar tests indicate that the major mechanism of coagulation is sweep floc coagulation, then a backmix-type mixer should be designed. On the other hand, if the tests indicate that adsorption–destabilization is the predominant mode of coagulation, then a high-intensity mixer of plug-flow configuration should be designed.

Two-stage mixing, where two or more chemicals are used, may prove beneficial in some instances. Two-stage mixing would be appropriate when a water of very low

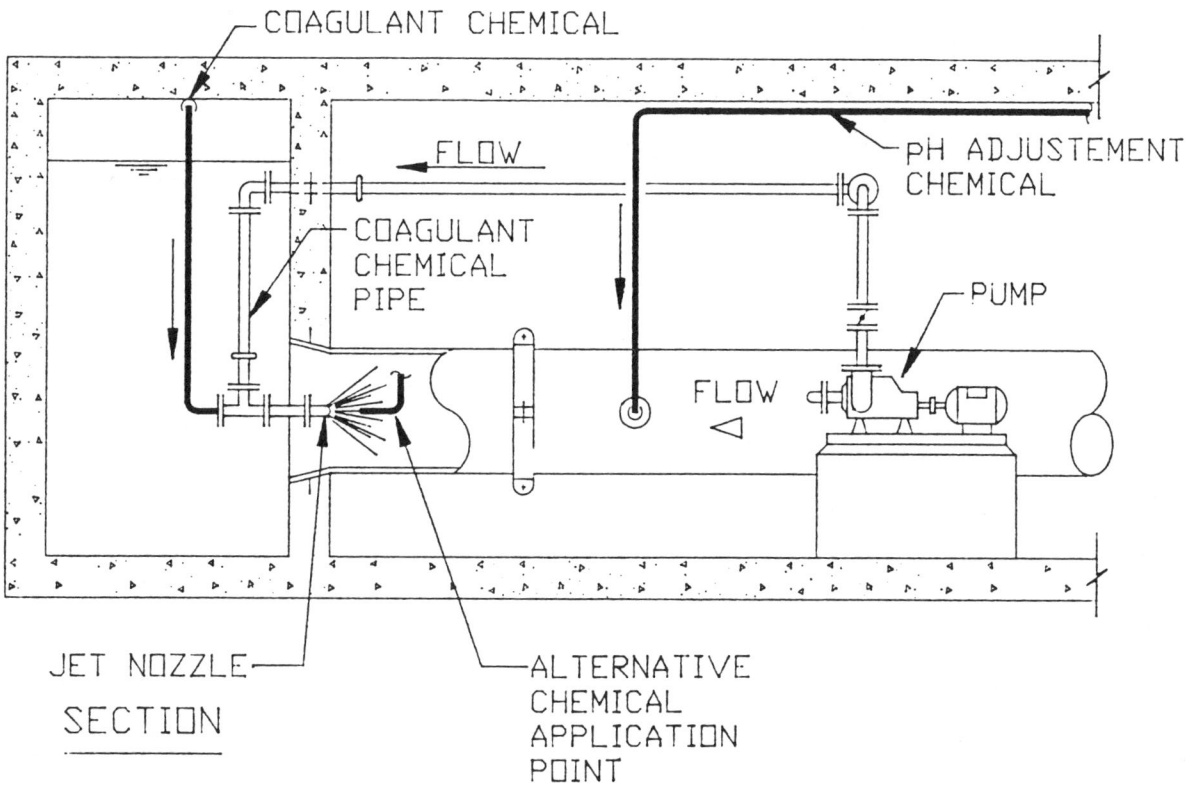

Figure 11-6 Pump injection rapid mixer.

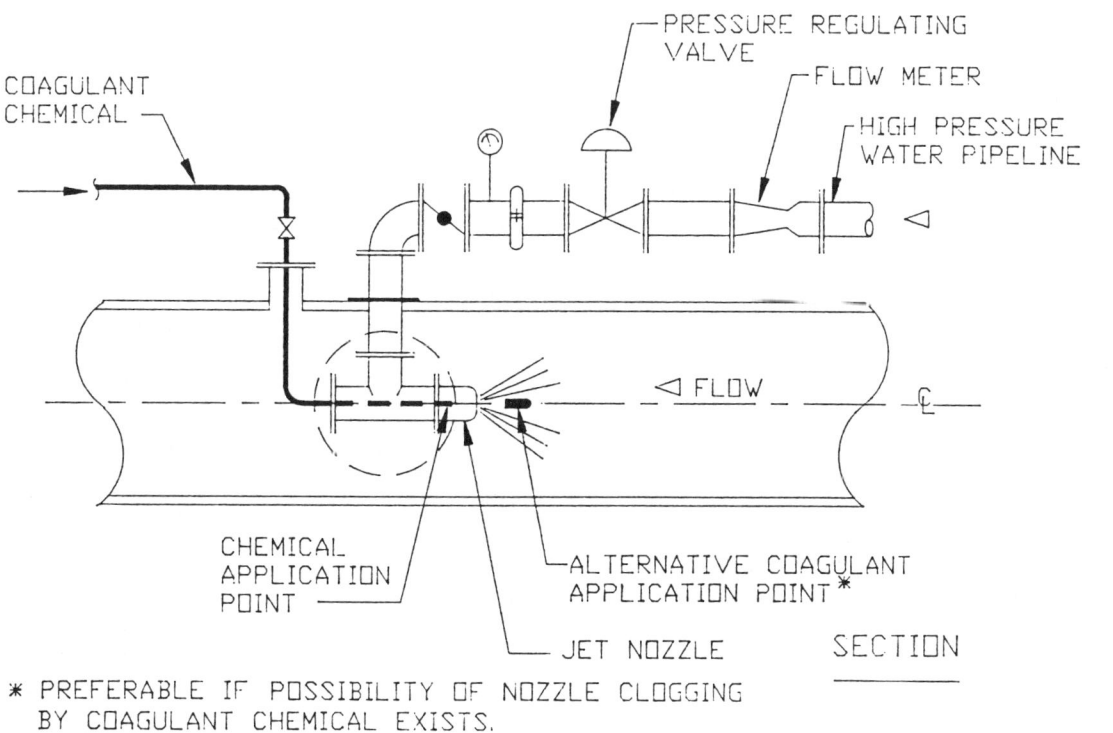

Figure 11-7 High-pressure water injection rapid mixer.

turbidity is to be treated by adding clay particles as nuclei for floc building. The clay should be added to the water in the first mixer, and the coagulant chemical to the second. Similarly, if a coagulant aid is to be used, then the aid may best be added at the second mixer. However, testing should be done to verify the best application point for some aids. Activated silica may be better added before the primary coagulant in treating low-turbidity water, as the activated silica can provide additional nuclei for flocculation. Added after the primary coagulant, activated silica can

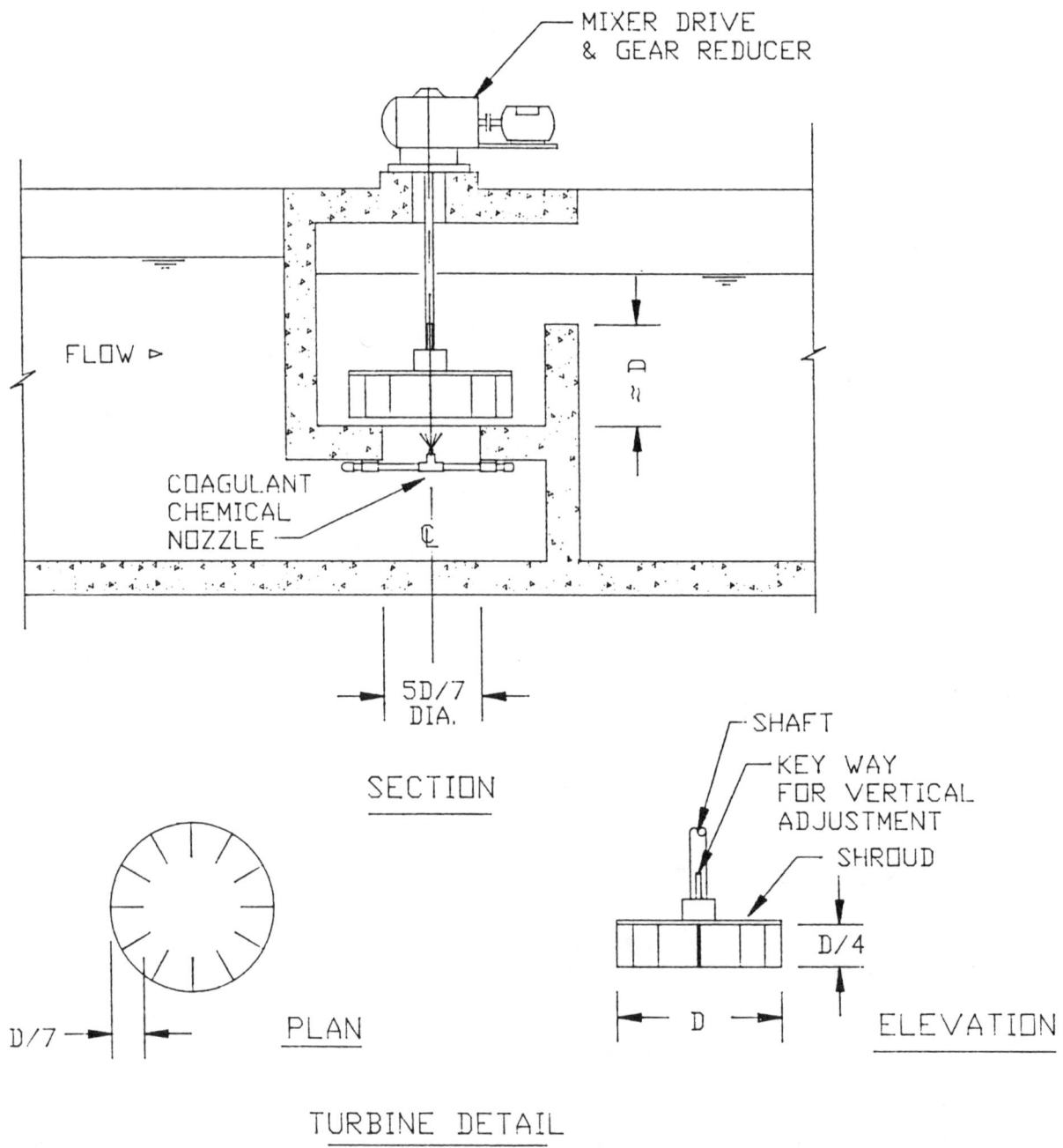

Figure 11-8 Radial turbine rapid mixer.

help in binding smaller flocs into larger and denser agglomerates. For the same reason, lime for pH adjustment added upstream may be more beneficial to the overall treatment process, provided the adjusted pH is appropriate for coagulation. Caution should be exercised when coagulating low-alkalinity water with alum.

Principles of Design

To effectively achieve the dispersion of chemical coagulants into raw water, intense turbulence must be generated in the water at the application point. The amount of energy dissipation and the detention time required to disperse the chemicals into the water have not been well-defined. The parameter most frequently used to express energy input is the mean velocity gradient G as given by Camp and Stein (1943):

$$G = \left(\frac{P_w}{\mu V}\right)^{1/2} s^{-1} \qquad \text{(Eq 11-1)}$$

Where:

P_w = the energy dissipated into the water in foot-pounds per second (kilogram-metre squared per cubic second [watts])

V = the volume, in cubic feet (cubic metres), of the basin where the energy is applied

μ = the absolute (dynamic) viscosity of the liquid in pound-seconds per square foot (kilograms per metre-second)

Whereas "the velocity gradient, or G-value, concept is a gross, simplistic, and totally inadequate parameter for design of mixers" (Amirtharajah 1978), as there are currently no other design approaches available, this parameter will continue to be used to design mixer units.

However, there are still wide ranges of G-values that are recommended for use, anywhere from 300 s^{-1} to 1000 s^{-1}, even greater than 6000 s^{-1} for motionless mixers, while recommended detention times range from fractions of a second (Hudson and Wolfner 1967; Vrale and Jorden 1971; Kawamura 1973) to 1 or more min (Letterman, Quon, and Gemmell 1973; Camp 1968). Often the product of G and t values are cited for the design of rapid mixers, and these recommended values can range from 10,000 to 40,000. Furthermore, often the design approach recommended is to provide 0.25 to 1.0 water hp/mgd (0.05 to 0.20 W/m^3/d) (Hudson 1981).

Letterman, Quon, and Gemmell (1973) have recommended, based upon their experimental work, an empirical relationship for optimizing the rapid-mix operation:

$$Gt_{opt} C^{1.46} = 5.9 \times 10^6 \qquad \text{(Eq 11-2)}$$

Where:

t_{opt} = the optimum rapid mix period in seconds
C = the alum coagulant concentration in milligrams per litre

Two things should be recognized about this formula: the dominant mechanism appeared to be sweep coagulation, and the equation was not proven under other conditions (Amirtharajah 1978).

As can be seen, the design of rapid mixers is still somewhat arbitrary. In plants where operators are competent and enthusiastic to experiment and optimize the treatment processes, flexibility should be built into the design of the mixers. For example, it may be prudent to provide mechanical rapid mixers with variable-speed drives to allow for variation in G-values for varying raw water flow and raw water characteristics, or when provision has been made in the plant for operating in the direct filtration mode (when the mechanism of adsorption–destabilization coagulation would be appropriate).

The energy requirements for mixing have also been expressed by Rushton (1952) by the following relationships:

$$P_w = \frac{k_2}{g} wN^3D^5 \text{ ft-lb/s } (= kwN^3D^5 \text{ kg-m}^2/\text{s}^3 \text{ [watts]}) \quad \text{(Eq 11-3)}$$

Where:

k_2 = a constant
g = the acceleration of gravity, 32.2 ft/s^2 (9.81 m/s^2)
w = the weight of fluid in pounds per cubic foot (kilograms per metre cubed)
N = the revolutions per second
D = the diameter of the impeller in feet (metres)

From Eq 11-1 and 11-3, it can be seen that:

$$G \propto N^{3/2} \quad \text{(Eq 11-4)}$$

Thus, the speed range of the mixer drive for possible velocity gradients for optimum Gt-values can be calculated from:

$$N_2 = N_1 \left(\frac{G_2}{G_1}\right)^{2/3} \quad \text{(Eq 11-5)}$$

Often it is desirable to have a degree of flexibility in the energy input to a mixer or the Gt-value. A variable-speed motor will give a full range of variability. However, a two- or three-speed motor may be a more economical alternative and give adequate flexibility.

Hydraulic mixing.

Weirs. For effective mixing, the head loss at the weir should be at least 4 in. (102 mm). Care must be taken to ensure that the coagulant chemical and aid are distributed uniformly across the weir width. If the head loss across the weir is excessive, large amounts of air can be entrapped in the water. An inadequate stilling area downstream of the weir will encourage the air to be carried to the next treatment unit and hinder the process (e.g., in a flocculation/clarifier unit).

Hydraulic jumps. For the formation of a hydraulic jump in a channel, the ratio of the upstream water depth (L_1) to the water depth downstream of the jump (L_2) must be greater than 2.4. The relative depths can be calculated from the following equation (Fair, Geyer, and Okun 1968):

$$\frac{L_2}{L_1} = \frac{1}{2}[(1 + 8F^2)^{1/2} - 1] \quad \text{(Eq 11-6)}$$

Where:

F (Froude number) $= \dfrac{v}{(gL_1)^{1/2}}$

v = the velocity of flow upstream of the jump in feet per second (metres per second)

g = the gravitational constant, 32.2 ft/s^2 (9.81 m/s^2)

To ensure the location of the hydraulic jump for siting the coagulant chemical diffuser, a drop in the floor must be provided. Standard hydraulic texts provide details of the configuration of Parshall and Palmer–Bowlus type flumes.

Static mixers. Static-mixer sizing should be based on the Gt-value required and selected from manufacturers' data. Mixers can be supplied with or without injectors and nozzles. Multiple elements can be furnished for adding different chemicals (e.g., primary coagulant, coagulant aid, and soda ash). The installation should be designed for removal of the complete mixer or the individual elements. Mixers can be custom fabricated from a variety of materials.

Mechanical mixing.

Completely mixed chamber. The water horsepower for a completely mixed chamber-type mixer can be calculated from Eq 11-1, water horsepower (hp) equals

$$\dfrac{P_w \text{ ft-lb/s}}{550 \text{ ft-lb/s/hp}} \quad \left(= \dfrac{P_w \text{ kg-m}^2/\text{s}^3}{1000 \text{ W/kW}} \right)$$

The motor horsepower is found by dividing the water horsepower by the efficiencies of both the gear reducer and the motor. These values can be obtained from equipment suppliers. The final motor selection will be the next standard size larger.

The velocity gradient G used in Eq 11-1 can range between 300 and 1000 s^{-1}. Here, engineering judgment must be used, and the choice will depend on the type of water to be treated. *Water Treatment Plant Design* (American Water Works Association 1969) suggests values of G and Gt for different types of raw water characteristics and treatment. The lowest anticipated water temperature must also be identified so that the corresponding value for dynamic viscosity can be used in Eq 11-1. The normal design approach is to select and specify the G and maximum motor horsepower. The selected mixer speed can be between 51 and 125 rpm, and the pumping capacity of the mixer Q, in gallons per minute (cubic metres per second), should be 1.5 times the flow through the basin:

$$Q = 7.48\, N_q N D^3 \text{ gpm} \quad \left(= \dfrac{N_q N D^3}{60} \text{ m}^3/\text{s} \right) \qquad \text{(Eq 11-7)}$$

Where:

N_q = the impeller pumping number
D = the impeller diameter in feet (metres)

The D/T_e ratio should be between 0.25 and 0.40:

$$T_e = 1.13\, \sqrt{L_L \times L_W} \qquad \text{(Eq 11-8)}$$

392　MIXING

Where:

T_e = the equivalent tank diameter
L_L = the length of the chamber
L_W = the width

Pump injection diffusion. The first step in the design of a pump injection (jet) diffusion system is to calculate the water horsepower at the lowest expected water temperature by rearranging Eq 11-1 and converting P_w from foot-pounds per second to horsepower (kg-m²/s³ to kW) as follows:

$$P_w = \frac{G^2 \mu V}{550} \text{ hp} \quad \left(= \frac{G^2 \mu V}{1000} \text{ kW} \right) \quad \text{(Eq 11-9)}$$

Where:

V = the mixing volume in cubic feet (cubic metres)

The volume will vary with the angle of the nozzle. The nozzle angle should be 90° or greater. The mixing zone volume (V) with a spray angle of 90° can be calculated as follows:

$$V = \frac{\pi}{24} D^3 \text{ ft}^3 \text{ (m}^3\text{)} \quad \text{(Eq 11-10)}$$

Where:

D = the diameter of the raw water pipeline in feet (metres)

The pump horsepower $= \dfrac{P_w}{E}$ hp (kW) (Eq 11-11)

Where:

E = the efficiency of the pump

The pump or mixing water capacity q (cubic feet per second [cubic metres per second]) should be between 2 and 4 percent of the raw water flow or should result in a coagulant dilution of at least 500 to 1.

The nozzle velocity head H (feet [metres]) can be calculated from the following equation:

$$H = \frac{P_w \, 550}{wq} \text{ ft} \quad \left(= \frac{1000 \, P_w}{wgq} \text{ m} \right) \quad \text{(Eq 11-12)}$$

and the orifice velocity $v_o = \sqrt{2gH}$ ft/s (m/s) (Eq 11-13)

Where:

w = the weight of fluid in pounds per cubic foot (kilograms per cubic metre)

Next, calculate the orifice size and then select a full-cone spray-pattern stainless-steel nozzle that meets the above criteria. Now the energy imparted by the nozzle can be found from the following expression:

$$P_w = \frac{C_d A v_o^3 w}{2g\,(550)} \text{ hp} \quad \left(= \frac{C_d A v_o^3 w}{2 \times 1000} \text{ kW} \right) \quad \text{(Eq 11-14)}$$

Where:

C_d = the coefficient of discharge (0.75)
A = the area of the nozzle in square feet (square metres)
v_o = the jet velocity in feet per second (metres per second)
w = the unit weight of water in pounds per cubic foot (kilograms per cubic metre)

P_w must be equal to or greater than the water horsepower of Eq 11-9. If the jet horsepower is less, then use the water horsepower to calculate the needed velocity and therefore the new mixing water capacity. From the nozzle manufacturer's tables, identify the pressure head needed for the selected nozzle and flow.

Next check that the velocity gradient G is equal to or greater than the selected value. The pump selection can be made using the total dynamic head based upon the pipe-friction head plus the nozzle-pressure head.

Shrouded radial turbine mixers. The design of a shrouded turbine mixer as shown in Figure 11-8 is based on the following formulae after Kalinske (Monk and Willis 1987):

$$D = \left(\frac{16tQ}{\pi} \right)^{1/3} \text{ ft (m)} \quad \text{(Eq 11-15)}$$

Where:

D = the turbine diameter in feet (metres)
t = the detention time of flow Q in the turbine in seconds
Q = the raw water flow in cubic feet per second (cubic metres per second)

Values for t should be small, say 0.25 s.

The height of the turbine blade $h = D/4$ ft (m).

The peripheral velocity $v_p = \dfrac{D}{2t}$ ft/s (m/s) \hfill (Eq 11-16)

The motor horsepower P_m (kW) (assuming 50 percent wire-to-water efficiency) can be found from the following:

$$P_m = \frac{Qwv_p^2}{550g} \text{ hp} \quad \left(= \frac{Qwv_p^2}{1000} \text{ kW} \right) \quad \text{(Eq 11-17)}$$

and the rotational speed of the turbine can be found as follows:

$$N = \frac{60v_p}{\pi D} \text{ rpm} \quad \text{(Eq 11-18)}$$

If a lower rpm is needed to match a standard gear reducer output, then go back and calculate the turbine size. Care must be taken to calculate head losses, and it is advisable to specify that the turbine be fixed to the shaft by means of a keyway that will allow up to a 12-in. (305-mm) vertical adjustment. The velocity gradient G can be calculated as follows:

$$G = \left[\frac{550P_mE}{\mu V}\right]^{1/2} = \left[\frac{1000P_mE}{\mu V}\right]^{1/2} \quad \text{(Eq 11-19)}$$

Where:

V = the volume of the turbine = $\frac{\pi D^3}{16}$ ft^3 (m^3)

E = the wire to water efficiency

The ratio D/T_e should be between 0.4 and 0.6.

Monitoring

There are five recognized methods of monitoring the effectiveness of chemical coagulation. Four methods, zeta potential meters, pilot filters, jar tests, and floc size observation, are used for monitoring the physical performance of chemical coagulation and flocculation. The fifth method, streaming current detection, produces an electrokinetic charge that can be used not only to measure the effectiveness of the process but to automatically control the amount of the chemical coagulant added to the raw water. Of course, in many plants "eyeballing" is still the operator's standard test for coagulation–flocculation–settling performance. A better procedure is to sample or continuously monitor the turbidity of the settled water.

Most operators still rely on the standard jar test to determine the optimum coagulant dosage on a daily basis, or in flashy river situations as often as every 2 h. Floc size is observed and the turbidity of decanted water taken after a 30-min standing period following flocculation. Floc size can also be observed by taking samples of flocculated water entering the sedimentation basin (or filter box in a direct filtration operation). The disadvantage of these methods is the time necessary to do the tests or make the observation and initiate changes.

Pilot columns are a comparative monitoring procedure. Coagulated water samples are continuously filtered through a pilot filter and the turbidity of the filtered water monitored. Two filters are normally used and each is backwashed frequently, every 45 min or so, to maintain uniform conditions. Any variation in the filtered water turbidity signals the operator to increase (or decrease) the coagulant chemicals. By this trial-and-error process, the "optimized" coagulant dosage is maintained. Pilot filters are primarily used in the direct filtration treatment process. In at least one large direct filtration water treatment plant (Monscvitz and Rexing 1983), a particle counter is also used in conjunction with laboratory coagulation, flocculation, and filtering to optimize the direct filtration process.

Zeta potential (ZP) monitoring is a controversial procedure. However, many operators use the system and swear by it (Neuman 1981). A zeta potential meter measures the negative charge of the colloidal particles in the raw water and effectiveness of the destabilization effects of cationic coagulant chemicals such as Al^{+3} and Fe^{+3}. "The control procedure requires monitoring the ZP of the coagulated water and changing the chemical dosage when the ZP varies outside a range known to produce the lowest turbidity. This range, though variable from plant to plant, is often −6 to −10 millivolts" (American Water Works Association 1984). Table 11-1 shows the degree of coagulation that occurs within various ZP ranges.

The streaming current detector (SCD) is a continuously sampling, on-line instrument that enables operators to know the optimum coagulant dosage needed. Streaming current is a measurable electric current that is generated when particles in water are temporarily immobilized, and the bulk liquid is forced to flow past the particles. Since the amount of current generated is proportional to the charge remaining after coagulants have been added, it is the algebraic sum of the currents generated, and more subjective than ZP measurements. An SCD samples continuously; therefore, the coagulant chemical dosage can be controlled automatically by means of a 4-20 mA output.

Practical Issues

Once the type of mixer has been selected, care must be taken as to the location and method of the chemical coagulant and aid application. The iron salt coagulants should be fed to the raw water at concentrations of 0.5 percent or less for best coagulation effects (Amirtharajah 1978). However, at this concentration, precipitation buildup will occur, and the final dilution should be done as close to the injection point as possible. To prevent pipe clogging, metal coagulants should not be diluted below 2 percent solution (Monk and Willis 1987). Provision must be made for easy cleaning of diffuser orifices and nozzles.

The addition of chemicals for pH adjustment can be made upstream or downstream of the rapid mixer. When possible, pilot-plant studies should identify the most effective location. Similarly, testing the advantages of two-stage mixing can best be done with pilot-plant studies.

In determining the motor horsepower of a mixer, the efficiency of both the gear reducer and the motor must be taken into account (i.e., motor horsepower will equal the water horsepower divided by the product of the gear reducer and motor efficiencies).

Specifications. The specification should clearly state whether the installation is indoors or outdoors; the ambient temperature extremes; and that the mixers are for 24-h/day and 7-day/week operation. The specification should include a manufacturer's minimum years of mixer design and manufacturing experience clause, and that the manufacturer shall be responsible for the supply of the entire unit and factory assembly and testing of the mixer.

Gear reducer. The gear reducer is the most critical part of the mixer, and the specification should spell out the quality required. Included should be: that the unit shall be designed and built for continuous 24-h/day, 7-day/week operation; a minimum American Gear Manufacturers Association (AGMA) service factor of 1.5 based

Table 11-1 Degree of Coagulation Within Different Zeta Potential Ranges

Average Zeta Potential	Degree of Coagulation
+3 to 0	Maximum
−1 to −4	Excellent
−5 to −10	Fair
−11 to −20	Poor
−21 to −30	Virtually none

Source: American Water Works Association. 1984. Introduction to Water Treatment: Principles and Practices of Water Supply Operations. *AWWA, Denver, Colo.*

on motor nameplate horsepower (2.5 for designs utilizing bearings in the gearbox for shaft support); a minimum B-10 life of 100,000 h based on Anti-Friction Bearing Manufacturers' Association (AFBMA) standards for bearings; an AGMA Standard 390.03 quality 10 or better for gears; and gear drives designed in accordance with AGMA Standard 420.04. Care must be taken to ensure that the lubrication system is adequate. Splash lubrication is normal and may not be adequate for low-speed units where an oil pump for lubrication may be needed.

Motors. Motors should be specified with a service factor of at least 1.15, and continuous-duty high-efficiency motors are advisable. Where plants are located above 3000 ft (914 m) the motor specification must spell out specific requirements for cooling the motor. Although the entire cost of high-efficiency motors may not be justified by energy savings, other advantages such as lower operating temperatures and noise levels should be evaluated. High-humidity and outdoor installations should dictate that motor heaters be specified. For lime softening applications, the impeller load should not exceed 80 percent of the motor nameplate horsepower to allow for calcium carbonate buildup.

Mixer shaft. The specification should require the operating speed to be equal to or less than 40 percent (60 percent for powdered activated carbon [PAC] and lime softening applications) of the first lateral critical speed; the stress in the impeller shaft should not exceed 8000 psi (55 MN/m^2) under maximum operating loads; and the mixer shaft should be connected to the reducer output shaft with a rigid flanged coupling.

Impeller. The specification must include: G-values; maximum tip speeds; type (and number) of impeller(s); the minimum superficial velocity at the maximum and minimum G-values; and the flow direction of the impeller (upward or downward). The stress in the impeller should not exceed 11,000 psi (76 MN/m^2) under maximum operating loads.

Special considerations. Care must be taken when mixers are used for applications other than "water." For example, for mixing and suspending PAC in a slurry tank, the specification must spell out the number of pounds (kilograms) of PAC per gallon (cubic metre) of water. The mixer drive unit should have a two-speed motor to allow for an initial feeding and mixing high speed (N = 56 to 84 rpm) and a lower speed (N = 45 to 63 rpm) for PAC suspension and be capable of resuspending PAC after several days of nonoperation of the mixer. The impeller should be sized to allow a superficial velocity (*SV*) as follows:

Pounds of PAC/gal	(Kilograms of PAC/m^3)	SV, ft/min	(m/s)
1.0	(121)	13	(4.0)
1.5	(182)	14	(4.3)
2.0	(243)	15	(4.6)

Superficial velocities can be calculated by use of Eq 11-39 (see p. 414). Two impellers should be specified. The upper impeller should be a downward-pumping, axial-flow, 32° or 45° pitched-blade turbine with a D/T_e ratio in the range of 0.20 to 0.25. The bottom impeller should be a radial-flow curved-blade turbine and have a D/T_e ratio in the range of 0.15 to 0.20.

The motor size should be based on the criteria of 1 hp per 1000 gal (197 W per m^3) of slurry at high speed. The center of the upper impeller should be two-thirds of the depth of the final slurry depth and the center of the lower impeller 12 to 18 in. (305 to 457 mm) off the floor level.

When mixers are used for lime slurry, the mixer design should be based on criteria for solids suspension, using the specific gravity of the lime slurry. The specific gravity can be calculated by the following expression:

$$SG = \frac{1}{\frac{C}{SG_L} + \frac{1-C}{SG_W}} \qquad \text{(Eq 11-20)}$$

Where:

- C = the lime slurry concentration by weight
- SG_L = the specific gravity of the lime (3.4 for limestone and 2.25 for lime or quicklime)
- SG_W = the specific gravity of the water (1.0)

The impeller must be sized for superficial velocities (SV) as follows:

Slurry Concentration, %	SV, ft/min	(m/s)
10	15	(4.6)
15	20	(6.1)
20	25	(7.6)
25	25	(7.6)
30	30	(9.1)
35	35	(10.7)

The impeller load should not exceed 80 percent of the motor horsepower to allow for calcium carbonate buildup on the impeller and shaft. The impeller should be located one-third of the impeller diameter (D) off the floor. If the ratio of L_D (depth of slurry) to T_e is equal to or greater than 0.60, add a second impeller. The recommended impellers are pitched-blade turbines, and D/T_e should be in the range of 0.28 to 0.33. The impeller speed N should be in the range of 45 to 84 rpm.

FLOCCULATION

Flocculation is the process step of particle agglomeration following the dispersion, hydrolysis, and polymerization of the coagulant by rapid mixing. In this step, the continuous agitation of the coagulated water occurs with less intensity. During this period, the minute coagulant ions are brought into contact with each other and colloids in the raw water, agglomerating progressively into larger and larger clusters, called floc, that can be separated by subsequent processes of sedimentation and/or filtration. The opposite-charged coagulant ions and natural colloids come together by adsorption and destabilization through contact, and the colloids may be entrapped within the aluminum and iron hydroxide precipitate. The reaction of the adsorption–destabilization mechanisms occur in less than 1 s, while the entrapped or sweep coagulation occurs in the range of 1 to 7 s.

The success of the flocculation process is first dependent on the effectiveness of the coagulation in the rapid mixer. For the flocculation process itself to be successful, the process must provide adequate time for the desired floc size and density to form. The stirring system must be adequate to mix the full volume of the basin to maximize collision opportunities for the floc and to prevent floc deposition on the basin floor, yet without localized high-intensity areas where the breakup of already formed floc can occur. The design must minimize short-circuiting through compartmentalization and incorporate appropriate inlet and outlet features. The detention time within the basin for the majority of the water should be as close as is practical to the theoretical contact time of flow divided by basin volume.

This chapter does not discuss mechanical flocculation in solids contact upflow clarifiers because of their proprietary nature. However, chap. 1 and Amirtharajah (1978) provide excellent treatment of the subject.

Flocculation Systems

The flocculation process requires energy to agitate the water to allow opportunity for floc contact and buildup. Energy can be provided by mechanical flocculation appliances; through energy dissipation by gravity flow of the coagulated water through specially designed hydraulic flocculation channels; or through energy dissipation of coagulated water into a sludge blanket. The alternative types of flocculators generally used are therefore grouped under hydraulic and mechanical flocculators.

Hydraulic flocculation. The most commonly used hydraulic flocculators are the baffled channel type. Less known and used in unindustrialized countries today is the hydraulic jet-action flocculator. While gravel-bed flocculators have been used in developing countries for many years, their application and evaluation in the United States are fairly recent.

Baffled channel flocculators. In baffled channel flocculators, stirring is achieved by changing the flow of the water through the channel with horizontal round-the-end or vertical under-and-over baffles. Because of the compartmentalized configuration, this type of flocculator operates close to plug-flow conditions with little or no short-circuiting problems. Figure 11-9 shows a typical tapered round-the-end hydraulic flocculation channel. MacDonald and Streicher (1977) record an innovative baffled channel flocculator that incorporates a tapered design and that

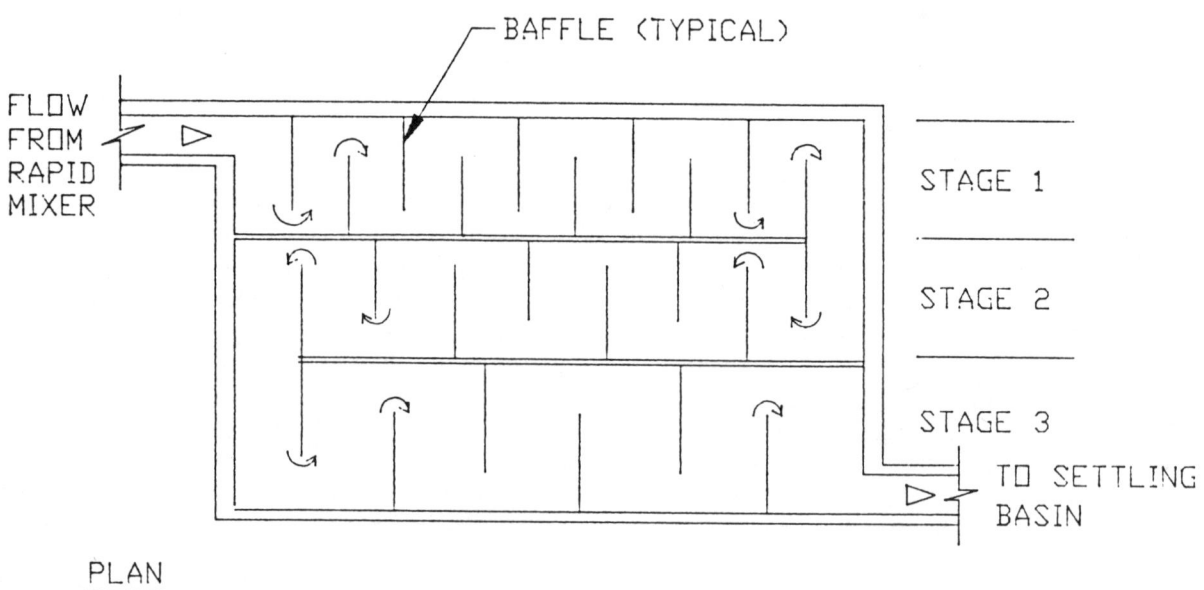

Figure 11-9 Tapered horizontal-flow hydraulic flocculator.

results in relatively little change in the Gt-value when wide swings in the flow rate occur.

Hydraulic jet-action flocculators. This type of flocculator uses the energy of the jet action of the inlet water to each of the multicompartmented flocculation basins. There are two types: the helicoidal-flow type, and the Alabama type. Both types have been used in the United States, although seldom if ever at present. Both types are still used in Latin America, while the helicoidal-flow type is now widely used in China (Schulz and Okun 1984). The helicoidal-flow type flocculator is shown in Figure 11-10; Figure 11-11 shows the Alabama-type flocculator.

Gravel-bed flocculation. Gravel-bed flocculators are a simple, effective, and inexpensive design for flocculation, and have been used successfully overseas. The process is being used in the United States by the Culligan International Company in two-stage filtration pressure filter package plants. Microfloc Products, Johnson Filtration Systems Inc., has also adopted the same concept using a buoyant plastic media in an upflow–downflow design and has numerous operating plants throughout the United States. A number of successful pilot plants in the United States have demonstrated the effectiveness of this hydraulic flocculation process (Kawamura 1985).

The large pore space in the gravel bed provides excellent conditions for the development and storage of agglomerated flocs within the interstices.

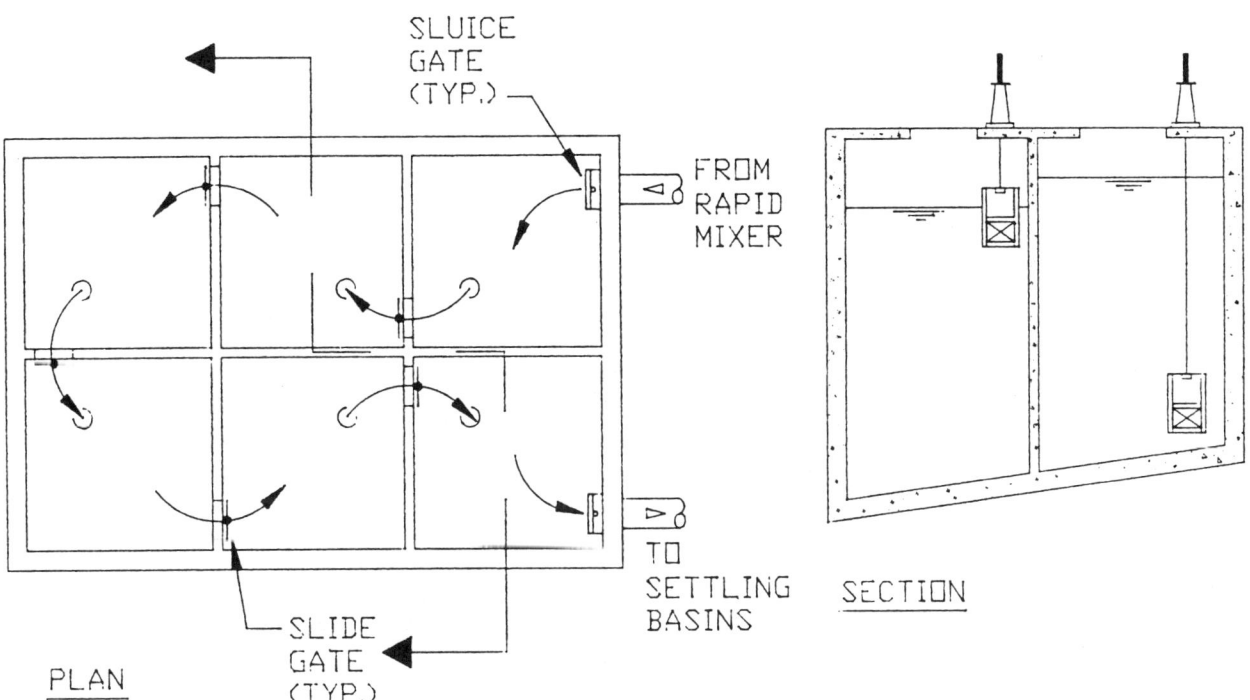

Figure 11-10 Helicoidal-flow flocculator.

Mechanical flocculation. A number of types of mechanical flocculation systems have been successfully used for various flocculation applications. The types of units that will be considered in this text include:

- horizontal shaft,
- vertical shaft, and
- reciprocating.

Horizontal shaft flocculators. Paddle-wheel flocculators are made up of shafts, drive units, and paddles. In some installations, the designs incorporate rotor paddles and starters. The shaft is positioned either transversely to the direction of flow or parallel to the flow. Horizontal paddle flocculators constructed parallel to the flow can be designed for stepped flocculation by varying the size and spacing of the paddles over the length of the shaft. Similarly, horizontal flocculators positioned transversely to the flow can be designed for different sized and spaced paddles on each of the three or four shafts; alternatively, the shafts may be operated at different speeds, each preceding shaft at a higher speed than the succeeding one. The motor for the horizontal shaft flocculators may be mounted in a dry well adjacent to the flocculation basin or on a platform above the water level. The horizontal shaft

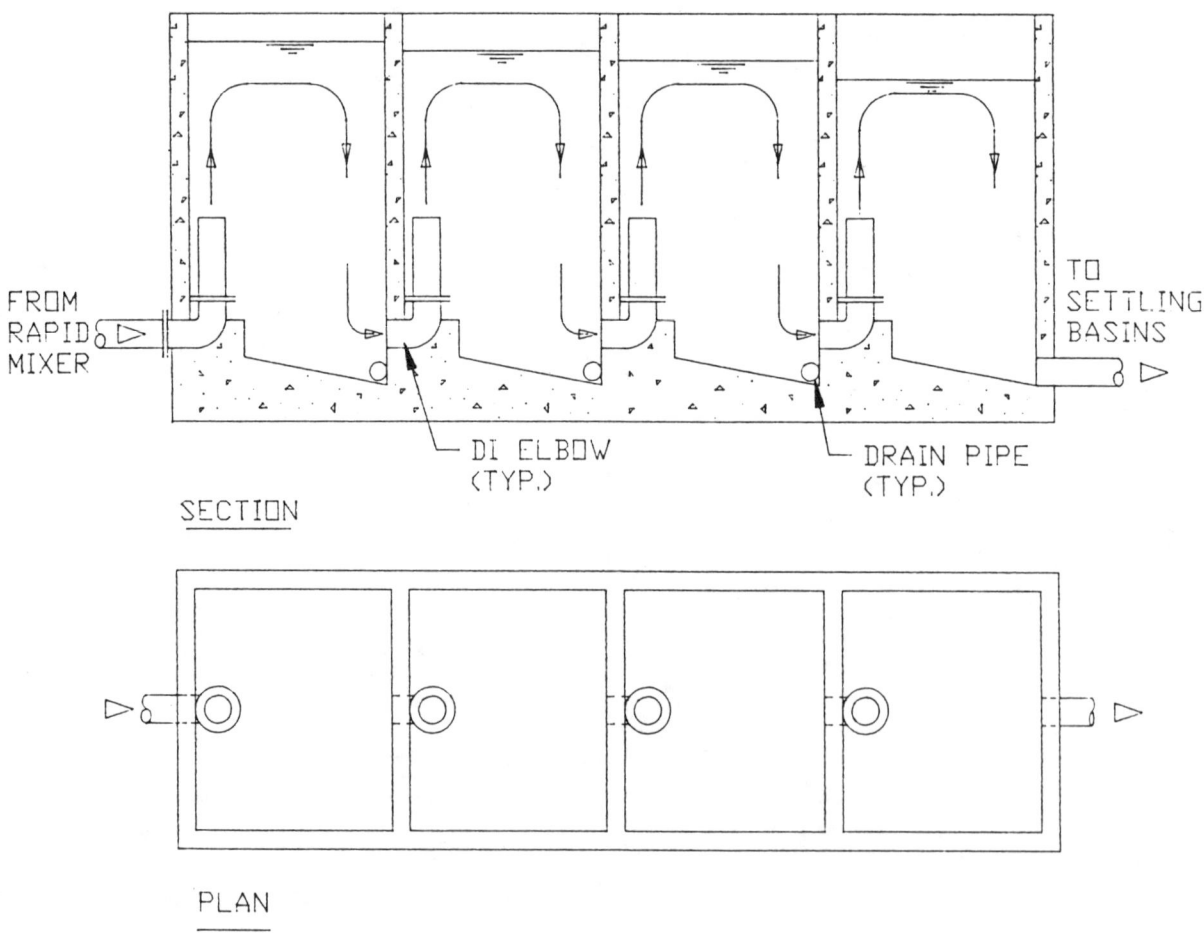

Figure 11-11 Alabama-type hydraulic flocculator.

units have underwater bearings. A typical transversely located paddle-wheel flocculator is illustrated in Figure 11-12.

In some instances, coarse wire mesh has been used instead of redwood or plastic slats, as it is claimed that the number of impingements or collisions between the growing floc and raw water colloids is increased manyfold (Riddick 1961). This type of flocculator is currently receiving attention in South American countries. Radial-flow type flocculators on horizontal shafts have been used in many installations, an example of which is shown in Figure 11-13.

Vertical shaft flocculators. Vertical shaft flocculators include radial-flow impellers and axial-flow impellers. The radial-flow type consists of vertical flat or curved blades mounted on a disk or a shroud (closed type) or to radial arms, or directly to

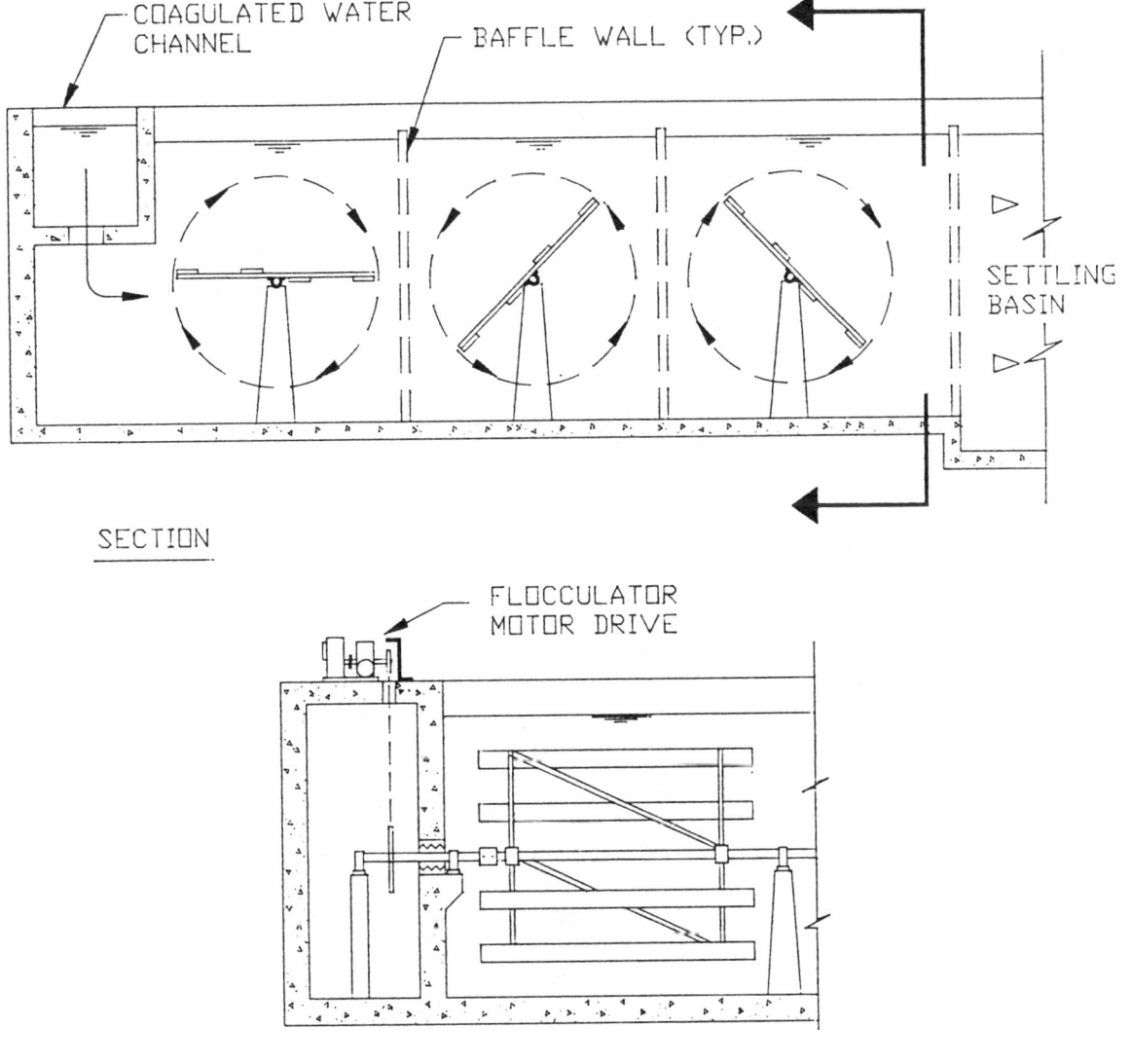

Figure 11-12 Transversely placed horizontal paddle-wheel flocculator.

the shaft hub (open type). Axial-flow impellers can be further classified as pitched-blade turbines or hydrofoils (propellers). Paddle-wheel flocculators have been used on vertical shaft installations but are not commonly used today. Figure 11-14 shows a modern high-efficiency hydrofoil type flocculator.

Tapered hydraulic flocculation has been successfully used in inverted truncated pyramid-shaped upflow solids-contact clarifiers (Amirtharajah 1978; Monk and Willis 1987). A flat-bottomed basin using the same principle (hydraulic tapered flocculation in a sludge blanket) was reported by Monk and Willis (1987).

Figure 11-13 Horizontal-shaft radial-flow flocculator.

MIXER DESIGN FOR TREATMENT PLANTS 403

Courtesy of Philadelphia Mixers, Palmyra, Pa.

Figure 11-14 Vertical shaft hydrofoil.

V-shaped support frame assembly. A number of these are fixed to a vertical connecting rod. The slats are moved up and down in a reciprocating motion by means of a drive motor, gear reducer, crank arms, connecting link lever, and beam. All the bearings, drive mechanisms, shafts, and guide devices are located above water level. Designs ensure full coverage of the basin and allow for compartmentalization. Figure 11-15 diagrammatically shows this type of floc conditioner, while Figure 11-16 shows an installed unit.

The oscillating flocculator has a pendulum motion. This type of flocculator is similar to the walking beam with slats on a frame suspended off a reciprocating horizontal shaft located above the water level. Figure 11-17 diagrammatically shows this type of flocculator. The shaft can be positioned either transversely to the direction of flow or parallel to the flow.

Selection Criteria

The selection of mechanical or hydraulic flocculation will be based more on the economics of facilities than on technical criteria. In locations where funds are limited or the importation of mechanical equipment is restricted, the selection of hydraulic flocculation is prudent. Hydraulic flocculation basins can be made from local materials, do not need a power supply, and require little operator attention.

Industrialized countries normally use mechanical flocculators because of their greater versatility and degree of agitation. This adds to the difficulty of selecting among the different types of mechanical flocculators. Often choices are based on the personal preference or experience of the engineer. Factors that should go into the final selection include: cost, both operating and capital; ease of maintenance; and the applicable treatment process. In lime softening, for example, the selected

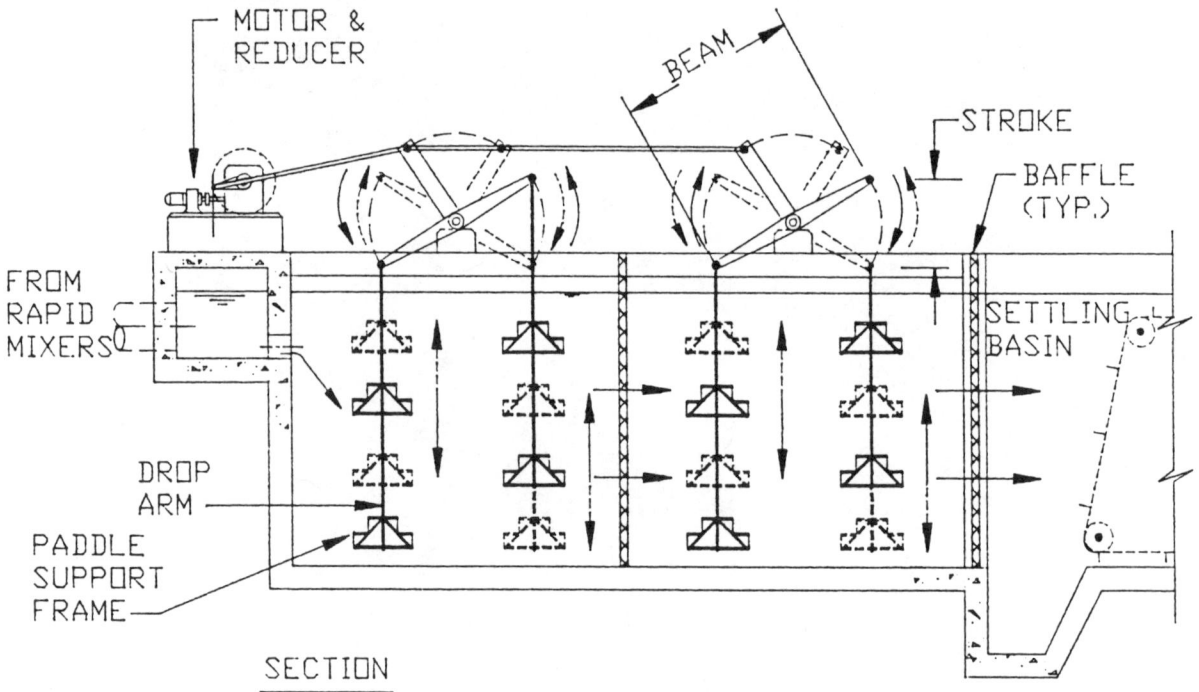

Figure 11-15 Walking beam flocculator diagram.

personal preference or experience of the engineer. Factors that should go into the final selection include: cost, both operating and capital; ease of maintenance; and the applicable treatment process. In lime softening, for example, the selected flocculators must be able to keep the heavy growing precipitate in suspension. This may be better achieved with an axial-flow flocculator than a horizontal paddle-wheel type.

Figure 11-18 schematically shows the estimated spatial distribution of velocity gradient values produced by paddle-wheel flocculators, radial-flow impellers, and axial-flow impellers throughout the volume of a flocculation basin. As can be seen, the variation from unduly low gradients to unduly high gradients is greater for paddle-wheel flocculators than for vertical shaft axial-flow flocculators. "Because too much mixing is as detrimental as too little, axial-flow impellers appear to be the preferred units for flocculation" (Hudson 1981).

Vertical shaft flocculators can and should be designed without underwater bearings. Mechanical flocculating impellers dissipate the energy imparted to them in two forms—liquid movement (pumping capacity) and shear. The perfect flocculator exhibits no shear—a ship's impeller is the closest design to this ideal. Axial-flow impellers better approach this ideal design, minimizing shear and local turbulence while maximizing bulk fluid motion.

Courtesy of Ralph B. Carter Company, Hackensack, N.J.

Figure 11-16 Installed walking beam flocculator.

values is by conducting pilot-plant studies. Ideally, multicompartmented flocculators with variable-speed agitators (and therefore variable G-values) should be used. Thus, by experimentation, the appropriate G-values for tapered flocculation, the number of compartments, and the flocculation time can be better identified for the selected process train. If pilot-plant studies are not conducted, the G- and t-values can be estimated from jar tests.

Principles of Design

Agitation of coagulated water accelerates the aggregation of colloidal particles. In agitated systems, fluid velocity varies both spatially and temporally. The spatial changes in velocity are characterized by the velocity gradient G. Particles that follow the fluid motion will have different velocities so that opportunities exist for

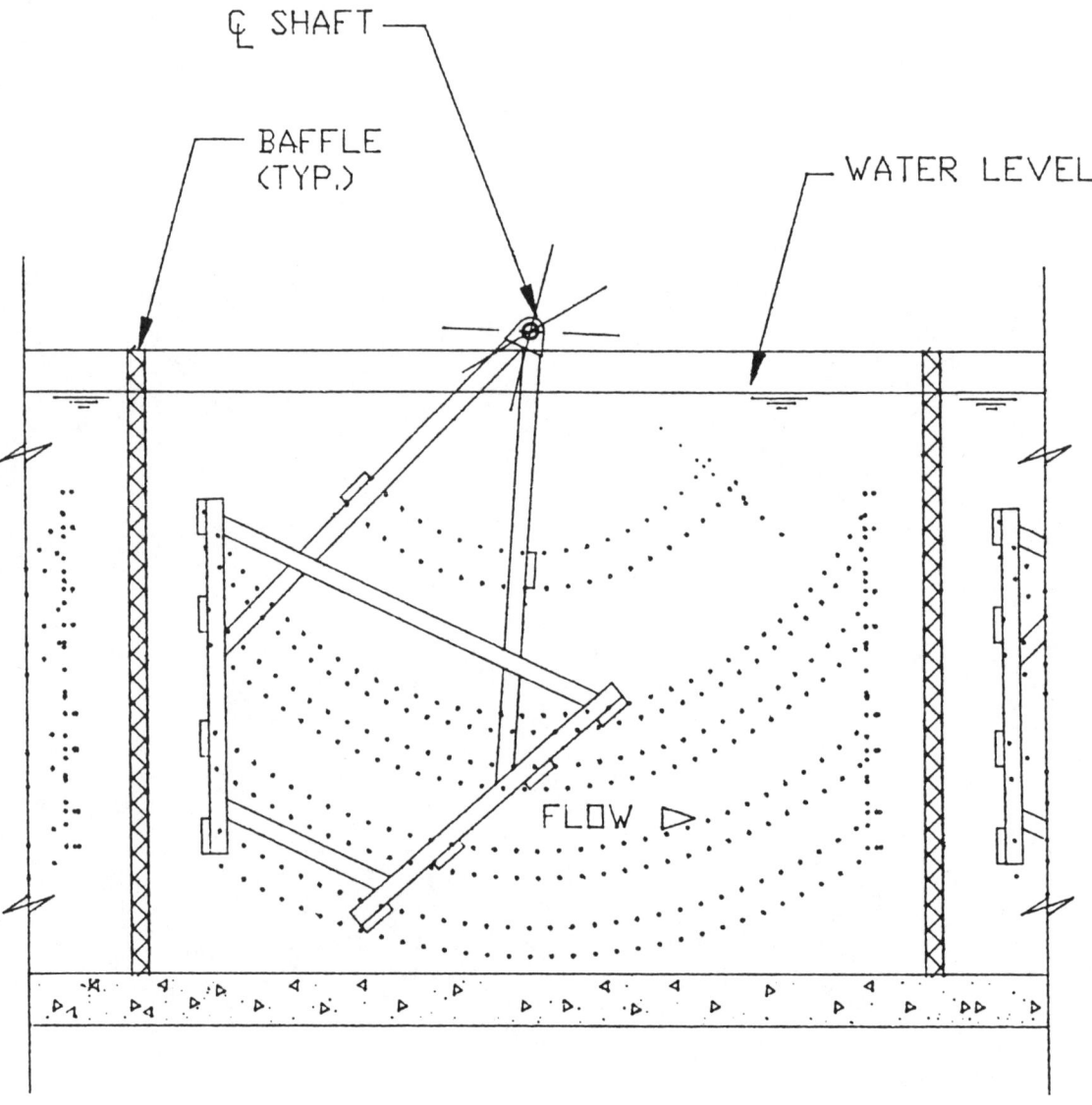

Figure 11-17 Oscillating type flocculator.

interparticle contacts. When contacts between particles are caused by fluid motion, the process is termed orthokinetic flocculation. The other mechanism of flocculation is perikinesis, which is aggregation resulting from random thermal motion of fluid molecules but is significant only for particles less than 1 or 2 µm. Orthokinetics is the predominant mechanism in water treatment flocculation.

For colloidal suspension of particles having a uniform particle size, the rate of collision between two particles with time due to orthokinetic flocculation can be characterized as follows:

$$\frac{dN}{dt} = \frac{G}{6} n_1 n_2 (d_1 + d_2)^3 \qquad \text{(Eq 11-21)}$$

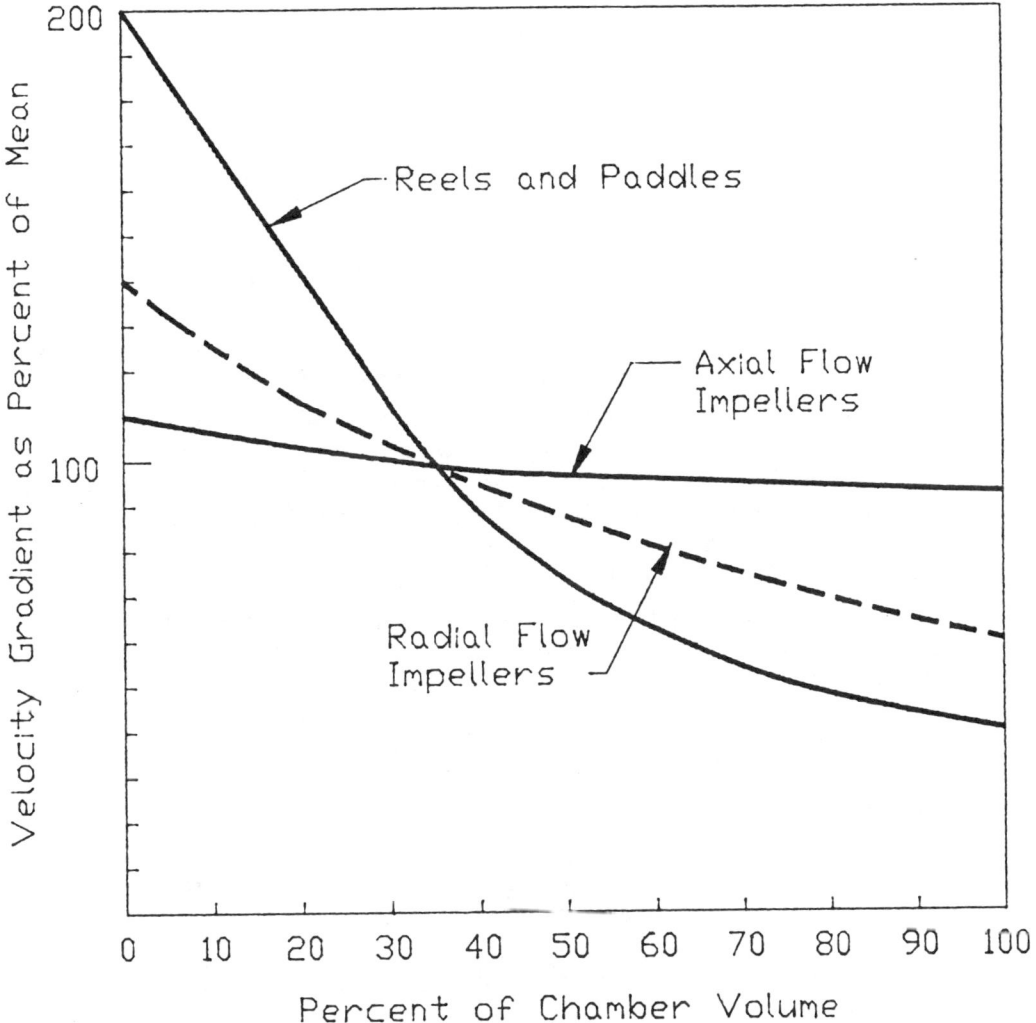

Source: Hudson, H.E. Jr. 1981. Water Clarification Processes: Practical Design and Evaluation, chap 6. Van Nostrand Reinhold, New York.

Figure 11-18 Assumed spatial distribution of velocity gradients for various impeller types.

Where:

G = the velocity gradient
n_1 and n_2 = the number density of the particles 1 and 2
d_1 and d_2 = the diameter of particles 1 and 2

Now the velocity gradient G for mechanical agitation can be calculated from Eq 11-1.

Similarly, the expression for G for hydraulic flocculation with a baffled basin can be determined from the following expression:

$$G = \left(\frac{wH}{\mu t}\right)^{1/2} \text{s}^{-1} \quad \left[= \left(\frac{wgH}{\mu t}\right)^{1/2} \text{s}^{-1} \right] \quad \text{(Eq 11-22)}$$

Where:

H = the head loss due to friction in feet (metres)
w = the weight of water in pounds per cubic foot (kilograms per cubic metre)
μ = the dynamic viscosity in pound-seconds per square foot (kilograms per metre-second)
t = the detention time in seconds
g = the acceleration of gravity, 32.2 ft/s² (9.81 m/s²)

Camp and Stein (1943) and others found the dimensionless product Gt to be a useful parameter for flocculator design and operation. Recommended values of Gt range from 2×10^4 to 2×10^5. However, consideration must be given to the type of treatment process. For example, for lime softening, the G- and t-values should both be higher; for a direct filtration process, the detention time will be less than for conventional treatment, whereas the G-values would be higher. For cold water, the detention time will be longer. Pilot-plant studies are advisable to identify these variables to produce more cost-effective designs, as detailed in chap. 10.

From bench-scale studies, Andreu-Villegas and Letterman (1976) developed an expression for an optimum value for G that is written as follows:

$$G_{opt} = \left(\frac{44 \times 10^5}{tC}\right)^{0.357} \quad \text{(Eq 11-23)}$$

Where:

C = the alum coagulant concentration in milligrams per litre in the 0 to 50-mg/L range

Many, for example, Camp (1955) and Kawamura (1976), have confirmed in practice that tapered flocculation with diminishing velocity gradients over the length of the flocculation basin is more effective than a uniform velocity gradient. Again, the value of the velocity gradient at the start of flocculation and degree of reduction over the length of the basin will depend on the overall treatment process.

Short-circuiting is an inherent problem in mechanical flocculation. To minimize this detrimental effect, compartmentalization of flocculation basins should be incorporated into any basin design. As can be seen from Figure 11-19, a single compartment results in approximately 40 percent of the inflow into a compartment passing through in less than half the theoretical detention time:

$$\frac{V}{Q}$$

Where:

Q = the rate of flow
V = the volume of the compartment

If the basin were divided into five compartments allowing for series flow, the arrangement would reduce this amount to about 10 percent. From a practical point of view, three or four compartments are recommended. Obviously, if short-circuiting is reduced, the detention time for flocculation can be reduced as more effective collision opportunities occur.

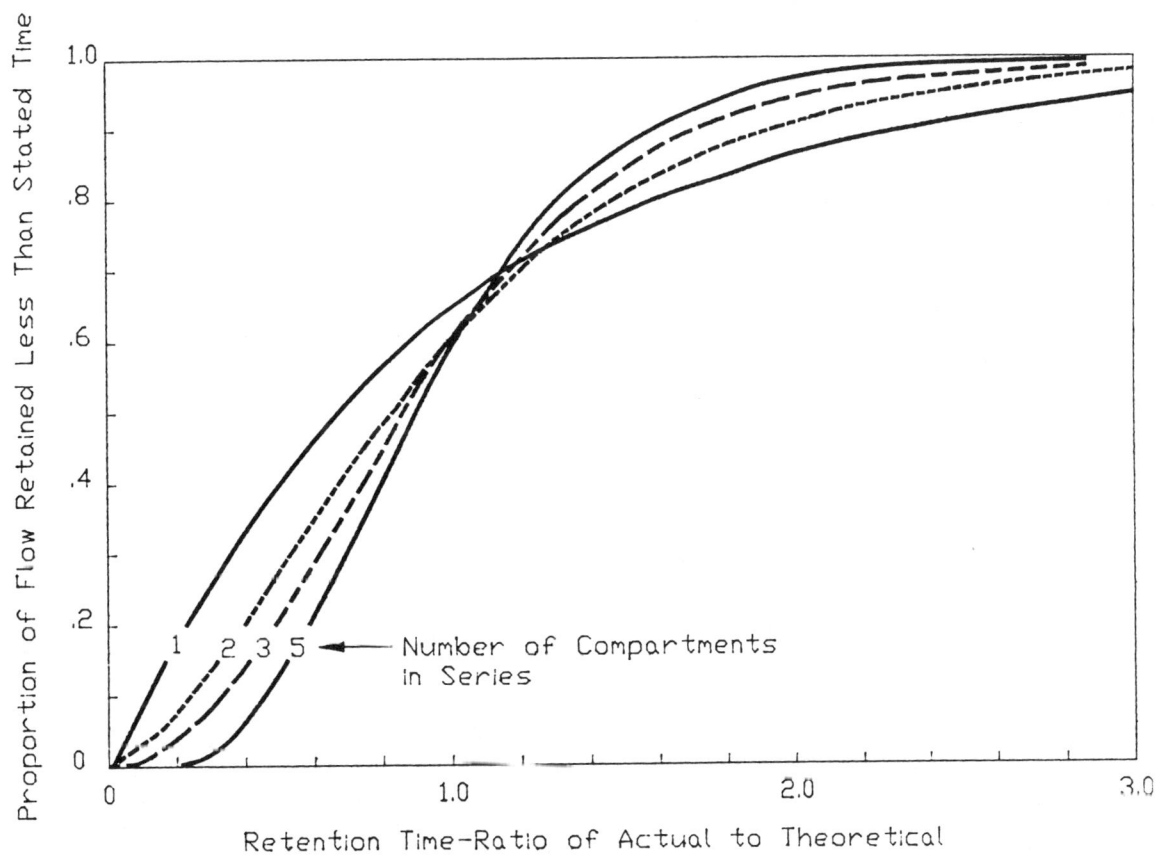

Source: American Water Works Association. 1969. Water Treatment Plant Design. McGraw-Hill, New York.

Figure 11-19 Short-circuiting in mixing compartments.

Hydraulic flocculation. For the design of hydraulic flocculation basins the total head loss per stage (for tapered flocculation) can be found from Eq 11-22. The head loss is the total of the head losses due to the water's 180° change of direction at the bends and frictional losses in the channels. The number of baffles needed to achieve a desired velocity gradient can be calculated from the following equations.

For horizontal units:

$$n = \left[\frac{2\mu t g}{w(1.44 + f)}\left(\frac{L_D L_L G}{Q}\right)^2\right]^{1/3} \qquad \text{(Eq 11-24)}$$

Where:

w	=	the weight of water in pounds per cubic foot (kilograms per cubic metre)
f	=	the coefficient of friction of the baffle material
L_L	=	the length of the basin in feet (metres)
Q	=	the plant flow in cubic feet per second (cubic metres per second)
L_D	=	the depth of water in feet (metres)
t	=	the time of flocculation in seconds

For vertical units:

$$n = \left[\frac{2\mu t g}{w(1.44 + f)}\left(\frac{L_W L_L G}{Q}\right)^2\right]^{1/3} \qquad \text{(Eq 11-25)}$$

Where:

L_W = the width of the basin

The head loss H, in feet, at each bend of either the horizontal or vertical units is approximated by the following formula:

$$H = K\frac{v^2}{2g} \text{ ft} \qquad \text{(Eq 11-26)}$$

Where:

v = fluid velocity in feet per second (metres per second)
K = an empirical constant (varying from 2.5 to 4)

Schultz and Okan (1984) give general guidelines for the design of the baffled channel, helicoidal, and Alabama-type hydraulic flocculators.

Gravel-bed flocculators. Gravel-bed flocculation has been successfully employed in many countries and more recently in the United States in proprietary package-type installations. There is not much design data available. The best approach for new facilities is to pilot the process to optimize the filtration rate, size, and depth of media, and the coagulant and aid chemical types and dosages.

However, enough is known about the dynamics of filtration to develop design data for initial medium selection. The intensity of the velocity gradient G in a filter bed depends on the filtration rate, the medium size, and the medium porosity. If changes in the medium porosity due to clogging in the filter are not accounted for, the value of the velocity gradient G in a filter can be calculated from the following expression:

$$G = \left(\frac{HwQ}{\mu\alpha V}\right)^{1/2} \quad \left[= \left(\frac{HwgQ}{\mu\alpha V}\right)^{1/2}\right] \quad \text{(Eq 11-27)}$$

Where:

H	=	the total head loss in feet (metres)
w	=	the weight of water per unit volume in pounds per cubic foot (kilograms per cubic metre)
Q	=	the filtration rate in cubic feet per second (cubic metres per second)
μ	=	the dynamic viscosity in pound-seconds per square foot (kilograms per metre-second)
α	=	the porosity of the media
V	=	the volume of the media in cubic feet (cubic metres)
g	=	the acceleration of gravity, 32.2 ft/s² (9.81 m/s²)

The head loss H can be calculated using the Kozency equation (Fair, Geyer, and Okun 1966) or, better, by observation in a pilot-plant study.

Gravel-bed flocculation is the first stage of the two-stage filtration process and takes the place of both "conventional" flocculation and sedimentation. G cannot be used to optimize the two-stage process; this can only be done through pilot-plant studies using the normal parameters of filtered water quality and production efficiency. The degree of flocculation and floc separation in the gravel-bed flocculator will affect the head loss rate and the load carried on to the second-stage filter.

The gravel-bed filter can be monomedium (from the authors' experience 6 mm effective size) or multimedia for tapered flocculation (Schulz and Okun 1984).

Mechanical flocculators.

Paddle-wheel type. For this type of flocculator, the useful power input (proportional to G^2) is a function of the drag of the paddles. This drag force (F_d) can be calculated from the following expression (Camp 1955):

$$F_d = C_D \frac{Awv^2}{2g} \text{ lb} \quad \left(= C_D \frac{Awv^2}{2} \text{ Newtons}\right) \quad \text{(Eq 11-28)}$$

Where:

C_D	=	the coefficient of drag
A	=	the area of the paddles in square feet (square metres)
w	=	the weight per unit volume of the water in pounds per cubic feet (kilograms per cubic metre)
v	=	the velocity of the paddles relative to the water in feet per second (metres per second)
g	=	the acceleration of gravity, 32.2 ft/s² (9.81 m/s²)

Therefore, P_w, the water power input, will equal force times velocity; i.e.,

$$P_w = \frac{C_D Awv^3}{2g} \text{ ft–lb/s} \quad \left(= \frac{C_D Awv^3}{2} \text{ kg–m}^2/\text{s}^3 \text{ [W]}\right) \quad \text{(Eq 11-29)}$$

If k_2 is the ratio of the fluid velocity to the blade velocity (v_1), the relative velocity of the blade $v = v_1 - k_2 v_1 = (1 - k_2)v_1 = 2\pi (1 - k_2) r_b N/60$ where r_b is the distance from the shaft to the center of the blade and N the shaft speed in revolutions per minute.

Therefore, P_w can be expressed as follows:

$$P_w = \frac{C_D A w}{2g} \left(\frac{2\pi (1 - k_2) r_b N}{60} \right)^3 \qquad \text{(Eq 11-30)}$$

$$= 1.1 \times 10^{-3} C_D (1 - k_2)^3 N^3 \Sigma A r_b^3 \qquad \text{(Eq 11-31)}$$

For a flocculator with several rotor paddles, at varying distances and from the shaft and/or varying paddle areas,

$$P_w = 1.1 \times 10^{-3} C_D (1 - k_2)^3 N^3 \Sigma A r_b^3 \qquad \text{(Eq 11-32)}$$

Where:

$\Sigma A r_b^3$ = the sum of the values of $A r_b^3$ for all paddles.

Now:
$$P_w = \mu V G^2 \text{ ft–lb/s} \qquad \text{(Eq 11-33)}$$

Therefore:

$$G = \left(\frac{1.1 \times 10^{-3} C_D (1 - k_2)^3 N^3 A r_b^3}{\mu V} \right)^{1/2} \text{ s}^{-1} \qquad \text{(Eq 11-34)}$$

and for multiple rotor paddles:

$$G = \left(\frac{1.1 \times 10^{-3} C_D (1 - k_2)^3 N^3 \Sigma A r_b^3}{\mu V} \right)^{1/2} \text{ s}^{-1} \qquad \text{(Eq 11-35)}$$

In practice, peripheral speeds of paddles range from 0.3 to 3 ft/s (0.09 to 0.91 m/s), C_D is approximately 1.8 for flat blades, k_2 is about 0.25 in the absence of stators, while the area of the paddle blades should not be more than 15 to 20 percent of the cross-sectional area of the basin (Beam 1953). Camp (1955) suggests G-values between 35 and 66 s^{-1}.

Vertical shaft flocculators. Radial-flow flat or curved-blade turbines can be designed using Eq 11-31 and 11-34. The turbine tip speed should not exceed 5 ft/s (1.52 m/s).

Impeller type flocculator sizing and selection technology must include the following criteria:

- velocity gradient (G),
- impeller tip speed (TS),
- ratio of impeller diameter (D) to equivalent tank diameter (T_e),

- superficial velocity (SV), and
- relative shear.

The velocity gradient G can be calculated from Eq 11-1, from which the water power input P_w in foot-pounds per second (kilogram-metre per second cubed [watts]) equals $\mu V G^2$. To convert to water horsepower, divide by 550 ft-lb/s/hp.

The impeller tip speed should not exceed 7 ft/s (2.1 m/s) for a 32° or 45° pitched-blade turbine and 8 ft/s (2.4 m/s) for a three- or four-blade hydrofoil.

The equivalent tank diameter T_e can be found from the following equation:

$$T_e = 1.13 \sqrt{L_L \times L_W} \qquad \text{(Eq 11-36)}$$

Where:

L_L = the tank length in feet (metres)
L_W = the tank width in feet (metres)

The object of flocculation is to provide bulk fluid motion and to dissipate the imported horsepower over a wide area. This calls for relatively large impellers and the optimum D/T_e ratio ranging from 0.35 to 0.40, with 0.30 considered the minimum ratio and 0.45 the maximum. A small-diameter impeller rotating at high speed will produce localized turbulence, reduced fluid motion, and therefore areas of little or no movement.

It is important that all floc particles in a flocculation basin be kept in gentle motion and that there are no dead areas where floc can settle out. To ensure this bulk fluid motion, the impeller must have sufficient primary pumping capacity to provide an optimum superficial velocity of approximately 6 ft/min (0.03 m/s) for conventional "water" flocculation. If variable-speed flocculators are used, then caution should be exercised to ensure that superficial velocity is not less than 3 ft/min (0.015 m/s) at the lowest operating speed. The maximum superficial velocity should not be greater than 10 ft/min (0.05 m/s).

Superficial velocity (SV) can be found from the following equation:

$$SV = \frac{Q}{7.48 \, L_L L_W} \text{ ft/min} \qquad \left(= \frac{Q}{L_L L_W} \text{ m/s} \right) \qquad \text{(Eq 11-37)}$$

Where:

Q = the primary pumping capacity in gallons per minute (cubic metres per second)

Now:

$$Q = 7.48 \, N_q N D^3 \text{ gpm} \qquad \left(= \frac{N_q N D^3}{60} \text{ m}^3/\text{s} \right) \qquad \text{(Eq 11-38)}$$

Where:

N_q = the impeller pumping number (dimensionless)
N = the revolutions per minute
D = the impeller diameter in feet (metres)

Values for N_q are given below, but these should be confirmed by flocculator equipment manufacturers.

Impeller Type	N_q
Flat-, pitched-, and curved-blade turbine	0.86
Four-blade hydrofoil	0.50
Three-blade hydrofoil	0.45

Therefore:

$$SV = \frac{N_q ND^3}{L_L W_L} \text{ ft/min} \quad \left(= \frac{N_q ND^3}{60 L_L W_L} \text{ m/s} \right) \quad \text{(Eq 11-39)}$$

Water or impeller horsepower may also be expressed as follows:

$$P_w = \frac{N_p D^5 N^3 SG}{6.12 \times 10^7} \text{ hp} \quad \text{(Eq 11-40)}$$

Where:

N_p = the impeller power number (dimensionless)
SG = the specific gravity of the fluid

Typical values for different types of impellers are given below.

Impeller Type	N_p
Flat-blade turbine (radial flow)	3.6
Curved-blade turbine (radial flow)	2.5
45° pitched-blade turbine (axial flow)	1.6
32° pitched-blade turbine (axial flow)	1.1
Four-blade hydrofoil	0.4
Three-blade hydrofoil	0.3

These values must be confirmed by flocculator manufacturers, as they may vary. Additional data on N_q and N_p values for impellers are given in chap. 9.

To select the correct impeller type and size for the flocculator, a series of calculations must be performed for comparative evaluation. First, select a velocity gradient (or range) for each stage of flocculation and the design water temperature and therefore the dynamic viscosity. Next, calculate the respective water horsepower required, the equivalent tank diameter (T_e) by Eq 11-36, and with the selected D/T_e ratio calculate the impeller diameter. Then calculate values of N from Eq 11-40. The impeller type or types must be narrowed down at this point or the number of alternatives will become too great.

The tip speed (TS) can be determined from the following expression:

$$TS = \frac{\pi ND}{60} \text{ ft/s} \quad \text{(Eq 11-41)}$$

Finally, the superficial velocity must be found using Eq 11-37 and 11-38. One further criterion that should be evaluated in comparing the alternative impellers is the relative shear values. Shear is proportional to the rotating speed squared and the diameter squared of the impellers under consideration (shear $\propto N^2D^2$). An impeller with the least shear is preferable.

Tabulating these results will help identify impellers that meet all the criteria. Final selection will be based on estimated costs.

When variable-speed drives are to be used to allow for a range of G-values, the ratio of N_{min}/N_{max} times maximum motor speed should not be less than 300 rpm.

Restraints exist for standard AGMA gear reducers. Standard AGMA gearbox output operating speeds (impeller speeds) for standard synchronous motor speeds are given in Table 11-2.

Normally, Eq 11-40 will result in a nonstandard speed (N); therefore, the speed should be reduced to the next lowest standard AGMA speed. However, if the next speed higher will not significantly increase the velocity gradient, tip speed, or superficial velocity, then select the higher speed.

Nonstandard gear reducer speeds are available, but these will result in higher costs than a standard unit.

If two velocity gradient values are to be used for each stage of flocculation, then the relationship between motor speeds and G-values can be used:

$$N_2 \approx N_1 \left(\frac{G_2}{G_1}\right)^{2/3} \quad \text{(Eq 11-42)}$$

Where:

G_1 and G_2 = the velocity gradient at high and low speed, respectively
N_1 and N_2 = the motor high and low speeds, respectively

Finally, the impeller must be located correctly relative to the basin bottom and water surface. The optimum distance that the impeller should be located below the water surface is 1.25 diameters, and the optimum distance above the floor level is

Table 11-2 Standard AGMA Gearbox Speeds

Synchronous Motor Speed, rpm		
1750	1165	875
350	230	175
280	190	140
230	155	115
190	125	94
155	100	77
125	84	63
100	68	51
84	56	42
68	45	34
56	37	28
45	30	23
37	25	18.6
30	20	15.2
25	16.5	12.4
20	13.5	10.1
16.5	11	8.3
13.5	9	6.7

1 diameter. The minimum distance above the floor for a down-pumping axial-flow impeller is 0.75 diameters. If the optimum D/T_e ratio equals 0.375, then the impeller diameter $D = 0.434 \sqrt{L_L L_W}$ (from Eq 11-36).

The optimum water depth $L_D = (1.25D + 1D)$, from which

$$D = \frac{L_D}{2.25} \qquad \text{(Eq 11-43)}$$

Therefore, the optimum water depth:

$$L_{D\,\text{opt}} = 0.95 \sqrt{L_L L_W} \qquad \text{(Eq 11-44)}$$

When conditions dictate—for example, when retrofitting an existing flocculation basin—the impeller cannot be placed above the recommended minimum space above the floor (0.75 diameters); therefore, an upward-pumping impeller should be specified. An upward-pumping impeller should be placed within 0.33 to 0.5 diameters of the floor and have at least 0.75 diameters of water above the impeller.

Walking beam flocculators. The manufacturers of walking beam flocculators design their flocculators with criteria "based upon intensive investigation and experience in the field or actual installation." These criteria are:

- peak water horsepower = 0.40 hp/mgd (0.08 W/m^3)
- detention time = 30 to 40 min
- an empirical formula, showing the relationship between peak horsepower and paddle length based upon stroke speed and stroke length.

Camp (1955) developed an expression for water power for vertically reciprocating blades:

$$P_w = 12.74\, C_D L_p^3 N^3\, \Sigma\, A \text{ lb-ft/s} \qquad \text{(Eq 11-45)}$$

Where:

C_D = the drag coefficient
L_p = the paddle stroke length in feet (metres)
N = the revolutions per second of the drive unit
A = the area of each paddle in square feet (square metres)

Camp (1955) recommended a value of 3 for the drag coefficient C_D.

Practical Issues

Tapered flocculation with diminishing velocity gradient has proved beneficial (Kawamura 1976; Camp 1955) and this flexibility should be designed into each flocculation system. Typical velocity gradients for three-stage flocculation basins for a conventional treatment plant are 70/40, 50/20, and 30/10 s^{-1}. However, velocity gradients will vary depending on the type of floc being formed and the treatment process adopted. It may be economical to specify all flocculators to have the same velocity gradient range (i.e., 70/10 when the required motor horsepower is small).

Care must be taken to see that a turbulence high enough to break up the formed floc is not generated downstream of the flocculators. The G-values of any openings, pipe bends, weirs, etc., should not be greater than the G-value of the last flocculation chamber.

In order to minimize short-circuiting in flocculation basins, a minimum of three compartments is recommended. Vertical shaft flocculation allows for easier compartmentalization, and positioning inlet and outlet openings of each compartment in a diagonally opposite location for serpentine flow, both horizontally and vertically, further helps minimize short-circuiting.

By specifying the mixer rotation opposite to the general flow pattern, short-circuiting can be reduced even further. Compartmentalized flocculation basins also allow more flexibility for varying G-values, detention times, and for taking units out of service for maintenance with very little effect on the performance of the process.

For added flexibility, it is advisable to include provision for alternative application points for coagulant chemical aids. These can be included at the first two stages of a three- or four-stage flocculation basin. This will allow full-plant tests to identify the optimum location for the addition of a coagulant aid, with the coagulant of the first (or only) rapid mixer, the second rapid mixer, the first flocculator, or second flocculator.

When vertical turbine type flocculators are to be used, avoid the use of bottom (underwater) bearings. Shafts and gearboxes can be designed to eliminate this difficult maintenance feature.

In order to determine the water horsepower of an installed flocculator, the efficiency of both the gear reducer and the motor must be taken into account (i.e., the water horsepower will equal the motor horsepower multiplied by the efficiency of both the gear reducer and motor). Chapter 9 details how these efficiencies can be determined.

Stators are not necessary in square basins when axial-flow impellers are used, particularly hydrofoil impellers.

Specifications. The following should be included in contract specifications.

Gear reducers. The gear reducer is the most critical part of the flocculator, and the specification should spell out the quality required. Included should be: that the unit shall be designed and built for continuous 24-h/day, 7-day/week operation; a minimum AGMA service factor of 1.5 based on motor nameplate horsepower (2.5 for designs utilizing bearings in the gearbox for shaft support); a minimum B-10 life of 100,000 h based on AFBMA standards for bearings; an AGMA Standard 390.03 quality 10 or better for gears; and gear drives designed in accordance with AGMA Standard 420.04. Care must be taken to ensure that the lubrication system is adequate. Splash lubrication is normal, but may not be adequate for low-speed units where an oil pump for lubrication may be needed.

Motors. Motors should be specified with a service factor of at least 1.15. For continuous-duty, high-efficiency motors are advisable. Where plants are located above 3000 ft (914 m), the motor specification must spell out specific requirements for cooling the motor. Although the entire cost of high-efficiency motors may not be justified by energy savings, other advantages such as lower operating temperatures and noise levels should be evaluated. For paddle-wheel and walking beam type flocculators, the motor size must be sized for the high starting loads (Hudson 1981; Camp 1955). For high-humidity and outdoor installations motor heaters should be specified.

Flocculator shaft. The specification should require the operating speed to be equal to or less than 70 percent of the first lateral critical speed. The stress in the impeller shaft should not exceed 8000 psi (55 MN/m^2) under maximum operating loads. The flocculator shaft should be connected to the reducer output shaft with ridged flanged coupling.

Impeller. The specification must include: G-values; maximum tip speeds; type (and number) of impeller(s); the minimum superficial velocity at the maximum and minimum G-values; the flow direction of the impeller (upward or downward); and location of the impeller(s) relative to the basin floor level. The stress in the impeller should not exceed 11,000 psi (76 MN/m^2) under maximum operating loads.

Submittals. The specifications should specify that the manufacturer submit calculations to substantiate the specified performance and construction criteria.

Hydraulic flocculation. In designing vertical baffled channel flocculators, care must be taken to allow for cleaning out accumulated settled matter on the basin floors. Holes should be allowed in the lower baffles. Schulz and Okun (1984) give further practical guidelines for the design and construction of baffled channel flocculators.

Special considerations. When the water treatment process includes softening using lime, the flocculators should be designed for the specific gravity of the coagulated water and for the buildup of calcium carbonate on the impeller or paddles and the flocculator shaft. The impeller load should not exceed 80 percent of the motor nameplate horsepower, and the operating speed of the shaft should be equal to or less than 60 percent of the first lateral critical speed to allow for lime buildup.

The addition of antiswirl baffles in the chambers of vertical shaft flocculators should be evaluated carefully. The most usual arrangement for water treatment is four baffles spaced at 90°, extending one-twelfth of the basin width into the basin, and down from the maximum water level to a point level with the lower edge of the impeller. The number and size will change with liquids of higher viscosities, and equipment manufacturers should be consulted.

References

AMERICAN WATER WORKS ASSOCIATION. 1969. *Water Treatment Plant Design.* McGraw-Hill, New York.

———. 1984. *Introduction to Water Treatment, Principles and Practices of Water Supply Operations.* AWWA, Denver, Colo.

AMIRTHARAJAH, A. 1978. Design of Rapid-Mix Units. In *Water Treatment Plant Design for the Practicing Engineer,* ed. by R.L. Sanks. Ann Arbor Science, Ann Arbor, Mich.

AMIRTHARAJAH, A. & MILLS, K.M. 1982. Rapid-Mix Design for Mechanisms of Alum Coagulation. *Jour. AWWA,* 74:4:210.

ANDREU-VILLEGAS, R. & LETTERMAN, R.D. 1976. Optimizing Flocculator Power Input. *Jour. Environ. Engrg. Div. ASCE,* 102:251.

BEAM, E.L. 1953. Study of Physical Factors Affecting Flocculation. *Wtr. Works Engrg.,* 106:1:33.

BOWERS, A.E. & BEARD II, J.D. 1981. New Concepts in Filtration Plant Design and Rehabilitation. *Jour. AWWA,* 73:9:457.

CAMP, T.R. 1955. Flocculation and Flocculation Basins. *Trans. Amer. Soc. Civ. Engrg.* 120:1.

———. 1968. Floc Volume Concentration. *Jour. AWWA,* 60:656.

CAMP, T.R. & STEIN, P.C. 1943. Velocity Gradients and Internal Work in Fluid Motion. *Jour. Boston Soc. Civ. Engrg.* 30:219.

FAIR, G.M.; GEYER, J.C.; & OKUN, D.A. 1966. *Water Supply and Wastewater Removal.* John Wiley and Sons, New York.

———. 1968. *Water and Wastewater Engineering.* John Wiley and Sons, New York (Vol. 2).

HUDSON, H.E. JR. 1981. *Water Clarification Processes: Practical Design and Evaluation,* chap. 6. Van Nostrand Reinhold, New York.

HUDSON, H.E. JR. & WOLFNER, J.P. 1967. Design of Mixing and Flocculating Basins. *Jour. AWWA,* 59:1257.

JM MONTGOMERY. 1985. *Water Treatment Principles and Design.* Wiley-Interscience, New York.

KAWAMURA, S. 1973. Coagulation Considerations. *Jour. AWWA,* 65:6:417.

———. 1976. Considerations on Improving Flocculation. *Jour. AWWA,* 68:328.

———. 1985. Two-Stage Filtration. *Jour. AWWA*, 77:12:42.

LETTERMAN, R.D.; QUON, J.E., & GEMMELL, R.S. 1973. Influence of Rapid-Mix Parameters on Flocculation. *Jour. AWWA*, 65:716.

MACDONALD, D.V. & STREICHER, L. 1977. Water Treatment Plant Design Is Cost Effective. *Public Works*, 108:8:86.

MONK, R.D.G. & WILLIS, J.F. 1987. Designing Water Treatment Facilities. *Jour. AWWA*, 79:2:45.

MONSCVITZ, J.T. & REXING, D.J. 1983. Direct Filtration Control by Particle Counting. AWWA Annual Conf., Las Vegas.

NEUMAN, W.E. 1981. Optimizing Coagulation With Pilot Filters and Zeta Potential. *Jour. AWWA*, 73:9:472.

RIDDICK, T.M. 1961. Zeta Potential and Its Application to Difficult Waters. *Jour. AWWA*, 53:8:1007.

RUSHTON, J.H. 1952. Mixing of Liquids in Chemical Processing. *Ind. Engrg. Chem.*, 44:2931.

SCHULZ, C.R. & OKUN, D.A. 1984. *Surface Water Treatment for Communities in Developing Countries*. John Wiley and Sons, New York.

VRALE, L. & JORDEN, R.M. 1971. Rapid Mixing in Water Treatment. *Jour. AWWA*, 63:52.

Index

NOTE: An *f*. following a page number refers to a figure; a *t*. refers to a table.; an *n*. refers to a footnote.

Aggregation, 217–218
Agitated-tank mixers, 44–48
ALT ratio, 260, 261*f*.
Alum, 4
  coagulation, 6*f*., 7*f*., 14
Aluminum, 4
  chemistry of, 202–205
  as a coagulant, 205
  conductivity, 204, 204*f*.
  electrophoretic mobility, 21*f*.
  monomers, 202, 235
  particle size distribution, 21–22, 22*f*.
  polymers, 202, 235
  precipitation of, 198–202
  speciation, 207, 208*f*., 209*f*.
  titration, 203–205, 203*f*.
Aluminum hydroxide flocs. *See* Floc
Aluminum sulfate. *See* Alum
Anemometry, 83
Anionic polymers, 24–25
  and metal coagulant, 25, 26*f*.
Axial dispersion model, 142–146, 144*f*., 168*t*.

Baffles, 45, 51
Batch jar test. *See* Jar test
Bench scale
  defined, 344
Binding force
  equation, 266
Bounding theorem, 159–160
Bypassing, 139–140

Camp's "G" concept, 317–322
Cationic polymers, 24
  and clay, 25, 26*f*.
Charge neutralization, 4, 6–7
Circulation time, 284
Closure
  conserved scalar approach, 100–103
  equilibrium assumption, 100
  higher-order moment, 103–104
Coagulation, 4–5, 4*f*., 80
  and constituent size, 20–21
  defined, 3
  first- and second-order kinetics of, 294*f*.
  and humic substances, 20
  mechanisms, 6*f*.
  mechanisms of particle destabilization, 4
  monitoring, 394–395
Coalescence–dispersion models, 110–112, 115*f*., 116*f*., 118*f*., 119*f*., 185–186
  costs, 120–121
Collision process, 26
  equation for collisions between particles, 27, 28–29, 30, 31
Collision rate
  equation, 10

Completely mixed chamber, 385, 386*f*., 391–392
Compression, 4
Concentration, equations, 108, 184
Concentration distribution
  coefficient of variation (equation), 233, 250
Concentration gradient
  standard deviation (equation), 227
Conserved scalar closure, 100–103
Crystal size distribution, 198, 199*f*.

Damköhler number, 290
Darcy equation, 219
David and Villermaux's model. *See* Four-stage mixing model
Derjaguin, Landau, Verwey, and Overbeek. *See* DLVO theory
Destabilization
  due to coagulants carried by microscale eddies, 11–12, 12*f*.
  equations for in inertial subrange, 13–14
  equations for in viscous subrange, 12–13
Diffusive time scales
  defined, 201
Discharge coefficient. *See* Pumping coefficient
Discharge rate
  equations, 53–54
DLVO theory, 4
Drop breakup rate
  equation, 295
Dynamic in-line mixers, 50–51, 50*f*., 76

Eddy size
  equation, 218
  and nonhomogeneity (equation), 56
Electrophoretic mobility
  of aluminum, 21, 21*f*.
  and velocity gradient, 15, 16*f*., 17*f*.
Equilibrium assumption closure, 100
Equivalent tank diameter
  equation, 413
Erosion model, 187, 188*f*.
Euler number
  equations, 250, 253
Eulerian modeling, 81, 95–96
  basic equation, 95
  and combustion engineering, 99
  necessary equations, 96–98
Eulerian view, 36
Exponential distribution, 167*f*.

Fanning friction factor (equation), 167
Ferric chloride coagulation, 14, 15, 16*f*., 17*f*.
Filterability number, 14
  and velocity gradient, 15, 16*f*., 17*f*.

Finite difference models, costs, 120
Floc, 256
  aging and density, 277, 277f.
  aging and electrophoretic mobility, 278–279
  aging and porosity, 277
  aging and regrowth ability, 278
  aging and strength, 278
  aging mechanism, 274–275
  aging process, 275–277
  area, 258, 259f.
  breakup, 267, 270–273
  breakup force (equations), 266–267
  clay, humic and aluminum, 24, 25f.
  combining clay and aluminum, 23–24, 23f.
  defined, 19
  density, 258–259, 260f., 263f., 266
  diameter, 257–258
  elements, 19–20
  equation for particle volume, 28
  equations for particle density, 28
  grown-floc, 29
  matrix with organic polymers, 24–25
  microfloc, 29
  porosity, 260–262
  random, 31
  shape, 258
  size, 257, 268, 294–295
  size and age, 275, 276f.
  size and energy dissipation, 268, 270f., 271t., 273t.
  size and rate of agitation, 268, 269f.
  strength and size, 266–270
  strength (equation), 268–270
  structure, 262–266, 263t.
  structure parameters, 278f.
  Vold model, 264
Flocculation, 80. See also Polymer flocculation, Slow mixing
  batch and continuous, 301–302, 304f.
  causes, 5
  contact, 29–31
  continuous-flow apparatus, 302, 303f.
  defined, 3, 368, 397
  design considerations, 362–364
  effects of scale-up, 282–283
  equation for, 10
  equation for in inertial subrange, 11
  equation for in viscous subrange, 11
  equation for under conditions of turbulent shear, 10
  evaluation of, 368–372, 371f., 373–374t., 375t., 376f.
  experiments with different impellers, 331–334, 333f., 334f., 335t., 336f., 336t., 337t.
  half time (equation), 292
  hydraulic, 398–399, 410–411
  under laminar and turbulent flow conditions, 26
  mechanical, 400–404, 408–409, 411–416
  pellet, 31–32
  random, and floc density, 27–29
  random, and floc size, 27
  settling of and initial mixing, 288f.
  short-circuiting, 408–409, 409f.
  steps, 327–329
  systems, 398–404
Flocculators
  baffled channel, 398–399, 398f.
  design considerations, 406–418
  gravel-bed, 399, 410–411
  horizontal-shaft, 400–401, 401f., 402f.
  hydraulic jet-action, 399, 399f., 400f.
  paddle-wheel, 400–401, 401f., 402f., 411–412
  vertical-shaft, 401–402, 403f., 412–416
Flow
  through tube (equations), 182–184
Flow coefficient. See Pump coefficient
Flow number. See Pumping coefficient
Fluctuating velocity, 35–36
Four-stage mixing model, 192–194, 193f., 195f., 196f.
  results for consecutive–competing reactions, 194–195
Fractional tubularity model, 135–137, 136f.
"Fractional unmixedness"
  equation, 40
Friction factor
  equations, 74, 75, 220
  and Reynolds number (equation), 42, 43f.
Froessling equation, 63
Froude number, 51, 65
Full scale
  defined, 344

G-value. See Velocity gradient
Gas–liquid mixing, 37, 66–67
Gearbox speeds, 415t.

Head, dimensionless, 54, 55f.
Head coefficient, 54
  equation, 43–44
Higher-order moment closure, 103–104
Humic substances
  and aluminum, 24
  removal by coagulation, 20
Hydraulic jumps, 382, 383f., 390–391
Hydrodynamic mixing model.
  See Interdiffusion model
Hydroxide flocs. See Floc

IEM model, 184, 185f., 186f., 188–189, 189f.
Impellers, 309–312
Impellers, 45f., 47f.
  axial-flow, 61, 315, 315f., 316f., 317, 317f.
  discharge characteristics, 52–54
  effects in microscale and macroscale, 312–317
  and flocculation experiments, 331–334, 333f., 335t., 336f., 336t., 337t.
  flow comparisons, 337, 338t., 339–340
  flow patterns, 46f., 47f.
  fluid foil, 311–312, 314f., 321f.
  power comparisons, 337–339, 338t., 339f., 340, 341f.

and pump mix, 53f.
pumping rate, 284
radial-flow, 315–317, 318f., 319f., 320f.
revolutions (equation), 40
Reynolds number vs. power number, 337, 338f.
rotation rate effects on stirred reactors, 103t.
scale-up principles, 322–325, 323t.
speed (equations), 61, 62
speed of and mixing performance, 326, 326f.
and tank bottoms, 61–62
types, 44–46, 310, 313t., 327, 328f., 329f., 331, 332f.
and velocity, 310f., 311–312, 312f.
vortices, 297f.
Initial mixing. *See* Rapid mixing
Inorganic coagulants, experimental results, 14–18
Interaction by exchange with the mean model. *See* IEM model
Interdiffusion model, 106–109, 108f., 109f., 110f., 111f., 112f., 113f.
equations, 108–109
Interparticle bridging, 4
Iron coagulation, 8, 9f.
Iron salts, 4

Jar test, 282–283, 288, 297–298, 301–302, 303–304, 331, 394
Jet mixers, 48–50, 49f.
axial jets, 71
coaxial, 72–73
diagonal jets, 71, 72
jet concentration (equation), 69
jet entrainment (equation), 69
jet velocity (equations), 69, 71
jet velocity and second fluid velocity (equations), 72
in pipes, 71–73, 231–234
pressure drops (equation), 253
Reynolds number (equation), 68
in tanks, 70–71
turbulent jet characteristics, 69f.
typical geometry, 70f.
Just-suspended condition. *See* Solid–liquid mixing—minimum mixing condition

Kolmogoroff length scale (equation), 88, 291
Kolmogoroff theory, 88

Lagrangian modeling, 81, 90, 93–95
level of segregation (equation), 93
micromixed model, 91–92, 92f.
multi-environment models, 93, 94t.
perfectly mixed model, 93
residence time distribution, 90–92, 91f.
segregated reactor model, 92, 92f., 93
Lamellar model, 114, 117f.
Laminar mixing, 38, 67–68
power consumption (equation), 67
power number and Reynolds number (equation), 41
Laminar shear, 10
Laser–Doppler technique, 53–54
Liquid–liquid mixing, 37–38, 66
Local diffusion model, 112–116
Log-log isoturbidity topogram, 368, 371f.

Macrofluid. *See* Segregation—complete
Macromixing, 171
Mass energy dissipation rate, 284–287
Mass transfer (equations), 63–64
Maximum mixedness, 161–163, 161f.
Mean velocity, 35–36
Means, 132–133
Metastable state and pellet flocculation, 31–32
Microfluid, 172–174
Micromixing, 163–164, 170–172
and chemical reactions, 174–176, 175f.
Danckwerts and Zwietering model of minimum mixedness, 183f.
empirical single-parameter models, 179–186
models, 181f., 186–190, 209–210
multistage semiempirical models, 190–191
and multistep reactions, 178–179, 180f.
physical multistage models, 192–196
and polymerization, 178
and precipitation of sparingly soluble salts, 198, 200f.
and single-step reactions, 176–178, 177f.
Zwietering model of maximum mixedness, 183f.
Mixed shear rate (equation), 68
Mixers. *See* Agitated-tank mixers, Dynamic in-line mixers, Jet mixers, Pipe mixers, Static mixers, Stirred-tank reactors. *See also* Mixing
scale-up and scale-down, 119–120
Mixing, 171f. *See also* Gas–liquid mixing, Laminar mixing, Liquid–liquid mixing, Macromixing, Micromixing, Mixers, Mixing models, Mixing time, Mixing time scales, Multiphase mixing, Rapid mixing, Solid–liquid mixing, Slow mixing, Turbulent mixing
and aluminum precipitation—modeling, 206–209
conserved scalar closure, 100–103
of disinfectants, 3
dynamic similarity of fluid forces, 325–326, 325t.
earliness, 181 184
effectiveness monitoring, 364–367
effects of scale-up, 282–283
energy dissipation, 172, 173f.
equilibrium assumption closure, 100
fast reactions, 100, 101f.
higher-order moment closure, 103–104
monitoring by conductivity probe, 367
monitoring by particle counting, 366
monitoring by photography, 367
monitoring by streaming-current detector, 367

424 MIXING

monitoring by testing electrophoretic mobility, 367
monitoring by testing zeta potential, 367
monitoring by tracer studies, 366
monitoring by turbidity measurement, 366
monitoring by velocity measurement, 367
monitoring by visible dye, 367
monitoring by visual inspection, 366–367
and pilot-plant studies, 343–344
rate (equation), 89
reaction rates (equation), 99–100
shear rate, 310, 313f.
space, 163f.
stages, 172, 173f., 187–190, 192–194
standard deviation (equations), 74, 76
and velocity, 35–36
zone volume (equation), 392
Mixing models, 89–90. *See also* Coalescence–dispersion model, Closure, Eulerian modeling, Finite difference models, Four-stage mixing model, Interdiffusion model, Lagrangian modeling, Lamellar model, Local diffusion model, Motionless-mixer models, Probability distribution functions, Similarity approximations, Tanks-in-series model
of complex chemical reactions, 116–118
correlations, 123
and experimental confirmation, 118–119
relative costs, 120–121
Mixing time, 38–40, 39f., 57–59
and progress of mixing (equations), 40–41, 59–61
Mixing time scales
defined, 201
and diffusive time scales, 200, 201f.
and reactive time scales, 200, 201f.
Models. *See* Axial dispersion model, Coalescence–dispersion model, Erosion model, Eulerian modeling, Finite difference models, Four-stage mixing model, IEM model, Lagrangian modeling, Lamellar model, Local diffusion model, Mixing models, Molecular diffusion model, Multistage semiempirical models, Piston-flow model, Interdiffusion model, Motionless-mixer models, Residence time, Stretching model, Tanks-in-series model
Molecular diffusion model, 188–190
Moments, 132–133
extracting from models, 152
Momentum flux ratio (equation), 233
Monodispersion decay rate (equations), 292, 293
Motionless-mixer models, 146
Multiphase mixing, 37–38
Multistage semiempirical models, 190–191, 191f.

N_h. *See* Head coefficient
N_q. *See* Pumping coefficient

Nonionic polymers, 24–25
Nozzle velocity head (equation), 392

Open systems and closed systems, 150f.
Organic polyelectrolytes, 4
Ozone contactors, design considerations, 364, 365f.

Palmer–Bowlus flumes, 391
Parshall flume, 382, 383f., 386, 391
Particle destabilization, 4–5, 217–218
and hydrolysis, 235, 237f.
by inorganic coagulants, 9–14
in turbulent pipe flow, 234–241
reaction rate, 238–239, 239f., 240f.
Particles
diameter, 257–258
shape, 258
size, 257
Pelleting. *See* Flocculation—pellet
Penney diagram, 287f.
Pilot columns, 394
Pilot plants
flocculator design considerations, 362–364, 405–406
hardware development, 360–361
operation of, 361–362
ozone contactor design considerations, 364, 365f.
process and instrumentation diagrams, 358–360, 359f.
rapid mixing design considerations, 362, 363f.
Pilot scale, defined, 344
Pilot studies. *See also* Pilot plants
alternative selection, 347
and bench-scale testing, 347–348, 349f.
design considerations, 345–346
duplicating source water variability, 354–355
effect of preseasoning on turbidity, 355, 356f.
flocculation evaluation, 368–372, 371f., 373–374t., 375t., 376f.
and full-scale testing, 347–348, 376–377, 377f., 378f.
literature reviews, 347
process selection, 354t.
program organization, 349
rapid-mixing evaluation, 367–368, 369f., 370t.
reviewing data, 347, 348f.
scaling, 356–357
scheduling, 348–349
selecting treatment process, 346
and source water quality, 350–354
testing and monitoring plan, 349–350, 350t., 351t.
Pipe mixers, 220–224, 221f., 222f., 223f., 224f., 225f., 226f., 227f., 228f., 229f., 230f.
agitator and ejector systems, 245–247, 247f.
comparative studies, 247–253, 249f., 251f., 252f.

high-velocity jet injection, 248
jet injection mixing, 231–234, 231f., 232f.
orifice plate, 248
particle aggregation, 241–244, 242f., 243f., 243t.
particle destabilization, 234–241, 235f.
particle destabilization, filtration efficiency, and chemical dose, 245, 246f.
performance criteria, 224–231
pipe expansion, 244–245
pipe expansion and filtration efficiency, 244, 245f.
pipe expansion with center injection, 248
pipe expansion with injection ring, 250–253
research needs, 253
turbo mixer, 248
Piston-flow model, 131–132
Plug-flow model. See Piston-flow model
Polymer flocculation, 18, 19f.
Polymers. See Anionic polymers, Cationic polymers, Nonionic polymers, Polymer flocculation
Power dissipation (equation), 253
Power number
  in different-sized vessels, 299, 301f.
  equation, 82, 390
  equation for Rushton turbines, 52
  in laminar mixing (equation), 67
  and Reynolds number, 41–42, 51, 52
Precipitation
  of aluminum, 198–202
  and micromixing, 198, 200f.
  process (equation), 196
  of sparingly soluble salts, 196–198
Pressure drop (equation), 253
Probability distribution functions, 104–110, 106f., 107f.
Process and instrumentation diagrams, 358–360, 359f.
Pump injection diffusion, 385, 387f., 392–393
Pump mixing, 385
Pumping capacity
  equation, 391
Pumping coefficient, 52
  and efficiency (equation), 55
  equation, 43–44
Pumping number. See Pumping coefficient

Rapid mixing, 3–5, 4f. See also Slow mixing
  for charge neutralization, 8–9
  competitive-consecutive reaction, 289, 290f.
  completely mixed chamber, 385, 386f., 391–392
  design considerations, 362, 363f.
  energy dissipation, 383, 384f.
  evaluation of, 367–368, 369f., 370t.
  and flocculation settling, 288f.
  hydraulic, 381–383
  hydraulic jumps, 382, 383f., 390–391
  in large-scale situations, 287–292
  mechanical, 385
  mixer design considerations, 389–397
  mixer selection criteria, 385–389
  mixing parameters, 14, 15t.
  pipe discharges, 383
  for polymer coagulants, 18–19
  pump injection diffusion, 385, 387f., 392–393
  pump mixing, 385
  shrouded radial turbine mixer, 385, 388f., 393–394
  static mixers, 383, 384f., 391
  for sweep coagulation, 7–8
  weirs, 382, 382f., 390
Reactive time scales, defined, 201
Residence time. See also Residence time density function, Residence time distribution
  differential distribution function, 128, 130f.
  distribution functions, 128–132
  frequency function, 128
  histogram, 129f.
  mean, 133–135
  means and moments, 132–133, 152
  normalized distributions, 133–135
  perfect mixing model, 130–131
  piston-flow model, 131–132, 168t.
  plug-flow model. See subhead piston-flow model
  washout function, 129, 130, 131f., 134, 135, 152
Residence time density function, 128, 130, 130f.
  calculations, 154–158, 156–157t.
  measurement by inert tracers, 146–147
  measurement in full-scale treatment facilities, 151–152
  measurement in multiport closed systems, 147–149
  measurement in open systems, 149–151
Residence time distribution
  achievement of mixing extremes, 164–165
  axial dispersion model, 142–146, 144f., 168t.
  design considerations, 164–168
  direct utilization of data, 154
  fractional-tank extension, 140–141
  fractional tubularity model, 135–137, 136f., 168t.
  gamma-function extension, 138–141
  in pilot-scale equipment, 164–165
  in simple closed system, 128, 129f.
  model formulation requirements, 135
  motionless-mixer models, 146
  parameter estimation in models, 152–154
  scale-up consideration, 165–168
  tanks-in-series model, 137–142, 138f., 168t.
Reynolds number
  and friction factor, 42, 43f.
  in jet mixing, 68
  and power number, 41–42, 51, 52
  in scale-up, 286

Scale-up
  basic laws, 284–287, 287f., 300f.
  experiments affecting flocculation, 297–301, 299t., 300f., 300t., 301f.
  and flocculation, 282–283, 292–297, 302–305
  and flow patterns, 327
  general effects, 283–284
  and impellers, 322–325, 323t.
  and initial mixing, 287–292, 302–305
  and mixing, 282–283
  and power input, 287
Scale-up factor (equation), 166
Segregation
  complete, 160, 161f., 174
  equations, 93, 95, 97
  intensity of (equation), 85
  microfluid, 172–174
  scale of (equations), 85
Segregation index (equation), 290
Separation coefficient, 64–65
Settled-water turbidity, 8, 8f.
Shear rate, 310, 313f., 317–322
  and detention time, 322
  diverging rates, 323, 324f.
Short-circuiting, 139
Shrouded radial turbine mixer, 385, 388f., 393–394
Similarity approximations, 121–123, 122f.
Slow mixing, 4f., 5–6. See also Rapid mixing
Smoluchowski expression, 10, 11
Smoluchowski theory of orthokinetic flocculation, 5, 26
Solid–liquid mixing, 37
  floating solids, 65–66
  homogeneous suspension, 64–65
  mass transfer, 63–64
  minimum mixing condition, 61–63
Source water
  algae in, 355
  characterization of, 351–354, 352t., 353f.
  duplicating variability of in pilot studies, 354–355
  turbidity of, 355, 356f.
SSE. See Sum-squared error
Standard deviation in mixing (equations), 74, 76
Static mixers, 48
  laminar mixing, 73–74
  multiphase mixing in, 38
  power dissipation (equation), 253
  rapid mixing in, 383, 384f., 391
  turbulent mixing, 36, 74–75
  types, 73
Steady bearings, 340–341
Stirred-tank reactors, 131n.
  design considerations, 362, 363f.

Streaming current detectors, 395
Stretching model, 187
Sulzer SMV (equations), 74–75
Sum-squared error (equation), 153, 158
Superficial velocity (equation), 413
Sweep coagulation, 4, 7–8, 8f.

Tanks-in-series model, 137–138, 138f.
  crossflow, 141–142, 142f.
  fractional-tank extension, 140–141
  gamma-function extension, 138–141
  stagnancy, 141–142, 142f.
Taylor scalar microscale, 89
Throughflow (equation), 148
Turbulence parameters (equations), 56, 57
Turbulent cloud diffusivity (equation), 291
Turbulent fields
  flocculation kinetics in, 9–11
  particle destabilization in, 11–14
Turbulent flow
  energy transfer, 218, 219f., 220f.
  and laminar flow, 35
  in pipes, 218–220
Turbulent mixing
  in agitated-tank mixers, 35–36
  analyzing, 37
  circulation time (equation), 58
  concentration, 83–85
  discharge rate, 52–54
  energy spectrum, 87f.
  head characteristics, 54–56
  mixing mechanisms, 36
  mixing time, 57–59
  modeling, 80
  in pipes, 36–37
  power consumption, 51–52
  power number and Reynolds numbers (equations), 41, 42f.
  spectral analysis (equations), 86–88
  stages, 82, 85
  in static mixers, 36, 74–75
  theory of, 85–89
  time parameters, 58–59
  turbulence, 56–57

Velocity gradient, 5–6, 407f.
  equation, 200, 389, 394, 408
Vold model, 264
Vortex incorporation time (equation), 291

Wadell sphericity, 258
Water power (equations), 390, 391, 392, 393, 411, 412, 416
Wave number (equation), 88
Weirs, 382, 382f., 390

Zeta potential meters, 394–395, 395t.
Zweitering correlation, 61, 62, 63